Cryptosporidium and Cryptosporidiosis

Edited by

Ronald Fayer

CRC Press
Boca Raton New York London Tokyo

Acquiring Editor: Marsha Baker
Project Editor: Sarah Fortener
Assistant Managing Editor: Gerry Jaffe
Marketing Manager: Susie Carlisle
Direct Marketing Manager: Becky McEldowney
Cover design: Denise Craig
PrePress: Carlos Esser
Manufacturing: Sheri Schwartz

Library of Congress Cataloging-in-Publication Data

Cryptosporidium and cryptosporidiosis / edited by Ronald Fayer.
p. cm.
Rev. ed. of: Cryptosporidiosis of man and animals. c1990.
Includes bibliographical references and index.
ISBN 0-8493-7695-5
1. Cryptosporidiosis. 2. Cryptosporidium. I. Fayer, R.
II. Cryptosporidiosis of man and animals.
[DNLM: 1. Cryptosporidiosis. 2. Cryptosporidium. WC 730 C956
1997]
RC136.5.C79 1997
616.9'36--dc20
DNLM/DLC
for Library of Congress 96-24303
CIP

International Standard Book Number 0-8493-7695-5
Library of Congress Card Number 96-24303
Printed in the United States of America 1 2 3 4 5 6 7 8 9 0
Printed on acid-free paper

PREFACE

Recognition of cryptosporidiosis and the organisms associated with the disease has evolved in the past few years from isolated observations of infections in animals to occasional pathogens in immunocompromised animals and humans to ubiquitous worldwide infections of large numbers of humans and animals. Concurrently, the literature and diversity of subject matter associated with this disease have grown enormously. Since the publication in 1990 of our first book, *Cryptosporidiosis of Man and Animals*, over 1000 new scientific articles have been published, making it difficult for experts and others interested in this area to keep current.

In this revision, an attempt has been made to summarize in the first chapter much of the data on taxonomy, life cycles, morphology, host species, and control methods from the first book. The following nine chapters reflect subject areas that have been emphasized in the scientific literature and that have aroused the greatest concern in the public health, medical, veterinary, and research communities: namely, diagnosis, epidemiology, waterborne events, prevention and treatment, immunity, biochemistry, cultivation, laboratory animal models, and molecular biology.

It is hoped that this book will serve as a guide for research biologists, public health workers, physicians, veterinarians, clinical laboratory technicians, and others concerned for human and animal health and for the safety of food, drinking water, and recreational water.

The editor acknowledges with special gratitude Eva Kovacs Nace for her many contributions to the preparation and editing of this book.

Ronald Fayer

THE EDITOR

Ronald Fayer, Ph.D., received his B.S. degree in 1962 from the University of Alaska and his M.S. and Ph.D. degrees in zoology from Utah State University in 1964 and 1968. From 1968 to 1972, he served as Zoologist with the Coccidiosis Project, Beltsville Parasitology Laboratory, U.S. Department of Agriculture (USDA), Agricultural Research Service (ARS), Beltsville, MD. He was Project Leader for the Sarcocystis Project from 1972 to 1978 at the Animal Parasitology Institute (API), USDA, and ARS and then Laboratory Chief for Ruminant Parasites Laboratory, API, USDA, ARS from 1978 to 1984. He was Director of API from 1984 to 1988. He served as Research Leader for the Zoonotic Diseases Laboratory, Livestock and Poultry Sciences Institute, from 1988 to 1994 and now serves as Senior Scientist in the Immunology and Disease Resistance Laboratory, USDA, ARS.

Dr. Fayer has spent over 25 years researching *Cryptosporidium* and related protozoan parasites of animals and humans. He has published three books and over 200 papers in scientific journals. He received the USDA Superior Service Medal, as well as the American Society of Parasitologists' H.B. Ward Medal. He served as President of the American Association of Veterinary Parasitologists and President of the Helminthological Society of Washington and was co-founder of the Federation of Societies for Parasitology. He is an internationally recognized scientist specializing in research on protozoan diseases.

CONTRIBUTORS

Michael J. Arrowood, Ph.D.
Research Microbiologist
Division of Parasitic Diseases
National Center for Infectious Diseases
Centers for Disease Control and Prevention
Public Health Service
U.S. Department of Health and Human Services
Atlanta, Georgia

Byron L. Blagburn, Ph.D.
Alumni Professor
Department of Pathobiology
College of Veterinary Medicine
Auburn University
Auburn, Alabama

David P. Casemore, Ph.D., M.R.C. Path.
Head, PHLS Cryptosporidium Reference Unit
Public Health Laboratory
Glan Clwyd District General Hospital
Rhyl
Clwyd, Wales

Robert L. Coop, Ph.D.
Moredun Research Institute
Edinburgh, Scotland

J. P. Dubey, Ph.D., M.V.Sc.
Senior Scientist
United States Department of Agriculture
Agriculture Research Service
Livestock and Poultry Science Institute
Parasite Biology and Epidemiology Laboratory
Beltsville, Maryland

Ronald Fayer, Ph.D.
Senior Scientist
United States Department of Agriculture
Agriculture Research Service
Livestock and Poultry Science Institute
Immunology and Disease Resistance Laboratory
Beltsville, Maryland

Mark C. Jenkins, Ph.D.
Microbiologist
United States Department of Agriculture
Agriculture Research Service
Immunology and Disease Resistance Laboratory
Beltsville, Maryland

Mark LeChevallier, Ph.D.
Director, Research
American Water Works Service Company
Voorhees, New Jersey

David S. Lindsay, Ph.D.
Senior Research Fellow
Department of Pathobiology
College of Veterinary Medicine
Auburn University
Auburn, Alabama

John T. Lisle, Ph.D.
Research Associate
Department of Microbiology
Montana State University
Bozeman, Montana

Carolyn Petersen, M.D.
Assistant Professor of Medicine and Infectious Diseases
Division of Infectious Diseases
San Francisco General Hospital
University of California
San Francisco, California

Michael W. Riggs, D.V.M., Ph.D.
Associate Professor
Department of Veterinary Science and Microbiology
University of Arizona
Tucson, Arizona

Joan B. Rose, Ph.D.
Associate Professor
Department of Marine Sciences
University of South Florida
St. Petersburg, Florida

Rosemary Soave, M.D.
Division of Infectious Diseases
Cornell University Medical College
New York, New York

C. A. Speer, Ph.D.
Professor
Department of Veterinary Molecular Biology
Montana State University
Bozeman, Montana

Michael Tilley, Ph.D.
Hitchings-Elion Postdoctoral Fellow
Department of Grain Science and Industry
Kansas State University
Manhattan, Kansas

Steve J. Upton, Ph.D.
Professor
Division of Biology
Kansas State University
Manhattan, Kansas

S. E. Wright, B.Sc.
Moredun Research Institute
Edinburgh, Scotland

CONTENTS

Chapter 1

The General Biology of *Cryptosporidium*

Ronald Fayer, C. A. Speer, and J. P. Dubey

CONTENTS

0-8493-7695-5/97/$0.00+$.50

I. INTRODUCTION AND HISTORY

Ernest Edward Tyzzer was the first person to recognize, clearly describe, and publish an account of a parasite he frequently found in the gastric glands of laboratory mice. In 1907, he described asexual and sexual stages and spores (oocysts), each with a specialized attachment organelle, and remarked that spores were excreted in the feces.[1] Recognizing the parasite as a sporozoan of uncertain taxonomic status, he named it *Cryptosporidium muris*. In 1910, in more detail, he proposed *Cryptosporidium* as a new genus and *C. muris* as the type species, added Japanese waltzing mice and English mice as hosts, and speculated that sporozoites from oocysts in the gastric glands might autoinfect the host.[2] Except for developmental stages as extracellular, Tyzzer's original description of the life cycle has been confirmed by electron microscopy.

In 1912, Tyzzer described a new species, *Cryptosporidium parvum*.[3] By experimentally infecting mice he demonstrated that *C. parvum* developed only in the small intestine and that its oocysts were smaller than those of *C. muris*. He remained ambiguous on the subject of whether stages were intracellular or extracellular and noted that stages similar to those of *C. parvum* in the mouse were present in the small intestine of the rabbit.

In 1929, Tyzzer illustrated but did not clearly describe developmental stages of *Cryptosporidium*, which he believed to be *C. parvum*, in chicken cecal epithelium.[4]

For 48 years after Tyzzer's first publication, because *Cryptosporidium* appeared to be of no economic, medical, or veterinary importance, it remained relatively obscure. Even in 1955, with the first report of a new species, *Cryptosporidium meleagridis*, associated with illness and death in young turkeys,[5] there remained little interest. And only slight interest was aroused when, in 1971, *Cryptosporidium* was found to be associated with bovine diarrhea.[6]

In 1976, two groups reported the first cases of cryptosporidiosis in humans.[7,8] Few human cases were reported thereafter until 1982. The report by the Centers for Disease Control that 21 males from six large cities in the U.S. had severe protracted diarrhea caused by *Cryptosporidium* in association with Acquired Immune Deficiency Syndrome (AIDS)[9] ushered in the present era of worldwide interest in and study of organisms in the genus *Cryptosporidium*.

In 1993, interest expanded dramatically following a massive waterborne outbreak in Milwaukee, WI, involving an estimated 403,000 persons.[10] The general public, public health agencies, agricultural agencies and groups, environmental agencies and groups, suppliers of drinking water, and others expressed concern and initiated studies on the basic biology of *Cryptosporidium* with emphasis on developing methods for recovery, detection, prevention, and treatment.

II. TAXONOMY

Cryptosporidium is one of several protozoan genera in the phylum Apicomplexa (Table 1). All are referred to as coccidia. Those that develop in the gastrointestinal tract of vertebrates through all of their life cycle include *Eimeria, Isospora, Cyclospora,* and *Cryptosporidium*. Those that are capable of or require extraintestinal development are referred to as cyst-forming coccidia and include *Besnoitia, Caryospora, Frenkelia, Hammondia, Neospora, Sarcocystis,* and *Toxoplasma*.

Several species of *Cryptosporidium* were named after the host in which they were found. Subsequent morphologic and cross-transmission studies have invalidated many species. *C. ameivae, C. anserinum,*

Table 1 Taxonomic Classification of *Cryptosporidium*

Classification	Name	Biological Characteristics
Empire	Eukaryota	Cells have nucleus containing most of cell's DNA, enclosed by double layer membrane; nearly all have mitochondria.
Kingdom	Protozoa (Goldfuss 1818)	Predominantly unicellular; most have cristate mitochondria, Golgi bodies, and peroxisomes.
Phylum	Apicomplexa (Levine 1970)	Unicellular endosymbiont or predator with apical complex typically composed of polar rings, rhoptries, micronemes, and usually a conoid; subpellicular microtubules and micropore common.
Class	Coccidea (Leuckart 1879)	Oocyst generally contains infective sporozoites that result from sporogony. Reproduction both asexual and sexual. Locomotion by flexing, gliding, or undulation.
Order	Eucoccidiorida (Leger & Duboscq 1910)	Merogony; infects vertebrates and invertebrates.
Family	Cryptosporidiidae (Leger 1911)	Develops beneath brush border and not in proper; meronts with knob-like attachment organelle; oocyst without sporocysts.

Source: Portions adapted from References 12a and 12b.

Table 2 Valid Named Species of *Cryptosporidium*

Species	Host	Author	Ref.
C. baileyi	*Gallus gallus* (chicken)	Current et al., 1986	42
C. felis	*Felis catis* (domestic cat)	Iseki, 1979	17
C. meleagridis	*Meleagris gallopavo* (turkey)	Slavin, 1955	5
C. muris	*Mus musculus* (house mouse)	Tyzzer, 1910	2
C. nasorum	*Naso literatus* (fish)	Hoover et al., 1981	24
C. parvum	*Mus musculus* (house mouse)	Tyzzer, 1912	3
C. serpentis	*Elaphe guttata* (corn snake)	Levine, 1980	12c
	E. subocularis (rat snake)		
	Sanzinia madagascarensus (Madagascar boa)		
C. wrairi	*Cavia porcellus* (guinea pig)	Vetterling et al., 1971	181

and *C. tyzzeri*, lacking adequate description, are *nomen nuda. C. crotali, C. ctenosauris, C. lampropeltis,* and *C. vulpis* were misidentified species of *Sarcocystis. C. agni, C. bovis, C. cuniculus, C. garnhami,* and *C. rhesi* are synonyms of *C. parvum. C. curyi* (cat source), with oocysts five to six times larger than those of *C. parvum,* are doubtful and unconfirmed. There are eight valid named species of *Cryptosporidium* (Table 2). There may be more than one species in fish. There are several unnamed species, including one from an amphibian,[11] five from reptiles,[12] one from bobwhite quail,[43] and possibly one from ostriches.[46]

C. parvum appears to be infectious for 79 species of mammals, including humans.[31] Although some isolates of *C. parvum* have been differentiated at the molecular level and others have been reported to vary in virulence, no traits of great enough significance have been reported to merit other taxon status.

III. HOST SPECIFICITY

In several studies, oocysts of *Cryptosporidium* were obtained from animals of one species and fed to or intubated into animals of another species. Generally, isolates from one class of vertebrates have not been infectious for animals of another class.[13] One study indicated that *C. parvum* oocysts from a human infected fish, amphibians, reptiles, birds, and mammals, based on finding oocysts in the feces of the recipients beginning 4 to 9 days after oral inoculation.[14] In contrast, *C. parvum* oocysts from a bovine did not infect fish, amphibians, or reptiles but simply passed through the digestive tract without infecting enterocytes.[15] In all other studies, oocysts from reptiles and birds did not infect mammals, with the possible exception of *C. baileyi* in an immunocompromised human,[59] and those from mammals did not infect birds. Even within a vertebrate class, oocysts of one species of *Cryptosporidium* did not always infect more than one host species. For example, *C. wrairi* infected only guinea pigs[16] and *C. felis* infected only cats.[17,18] In contrast, oocysts of *C. parvum* have been found infectious for virtually all mammals.

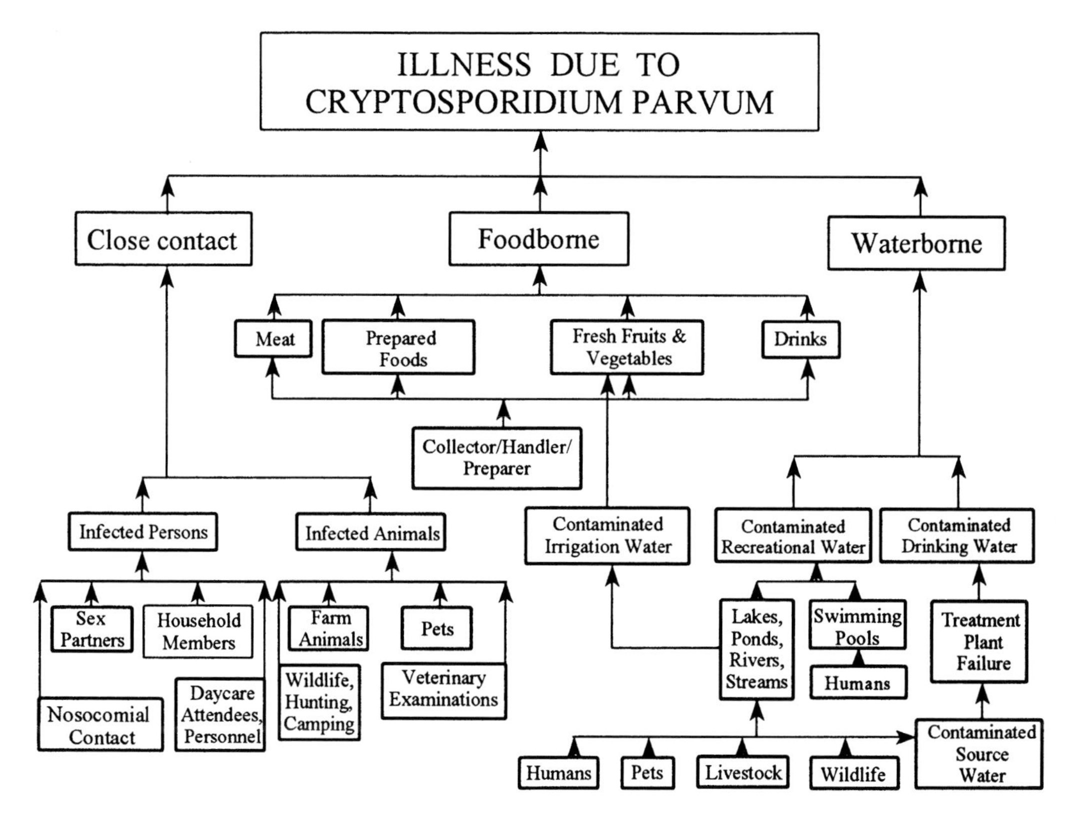

Figure 1 A "Fault Tree" depicting sources of oocysts and routes of transmission. (Courtest of Nancy N. Subat.)

IV. TRANSMISSION

Cryptosporidiosis is transmitted by the fecal-oral route via the oocyst stage. Sources and rates of infection are discussed in detail in Chapter 3. *C. parvum* is zoonotic, apparently lacking host specificity among mammals. Opportunities for such infection increase with close human-to-human contact and during care of infected livestock, zoo animals, or companion animals. Drinking or recreational waters serve as vehicles for transmission, and several outbreaks have been reported from both sources (see Chapter 4). The only published case of foodborne infection has been via cider made from apples that may have fallen to the ground and become contaminated with animal feces.[19] A "Fault Tree" indicating known and potential sources of infection is presented as Figure 1.

V. LIFE CYCLE

The life cycle is shown diagrammatically in Figure 2. The sporulated oocyst is the only exogenous stage. Consisting of four sporozoites within a tough two-layered wall, it is excreted from the body of an infected host in the feces.

The endogenous phase begins after the oocyst is ingested by a suitable host. Sporozoites excyst from the oocyst and parasitize epithelial cells of the gastrointestinal or respiratory tract. For most coccidia, excystation of sporozoites requires exposure to reducing conditions followed by exposure to pancreatic enzymes and/or bile salts. For *Cryptosporidium,* such exposure may enhance excystation, but sporozoites can excyst in warm aqueous solutions alone, possibly enabling infection and autoinfection of extraint-estinal sites: the conjunctiva of the eye, respiratory tract,[13] gall bladder, lymph nodes, testicle, ovary, uterus, and vagina.[20]

The anterior end of each excysted sporozoite adheres to the luminal surface of an epithelial cell until microvilli surround it, making it intracellular but extracytoplasmic. A unique organelle, referred to as an attachment or feeder organelle, develops between the parasite and host cell cytoplasm. Its function is unknown. Except for merozoites and microgametes, which leave host cells to invade other cells, all endogenous stages are located on the epithelial surface. Each sporozoite differentiates into a spherical

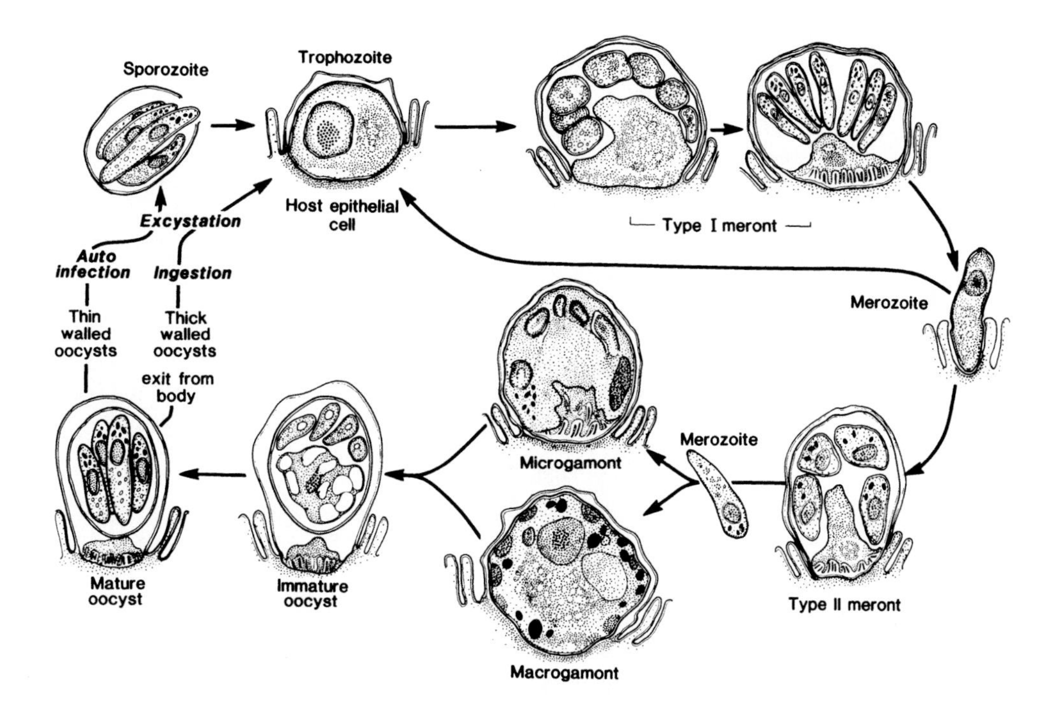

Figure 2 Diagrammatic representation of the life cycle of *C. parvum*. (From Fayer, R., Speer, C. A., and Dubey, J. P., in *Cryptosporidiosis of Man and Animals*, Dubey, J. P., Speer, C. A., and Fayer, R., CRC Press, Boca Raton, FL, 1990, 1. With permission)

trophozoite. Asexual multiplication, called schizogony or merogony, results when the trophozoite nucleus divides. *C. baileyi* has three types of schizonts or meronts, and *C. parvum* has two types. For *C. parvum*, type I schizonts develop six or eight nuclei, and each is incorporated into a merozoite, a stage structurally similar to the sporozoite. Each mature merozoite, theoretically, leaves the schizont to infect another host cell and develop into another type I or into a type II schizont which produces four merozoites. It is thought that only merozoites from type II schizonts initiate sexual multiplication (gametogony) upon infecting new host cells by differentiating into either a microgamont (male) or macrogamont (female) stage. Each microgamont becomes multinucleate, and each nucleus is incorporated into a microgamete, a sperm cell equivalent. Macrogamonts remain uninucleate, an ovum equivalent. It is assumed that only fertilized macrogamonts develop into oocysts that sporulate *in situ* and contain four sporozoites. Oocysts in the gastrointestinal tract are excreted with feces, whereas those in the respiratory tract exit the body with respiratory or nasal secretions. Some reports suggest that oocysts with thin walls release sporozoites that autoinfect the host, whereas those with thicker walls leave the body to infect other hosts.[21,22]

Autoinfection, in which the sequential asexual and sexual phases of the life cycle are repeated within the same host, is uncommon among other coccidian genera. Oocysts of *Eimeria, Hammondia, Isospora,* and *Toxoplasma* must sporulate outside the body to become infectious. Oocysts of *Frenkelia* and *Sarcocystis* sporulate internally but are infective only for other host species. *Cryptosporidium* and *Caryospora* are the only coccidia whose oocysts sporulate *in situ* and autoinfect.

The prepatent period is the shortest time after ingestion of infective oocysts to complete the endogenous life cycle and excrete newly developed oocysts. This time varies with the host and species of *Cryptosporidium*. Experimentally determined prepatent periods for *C. parvum* range from 2 to 7 days for calves, 2 to 14 days for dogs, 3 to 6 days for pigs, 2 to 5 days for lambs,[13] and 4 to 22 days for humans.[23] The patent period is the duration of oocyst excretion. Experimentally determined patent periods for *C. parvum* range from 1 to 12 days for calves, 3 to 33 days for dogs, 5 to 14 days for pigs,[13] and 1 to 20 days for humans.[23]

A. ELECTRON MICROSCOPY

The oocyst wall of *Cryptosporidium* spp., like that of other coccidia, has distinct inner and outer layers but is unique in having a suture at one end. The suture dissolves during excystation, opening the wall

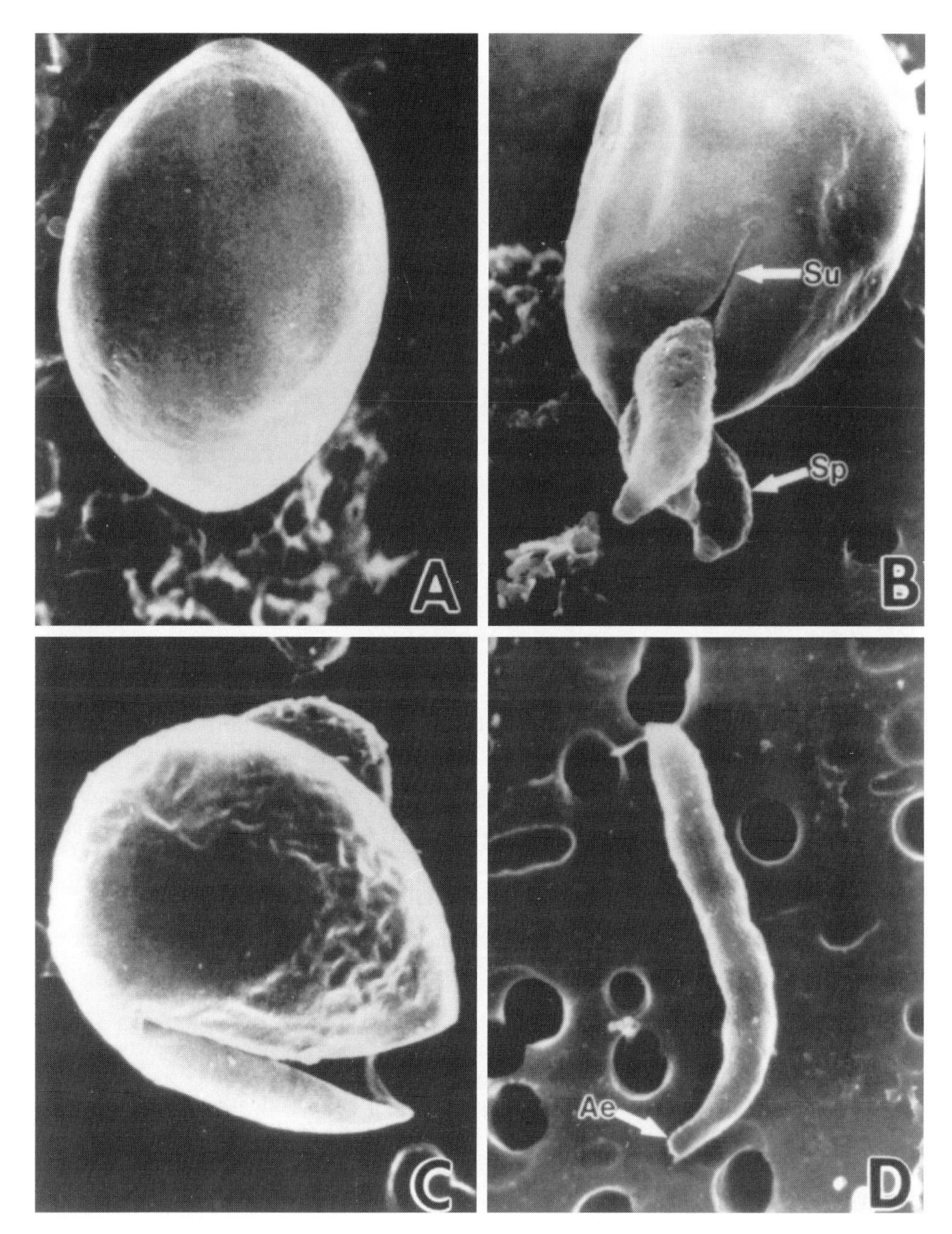

Figure 3 Scanning electron micrograph (SEM) of *C. parvum* oocysts and excysting sporozoites. **(A)** Intact oocyst before excystation (original magnification × 16,000). **(B)** Three sporozoites (Sp) excysting from an oocyst simultaneously via the cleaved suture (Su) (original magnification × 16,000). **(C)** Empty oocyst (original magnification × 16,000). **(D)** Excysted sporozoite; Ae = apical end (original magnification × 14,000). (A, B, and C from Reduker, D. W., Speer, C. A., and Blixt, J. A., *J. Protozool.,* 32, 708, 1985. D from Fayer, R., Speer, C. A., and Dubey, J. P., in *Cryptosporidiosis of Man and Animals*, Dubey, J. P., Speer, C. A., and Fayer R., Eds., CRC Press, Boca Raton, FL, 1990, 1. With permission.)

through which the sporozoites leave the oocyst (Figure 3). A trypsin-bile sensitive suture similar in appearance to that of *C. parvum* is found in the sporocyst wall of *Sarcocystis*, *Toxoplasma*, *Isospora*, and *Goussia*, suggesting a close phylogenetic relationship between *Cryptosporidium* and the sarcocystids or the calyptosporids (i.e., *Goussia*, *Calyptospora*).

Sporozoites and merozoites of *Cryptosporidium* spp. appear similar to those of other coccidia with organelles typical of the phylum, such as the pellicle, rhoptries, micronemes, electron-dense granules, nucleus, ribosomes, subpellicular microtublues, and apical rings. However, they lack other organelles such as typical polar rings, mitochondria, micropores, and the conoid. Posterior to the apical rings, sporozoites and merozoites have a cylindrical collar that appears to be the site of origin for the inner membrane complex and the subpellicular microtubules (Figure 4).

Sporozoites and merozoites bound to epithelial cells (possibly by specific receptors) become enveloped by microvilli until they reside within a parasitophorous vacuole (Figure 4). Changes in the apex of the host cell and in the parasite result in formation of an attachment or feeder organelle (Figure 4). At the contact site, an electron-dense layer forms in the host cell immediately above the terminal web of microfilaments, while the parasite plasmalemma invaginates at the apex just below the apical rings, which disappear. This invagination enlarges as an electron-dense collar expands laterally, eventually locating close to the lateral margin of the electron-dense layer. Concurrently, a fibrous layer that spans the area within the electron-dense collar forms between the invaginated parasite plasmalemma and the electron-dense layer of the host cell. The plasmalemma in this region becomes highly convoluted, increasing its surface area (Figure 4). The attachment organelle is separable from the schizont residual body only after merozoites are fully formed. Similarly, microgametocytes separate only after microgametes mature (Figure 8), and the plasmalemma invaginates until the mature zygote separates from the attachment organelle (Figure 11).

Trophozoites contain a prominent nucleolus within a single nucleus surrounded by cytoplasm, and a well-developed attachment/feeder organelle (Figure 5). During nuclear division of schizogony, division spindles, nuclear plaques, and centrioles have not been observed. After nuclear division, merozoites develop simultaneously around the margin of the schizont. Merozoite anlagen including a double inner-membrane complex, an electron-dense collar, electron-dense granules, a few micronemes, ribosomes, and cytoplasm form immediately beneath the schizont plasmalemma just above each schizont nucleus. Merozoites remain attached to a residuum at their posterior end and become more elongate during maturation. During this time a pair of rhoptries, more micronemes, numerous electron-dense granules, and many ribosomes form in the cytoplasm (Figure 6). Micronemes often form in rows perpendicular to the long axis of the merozoite (Figure 6). Upon maturity, merozoites separate from the residual body, the host cell membrane surrounding the schizont lyses, and merozoites become extracellular (Figure 7), able to infect other host cells.

Microgamonts have been found less frequently than other stages. Immature microgamonts resemble schizonts but contain small, compact nuclei. The single surface membrane later doubles at sites around the margin where microgametes form (Figure 8). Each microgamete forms as a nuclear protrusion at the gamont surface. Midway in microgamete formation, the nucleus in the microgamete bud contains equal amounts of electron-lucent and electron-dense chromatin, and an apical cap consisting of two closely applied membranes forms over the anterior end (Figure 9). Mature microgametes separate from the gamont surface, leaving the residual body surrounded by a single membrane and containing numerous ribosomes, endoplasmic reticulum, and a few micronemes (Figure 8).

Microgametes are rod shaped (1.4×0.5 µm for *C. parvum*), with a flattened anterior end, and lack both flagellae and mitochondria typically observed in microgametes of other coccidia. Most of the microgamete consists of a condensed nucleus (Figure 9). A plasmalemma completely surrounds the body. Beneath it a single membrane extends approximately two thirds the body length. Originating at an anterior conical structure, eight microtubules extend posteriad in close proximity to the surface of the nucleus (Figure 9). Three to five concentric lamellae extend outward at 90 degrees to the long axis at the posterior margin of the apical cap (Figure 9). Electron-dense granules of undetermined function are found in the cytoplasm at midbody (Figure 9).

Macrogamonts of *C. parvum* are approximately 4 to 6 µm and spherical to ovoid, have a large central nucleus with a prominent nucleolus, and contain lipid bodies, amylopectin granules, and unique wall-forming bodies in the cytoplasm (Figure 10). Little of the fertilization process of a macrogamont by a microgamete has been recorded, suggesting that the process is rapid. Microgametes attach at their apical cap to the surface of host cells harboring macrogamonts (Figure 9). Only the microgamete nucleus and associated microtubules have been observed within macrogamonts (Figure 10). Fusion of nuclei has not been observed.

The fertilized macrogamont, or zygote, develops into an oocyst with either a thin or a thick wall (Figure 11). Those that develop into thick-walled oocysts have type I and II wall-forming bodies similar

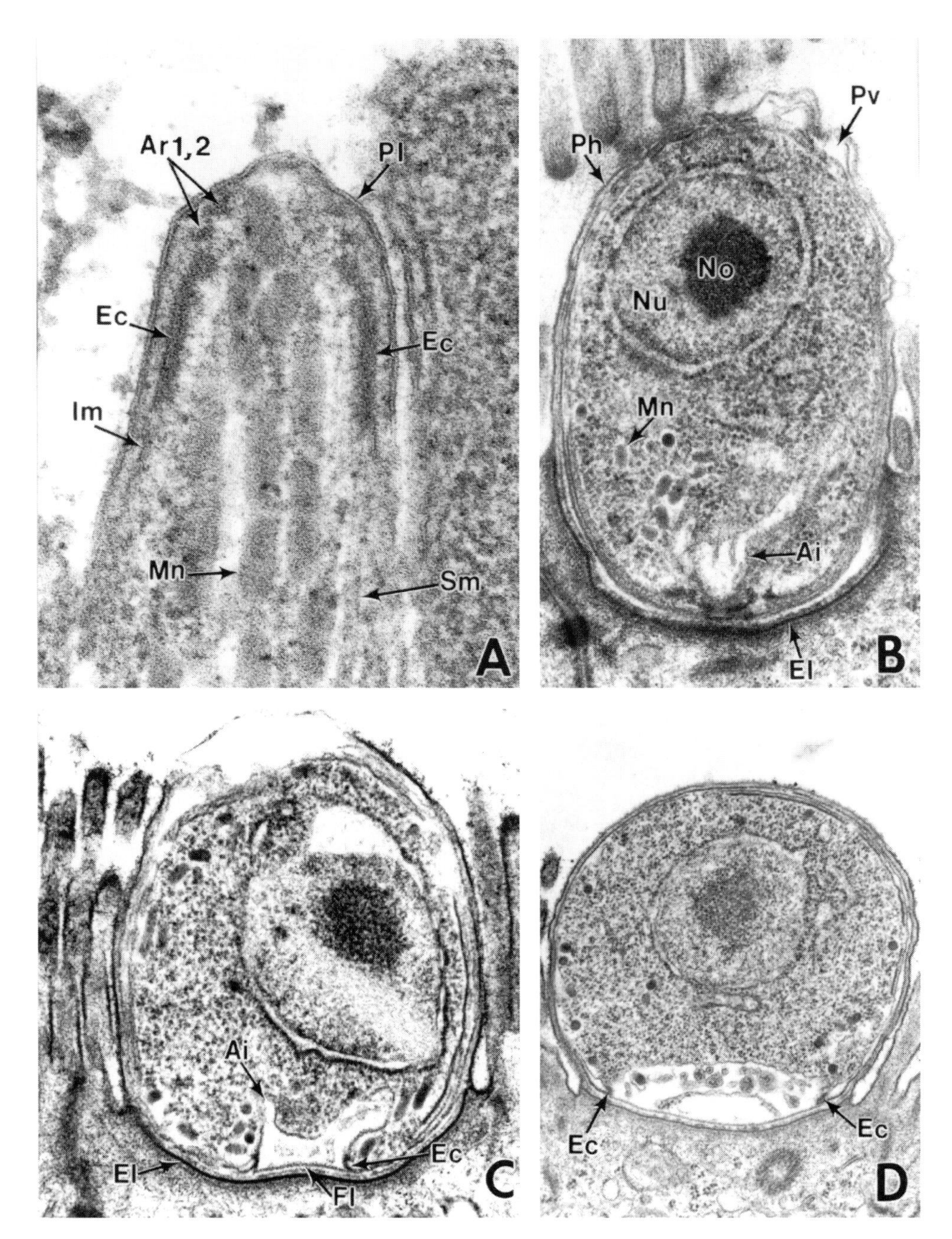

Figure 4 Transmission electron micrographs (TEMs) of *C. parvum*. **(A)** High magnification of apical complex of merozoite showing apical rings (Ar1, 2), micronemes (Mn), plasmalemma (Pl), inner membrane complex (Im), subpellicular microtubule (Sm), and electron-dense collar (Ec) (original magnification × 99,000). **(B)** Merozoite in early stage of attachment to an epithelial cell; Ai = anterior invagination; El = electron-dense layer of attachment zone; Mn = microneme; No = nucleolus; Nu = nucleus; Ph = plasmalemma of host cell; Pv = parasitophorous vacuole (original magnification × 32,500). **(C)** Slightly more advanced stage of attachment; note the enlarged anterior invagination (Ai), electron-dense collar (Ec), electron-dense layer (El), and fibrous layer (Fl) of attachment organelle (original magnification × 34,000). **(D)** Trophozoite with electron-dense collar located at margins of attachment organelle (original magnification × 19,500). (B and C from Fayer, R., Speer, C. A., and Dubey, J. P., in *Cryptosporidiosis of Man and Animals*, Dubey, J. P., Speer, C. A., and Fayer, R., Eds., CRC Press, Boca Raton, FL, 1990, 1. With permission)

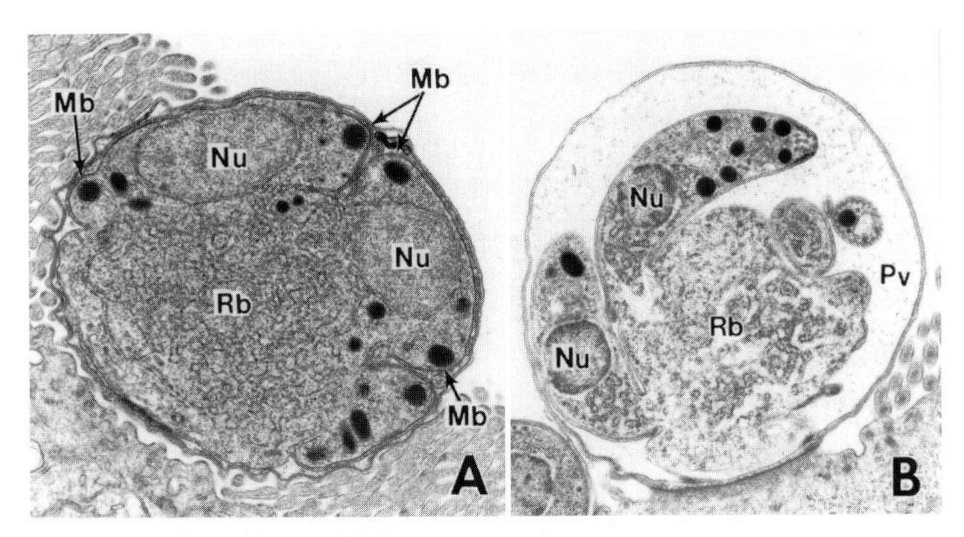

Figure 5 TEMs of schizogony of *C. parvum.* **(A)** Schizont showing early stages of merozoite formation at margin of schizont; note that a merozoite bud (Mb) is developing above each pole of the two nuclei (Nu) visible in this section; Rb = residual body (original magnification × 15,000). **(B)** Merozoites in an advanced stage of budding from the schizont residual body (Rb); Pv = parasitophorous vacuole (original magnification × 14,000). (Part B from Fayer, R., Speer, C. A., and Dubey, J. P., in *Cryptosporidiosis of Man and Animals*, Dubey, J. P., Speer, C. A., and Fayer, R., Eds., CRC Press, Boca Raton, FL, 1990, 1. With permission.)

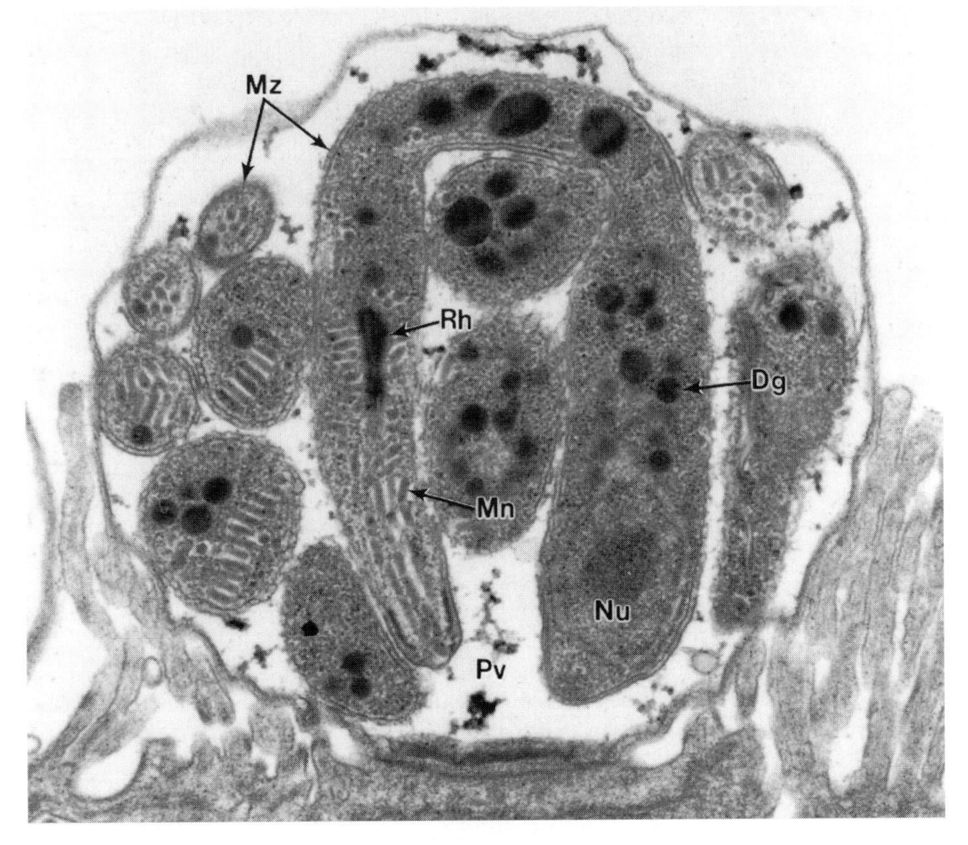

Figure 6 TEM of mature *C. parvum* schizont showing merozoites (Mz). Note dense granules (Dg), electron-dense collar (Ec), micronemes (Mn), nucleolus (No), nucleus (Nu), parasitophorous vacuole (Pv), and rhoptry (Rh) (original magnification × 28,000).

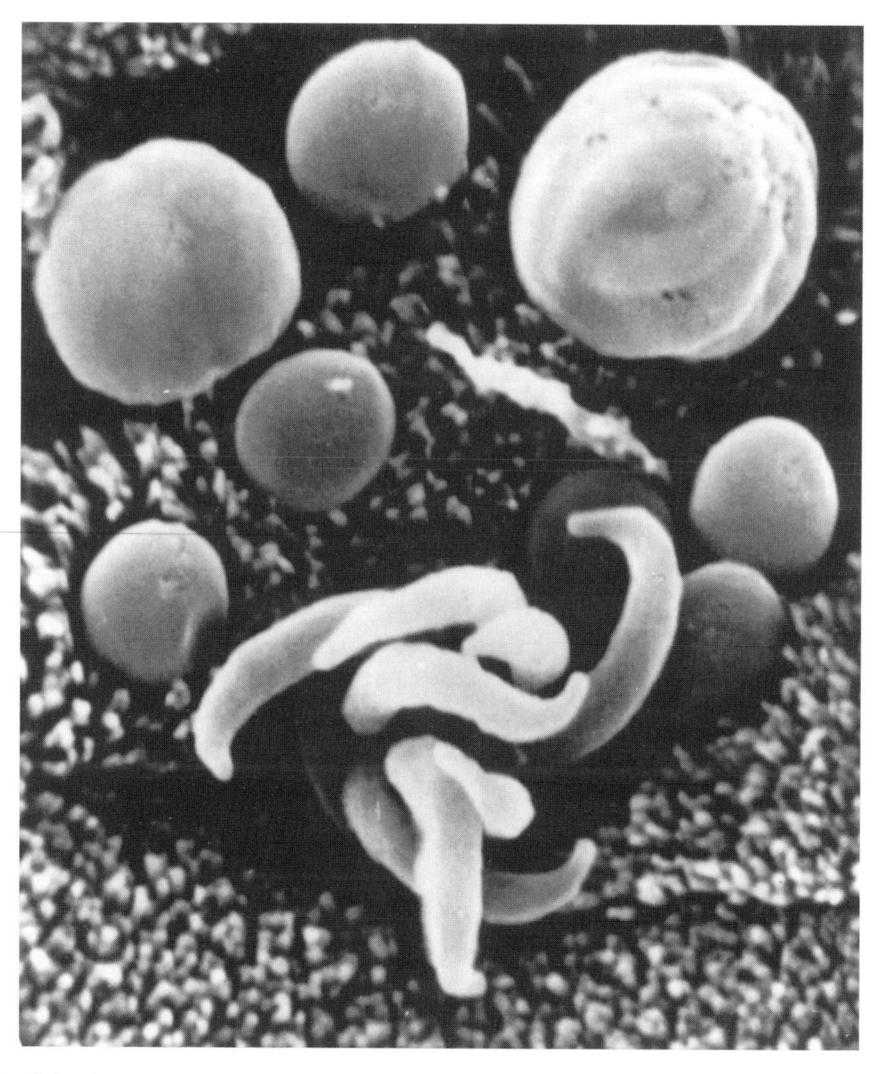

Figure 7 SEM of cryptosporidia on sheep intestinal epithelial cells. Eight merozoites are escaping from one host cell, and the impressions of merozoites are visible in the host cell at the upper right (original magnification × 15,750). (From Fayer, R., Speer, C. A., and Dubey, J. P., in *Cryptosporidiosis of Man and Animals*, Dubey, J. P., Speer, C. A., and Fayer, R., Eds., CRC Press, Boca Raton, FL, 1990, 1. With permission.)

to other coccidia. Those that develop into thin-walled oocysts lack the characteristic wall-forming bodies. Initially, two unit membranes form simultaneously external to the plasmalemma, while the sporont separates from the feeder/attachment organelle (Figure 12). Then, wall-forming body material is transported or exocytosed across the oocyst pellicle (i.e., plasmalemma and inner membrane), where it forms a thin, moderately coarse outer layer and a finely granular inner layer (Figure 12). Between these two layers of the oocyst wall is an electron-lucent zone that consists of the two oocyst membranes sandwiched between the outer and inner layers of the oocyst wall (Figure 12). The outer layer of the wall is continuous and of uniform thickness. The inner layer contains a suture at one pole which spans ⅓ to ½ the circumference of the oocyst (Figures 4, 11, 12).

VI. HOSTS

A. PISCINE CRYPTOSPORIDIOSIS

Cryptosporidium has been reported in nine species of freshwater and marine fish (Table 3) worldwide. The first report described a progressive illness lasting 2 months in a tropical marine fish (*Naso lituratus*).[24]

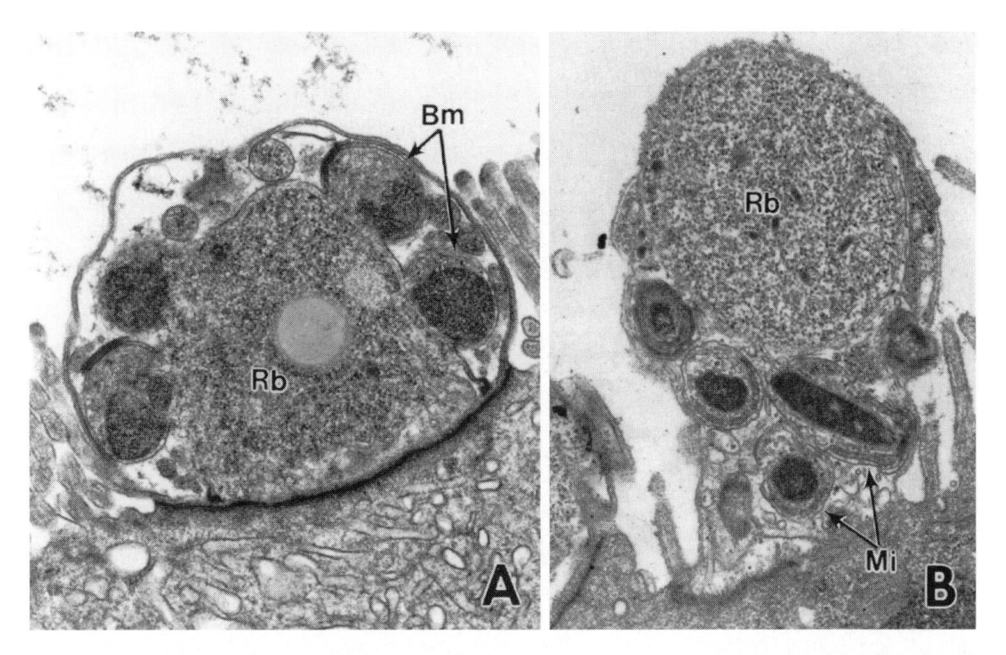

Figure 8 TEMs of *C. parvum* microgamonts. **(A)** Microgamont with microgametes budding (Bm) from a residual body (Rb) (original magnification × 16,000). **(B)** Mature microgamont with portions of six microgametes (Mi) separated from the residual body (Rb) (original magnification × 19,000). (From Fayer, R., Speer, C. A., and Dubey, J. P., in *Cryptosporidiosis of Man and Animals*, Dubey, J. P., Speer, C. A., and Fayer, R., Eds., CRC Press, Boca Raton, FL, 1990, 1. With permission.)

Intestinal morphology was normal except for endogenous parasite stages in the microvillar surface. Named *Cryptosporidium nasorum*, it is the only valid species of *Cryptosporidium* in fish.

Cryptosporidium sp. was found as developmental stages in intestinal villi in 5 of 35 carp in Czechoslovakia[25] and in intestinal contents from five brown trout from a reservoir near Sheffield, England.[26] Neither carp nor trout were reported ill.

Cryptosporidium sp. was found in stomachs of hatchery-reared fry and fingerling cichlids from a lake in Israel.[27,28] Electron microscopy revealed a unique membrane covering developmental stages.[28] Oocysts were found intracellularly in foci of necrotic epithelial cells. *Cryptosporidium* sp. was found in barramundi in association with inflammatory cells in the intestinal lamina propria.[29]

Although *C. parvum* oocysts from a human were reported infectious for fish based on gut histology,[14] attempts to experimentally infect 1.5-cm guppies[30] and bluegills[15] with *C. parvum* oocysts from a batch found infectious for suckling mice were unsuccessful. Slow passage retention of the oocyst inoculum in the gut or a prior undetected infection might result in misdiagnosis of an experimental infection.[15] No attempts have been made to determine infectivity of piscine cryptosporidian oocysts among fish species or for other vertebrate species.

It is evident that little is known of the species of *Cryptosporidium* infecting fish. Nothing is known of the host range of such species for fish or other animals, the prevalence or geographic distribution, nor the pathogenicity. Care must be taken in future reports to clearly distinguish *Cryptosporidium* sp. from *Epieimeria* sp., a superficially similar but clearly distinguishable genus parasitizing fish.

B. AMPHIBIAN CRYPTOSPORIDIOSIS

There is only one report of a natural infection in an amphibian. Feces collected from *Ceratophrys ornata* (Bell's horned frog) at the Metropolitan Toronto Zoo and placed in sodium acetate/formalin solution were stained by modified Ziehl-Neelsen stain.[11] Although oocysts were detected, the authors suggested these might have come from ingestion of infected mice. No other information was provided.

One group of investigators reported experimental infection of amphibians with *C. parvum* oocysts of human origin,[14] whereas another group was unable to infect African clawed frogs or poison dart frogs with *C. parvum* oocysts, which were infectious for suckling mice.[15]

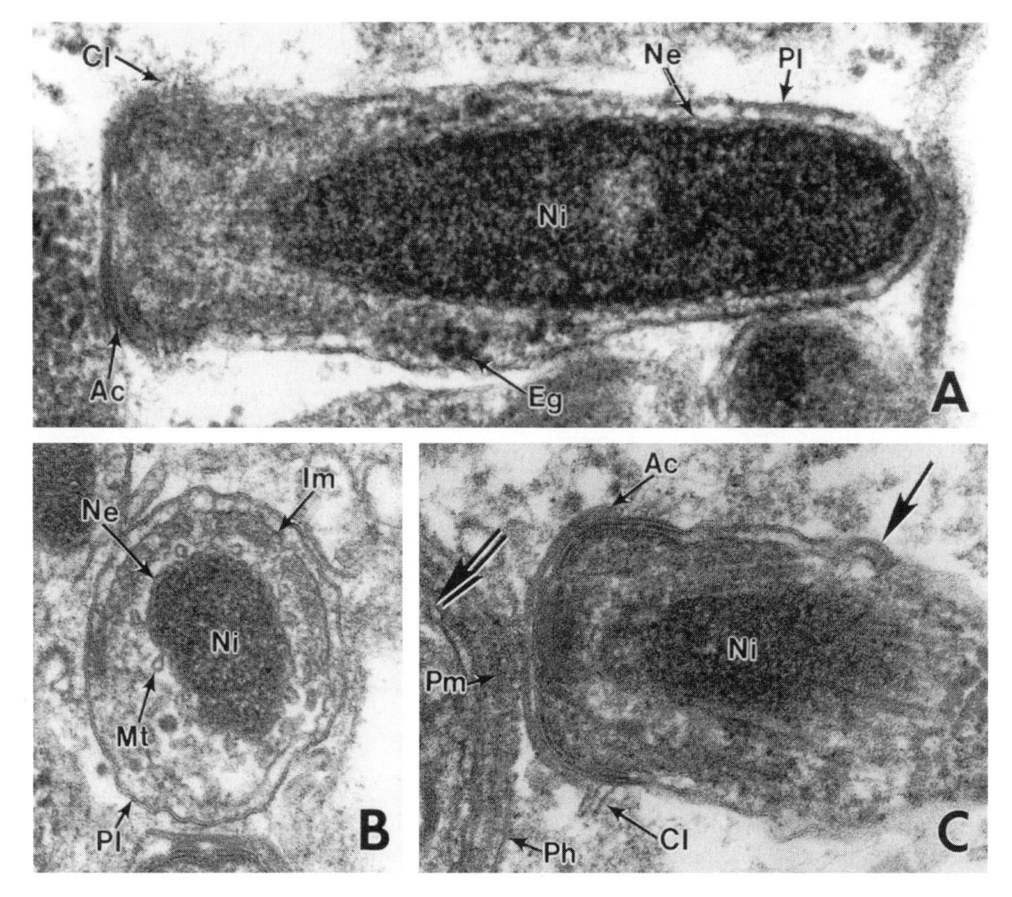

Figure 9 TEMs of *C. parvum* microgametes. **(A)** Longitudinal section. Note concentric lamellae (Cl) just posterior to the apical cap (Ac); Eg = electron-dense granules; Mt = microtubule; Ni = nucleus of microgamete; Pl = plasmalemma of microgamete (original magnification × 75,600). **(B)** Cross-section showing microtubules (Mt) in close proximity to the nuclear envelope (Ne), microgamete nucleus (Ni), plasmalemma (Pl), and inner membrane (Im) of microgamete (original magnification × 66,000). **(C)** High magnification of a microgamete in close proximity to a host cell containing a macrogamont; plasmalemmae of microgamete = single arrow and macrogamont = double arrow; Ac = apical cap; Cl = concentric lamellae; M = nucleus of microgamete; Ph = plasmalemma of host cell; Pm = parasitophorous vacuolar membrane (original magnification × 80,000). (From Fayer, R., Speer, C. A., and Dubey, J. P., in *Cryptosporidiosis of Man and Animals*, Dubey, J. P., Speer, C. A., and Fayer, R., Eds., CRC Press, Boca Raton, FL, 1990, 1. With permission.)

C. REPTILIAN CRYPTOSPORIDIOSIS

Cryptosporidium has been reported in 57 species of reptiles, including 40 species of snakes, 15 species of lizards, and 2 species of tortoises.[31] Little is known of the distribution or prevalence. Nearly all reports are from the U.S., Canada, and Australia. Most reports have described clinical and subclinical infections in captive reptiles, whereas only subclinical infections have been reported in wild reptiles.

Only one of five named species of *Cryptosporidium* from reptiles is valid. Because descriptions of *Cryptosporidium crotali*, *Cryptosporidium lampropeltis*, *Cryptosporidium ameivae*, and *Cryptosporidium ctenosauris* were consistent with the morphology of *Sarcocystis* sp., they are invalid. *Cryptosporidium serpentis*, described in snakes, is the only valid species, although as many as five species of *Cryptosporidium* infecting reptiles might exist based on oocyst morphology.[30]

1. Snakes

In boids, colubrids, elapids, and viperids, *C. serpentis* infects the gastric mucosa, primarily in mature animals.[30] Oocysts have been excreted in feces for years without clinical signs.[32] Oocysts also leave the body with regurgitated, partially digested food. Although surface antigen on oocysts of *C. serpentis* was recognized by antibody to oocysts of *C. parvum* from a commercial diagnostic kit,[33] the electrophoretic

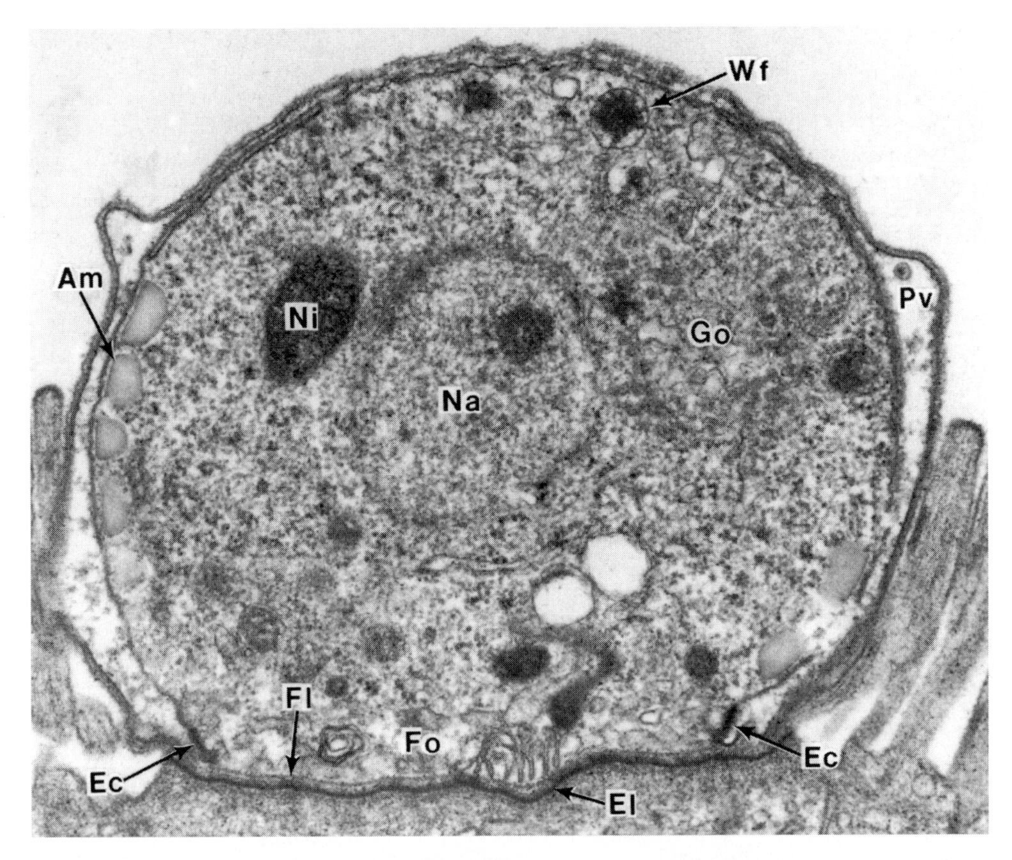

Figure 10 TEM of a *C. parvum* macrogamont containing a microgamete nucleus (Ni); Am = amylopectin granules; Ec = electron-dense collar; Fo = feeding organelle; Go = Golgi complex; Na = nucleus of macrogamont; No = nucleolus of macrogamont; Pv = parasitophorous vacuole; Wf = wall-forming body (original magnification × 34,000).

protein patterns from oocysts of *C. serpentis* were clearly different from those of *C. parvum*.[34] Diagnosis of most infections has followed recognition of clinical signs.

Disease is characterized by anorexia, postprandial regurgitation, lethargy, firm midbody swelling, weight loss, and death. Illness is chronic and progressive and self cure has never been observed.[35] Gastric hypertrophy, marked by a narrowed gastric lumen and thickened stomach wall, has been diagnosed by gastrography and fluoroscopy and confirmed by postmortem examination. The mucosal surface may exhibit petechiae, enlarged rugae, excess mucous, and multiple foci of necrosis. Microscopic examination of gastric tissue has revealed hyperplasia and hypertrophy of gastric glands, atrophy of granular cells, edema of the submucosa and lamina propria, and inflammation of the gastric mucosa characterized by infiltration with lymphocytes and heterophils. Gram-negative bacteria and adenovirus-like intranuclear inclusions also have been found in some infected snakes.

Little is known of the epidemiology, species specificity, or host range of *C. serpentis* or other species of *Cryptosporidium* from snakes. Attempts by two groups to experimentally infect suckling mice with numerous isolates of oocysts of *C. serpentis* have failed.[34,36] One group of investigators reported experimental infection of snakes with *C. parvum* oocysts of human origin,[14] whereas another was unable to infect corn snakes with *C. parvum* oocysts of bovine origin that were infectious for suckling mice.[15]

Neither halofuginone nor spiramycin produced satisfactory therapeutic outcomes for clinical infections in snakes.[36a] The only successful treatment of snakes has been supportive care including liquid diet and trimethoprim-sulfa.[37] Control in zoological collections can be managed by quarantine of new acquisitions for several weeks while screening feces for oocysts and rejecting all infected snakes.

2. Lizards

There are no named species of *Cryptosporidium* in lizards. Most infected lizards have no clinical signs of cryptosporidiosis; however, severe disease was reported in a pre-adult chameleon in a zoo in Birmingham,

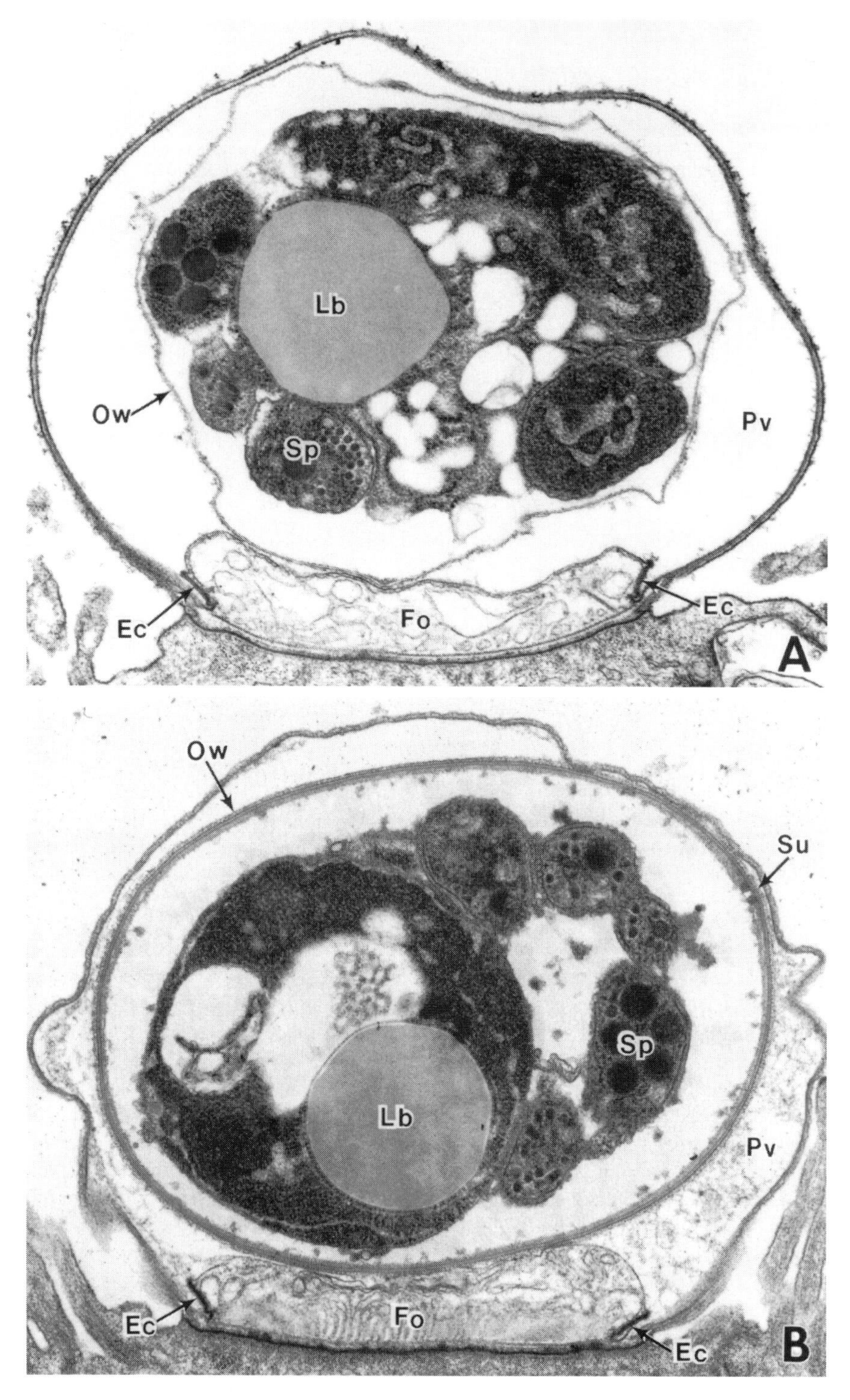

Figure 11 TEMs of *C. parvum* oocysts. **(A)** Thin-walled oocyst; Am = amylopectin; Fo = feeding (attachment) organelle; Lb = lipid body; Ow = thin oocyst wall; Pv = parasitophorous vacuole; Sp = sporozoite (original magnification × 23,240). **(B)** Thick-walled oocyst with suture (Su) at one pole; Lb = lipid body; Ow = thick oocyst wall; Pv = parasitophorous vacuole; Sp = sporozoite (original magnification × 23,240). (From Fayer, R., Speer, C. A., and Dubey, J. P., in *Cryptosporidiosis of Man and Animals*, Dubey, J. P., Speer, C. A., and Fayer, R., Eds., CRC Press, Boca Raton, FL, 1990, 1. With permission.)

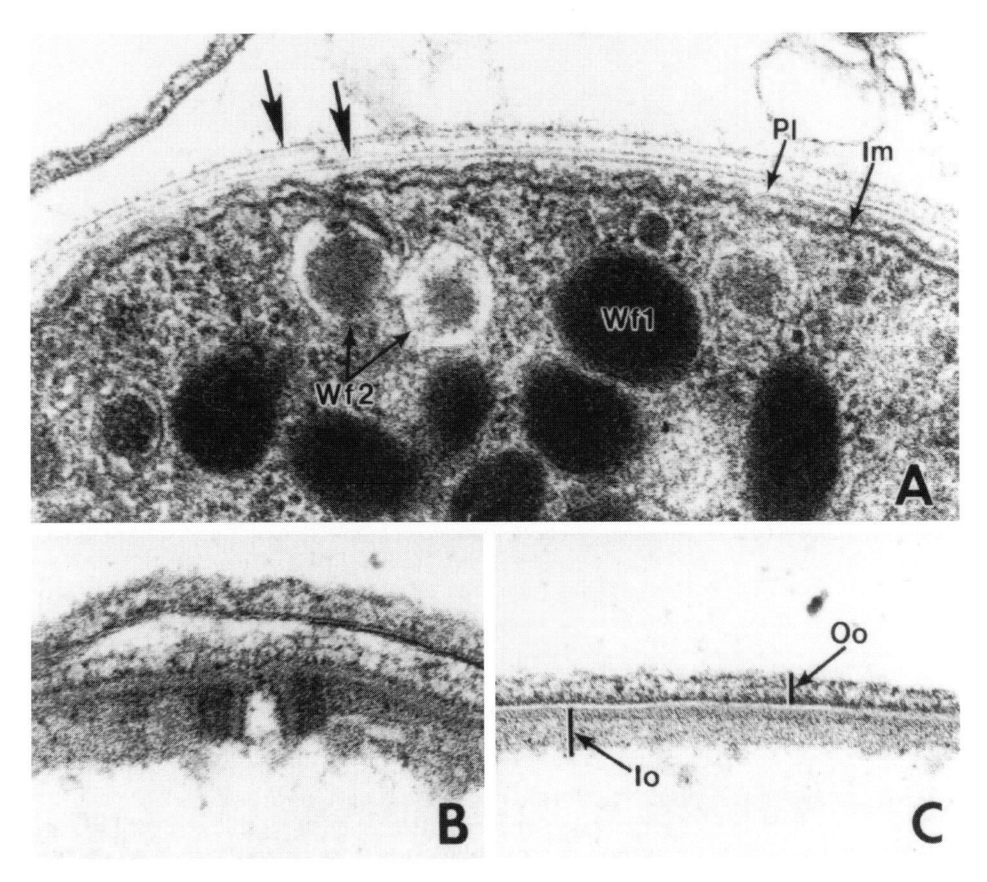

Figure 12 High magnification *C. parvum* TEMs. **(A)** Margin of zygote showing two membranes (arrows) external to the parasite plasmalemma (Pl) and two types of wall-forming bodies (Wf1 and Wf2); Im = inner membrane of zygote (original magnification × 80,000). **(B)** Portion of Figure 11B showing suture in oocyst wall (original magnification × 118,000). **(C)** Cross-section of completely formed oocyst wall showing inner (Io) and outer (Oo) layers (original magnification × 118,000). (From Fayer, R., Speer, C. A., and Dubey, J. P., in *Cryptosporidiosis of Man and Animals*, Dubey, J. P., Speer, C. A., and Fayer, R., Eds., CRC Press, Boca Raton, FL, 1990, 1. With permission.)

Table 3 *Cryptosporidium* Infections Reported in Fish

Fish	Locality	Ref.
Order Cypriniformes		
Bagrus bayad (catfish)	Egypt	203
Clarias lazera (catfish)	Egypt	203
Cyprinus carpio (carp)	Bohemia	26
Tilapia nilotica (tilapia)	Egypt	203
Order Perciformes		
Lates calcarifer (barramundi)	Australia	29
Naso lituratus (naso tang)	Canada	24
Oreochromis aureus (cichlid)	Israel	27
Oreochromis hybrids (cichlids)	Israel	28
Order Salmoniforms		
Salmo trutta (brown trout)	U.K.	26
Unidentified species		
Tropical marine fish	Canada	24

Source: Modified from References 30 and 31.

AL.[38] After exhibiting anorexia, weight loss, and lethargy, the chameleon died. Histological sections revealed numerous organisms in the gastric mucosa with lesions similar to those observed in snakes. An adult ocellated lacerta at the National Zoological Park in Washington, D.C., presented with postprandial regurgitation and passage of undigested food in its feces, lost weight, and died 3 weeks later.[39] Initial fecal examination revealed numerous oxyurid nematode ova and flagellated protozoa. Cytological preparations from gastric lavage and sucrose flotation of feces revealed round oocysts 4 to 6 μm in diameter. Major histologic findings included atrophied gastric mucosa and mononuclear cell infiltration of the lamina propria; most remaining gastric glands had few mucous cells and loss of granular cells. Although many geckos imported illegally into the U.S. from Madagascar died or were killed while moribund, the role of cryptosporidiosis could not be determined because of concurrent infections with other coccidia and lack of information on viral and bacterial pathogens.[30] This report, in which oocysts and developmental stages of *Cryptosporidium* were found in the cloaca of several geckos, is the only report of *Cryptosporidium* infecting the cloaca of a reptile, suggesting this might be a separate species.[30]

Little is known of the specificity or host range of *Cryptosporidium* from lizards. Three-month-old oocysts from *Varanus* lizards in Texas were not infectious for suckling ICR outbred mice.[30]

3. Tortoises

Among the Testudines, *Cryptosporidium* has been reported only in tortoises. Oocysts were observed in feces from a *Geochelone elegans* (star tortoise) at the San Diego Wild Animal Park.[40] Another zoo specimen, *Geochelone carbonaria* (red-footed tortoise), underwent a steady decline in condition and was euthanized soon after cryptosporidiosis was diagnosed.[37] No description of the parasite was provided in either case.

D. AVIAN CRYPTOSPORIDIOSIS

1. Species

Of four named species of *Cryptosporidium* from birds, only two are valid. Because the original descriptions of *C. tyzzeri* Levine 1961 from the chicken and *C. anserinum* Proctor and Kemp 1974 from the goose lacked sufficient detail for subsequent identification, they are considered invalid. The two valid species are *C. meleagridis* Slavin 1955 from turkey poults described by endogenous intestinal stages[5] and later by the oocyst and sporozoite morphology,[41] and *C. baileyi* Current, Upton and Haynes 1986 from broiler chickens for which the entire life cycle and oocyst morphology were described.[42] A possible third species, from bobwhite quail, with oocysts similar to those of *C. meleagridis,* differed from *C. meleagridis* by infecting the entire small intestine and by causing severe morbidity and mortality.[43-45] A possible fourth species, from ostriches, with oocysts similar in size to those of *C. meleagridis,* were not infectious for freshly hatched chickens, turkeys, or quail.[46]

2. Hosts

Infections with *Cryptosporidium* have been reported in over 30 species of birds,[31] including domesticated chickens, turkeys, ducks, geese, quail, pheasants, peacocks, and a wide variety of wild and captive birds. An 8-week-old jungle fowl (*Gallus sonneratii*) died with renal and respiratory cryptosporidiosis.[47] A 4-month-old black-throated finch (*Poephila cincta*), found dead, had renal cryptosporidiosis.[48] Nearly all lovebird (*Agapornis* sp.) chicks less than 1 month of age died from intestinal cryptosporidiosis.[49] A young pigeon died of intestinal cryptosporidiosis.[50] A budgerigar (*Melopsittacus undulatus*), found dead, had respiratory cryptosporidiosis; another had intestinal cryptosporidiosis.[51] Cockatiels (*Nymphocus hollandicus*) had intestinal cryptosporidiosis.[51] Black-headed gull chicks (*Larus ridibundus*) had respiratory and intestinal cryptosporidiosis.[52] Four-week-old canaries[53] and an Australian diamond firetail finch (*Staganoplura bella*)[54] had proventricular cryptosporidiosis. A red-lored parrot (*Amazona autumnalis*)[55] had cloacal cryptosporidiosis. Oocysts were found in feces of a macaw (*Ara* sp.) and a Tundra swan (*Cygnus* sp.).[56] Oocysts were found in feces and in tissue sections of the cloaca of ostriches.[46,57]

All attempts to experimentally infect rodent, porcine, caprine, and bovine hosts with *C. baileyi* have been unsuccessful.[58] However, there has been one report of *C. baileyi* oocysts excreted by an immunocompromised human and endogenous stages found in organs taken at autopsy.[59]

3. Geographic Distribution and Prevalence

C. meleagridis or unnamed species have been reported in turkeys in Canada, the Czech Republic, Romania, Scotland, and the U.S. The prevalence is unknown.

C. baileyi or unnamed species have been reported in chickens in Argentina, the Czech Republic, Denmark, Greece, Hungary, Japan, Korea, the Netherlands, Romania, Scotland, Spain, Taiwan, Turkey, and the U.S. Information on prevalence is scarce. Serological data from 18 flocks in Delaware, Maryland, and Virginia indicated 22 to 50% of 49- to 63-day-old birds had antibodies to *Cryptosporidium*.[60] Neither the species nor the prevalence in wild birds is known.

4. Life Cycle and Morphology

As first reported, unsporulated oocysts of *C. meleagridis* were 4.5 × 4.0 μm, young schizonts were attached to villar epithelium, some forms were in goblet cells and in depressions between epithelial cells, mature schizonts contained eight falciform merozoites 5 × 1 μm, microgamonts were round or oval and contained 16 microgametes with no flagella, and macrogametes were roughly oval, 4.5–5.0 × 3.5–4.0 μm.[5] Fully sporulated oocysts in feces contained four sporozoites and a residuum, but no micropyle, and measured 5.2 × 4.6 μm.[41] The factors influencing the site of endogenous development are not known. Oral infection of turkey poults with oocysts produced infections in the ileum, ceca, colon, cloaca, and bursa of Fabricius.[41] Natural infections in turkeys with *Cryptosporidium* sp. have been found in enteric and respiratory sites.

As first reported, *C. baileyi* had three structural types (generations) of schizonts and two types (thin- and thick-walled) of oocysts.[42] Thin-walled oocysts were thought to excyst *in situ* and to reinitiate the endogenous cycle, whereas thick-walled oocysts were excreted in the feces. Unsporulated oocysts were 6.3 × 5.2 μm, and sporulated oocysts were 6.2 × 4.6 μm.[42] First generation schizonts were 5.0 × 4.9 μm and contained eight 6.9 × 1.1-μm merozoites; second generation schizonts were 5.1 × 5.1 μm and contained four 5.0 × 1.3-μm merozoites; third generation schizonts were 5.2 × 5.1 μm and contained eight 3.6 × 1.1-μm merozoites. Microgamonts were 4.0 μm in diameter and contained 16 microgametes. Macrogamonts were 5.8 μm in diameter. The primary sites of development were the bursa of Fabricius and the cloaca, with very few birds harboring organisms in the respiratory tract. Prepatent and patent periods in chickens were 2 to 5 days and 24 to 30 days, respectively, and in mallards 3 to 4 and 10 to 18 days, respectively.[61,62] Intratracheal inoculation of chickens with oocysts resulted in extensive infection of the respiratory tract, including the nasopharynx, larynx, trachea, bronchi, and air sacs. Placing oocysts on the conjunctiva resulted in conjunctival, bursal, and cloaca infections. These and other routes of inoculation have been reviewed by Lindsay and Blagburn.[58]

5. Age, Disease, and Lesions

Natural infections with *C. baileyi* occur in chickens under 11 weeks of age.[63] In experimentally infected chickens, susceptibility to infection was age related with an inverse relationship between the age of chickens and length of the patent period.[42,64,64a]

Avian cryptosporidiosis can be respiratory, enteric, or renal. Usually only one location is affected in an outbreak. Cryptosporidiosis in a flock of broiler chickens was reported concurrent with chicken anemia virus infection.[65] Neither the species nor the exact relationship between cryptosporidiosis and acute transmissible enteritis are known. This severe illness of turkeys has resulted in losses of approximately $25 million in North Carolina in 1994 and 1995. *Cryptosporidium,* as well as enteric viruses and bacteria, are associated with this disease.

Respiratory disease has been reported in chickens, turkeys, quail, mallard ducks, ring-necked pheasants, budgerigars, black-headed gulls, and a jungle fowl.[58] Generally, clinical signs include coughing, sneezing, dyspnea, and rales.[58] Excess mucus may be present in the trachea, sinuses, and nasal passages, and fluid may be present in air sacs. Microscopically, infected epithelium may be hypertrophic and hyperplastic with macrophage, heterophil, lymphocyte, and plasma cell infiltrates. Cilia may be reduced or absent, and microvilli may be branched, blunted, or atrophied.[58a]

Enteric disease has been reported in turkeys, bobwhite quail, budgerigars, cockatiels, lovebirds, pigeons, and black-headed gulls. High morbidity and mortality have been seen in bobwhite quail and turkeys. Gross lesions include small intestinal and cecal distension with mucoid contacts and gas. Microscopic lesions include villous atrophy and fusion, and crypt hyperplasia (Figure 13). Inflammatory cell infiltrates may be present.

Renal disease has been reported in finches, jungle fowl, and chickens.[58] Kidneys were pale and enlarged. Epithelial cells of collecting ducts, collecting tubules, distal convoluted tubules, and ureters were hypertrophic and hyperplastic. Inflammatory cell infiltrates were often present.

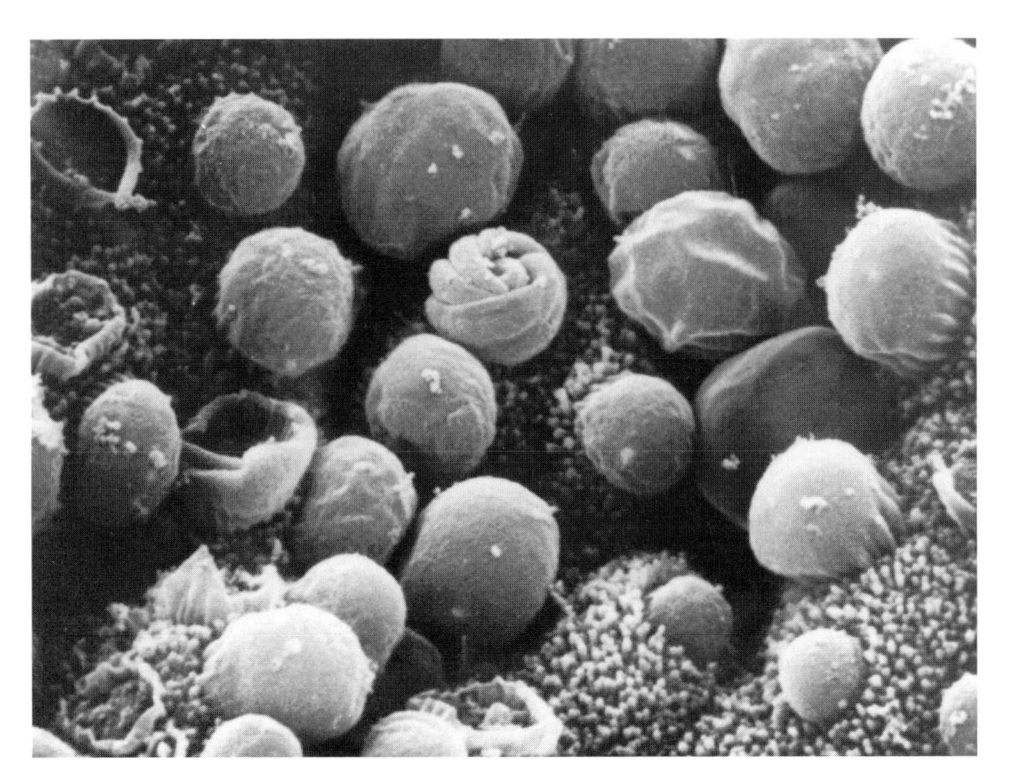

Figure 13 SEM of the cloaca of a chicken experimentally infected with *C. baileyi.* Some villi are fused near the surface of the parasites. The host cell membrane covering one schizont is gone, exposing merozoites. Note the craters that remain after parasites have ruptured host cells at the attachment site. (Courtesy of W. L. Current and S. L. White, Lilly Research Laboratories; Indianapolis, IN.)

6. Treatment and Control

Few reports of avian cryptosporidiosis mention whether anticoccidials drugs were present in feed or if they were used for therapy.[58] However, of those mentioned, none has been effective against natural respiratory or enteric infections. Chickens develop immunity to *C. baileyi* that clears the parasite from intestinal and respiratory sites and leaves birds resistant to reinfection.[64a,65a] Cell-mediated immunity plays the primary role in resistance, as opposed to antibody-mediated mechanisms.[65b] No specific immune prevention or therapy is available at the present time. In experimental studies, the most widely used anticoccidials were ineffective for prevention and treatment. Control methods have relied primarily on sanitation and disinfectants such as ammonia compounds.[44] Metal brooders, feeders, and waterers were disinfected and exposed to direct sunlight for 3 days, concrete floor pens were cleaned, and fresh shavings were applied. Such sanitation methods are most applicable to small farms and pet bird facilities. Because darkling beetles (litter beetles) serve as vectors for many common avian pathogens, including *Eimeria,* abatement programs should be incorporated in poultry disease control programs.[58b]

E. MAMMALIAN CRYPTOSPORIDIOSIS

Infections have been reported in 79 mammalian species.[31]

1. Ruminants

Two species of *Cryptosporidium* have been reported in ruminants. The intestinal species, *C. parvum,* most commonly causes acute cryptosporidiosis in young ruminants. An abomasal species resembling *C. muris* has been reported in chronically infected adults.

a. Bovine (Bos taurus)

i. History and Life Cycle

In 1971, the first report of bovine cryptosporidiosis described endogenous stages in the jejunum of an 8-month old heifer with chronic diarrhea.[6] Reports followed of diarrheic calves 2 weeks old or less from beef and dairy herds.[66-68] *C. parvum* was found in histological sections of lower jejunum and ileum.

Other enteropathogens often associated with calfhood diarrhea were not found in some calves, leaving *C. parvum* as the sole pathogen.[68] A study in the U.S. confirmed these Canadian findings and suggested that cryptosporidia were common enteropathogens of calves.[69] Studies worldwide have supported this.[70] Experimental infections with purified oocysts confirmed that *C. parvum* alone induced clinical illness.[71,72] The prepatent period of *C. parvum* in 20 experimentally infected calves is 3 to 6 days (Fayer, unpublished).

ii. Prevalence, Geographic Distribution, and Clinical Signs

C. parvum is highly prevalent in young calves worldwide. Natural infection has been reported in calves as young as 4 days[73] but is most common around 2 weeks of age and rare in calves under 1[69] and over 4 weeks of age.[10] Observations of large numbers of *C. parvum* oocysts per gram of diarrheic feces from preweaned calves and few or no oocysts in normal feces of older cattle has led to assumptions that older cattle are refractory to infection. Small numbers of oocysts per gram of feces and insensitive detection methods may have influenced findings. In France,[74] Canada,[74a] the U.S.,[75,76] Spain,[76a,77] the U.K.,[78,79,79a] and Brazil,[79b] adult cattle or calves over 20 weeks of age excreted oocysts considered *C. parvum.* (For details, see Chapter 3.)

The most prominent signs of cryptosporidiosis are seen in preweaned calves and include diarrhea accompanied by lethargy, inappetence, fever, dehydration, and/or poor condition. Peak numbers of oocysts in diarrheic stools from natural and from experimental infections can range from 10^5 to 10^7 oocysts per gram of feces.[80,81] Experimental infections with as many as 1.5×10^6 oocysts of *C. parvum* alone caused diarrhea of 4 to 18 days' duration (usually 6 days) but seldom caused the severe dehydration, extreme morbidity, and high mortality of ETEC-K99+ or rotavirus infections. Severe outbreaks with prolonged and intractable diarrhea or high mortality where *C. parvum* was the only enteropathogen found[82,83] could have resulted from infection with highly virulent isolates of *C. parvum*, an immunodeficient host population, or misdiagnosis.[71] Because calves are highly vulnerable to ETEC-K99+ in the first few days of life and to rotavirus and coronavirus infections at 1 and 2 weeks of age,[71] investigations of calfhood diarrhea should always consider the possibility of multiple infections.

In 1985, cryptosporidiosis with endogenous stages in the abomasum and an oocyst stage indistinguishable from *C. muris* was found in feces from older calves and mature cattle in the U.S.[84,85] The finding of two 2-week-old calves excreting *C. muris*-like oocysts was a rare event (Anderson, personal communication). In 150 dairies in 12 states, 1.6% of 48,810 fecal smears were positive for *C. muris*-like oocysts by acid-fast staining.[86] Of 1746 fecal specimens examined from 486 to 693 cows at a dairy in California, 7.0 to 9.7% were excreting *C. muris*-like oocysts.[86a] Infected cows appeared to excrete oocysts continually for a prolonged time.[86a] Milk production was significantly reduced in those affected. Cattle in feedlot pens from 10 beef-producing states in the U.S. and in Mexico also were positive.[86] Infected cattle had significantly elevated plasma pepsinogen levels and weighed significantly less than uninfected cattle of comparable age. Cattle with abomasal cryptosporidiosis had no diarrhea, though they excreted oocysts for several months.

iii. Diagnosis

Finding oocysts in diarrheic feces of cattle is indicative of infection. For detailed methods of detection, Chapter 2 and Reference 71. Oocysts of *C. parvum* range from 4.5–5.4 × 4.2–5.0 μm with a mean size of 5.0 × 4.5 μm and a shape index of 1.1[86b] but can vary, depending on whether they are fresh or fixed and how they are processed, whereas *C. muris*-like oocysts are larger: 6.6–7.9 × 5.3–6.5 μm with a mean size of 7.4 × 5.6 μm and a shpae index of 1.3.[86b] Necropsy findings are nonspecific, because congested mucosae and alterations in the color, odor, and consistency of gut contents are not pathognomonic.[71]

Endogenous stages and oocysts can be detected in stained impression smears from the ileal mucosa and in histological sections of ileum fixed before autolysis begins, within 20 minutes after death.[88,89] Euthanasia of a moribund calf with immediate fixation or freezing of intestine for histology, immunocytochemistry, or identification of agents in feces can yield excellent results.[90] Because calves acquire antibody via colostrum and because nearly all calves become exposed to the organism, serological findings are not diagnostic of acute infection.[91]

iv. Lesions

C. parvum is most often in the distal small intestine associated with villous atrophy, villous fusion, metaplasia of the surface epithelium to low columnar or cuboidal cells, degeneration or sloughing of individual enterocytes, and shortening of microvilli.[71] Mononuclear cells and neutrophils infiltrate the lamina propria.[83] Cecum, colon, and duodenum can also be infected.[71,83] Crypts in all sites become dilated and contain necrotic debris or dead leukocytes.[83] Such lesions have reduced absorption of vitamin

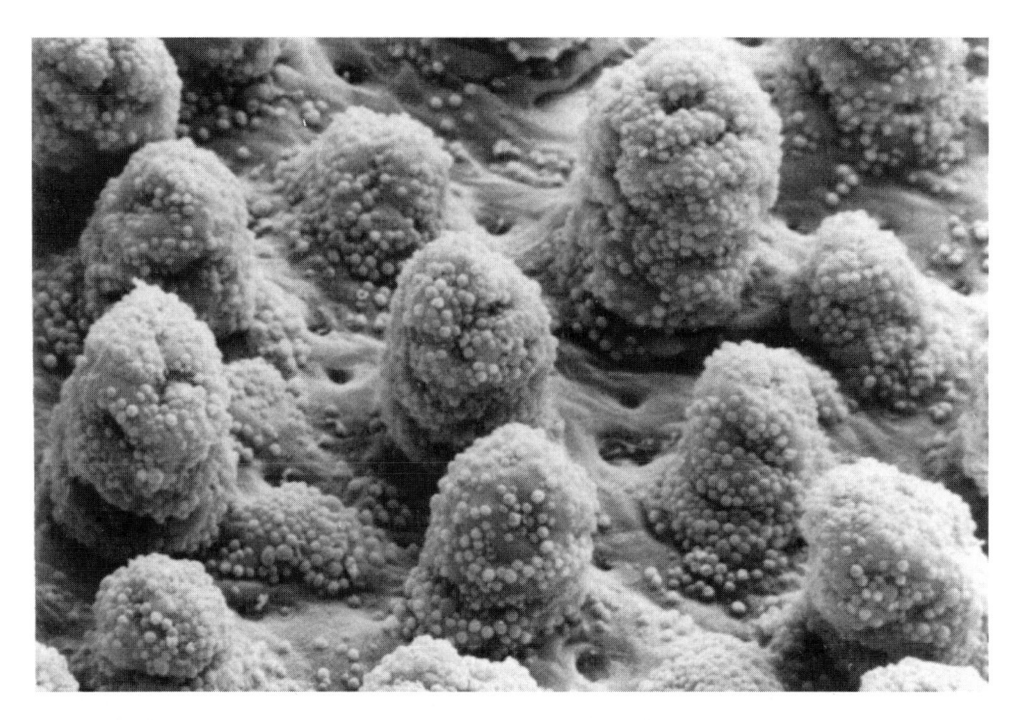

Figure 14 SEM of stages covering the villar and intervillar surface of the small intestine of an experimentally *C. parvum* infected mouse (original magnification × 225). (From Vitovec, J. and Koudela, B., *J. Vet. Med.*, 35, 515, 1988. With permission.)

A[92] and carbohydrates.[93] The lining of the urinary tract in diarrheic calves excreting oocysts in feces and urine[94] and in the lung of a diarrheic calf[95] also have been infected.

Mechanisms by which *C. parvum* damages the gut and produces clinical effects are somewhat speculative.[71] Parasites grow and derive energy from the host by redirecting activities of parasitized cells or by utilizing nutrients absorbed from the gut lumen. If microvilli reduce in number and size by displacement with endogenous stages (Figures 14 and 15) and remaining mature enterocytes are further reduced by sloughing (Figure 16), microvillous disaccharidase activity decreases, enabling lactose and other sugars to enter the large bowel undegraded. These sugars could promote bacterial overgrowth with formation of volatile free fatty acids which change osmotic pressure,[81,91,96] or accumulated nonabsorbed hypertonic nutrients in the lumen[96] could induce diarrhea. Hypersecretion of fluids and electrolytes in the ileum suggests a parasite-produced, cholera-like toxin that stimulates cyclic AMP production through the adenyl-cyclase system in the enterocytes.[81]

Endogenous stages of the *C. muris*-like species have been found from apex to orifice on flattened epithelial cells lining dilated digestive glands of the abomasum with lymphocyte and plasmacyte infiltrates between infected glands.[97] Elevated plasma pepsinogen concentrations probably result from destruction of microvilli lining these peptic glands.[71]

v. Epidemiology

Potential sources of infection for calves are (1) other infected calves or mature cattle, especially in communal housing or in permanent outdoor pens; (2) other infected animals — rodents, cats, dogs, wild animals; (3) mechanical carriers such as insects, birds, and humans; and (4) fecal-contaminated cows' teats, feed, water, bedding, feed utensils, brushes, shovels, wheels of vehicles. Although subclinical infections have been confirmed in adult cattle, a periparturient rise in *C. parvum* oocyst excretion has not been demonstrated in cows. See Chapter 3 for additional comments.

vi. Treatment and Control

No antibiotics or antiprotozoal drugs licensed for animal use have been approved for prophylaxis or therapy of cryptosporidiosis (see Chapter 5). For experimentally infected calves under controlled conditions, anti-cryptosporidial activity was reported for lasalocid,[98,99] halofuginone,[100,101] decoquinate,[102] and paromomycin.[103] Prophylaxis with paromomycin was highly efficacious and nontoxic, but it has not been tested under field conditions.

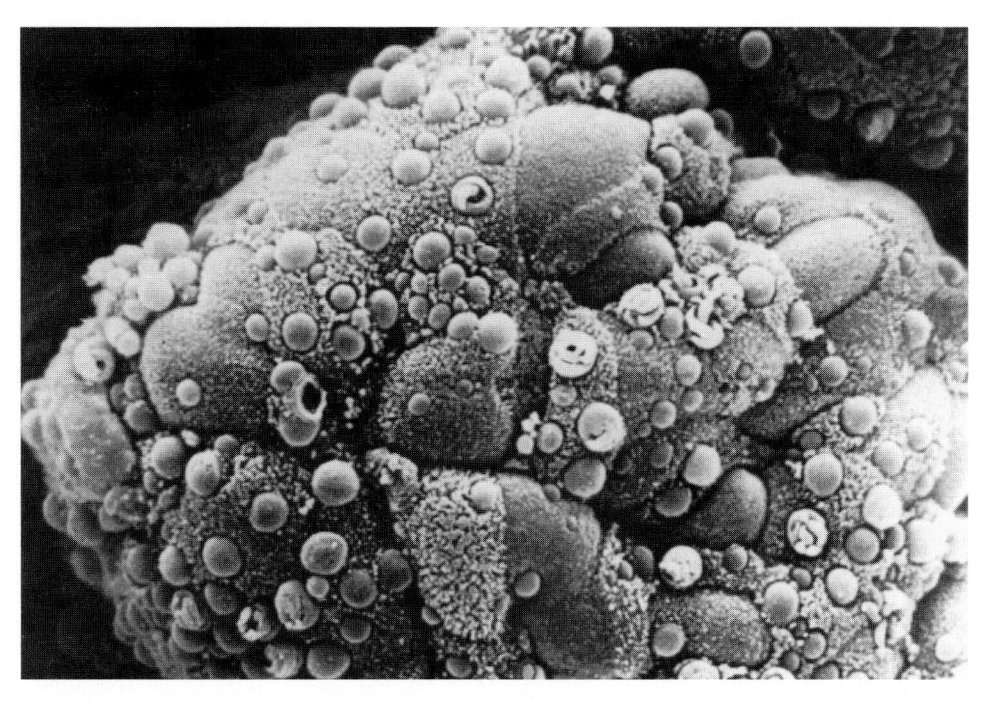

Figure 15 SEM showing erosion of villi in an experimentally *C. parvum*-infected mouse (original magnification × 925). (From Vitovec, J. and Koudela, B., *J. Vet. Med.,* 35, 515, 1988. With permission.)

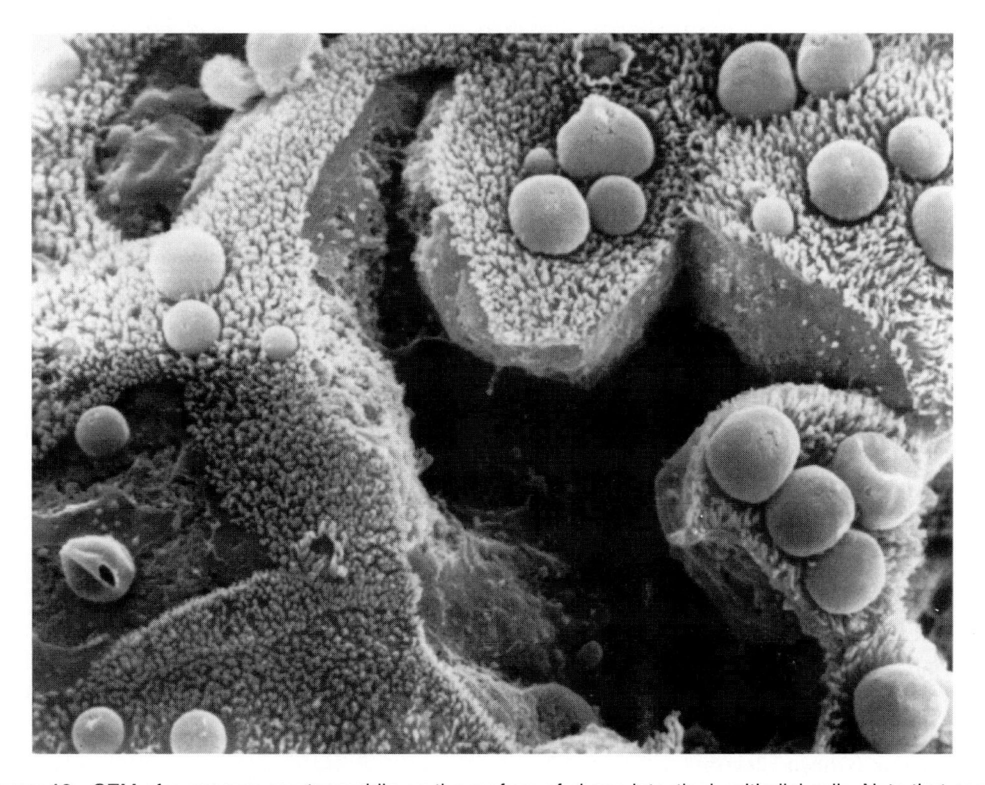

Figure 16 SEM of numerous cryptosporidia on the surface of sheep intestinal epithelial cells. Note that some of the epithelial cells have sloughed (original magnification × 3900).

Protective passive immunity from prophylactic feeding of hyperimmune bovine colostrum reduced the severity of diarrhea and the number of oocysts excreted by calves experimentally infected with *C. parvum* (see Chapter 6).[104]

Initiate oral or parenteral rehydration with electrolyte solutions even before pathogen status can be established in cases of severe diarrhea. Couple rehydration with kaolin-based absorbent and demulcent suspensions or antispasmodic drugs, followed by isolation and hygiene measures to combat the spread of infection.

Control measures based on housing and management, disinfection, and hygiene can significantly reduce morbidity and spread. If infection can be prevented in the first 2 to 3 weeks of life by reducing oocysts in the environment, clinical effects of *C. parvum* are likely to be innocuous and transient. However, oocysts are remarkably resistant to nontoxic or nonirritating chemical disinfectants, making it virtually impossible to keep calving areas completely free of oocysts. Even calves taken in plastic bags at birth to brand-new wooden stalls still excreted oocysts within a few days.[71] Where possible, rearing areas should be cleaned of feces daily. In well managed rearing units, oocyst excretion was delayed when calves were individually penned, and few developed clinical disease until moved to communal housing.[71] The following procedures are recommended: (1) to ensure thorough cleaning of housing, all calves should enter and leave at the same time, not in continual replacement; (2) individual hutches or pens should be scrupulously cleaned and thoroughly dried between batches of calves; (3) calves should be born and raised in a clean, dry environment; (4) newborn calves should be penned alone for 2 to 3 weeks; (5) sick calves should be isolated from healthy calves and have different attendants; (6) attendants should keep hands, boots, and protective clothing as free of feces as possible; (7) utensils should be heat-sterilized daily; (8) vermin, dogs, and cats should be controlled; (9) colostrum and nutritional supplements should be provided; and (10) prophylaxis against other agents should be used, such as rotavirus and ETEC-K99+ vaccines.

b. *Ovine* (Ovis aries)

Ovine cryptosporidiosis has been reported worldwide (see Chapter 3). It was first diagnosed in 1974 in 1- to 3-week-old diarrheic lambs on a farm that produced ovine cheese and milk products in Australia.[67] In 1981, 40% of 40 lambs died with acute diarrhea that began at 5 to 12 days of age and lasted for 7 to 16 days.[105] Numerous cryptosporidia were found in the lower jejunum, ileum, cecum, and colon, but the only other enteropathogen found was rotavirus in one lamb. Where ewes were brought indoors to lamb, the earliest lambs were healthy, but lambs born later had diarrhea associated with *Cryptosporidium* sp. oocysts, and 78% of 37 feces contained oocysts. Again, only a single specimen contained another enteropathogen.[106] Oocysts free from viruses and bacteria induced diarrhea and pathologic changes in neonatal germ-free lambs, confirming that *Cryptosporidium* sp. was pathogenic.[107]

In contrast, from diarrhea outbreaks in lambs in Idaho it was concluded that uncomplicated cryptosporidiosis in lambs was mild, but environmental stress and concurrent infections with other enteropathogens exacerbated illness and caused mortality.[108] Contrasting findings suggest exposure to *Cryptosporidium* isolates of different virulence, but because many factors influence the severity of infection (see Chapter 3), this suggestion must be regarded as speculative.

As parasite-free lambs experimentally infected with *C. parvum* increased in age, the severity of clinical signs decreased, suggesting age-related resistance to cryptosporidiosis.[109] However, asymptomatic adult sheep excreted oocysts of *C. parvum* before and after parturition but reached peak numbers at or shortly after parturition, indicating they serve as a source of oocysts for lambs.[110]

In a study designed to simulate natural waterborne transmission using low-level accumulative dosing, 10 out of 10 gnotobiotic neonatal lambs became clinically infected after receiving a mean of 3.1 oocysts per liter.[110a] Clinical features of ovine cryptosporidiosis, diagnosis, lesions, interactions with other pathogens, epidemiology, treatment, and control measures are similar to bovine cryptosporidiosis and have been reviewed in depth.[71] Oral dosing with halofuginone for 3 days reduced oocyst excretion and was well tolerated.[111] More recently, untreated control lambs inoculated with 1×10^6 *C. parvum* oocysts had diarrhea, oocyst excretion, and mortality, whereas normal colostrum prevented mortality and clinical cryptosporidiosis, and hyperimmune ovine or bovine colostrum gave the best results with no mortality, no diarrhea, and low oocyst shedding.[111a]

c. *Caprine* (Capra hircus)

Cryptosporidiosis is found in goats worldwide. First reported in Tasmania, Australia, in 1981, a 2-week-old Angora kid with acute diarrhea died; cryptosporidia but no other enteropathogens were found on histological and electron microscopic examination of jejunum and ileum.[112] The following year in

Australia during an outbreak of acute diarrhea in young goats, 21 of 29 kids under 21 days of age had acute diarrhea, for 3 to 7 days, many became dehydrated, and 3 died.[113] Feces from nine diarrheic kids contained *Cryptosporidium* sp. oocysts, but no other enteropathogens were found. In Europe, it was first reported in 1983 in France,[114] where the parasite was considered a common cause of diarrhea and where diarrhea was the most common cause of death in young goats.[115] *C. parvum* was thought to be the dominant pathogen.[116] In Hungary, multiple infections including *Cryptosporidium* sp., rotavirus, coronavirus, and ETEC-K99+ were reported in young goats.[117] In Grand Canary Island, diarrheic kids of a Canadian breed excreted oocysts of *Cryptosporidium* sp.[118]

i. Lesions

Although acute diarrhea and dehydration are the main clinical signs of cryptosporidiosis in 1- to 3-week-old goat kids,[113] cryptosporidiosis without diarrhea but with progressive emaciation and high mortality was detected histologically in 5- to 8-day-old kids in two herds.[119] Oocyst excretion (in the absence of other enteropathogens) with diarrhea and progressive loss of condition were observed in goats about 6 weeks of age,[119] and there is a report of cryptosporidiosis and intermittent diarrhea in a 6-month-old goat, in the absence of viral or bacterial enteropathogens.[120] Subclinical infection has also been confirmed by ileal biopsy in 6-week-old goats.[121] For additional clinical features, diagnosis, epidemiology, treatment, and control of caprine cryptosporidiosis, see Angus.[70]

ii. Treatment

Administration of paromomycin to two groups of 18 and 12 kids from days 2 to 13 after birth strongly reduced but did not prevent oocyst excretion while preventing clinical signs and mortality.[121a]

d. Cervine (Capreolus, Cervus, Dama, Odocoileus)

The most commonly farmed species are red deer (*Cervus elephus*) and fallow deer (*Dama dama*). Reports of *C. parvum* in red deer from Scotland[122] and New Zealand[123] described diarrhea outbreaks in domestically reared deer calves with high mortality. Diarrhea lasting 2 to 14 days began 1 to 2 weeks after calves were housed. Field studies also have revealed cryptosporidiosis in naturally reared red deer calves.[82,124] Clinical signs, diagnosis, lesions, epidemiology, treatment, and control of cryptosporidiosis in farmed red deer have been reviewed.[71]

Among wild deer, cryptosporidial infections have been found in roe deer (*Capreolus capreolus*),[125] fallow deer, sika deer (*Cervus nippon*), mule deer (*Odocoileus hermionus*), Eld's deer (*Cervus eldi thamin*), axis deer (*Axis axis*), and barasingha deer (*Cervus duvauceli*).[40] Most cases were diarrheic neonatal deer hand-raised in a wildlife park and diagnosed by finding oocysts in feces. A white-tailed deer (*Odocoileus virginianus*) doe from a captive herd was housed in an isolated building where she fawned; she and her fawn, as well as twin fawns in the same building, excreted oocysts of *C. parvum* but none had signs of illness.[126] Despite exceedingly high population densities throughout the U.S., the prevalence of *Cryptosporidium* in wild white-tailed deer is unknown.

e. Other Ruminants

Cryptosporidiosis has been reported in the water buffalo (*Bubalus bubalis*) in Italy,[127,128] Egypt,[129] Cuba,[130] and India.[130a] In Italy, 7.4% of 229 diarrheic calves under 15 days of age excreted oocysts.[127] In Egypt, 21% of 87 calves under 12 days of age were infected, and 4 died.[129] In Cuba, none of 416 adults and 20% of 200 calves under 4 months of age were infected.[130] Oocysts were found in 11 of 78 buffalo calves in India.[130a]

In neonatal exotic ruminants, *C. parvum* was confirmed by histology or by oocyst detection in blackbuck (*Antelope cervicapra*), sable antelope (*Hippotragus niger*), scimitar-horned (*Oryx gazella dammah*) and fringe-eared (*Oryx gazella callotys*) oryx, and in addax (*Addax nasomaculatus*).[40,131] One neonatal Persian gazelle (*Gazella subgutterosa*) died with cryptosporidiosis,[40] others had diarrhea and excreted oocysts.[132] Oocysts were found in feces of diarrheic neonatal impala (*Aepyceros melampus*), springbok (*Antidorcu marsupialis*), nilgai (*Boselaphus tragocamelus*), Turkomen markhor (*Capra falconeri*), Addra gazelle (*Gazella dama ruficollis*), slender-horned gazelle (*Gazella leptoceros*), eland (*Taufotragus oryx*), and Armenian mouflon (*Ovis orientalis gmelini*).[40] Mountain gazelles (*Gazella cuvieri*) from the Munich Zoo had *C. muris*-like parasites in the abomasum.[133]

A bactrian camel (*Camelus bactrianus*) with a history of parasitic and bacterial infections at the National Zoological Park in Washington, D.C., excreted oocysts resembling *C. muris* for over a year without notable clinical signs.[134] These oocysts were not infectious for calves[134] but were infectious for mice.[135]

A neonatal full-term female llama developed profuse watery diarrhea 4 days after a surgical procedure at a veterinary teaching hospital.[135a] *Cryptosporidium* sp. was isolated from the stool, which was negative for *Salmonella, Campylobacter,* rotavirus, coronavirus, and other parasites. The llama became febrile, dehydrated, and anorectic and lost weight. The hemogram indicated neutropenia and left shift. Serum electrolyte analysis indicated hyperosmolar syndrome. Despite plasma transfusion and vigorous intravenous fluid and electrolyte administration, the llama became severely debilitated. Total parenteral nutrition was supplied through a jugular catheter, and 3 days later the diarrhea resolved; weight loss, cachexia, and electrolyte inbalances were reversed.

2. Nonruminants

a. *Porcine* (Sus scrofa)

Porcine cryptosporidiosis, first reported in the U.S. in 1977,[136,137] has been found in Australia, Bulgaria, Canada, Chile, Czechoslovakia, Germany, Italy, Japan, and Vietnam (in Reference 138). Although *C. parvum* oocysts from bovine and human sources infected piglets,[139-141] little is known of the prevalence of any species of *Cryptosporidium* in pigs. In Ontario, Canada, based on histologic specimens, 5.3% of 3491 pigs were infected.[142] Most were 1- to 12-week-old piglets, but pigs as old as 30 weeks were infected. Based on rectal swabs, 5% of 200 market pigs in California were excreting oocysts.[143] In Ohio, no oocysts were found in 20 litters of nursing pigs or their sows on one farm, but 8 of 30 weanlings excreted oocysts.[143a] On a second farm, 7 of 24 nursing litters and 6 of 32 weanlings, but no sows, were excreting oocysts.[143a] The first farm kept pigs on woven-wire floors, whereas the second farm kept pigs on old porous concrete floors.

Endogenous stages have been found in the jejunum, ileum, cecum, and colon in the microvillous surface of epithelial cells of villi and crypts and within leukocytes in crypts.[142] Most parasites have been found in the terminal ileum.

Villi have been only mildly to moderately atrophied in the ileum, where large numbers of mononuclear cells and a few eosinophils infiltrated the lamina propria.[142] Although diarrhea has been the major clinical sign, many infected pigs have had one or more additional enteropathogens[142] and many pigs apparently remain asymptomatic.

b. *Equine* (Equus caballus)

The first review article that addressed cryptosporidiosis in equines indicated that data were sparse.[138] Presently, little more is known. Clinical cryptosporidiosis in horses, first reported in immunodeficient Arabian foals in the U.S.[144] and then Australia,[145] later was reported in immunodeficient Thoroughbred foals in the U.S.,[146] and a year later it was reported in an immunodeficient Arabian foal in the U.K.[147] These early reports and the absence of any reports of cryptosporidiosis in other horses suggested that only immunodeficient horses became infected. But when antibodies against the parasite were found in 91% of 22 immunologically normal horses in Scotland[148] and then the parasite was found in feces from normal foals in Louisiana,[149] Illinois,[150] Canada,[151] France,[152,153] Spain,[154] Italy,[155] and Iceland,[156] it became apparent that all horses worldwide were susceptible to infection. In reports that provided a detailed description of the parasite, the only named species with supporting characteristics was *C. parvum.*

Neither the prevalence nor the clinical significance of natural cryptosporidiosis in equines is clearly understood. Immunologically normal foals have been studied most. Of 82 3-week to 3½-month-old foals in France, 13 were infected but none appeared clinically ill.[153] Of 14 Quarter Horse foals in Ohio, none were found infected from birth to 28 days of age.[157] Of 19 infected foals in Spain, 16 died.[154] Of two 1- and 6-week-old foals in Canada, the younger foal died.[151] Of 22 9- to 28-day-old pony foals in Louisiana, 64% were ill and 3 died.[149] Also in Louisiana, 2 of 17 Thoroughbred foals, 3 of 14 Quarter Horse foals, and 3 of 26 pony foals on pasture with their dams were infected. A similar survey conducted during foaling one year later detected no infections in 58 Quarter Horse, Arabian, and pony foals. All of 22 immunologically normal 5- to 56-day-old pony foals in Illinois were infected and ill, and 7 died.[150] Of 66 foals in Ohio and Kentucky, 15 to 31% were positive.[158] The major sign in symptomatic foals was diarrhea. Possible concurrent foal heat diarrhea or infection with additional organisms was not always determined or reported.

Because studies involving older horses are relatively rare, their possible role in disseminating oocysts remains unresolved. In Ohio and Kentucky, 0 of 71 mares were positive.[158] In another study in Ohio, 0 of 14 mares were positive.[157] In Louisiana, three 100- to 123-day-old Quarter Horses were found infected.[149] In France, for 3 consecutive years mares were observed around the time they had their foals.[159] Of 36, 21, and 22 mares examined, 29, 20, and 19, respectively, were detected excreting oocysts. Although this and other studies in France suggested that mares might provide infectious oocysts for their foals, yearlings had an infection rate similar to mares. In another study in France, 48 mares were not found infected with *Cryptosporidium.* In Ohio, 9-day-old foals pastured with *Cryptosporidium*-negative mares did not become infected until 4 weeks of age, suggesting that mares were not the source of infection for foals.[158] In the foregoing and other studies, it is important to interpret findings on the basis of the diagnostic test used to detect oocysts and to consider its sensitivity. See Chapter 2 for more information.

Diarrhea, the most common sign of cryptosporidiosis, has not always been found associated with excretion of oocysts, so the role of the parasite in the cause of diarrhea in immunocompetent foals is not fully understood. *Cryptosporidium* has been found as the sole pathogen in some cases, while associated with infectious agents such as rotavirus, coronavirus, adenovirus, *Giardia,* and *Salmonella* in other cases. Infection in immunodeficient foals reported during the first month of life caused acute diarrhea lasting 1 or 2 weeks, and then foals died. Infection in immunologically normal foals most often has been reported to begin in the first month with soft or diarrheic feces lasting a few days to about a week, and the infection has cleared. Severe illness and high mortality probably result from concurrent infection with other pathogens. Infection in older horses has been asymptomatic or chronic with soft feces or diarrhea.

Because present data indicate that young cattle have a high prevalence of *C. parvum* infection and some adult cattle may be asymptomatic excretors of oocysts, the management practice of alternate pasturing of horses and cattle could unknowingly place horses in highly contaminated environments.

c. *Feline* (Felis catus)

There is confusion concerning the species of *Cryptosporidium* in cats. In 1979, oocysts were reported in feces of 5 of 13 cats in Japan.[17] Oocysts from a naturally infected cat were infectious to other cats but not to mice or guinea pigs. A later study in Japan confirmed that oocysts from a naturally infected cat were not infectious to suckling and adult mice, rats, guinea pigs, dogs, and immunosuppressed mice, but they did infect cats.[18] Cats inoculated with the feline isolate of *Cryptosporidium* excreted oocysts for several weeks without clinical signs. Because oocysts were approximately 5 μm in diameter with endogenous stages in intestine and not in gastric mucosa, this organism was not *C. muris.*[18] A *C. muris*-like organism from rats was experimentally transmitted to cats.[160,161] *C. parvum* from human feces[162] and from bovine feces[163] was transmitted to cats (1 to 100 days old);[163] the recipients excreted oocysts but remained clinically normal.[163] Another cat fed *C. parvum* from bovine feces had diarrhea and excreted *C. parvum*-like oocysts; therefore, cats may serve as hosts to *C. felis, C. muris,* and *C. parvum.*[164]

Cryptosporidial antibodies have been found in 20 of 23 (87%) cats, suggesting widespread exposure of cats to *Cryptosporidium.*[148] There is only one report of of natural *C. muris* infection in a cat.[204a] See Table 4 for summary. There is only one report of successful treatment of a chronically infected cat with paromomycin.[165]

d. *Canine* (Canis familaris)

Cryptosporidiosis was first reported in a dog in the U.S. in 1983.[166] This and other natural infections in dogs were considered to be due to *C. parvum.*[167] Most infections involved asymptomatic young dogs and have been asymptomatic unless dogs were concurrently infected with canine distemper virus infection, which is known to be immunosuppressive (Table 5). Experimentally, dogs have been infected with *C. parvum* oocysts from humans[162] and calves.[163] Although antibodies to *Cryptosporidium* were found in 16 of 20 (80%) dogs from Scotland,[148] surveys in Finland, Germany, and Scotland failed to detect oocysts in feces of healthy dogs, and the prevalence remains unknown.[167] Dogs are not considered a good host for *C. muris.* Dogs fed *C. muris* oocysts from rats excreted a few oocysts and remained asymptomatic.[160,161]

e. *Raccoon* (Procyon lotor)

There are three reports of cryptosporidial infections in raccoons from the U.S.[168-170] Cryptosporidial bodies were found in histologic section of small intestine of a young apparently healthy raccoon trapped near Waterford, CN,[168] in histologic sections of intestine of a juvenile raccoon found moribund near Fort Collins, CO,[169] and in feces of 13 juveniles of 100 trapped or shot raccoons.[170] Clinical signs of infection have not been observed.

f. *Ferret* (Mustela furo)

There is one report of cryptosporidiosis in ferrets. *Cryptosporidium parvum*-like oocysts were seen in 40 of 66 (60.6%) ferrets from a farm in New York.[171] Cryptosporidia persisted up to 20 weeks in ferrets given corticosteroids.

Table 4 Summary of Cryptosporidial Reports in Cats

Country	No. of Cats	Remarks	Ref.
Germany	4/300 (0.75%)	Subclinical	163
Germany	3/70 (4.3%)	Litters of kittens kept outdoors and 1 of 30 litters kept indoors	203
Italy (Bologna)	40/88	Also one asymptomatic with *C. muris*	204
Japan (Osaka City)	5/13	Subclinical, considered *C. felis*	17
Japan (Hyogo Prefecture)	20/507 (3.9%)	6.1% of cats with diarrhea and 3.6% of cats without diarrhea	205
Japan (Tokyo)	23/608 (3.8%)	Infection more common in kittens, and most were subclinical	206
Switzerland	1	Kitten with oocysts in feces considered to be source of infection for cryptosporidiosis in an 8-year-old boy	207
U.K. (Liverpool)	3	4- to 6-month old cats with diarrhea, infection verified by histology	208
U.K. (Glasgow)	19/235 (8.1%)	Prevalences were similar in feral and domestic cats; more kittens were infected and endogenous stages were found in intestines of eight cats	209
U.K. (Glasgow)	7/57 (12.2%)	Cats were from eight farms; 33.3% (3 of 9) < 6-month-old vs. 83% (4 of 48) > 6-month-old cats were infected.	210
U.S. (Kentucky)	1	5-year-old cat with persistent diarrhea, infection verified by histology	211
U.S. (N. Carolina)	1	4-year-old cat with intermittent diarrhea for 1 year, concurrent feline leukemia virus infection	212
U.S. (Georgia)	1	Cat concurrently infected with feline leukemia virus	213
U.S. (Indiana)	1	13-year-old cat with diarrhea, concurrently infected with lymphosarcoma	214
U.S. (New York)	1	6-month-old cat with persistent diarrhea; treated successfully with paromomycin	165

Table 5 *Cryptosporidium parvum*-Like Infections in Dogs

Country	No. of Dogs	Remarks	Ref.
Argentina	1	Oocysts found in 1½-year-old mixed breed dog with diarrhea	215a
France	13/29	Oocysts found in asymptomatic dogs	215
Japan	3/213	Oocysts in a routine survey; dogs were asymptomatic	205
U.S. (Tennessee)	1	1-week-old Pomeranian dog with diarrhea, confirmed by histology	166
U.S. (Tennessee)	1	3-month-old Cockapoo, concurrent distemper virus infection, confirmed histologically	216
U.S. (Louisiana)	2	6-week-old mixed breed pups, one with concurrent toxaphene intoxication, confirmed histologically	217
U.S. (Louisiana)	1	6-month-old Siberian Husky, concurrently infected with canine distemper virus, diarrhea; immunosuppression confirmed	218
U.S. (Georgia)	1	5-year-old Pointer with intermittent diarrhea for 1 year, confirmed by histology of intestinal biopsy	219
U.S. (Georgia)	5/49 (10.2%)	No clinical signs	220
U.S. (California)	4/200 (2%)	Impounded in animal shelter	220a

g. *Chinchilla* (Chinchilla laniger)

Cryptosporidial organisms were found in histologic sections of stomach and small and large intestines of an 8-month-old diarrheic chinchilla from a pet shop in Michigan.[173]

h. *Gray squirrel* (Sciurus carolinensis)

Cryptosporidial schizonts and gamonts were identified in cecal epithelium of a young squirrel trapped near Storrs, CN.[174]

i. *Rabbit* (Oryctolagus cuniculus; Sylvilagus floridanus)

Cryptosporidium in a rabbit was first reported by Tyzzer in 1912.[3] It was next reported in a healthy rabbit from Washington, D.C., in 1979 and was named *C. cuniculus* only because cryptosporidia were considered to be host specific at that time.[175] Retrospectively, *C. cuniculus* is synonymous with *C. parvum*. Since that time, cryptosporidiosis has been reported in laboratory-raised and farmed rabbits in the U.S., Belgium, and Germany (in Reference 167); in laboratory and wild rabbits in Brazil;[175a] and in a wild cottontail rabbit (*S. floridanus*).[176] Prevalence in rabbits is unknown. Although diarrhea was associated with *Cryptosporidium* in 4- to 5-week-old farmed rabbits, its role could not be determined because other enteropathogens were present.[177] Three New Zealand white rabbits, experimentally exposed to *C. muris* strain RN 66 developed gastric infections and excreted oocysts for 11 to 43 days without clinical signs.[160] (See Chapter 9 for more information.)

j. Cotton Rat (Sigmodon hispidus)

Cryptosporidium stages were seen in sections of large intestines (but not small intestine) of 1 of 9 cotton rats trapped from Pryor, OK.[178]

k. Hamster (Mesocricetus auratus)

A litter of 6 hamsters developed diarrhea starting at 18 days of age.[179] All hamsters died within 1 month of age and one was necropsied. Cryptosporidia were identified in sections of large intestine. (See Chapter 9 for more information.)

l. Guinea Pig (Cavia porcellus)

Cryptosporidiosis, first recognized as a cause of chronic enteritis in a colony of guinea pigs at the Walter Reed Army Medical Center, Washington, D.C.,[180] was characterized and named *Cryptosporidium wrairi*.[181,182] Unlike *C. parvum, C. wrairi* was not infective for mice and rabbits.[16] It caused anorexia, weight loss, diarrhea, abdominal distension, and even death in guinea pigs.[16,183] (See Chapter 9 for more information.)

m. Vole (Microtus *sp.*), *Mouse* (Mus musculus), *Rat* (Rattus *sp.*)

In healthy voles in Finland, *Cryptosporidium* sp. oocysts were found in 1 of 13 *Microtus agrestis*, 1 of 41 *Clethrionomys glareolus,* and 0 of 43 *Microtus oeconomus*.[184] There are few reports of natural infections in mice and no reports of clinical cryptosporidiosis in wild mice since the first reports of *C. muris* and *C. parvum* by Tyzzer in 1907[1] and 1912.[3] Recently, *C. parvum* oocysts were found in feces of 19 of 58 (32.7%) and *C. muris* oocysts were found in feces of 115 of 58 (25.8%) wild mice trapped near Moreton Morrell, U.K.[185] Natural *C. muris* infections were reported from Turkey[161] and from Alabama,[186] where oocysts were found in 30 of 115 mice. The prevalence of either species in mice is not known.

In house rats (*Rattus norvegicus*) in Osaka City, Japan, *C. parvum* oocysts were found in feces of 6 of 61 (14.8%) and *C. muris* oocysts were found in 3 of 61 (4.9%).[187] Both species were experimentally transmitted to SPF rats. Cryptosporidial oocysts (not speciated) were found in 48.5% of 171 *Rattus rattus* and 21.3% of 47 *R. norvegicus* around the Tokyo and Chiba districts of Japan.[188] In a subsequent study, prevalence rates of 17.7% in *R. rattus* and 2.1% in *R. norvegicus* were reported for 231 house rats trapped around Tokyo.[189] Rats in all three Japanese surveys were clinically normal.

The only clinical cryptosporidiosis in naturally infected rats was in a colony of Rapp hypertensive rats *(R. norvegicus*) in New York.[190] After a sudden onset of diarrhea, 32 and 43 offspring died within 3 weeks of birth. Histological lesions were confined to the mucosa of small intestine, and cryptosporidia were found in microvillus border. (See Chapter 9 for more information.)

n. Marsupials: Opossum (Didelphis virginiana), *Antechinus* (Antechinus stuartii), *Bandicoot* (Isoodon obesulus), *Kangaroo* (Macropus rufus), *Koala* (Phascolarctos cinereus), *and Pademelon* (Thylogale billardierii)

Natural infection has not been reported in the opossum (*Didelphis virginiana*); however, 11-week-old nursing opossums, orally inoculated with *C. parvum* oocysts from a bovine source, excreted oocysts in their feces and had developmental stages in epithelial cells of ileum, cecum, and colon 6 to 10 days after inoculation.[173] Developmental stages of *Cryptosporidium* sp. were observed in histological sections of small intestine from several male brown antechinus (*Antechinus stuartii*), a murine-like dasyurid marsupial found in southeastern Australia, which died (like all other males of this species) shortly after mating.[173a]

Oocysts resembling those of *C. parvum* were found in feces from two southern brown bandicoots (*Isoodon obesulus*), a rat-like marsupial, in the Mt. Lofty area of South Australia (O'Donoghue, personal communication).[31] Oocysts resembling those of *C. parvum* were found in feces from a hand-reared orphan juvenile red kangaroo (*Macropus rufus*) in South Australia (O'Donoghue, personal communication).[31]

Developmental stages of *Cryptosporidium* sp. were observed in the small intestines of five koalas (*Phascolarctos cincereus*) from a wildlife park in South Australia (O'Donoghue, personal communication).[31] Developmental stages were found in the intestine of a hand-reared juvenile tasmanian pademelon (*Thylogale billardierii*), a small compact-bodied macropod from Tasmania. These wallabies often graze at night in pastures occupied by sheep and cattle (O'Donoghue, personal communication).[31]

o. *Echidna* (Tachyglossus aculeatus)

Oocysts were found in the feces and developmental stages were found in the intestine of a captive short-beaked echidna (*Tachyglossus aculeatus*), a toothless porcupine-like monotreme that ingests insects found in soil and a considerable amount of soil itself (O'Donoghue, personal communication).[31]

3. Zoo Animals

For additional information on cryptosporidial infections in zoo animals not found in the Cervine, Other Ruminants, or Nonhuman Primates sections, see reviews by Crawshaw and Mehren,[11] Heuschele et al.,[40] and O'Donoghue.[31]

4. Primates

a. *Nonhuman Primates*

Although the name *Cryptosporidium rhesi* was given to the species infecting rhesus monkeys, all available evidence from intestinal, hepatopancreatic, and pulmonary infections indicates the presence of organisms indistinguishable from *C. parvum*.[167,190a] Numerous reports of natural and experimentally induced infections in macaques as well as reports of cryptosporidiosis in a squirrel monkey (*Saimiri sciureus*), a red-ruffed lemur (*Varecia variegata rubra*), and a cotton-topped tamarin (*Saguinis oedipus*) have been reviewed extensively.[167] Subsequently, 81 cases of clinical cryptosporidiosis were diagnosed during 11 months at the Regional Primate Research Center, Seattle, WA.[191] Cryptosporidiosis was found in 66 of 128 (52%) *Macaca nemestrina,* 10 of 22 (45%) *M. fascicularis*, 2 of 3 *C. fiscata*, and 3 of 4 *Papio* sp. Thirteen to 14 new cases were detected every month with no seasonal bias. Although diarrhea persisted for 3 to 79 days, only one death was attributed to cryptosporidiosis. Oocysts were seen even after cessation of diarrhea. The mean duration of oocyst excretion was 36 days.

C. parvum-like oocysts were detected in 8 of 29 monkeys in the Barcelona Zoo, Spain.[192] Infected monkeys were *Aetles belzebuth, Cercocebus torquatus lunulatus, Cercopithecus aethiops, C. campbelli, C. talapuin, Erethrocebus patas, Macaca mulatta*, and *Lemur macaco-mayotensis*.

In Germany, cryptosporidiosis was found in 4 of 11 rhesus monkeys in a vaccination experiment with simian immunodeficiency virus.[193] One monkey had severe cryptosporidiosis involving the intestine; three others had severe intestinal, pancreatic, and hepatic cryptosporidiosis. In a captive population of 230 common marmosets in Wisconsin, oocysts were detected in 1 of 25 animals under 1 year of age and in 4 of 25 animals over 1 year of age, suggesting that captive marmosets of all ages are potential carriers.[193e] (See Chapter 9 for more information.)

b. *Humans* (Homo sapiens)

Except for a single case in which an immunocompromised person in Czechoslovakia was reported to be infected with *C. baileyi*,[59] all other human infections are considered to be due to *C. parvum*. The first two cases of human cryptosporidiosis were reported in 1976.[7,8] Until 1982 when the Centers for Disease Control issued reports of widespread cryptosporidiosis in AIDS patients[9] and an outbreak in immunocompetent persons,[194] only 11 additional cases had been reported.[195] Over 1000 subsequent reports have documented cryptosporidiosis in humans in 95 countries in all continents except Antarctica (Table 6). Infections have been documented based on age, sex, occupation, social group, immune status, country, locality, source of infection, and other variables (reviewed in Chapter 2, Chapter 3, and by others[31,87,195]). *C. parvum* oocysts measured from human feces ranged from 3.8–6.0 × 3.0–5.3 μm with a mean size of 5.0 ± 0.25 × 4.5 ± 0.26 μm and a shape index (major diameter divided by minor diameter) of 1.1.[195a] Compilations from numerous coprologic surveys through 1991 excluding AIDS patients and specific outbreaks indicate prevalence rates of 0.1 to 27.1% (mean 4.9%) in industrialized countries vs. 0.1 to 37.5% (mean 7.9%) in lesser developed countries.[31] Seroprevalence data indicate a much higher level of exposure to *Cryptosporidium* than data from coprologic surveys.[31] Higher prevalence has been associated with poor sanitation, malnutrition, contaminated drinking water, close contact with infected persons, and contact with animals.

Typically, diarrhea is the most common clinical sign. Stools may be watery and may contain mucus but rarely blood. Bowel movements can be copious and numerous, resulting in weight loss and dehydration. As many as 71 bowel movements and fluid loss of 12 to 17 liters per day has been reported.[195] Other common clinical signs include abdominal cramps, fever, nausea, and vomiting. Less common general signs include headache, weakness, fatigue, myalgia, and inappetence. (See Chapter 2 for more information.)

Severity and duration of infection is influenced greatly by the immune response. Most immunocompetent persons present with acute enteritis, mild to moderate, lasting 1 to 2 weeks. Most immunocompromised persons present with chronic enteritis, moderate to severe, lasting as long as the immune impairment,

Table 6 Geographic Distribution of Reported Human Cryptosporidiosis

Africa	**Central/South America**	**Middle East**
Algeria	Argentina	Egypt
Burundi	Brazil	Iran
Cameroon	Colombia	Israel
Ethiopia	Chile	Kuwait
Gabon	Costa Rica	Saudi Arabia
Ghana	Ecuador	
Guinea	El Salvador	**North America**
Guinea-Bissau	Guatemala	Canada
Ivory Coast	Mexico	United States
Kenya	Panama	
Liberia	Peru	**Pacific**
Mauritania	Uruguay	Australia
Mauritius	Venezuela	Malaysia
Morocco		Mali
Nigeria	**Europe**	New Zealand
Rwanda	Austria	Papua-
South Africa	Belgium	New Guinea
Sudan	Czechoslovakia	Philippines
Togo	Denmark	Singapore
Tunisia	England	
Uganda	Finland	**Caribbean**
Zaire	France	Cuba
Zambia	Germany	Haiti
Zimbabwe	Greece	Jamaica
	Hungary	Puerto Rico
Asia	Ireland	St. Lucia
Bangladesh	Italy	Tobago
Belarus	Lithuania	Trinidad
Cambodia	Netherlands	Virgin Islands
China	Poland	
India	Portugal	
Japan	Romania	
Korea	Serbia	
Myanmar Republic	Spain	
Pakistan	Sweden	
Russia	Switzerland	
Sri Lanka	Turkey	
Taiwan	Wales	
Thailand		

sometimes resolving and then recurring, other times persisting and becoming life threatening. This group includes persons receiving immunosuppressive drugs for treatment of a variety of neoplastic diseases, organ or tissue transplants, and skin lesions; malnourished persons; those with inherited immune deficiencies; and persons with concurrent infectious diseases such as measles, chickenpox, or AIDS. However, the clinical course and severity of illness can vary greatly from one individual to another and some immunocompetent patients can present with symptoms more severe than immunocompromised patients. (See Chapter 2 for more information.) In general, when CD4+ cells are greater than 200/μl, infections are acute and resolve; when CD4+ cells are less than 200/μl, infections may be chronic and may not resolve.[196]

Although 37% of 54 human fecals containing *C. parvum* oocysts also contained an atypical picobirnavirus detected by polyacrylamide gel electrophoresis, any relation between the virus and cryptosporidiosis remains unclear.[197] Little information is available on asymptomatic carriers. Of 78 immunocompetent and 50 immunodeficient children, 6.4 and 22%, respectively, had asymptomatic cryptosporidiosis.

Infections have not always been confined to the small intestine, especially in immunocompromised persons. Developmental stages have been found in the lungs, esophagus, stomach, liver, pancreas, gall bladder, appendix, colon, rectum, and conjunctiva of the eye. Persons with extra-intestinal infections have presented with clinical signs associated with the affected organ. Those with respiratory infections have had coughing, wheezing, croup, hoarseness, and shortness of breath. Infections in other sites have resulted in hepatitis, pancreatitis, cholecystitis, cholangitis, and conjunctivitis.

Table 7 Physical Disinfection of *Cryptosporidium* Oocysts

Agent	Conditions	Results	Test	Ref.
Heat	121°C, 10 minutes	Protein changes	DEP	221
Heat	50–55°C, 5 minutes	NI	*In vivo*	199
Heat	45°C, 20 minutes	NI	*In vivo*	222
	60°C, 6 minutes	NI	*In vivo*	
Heat	59.7°C, 5 minutes	I	*In vivo*	200
	64.2°C, 5 minutes	NI	*In vivo*	
	67.5°C, 1 minutes	I	*In vivo*	
	72.4°C, 1 minute	NI	*In vivo*	
Freezing	–196°C, 10 minutes	NI	*In vivo*	223
	–20°C, 3 days	NI	*In vivo*	
Freezing	–70°C, 1 h	NI	*In vivo*	201
	–20°C, 8 hours; 1 day	I; NI	*In vivo*	
	–15°C, 24 hours; 1 week	I; NI	*In vivo*	
	–10°C, 1 week	I	*In vivo*	
Freezing	Liquid nitrogen,	100% reduced	Ex/dyes	224
	–22°C, ≤32 days	98% reduced	Ex/dyes	
Ultraviolet	15,000 mW/sec for 2 hours	I	*In vivo*	225
	15,000 mW/sec for 2.5 hours	NI	*In vivo*	
Ultraviolet	80 mW/sec cm^{-2}	90% reduced	Ex	226
	120 mW/sec cm^{-2}	99% reduced	Ex	
Ultraviolet	8748 mW/sec cm^{-2}	100% reduced	Ex/dyes	227
Pulsed light	1 J/cm^2	100% reduced	*In vivo*	227a
Drying	Air dried, 2 hours	97% reduced	Ex/dyes	224
	Air dried, 4 hours	100% reduced	Ex/dyes	
Drying	Air dried in feces, 1–4 days	NI	*In vivo*	228

Note: I = infectious; NI = noninfectious; *in vivo* testing performed in mice; DEP = dielectrophoresis; Ex = excystation.

Even when infection has been confined to the intestine, clinical manifestations have not always been associated with the gastrointestinal tract. Six cases have been reported for individuals from 7 to 27 years of age who have presented with arthritis, tenosynovitis, plantar fasciitis, conjunctivitis, urethritis, and/or oral mucosal erythema (in Reference 198). Several of these signs are indicative of Reiter's syndrome, a reaction most commonly observed after infection with *Salmonella, Shigella, Yersinia,* or *Campylobacter.*[198]

Histologically, enteric infections have been characterized by villous atrophy, enlarged crypts, and infiltration of inflammatory cells into the lamina propria. Respiratory infections have been characterized by infiltration of inflammatory cells into the lamina propria, loss of cilia, hyperplasia, and hypertrophy of epithelial cells. Similar inflammatory cell infiltration has been reported for infections in other organs.

VII. PREVENTION, CONTROL, AND TREATMENT

As the sole mechanism for transmission, oocysts have evolved to be dispersed and survive in harsh environments for long periods of time. They are unusually resistant to natural stresses and many man-made chemical disinfectants. As Blewett[199] noted, oocysts probably evolved in dispersed mobile populations where there was strong selective pressure for long-term survival, whereas modern times are characterized by concentrated, fixed populations which become exposed to extremely high levels of infective organisms. To control infections in animal populations, the current best strategy is to modify this imbalance by moving animals to clean areas. For human populations, disinfection procedures are sought to minimize person-to-person transmission in domestic and institutional settings and to deal effectively with contamination of recreational and drinking water.

Because all infections with *Cryptosporidium* are initiated by ingestion or inhalation of the oocyst stage, measures to prevent or limit the spread of infection must be targeted to eliminate or reduce infectious oocysts in the environment. There are no drugs approved for prophylaxis or therapy for humans or animals that will prevent or stop oocyst production by infected individuals. Hygiene, including disinfection, remains the most effective management tool. To determine the effectiveness of any anti-oocyst activity, there must be an assay to determine viability and infectivity after treatment. Two principal methods are currently used: (1) animal infectivity, which is expensive and laborious, requires special

159. Chermette, R., Tarnau, C., Boufassa-Ouzrout, S., and Guderc, O., Survey on equine cryptosporidiosis in Normandy, *5th Int. Coccidiosis Conf. Tours (France),* 17–20 October 1989, INRA Publ., 49, 493, 1989.
160. Iseki, M., Maekawa, T., Moriya, K., Uni, S., and Takada, S., Infectivity of *Cryptosporidium muris* (strain RN 66) in various laboratory animals, *Parasitol. Res.,* 75, 218, 1989.
161. Ozkul, I. A. and Aydin, Y., Infectivity of *Cryptosporidium muris* directly isolated from the murine stomach for various laboratory animals, *Vet. Parasitol.,* in press.
162. Current, W. L., Reese, N. C., Ernst, J. V., Bailey, W. S., Heyman, M. B., and Weinstein, W. M., Human cryptosporidiosis in immunocompetent and immunodeficient persons. Studies of an outbreak and experimental transmission, *N. Engl. J. Med.,* 308, 1252, 1983.
163. Augustin-Bichl, G., Boch, J., and Henkel, G., Kryptosporidien-Infektionen bei Hund und Katze, *Berl. Muench. Tieraerztl. Wochenschr.,* 97, 179, 1984.
164. Pavlasek, I., Experimental infection of cat and chicken with *Cryptosporidium* sp. oocysts isolated from a calf, *Folia Parasitol. (Prague),* 30, 121, 1983.
165. Barr, S. C., Jamrosz, G. F., Hornbuckle, W. E., Bowman, D. D., and Fayer, R., Use of paromomycin for treatment of cryptosporidiosis in a cat, *J. Am. Vet. Med. Assoc.,* 205, 1742, 1994.
166. Wilson, R. B., Holscher, M. A., and Lyle, S. J., Cryptosporidiosis in a pup, *J. Am. Med. Assoc.,* 183, 1005, 1983.
167. Riggs, M. W., Cryptosporidiosis in cats, dogs, ferrets, raccoons, opossums, rabbits, and nonhuman primates, in *Cryptosporidiosis in Man and Animals,* Dubey, J. P., Speer, C. A., and Fayer, R., Eds., CRC Press, Boca Raton, FL, 1990, 113.
168. Carlson, B. L. and Neilsen, S. W., Cryptosporidiosis in a raccoon, *J. Am. Vet. Med. Assoc.,* 181, 1405, 1982.
169. Marion, H. D. and Zeidner, N. S., Concomitant cryptosporidia, coronavirus, and parvovirus in a raccoon (*Procyon lotor*), *J. Wildl. Dis.,* 28, 113, 1992.
170. Snyder, D. E., Indirect immunofluorescent detection of oocysts of *Cryptosporidium parvum* in the feces of naturally infected raccoons (*Procyon lotor*), *J. Parasitol.,* 74, 1050, 1988.
171. Rehg, J. E., Gigliotti, F., and Stokes, D. C., Cryptosporidiosis in ferrets, *Lab Anim. Sci.,* 38, 155, 1988.
172. Lindsay, D. S., Hendrix, C. M., and Blagburn, B. L., Experimental *Cryptosporidium parvum* infection in opossums (*Didelphis virginiana*), *J. Wildl. Dis.,* 24, 157, 1988.
173. Yamini, B. and Raju, N. R., Gastroenteritis associated with a *Cryptosporidium* sp. in a chinchilla, *J. Am. Vet. Med. Assoc.,* 189, 1158, 1986.
173a. Barker, I. K., Beveridge, I., Bradley, A. J., and Lee, A. K., Observations on spontaneous stress-related mortality among males of the dasyurid marsupial, *Antechinus stuartii* Macleay, *Aust. J. Zool.,* 26, 435, 1978.
174. Sundberg, J. P., Hill, D., and Ryan, M. J., Cryptosporidiosis in a gray squirrel, *J. Am. Vet. Med. Assoc.,* 181, 1420, 1982.
175. Inman, L. R. and Takeuchi, A., Spontaneous cryptosporidiosis in an adult female rabbit, *Vet. Pathol.,* 16, 89, 1979.
175a. DaSilva, N. R. S., Bigatti, L. E., Amado, R. K., DeAraujo, F. A. P., and Chaplin, E. L., *Eimeria* and *Cryptosporidium* infections in rabbits in the municipality of Porto Alegre, Rio Grande, Brazil, *Arq. Faculd. Vet. UFRGS,* 20, 235, 1992.
176. Ryan, M. J., Sundberg, J. P., Sauerschell, R. J., and Todd, K. S., *Cryptosporidium* in a wild cottontail rabbit (*Sylvilagus floridanus*), *J. Wildl. Dis.,* 22, 267, 1986.
177. Peeters, J. E., Geeroms, R., Carman, R. J., and Wilkins, T. D., Significance of *Clostridium spiroforme* in the enteritis complex of commercial rabbits, *Vet. Microbiol.,* 12, 25, 1986.
178. Elangbam, C. S., Qualis, C. W., Ewing, S. A., and Lochmiller, R. L., Cryptosporidiosis in a cotton rat (*Sigmodon hispidus*), *J. Wildl. Dis.,* 29, 161, 1993.
179. Orr, J. P., *Cryptosporidium* infection associated with proliferative enteritis (wet tail) in Syrian hamsters, *Can. Vet. J.,* 29, 843, 1988.
180. Jervis, H. R., Merrill, T. G., and Sprinz, H., Coccidiosis in the guinea pig small intestine due to a *Cryptosporidium, Am. J. Vet. Res.,* 27, 408, 1966.
181. Vetterling, J. M., Jervis, H. R., Merrill, T. G., and Sprinz, H., *Cryptosporidium wrairi* sp. n. from the guinea pig *Cavia porcellus* with an emendation of the genus, *J. Protozool.,* 18, 243, 1971.
182. Vetterling, J. M., Takeuchi, A., and Madden, P. A., Ultrastructure of *Cryptosporidium wrairi* sp. n. from the guinea pig, *J. Protozool.,* 18, 248, 1971.
183. Chrisp, C. E., Reid, W. C., Suckow, M. A., Rush, H. R., and Thomann, M., Cryptosporidiosis in guinea pigs: an animal model, *Infect. Immun.,* 58, 674, 1990.
184. Laakhonen, J., Soveri, Y., and Henttonen, H., Prevalence of *Cryptosporidium* sp. in *Microtus oeconomus* and *Clethrionomys glareolus* populations, *J. Wildl. Dis.,* 30, 110, 1994.
185. Chalmers, R. M. and Sturdee, A. P., *Cryptosporidium muris* in wild house mice (*Mus musculus*): first report in the U.K., *Eur. J. Protistol.,* 30, 151, 1994.
186. Klesius, P. H., Haynes, T. B., and Malo, L. K., Infectivity of *Cryptosporidium* sp. isolated from wild mice for calves and mice, *J. Am. Vet. Med. Assoc.,* 189, 192, 1986.
187. Iseki, M., Two species of *Cryptosporidium* naturally infecting house rats, *Rattus norvegicus, Jpn. J. Parasitol.,* 35, 521, 1986.
188. Miyaji, S., Tanikawa, T., and Shikata, J., Prevalence of *Cryptosporidium* in *Rattus rattus* and *R. norvegicus* in Japan, *Jpn. J. Parasitol.,* 38, 368, 1989.

189. Yamaura, H., Shirasaka, R., Asahi, H., Koyama, T., Motoki, M., and Ito, H., Prevalence of *Cryptosporidium* infection among house rats, *Rattus rattus* and *R. norvegicus*, in Tokyo and experimental cryptosporidiosis in roof rats, *Jpn. J. Parasitol.*, 39, 439, 1990.
190. Moody, K. D., Brownstein, D. G., and Johnson, E. A., Cryptosporidiosis in suckling laboratory rats, *Lab. Animal Sci.*, 41, 625, 1991.
190a. Lackner, A. A. and Wilson, D. W., Cryptosporidiosis, intestines, pancreatic duct, bile duct, gall bladder, *Macaca mulatta,* in *Nonhuman Primates, II,* Jones, T. C., Mohr, U., and Hunt, R. D., Eds., Springer-Verlag, Berlin, 1993, 41.
191. Miller, R. A., Bronsdon, M. A. and Morton, W. R., Experimental cryptosporidiosis in a primate model, *J. Infect. Dis.*, 161, 316, 1990.
192. Gomez, M. S., Gracenea, M., Gosalbez, P., Feliu, C., Ensenat, C., and Hidalgo, R., Detection of oocysts of *Cryptosporidium* in several species of monkeys and in one prosimian species at the Barcelona Zoo, *Parasitol. Res.*, 78, 619, 1992.
193. Kaup, F. J., Kuhn, E. M., Makoschey, B., and Hunsmann, G., Cryptosporidiosis of liver and pancreas in rhesus monkeys with experimental SIV infection, *J. Med. Primatol.*, 23, 304, 1994.
193a. Kalishman, J., Paul-Murphy, J., Scheffler, J., and Thomson, J. A., Survey of *Cryptosporidium* and *Giardia* spp. in a captive population of common marmosets, *Lab Anim. Sci.*, 46, 116, 1996.
194. Anon., Human cryptosporidiosis — Alabama, *Morbid. Mortal. Wkly. Rpt.*, 31, 252, 1982.
195. Ungar, B. L. P., Cryptosporidiosis in humans (*Homo sapiens*), in *Cryptosporidiosis of Man and Animals,* Dubey, J. P., Speer, C. A., and Fayer, R., Eds., CRC Press, Boca Raton, FL, 1990, 59.
195a. Mercado, R. and Santander, F., Size of *Cryptosporidium* oocysts excreted by symptomatic children of Santiago, Chile, *Rev. Inst. Med. Trop. Sao Paulo,* 37, 473, 1995.
196. Gazzard, B. G., Opportunistic infections in HIV seropositive individuals, *J. Roy. Coll. Phys. Lond.*, 29, 335, 1995.
197. Gallimore, C. I., Green, J., Casemore, D. P., and Brown, D. W. G., Detection of a picobirnavirus associated with *Cryptosporidium* positive stools from human, *Arch. Virol.*, 140, 1275, 1995.
197a. Pettoello-Mantavani, M., DiMartino, L., Dettori, G., Vajro, P., Scotti, S., Ditullio, M. T., and Guandalini, S., Asymptomatic carriage of intestinal *Cryptosporidium* in immunocompetent and immunodeficient children: a prospective study, *Pediatr. Infect. Dis. J.*, 14, 1042, 1995.
198. Cron, R. Q. and Sherry, D. D., Reiter's syndrome associated with cryptosporidial gastroenteritis, *J. Rheumatol.*, 22 1962, 1995.
199. Blewett, D. A., Disinfection and oocysts, in *Proc. 1st Int. Workshop on Cryptosporidiosis,* Angus, K. W. and Blewett, D. A., Eds., Sept. 7–8, 1988, Animal Diseases Research Institute, Edinburgh, 1989, 107.
200. Fayer, R., Effect of high temperature on infectivity of *Cryptosporidium parvum* oocysts in water, *Appl. Environ. Microbiol.*, 60, 2732, 1994.
200a. Harp, J. A., Fayer, R., Pesch, B. A., and Jackson, G. J., Effect of pasteurization on infectivity of *Cryptosporidium parvum* oocysts in water and milk, *Appl. Environ. Microbiol.*, 62, 197, 1996.
201. Fayer, R. and Nerad, T., Effects of low temperatures on viability of *Cryptosporidium parvum* oocysts, *Appl. Environ. Microbiol.*, 62, x, 1996.
202. Fayer, R., Nerad, T., Rall, W., Lindsay, D. S., and Blagburn, B. L., Studies on cryopreservation of *Cryptosporidium parvum, J. Parasitol.*, 77, 357, 1991.
203. Beelitz, P., Gobel, E., and Gothe, R., Fauna und Befallschaufigkeit von Endoparasiten bei Katzenwelpen und ihren Muttern unterschiedlicher Haltung in Suddeutschland, *Tierarztl. Prax.*, 20, 297, 1992.
204. Canestri-Trotti, G. and Arnone, B., *Cryptosporidium muris* nel gatto a Bologn, *Parassitologie*, 28, 212, 1986.
205. Uga, S., Matsumura, T., Ishibashi, K., Yoda, Y., Yatomi, K., and Katoaoka, N., Cryptosporidiosis in dogs and cats in Hyogo Prefecture, Japan, *Jpn. J. Parasitol.*, 38, 139, 1989.
206. Arai, H., Fukuda, Y., Hara, T., Funakoshi, Y., Kaneko, S., Yoshida, T., Asahi, H., Kumada, M., Kato, K., and Koyama, T., Prevalence of *Cryptosporidium* infection among domestic cats in the Tokyo metropolitan district, Japan, *Jpn. J. Med. Sci. Biol.*, 43, 7, 1990.
207. Egger, M., Nguyen, M., Schaad, U. B., Krech, T., Intestinal cryptosporidiosis acquired from a cat, *Infection,* 18, 177, 1990.
208. Bennett M., Baxby, D., Blundell, N., Gaskell, C. J., Hart, C. A., and Kelly, D. F., Cryptosporidiosis in the domestic cat, *Vet. Rec.*, 116, 73, 1985.
209. Mtambo, M. M. A., Nash, A. S., Blewett, D. A., Smith, H. V., and Wright, S., *Cryptosporidium* infection in cats: prevalence of infection in domestic and feral cats in the Glasgowarea, *Vet. Rec.*, 129, 502, 1991.
210. Nash, A. S., Mtambo, M. M. A., and Gibbs, H. A., *Cryptosporidium* infection in farm cats in the Glasgowarea, *Vet. Res.,* 133, 576, 1993.
211. Poonacha, K. B. and Pippin, C., Intestinal cryptosporidiosis in a cat, *Vet. Pathol.*, 19, 708, 1982.
212. Monticello, T. M., Levy, M. G., Bunch, S. E., and Fairley, R. A., Cryptosporidiosis in a feline leukemia virus-positive cat, *J. Am. Vet. Med. Assoc.*, 191, 705, 1987.
213. Goodwin, M. A. and Barsanti, J. A., Intractable diarrhea associated with intestinal cryptosporidiosis in a domestic cat also infected with feline leukemia virus, *J. Am. Anim. Hosp. Assoc.*, 26, 365, 1990.
214. Lent, S. F., Burkhardt, J. E., and Bolka, D., Coincident enteric cryptosporidiosis and lymphosarcoma in a cat with diarrhea, *J. Am. Anim. Hosp. Assoc.*, 29, 492, 1993.
215. Chermette, R. and Blonde, S., Cryptosporidiose des carnivores domestiques, resultats preliminaires en France, *Bull. Soc. Fr. Parasitol.*, 7, 31, 1989.

215a. Dominguez, S. and Almarza, A., El primer hallazgo de *Cryptosporidium* en un canino en la Republica Argentina, *Pet's Ciencia,* 4, 244, 1988.

216. Fukushima, K. and Helman, R. G., Cryptosporidiosis in a pup with distemper, *Vet. Pathol.,* 21, 247, 1984.

217. Sisk, D. B., Gosser, H. S., and Styer, E. L., Intestinal cryptosporidiosis in two pups, *J. Am. Vet. Med. Assoc.,* 184, 835, 1984.

218. Turnwald, G. H., Barta, O., Taylor, H. W., Greeger, J., Coleman, S. U., and Pourciau, S. S., Cryptosporidiosis associated with immunosuppression attributable to distemper in a pup, *J. Am. Vet. Med. Assoc.,* 192, 79, 1988.

219. Greene, C. E., Jacobs, G. J., and Prickett, D., Intestinal malabsorption and cryptosporidiosis in an adult dog, *J. Am. Vet. Med. Assoc.,* 197, 365, 1990.

220. Jafri, H. S., Moorhead, A. R., Reedy, T., Dickerson, J. W., Wahlquist, S. P., and Schantz, P. M., Detection of pathogenic protozoan in fecal specimens from urban dwelling dogs, *Am. J. Trop. Med. Hyg. (Suppl.),* 49, 269, 1993.

220a. El-Ahraf, A., Tacal, J. V., Sobih, M., Amin, M., Lawrence, W., and Wilcke, B. W., Prevalence of cryptosporidiosis in dogs and human beings in San Bernardine County, California, *J. Am. Vet. Med. Assoc.,* 198, 631, 1991.

221. Archer, G. P., Betts, W. B., and Haigh, T., Rapid differentiation of untreated, autoclaved and ozone-treated *Cryptosporidium parvum* oocysts using dielectrophoresis, *Microbios,* 73, 165, 1993.

222. Anderson, B. C., Moist heat inactivation of *Cryptosporidium* sp., *Am. J. Public Health,* 75, 1433, 1985.

223. Sherwood, D., Angus, K. W., Snodgrass, D. R., and Tzipori, S., Experimental cryptosporidiosis in laboratory mice, *Infect. Immun.,* 38, 471, 1982.

224. Robertson, L. J., Campbell, A. T., and Smith, H. V., Survival of *Cryptosporidium parvum* oocysts under various environmental pressures, *Appl. Environ. Microbiol.,* 58, 3494, 1992.

225. Lorenzo-Lorenzo, M. J., Ares-Mazas, M. E., Villa Corta, I., and Duran-Oreiro, D., Effect of ultraviolet disinfection of drinking water on the viability of *Cryptosporidium parvum* oocysts, *J. Parasitol.,* 79, 67, 1993.

226. Ransome, M. E., Whitmore, T. N., Carrington, E. G., Effect of disinfectants on the viability of *Cryptosporidium parvum* oocysts, *Water Supply,* 11, 75, 1993.

227. Campbell, A. T., Robertson, L. J., Snowball, M. R., Smith, H. V., Inactivation of oocysts of *Cryptosporidium parvum* by ultraviolet irradiation, *Water Res.,* 29, 2583, 1995.

227a. Dunn, J., Ott, T., and Clark, W., Pulsed light treatment of food and packaging, *Food Technol.,* 49, 95, 1995.

228. Anderson, B., Effect of drying on the infectivity of cryptosporidia-laden calf feces for 3- to 7-day-old mice, *Am. J. Vet. Res.,* 47, 2272, 1986.

229. Pavlasek, I., Effect of disinfectants in infectiousness of oocysts of *Cryptosporidium* sp., *Cs. Epidem. Mikrobiol. Immunol.,* 33, 97, 1984.

230. Holton, J., Nye, P., and McDonald V., Efficacy of selected disinfectants against mycobacteria and cryptosporidia, *J. Hosp. Infect.,* 27, 105, 1994.

231. Sundermann, C. A., Lindsay, D. S., and Blagburn, B. L., Evaluation of disinfectants for ability to kill avian *Cryptosporidium* oocysts, *Comp. Anim. Pract.,* 36, 1987.

232. Fayer, R., Graczyk, T. K., Cranfield, M. R., and Trout, J., Effect of low molecular weight gases on disinfection of *Cryptosporidium* oocysts, *Appl. Environ. Microbiol.,* 1996.

233. Quinn, C. M. and Betts, W. B., Longer term viability status of chlorine-treated *Cryptosporidium* oocysts in tap water, *Biomed. Lett.,* 48, 315, 1993.

234. Smith, H. V., Smith, A. L., Girdwood, R. W. A., and Carrington, E. G., The effect of free chlorine on the viability of *Cryptosporidium* oocysts isolated from human feces, in *Cryptosporidium in Water Supplies,* Badenoch, J., Ed., Her Majesty's Stationery Office, London, 1990, 185.

235. Korich, D. G., Mead, J. R., Madore, M. S., Sinclair, N. A., and Sterling, C. R., Effects of ozone, chlorine dioxide, chlorine, and monochloramine on *Cryptosporidium parvum* oocyst viability, *Appl. Environ. Microbiol.,* 56, 1423, 1990.

236. Parker, J. and Smith H., Destruction of oocysts of *Cryptosporidium parvum* by sand and chlorine, *Water Res.* 27, 729, 1993.

237. Campbell, I., Tzipori, S., Hutchison, G., and Angus, K. W., Effect of disinfectants on survival of *Cryptosporidium* oocysts, *Vet. Rec.,* 111, 414, 1982.

238. Fayer, R., Effect of sodium hypoclorite exposure on infectivity of *Cryptosporidium parvum* oocysts for neonatal BALB/c mice, *Appl. Environ. Microbiol.,* 61, 844, 1995.

239. Peeters, J. E., Ares Mazas, E., Masschelein, W. J., Villacorta, I., and Debacker, E., Effect of disinfection of drinking water with ozone or chlorine dioxide on survival of *Cryptosporidium parvum* oocysts, *Appl. Environ. Microbiol.,* 55, 1519, 1989.

240. Campbell, A T., Robertson, L. J., and Smith, H. V., Effects of preservatives on viability of *Cryptosporidium parvum* oocysts, *Appl. Environ. Microbiol.,* 59, 4361, 1993.

241. Angus, K., Sherwood, D., Hutchison, G., and Campbell, I., Evaluation of the effect of two aldehyde-based disinfectants on the infectivity of faecal cryptosporidia for mice, *Res. Vet. Sci.,* 33, 379, 1982.

Chapter 2

Diagnosis

Michael J. Arrowood

CONTENTS

I. INTRODUCTION

With the exception of a single case report,[1] *Cryptosporidium parvum* is the only species of the genus considered to produce infection and disease in humans. While biological and molecular differences have been noted among isolates of *C. parvum* infecting various mammals, subspecies or strains are not yet recognized. The diagnostic assays described below will range from those that can be widely applied to all *Cryptosporidium* species to those that are species specific. The major focus of the following discussion will be directed to the detection of infections caused by *C. parvum*.

Clinical signs associated with cryptosporidiosis, including profuse, watery diarrhea, are generally not pathognomonic in the absence of detecting the parasite or parasite antigens. Great variability in presentation of clinical signs makes diagnostic interpretation subjective at best. Some utility is afforded by the observation of chronic symptoms in immunocompromised hosts, but parasitologic confirmation is yet necessitated.

Diagnosing infections caused by *Cryptosporidium* spp. requires laboratory tests to detect the parasite or specific antigens in the feces, body fluids, or tissues of the host. Considerable labor and expertise in microscopy were formerly required for the identification of this protozoan pathogen. In laboratories where conventional microscopic examination methods continue to be used for routine specimen processing, commercial immunofluorescent assays (IFA) and, more recently, enzyme immunoassays (EIA) have provided invaluable tools to identify and confirm the presence of cryptosporidial oocysts or antigens in clinical specimens.

Detection of parasite-specific antibodies has proven to be an excellent aid to the clinical diagnosis of some parasitic diseases when parasite detection is not possible.[2] Recent studies suggest that serologic

0-8493-7695-5/97/$0.00+$.50

assays for demonstrating recent cryptosporidial infections may be practical, and there are great opportunities for the development of sensitive and specific assays for different species of *Cryptosporidium* and possibly different strains of *C. parvum* using biochemical assays or nucleic acid-specific molecular probes.

An important caveat to remember while discussing diagnostic methodologies is the absence of effective therapies for resolving cryptosporidial infections. This fact more than any other (including expense) probably accounts for the reluctance of physicians and veterinarians to request specific tests for the diagnosis of cryptosporidiosis. Nevertheless, diagnosis is valuable since it provides the opportunity to avoid prescribing antibiotics that may disrupt the normal intestinal flora and possibly exacerbate the cryptosporidial infection.

II. CLINICAL SIGNS OF CRYPTOSPORIDIOSIS

A. INCUBATION PERIOD AND DURATION OF INFECTION IN HUMANS

Cryptosporidial infections result from oral ingestion of oocysts which may be encountered in contaminated food or water, through direct contact with infected animals or humans, or on contaminated surfaces. Several case reports of human infections have provided sufficient data to reasonably estimate the incubation period: 2 to 10 days. The most frequently reported period preceding the onset of symptoms is 7 days,[3,4] which is consistent with the onset reported among human volunteers enrolled in a *C. parvum* infectious dose study.[5] The latter report presented a mean of 9 days and a median of 6.5 days preceding development of clinical signs. These data are consistent with the reported timing of the *C. parvum* life cycle *in vitro*[6-9] and in animal models.[10]

The duration of clinical illness and oocyst excretion in immunocompetent individuals has been described in many case reports and surveys.[3,4,11] Symptoms, including diarrhea, generally persist for 1 to 2 weeks. Oocyst excretion persists for 1 to 4 weeks or more after symptoms are resolved. A median period of 18 days lapsed between the onset of clinical symptoms and the disappearance of oocysts from stool specimens among 49 patients with symptomatic cryptosporidiosis.[12] The median periods were 19 and 17 days, respectively, for adults (19 patients) and children (30 patients). The period between the appearance of symptoms and the disappearance of oocysts ranged from 9 to 50 days in children and 11 to 28 days in adults. Oocyst excretion has been estimated to persist for approximately twice as long as the duration of symptoms,[13] but in most reports the range is quite variable.[12,14-17] Despite paromomycin treatment of symptomatic, experimentally infected human volunteers, the onset of symptoms, oocyst excretion, and oocyst excretion duration were consistent with patterns observed in naturally infected persons.[5,18]

Conclusive studies estimating the prevalence of asymptomatic infections and the corresponding period of oocyst excretion have not been performed. Bile samples from 169 patients undergoing routine endoscopy revealed 12.7% were positive for oocyst presence.[19] While half of these patients had demonstrable oocysts in their stools, none were diarrheic. The authors suggest that the potential for asymptomatic infections may exceed prior expectations. Other studies, especially of children in daycare centers, suggest that asymptomatic excretion of oocysts is not uncommon.[20-23] Oocyst excretion patterns were intermittent among symptomatic and asymptomatic individuals experimentally infected with *C. parvum*.[18] The study revealed positive stools alternating with negative using an immunofluorescent assay[24] on aliquots of the unconcentrated stools. As for other parasite infections, this observation reinforces the advocacy of examining up to three independent stools to avoid false negatives (parasitologists' rule of thumb: O&P times 3).

Cryptosporidial infections in immunosuppressed or immunodeficient individuals can be chronic, with persistent clinical symptoms and oocyst excretion. The resolution of infections in these patients is directly related to the withdrawal of immunosuppressive treatments (e.g., cancer and organ transplant patients), the resolution of immunosuppressive intercurrent infections (e.g., measles, chickenpox, cytomegalovirus), or the correction of immunosuppression resulting from malnutrition. Congenitally immunodeficient individuals (e.g., hypo- and agammaglobulinemia) and those with untreatable or nonresolving immunosuppressive infections (e.g., HIV-infected patients) generally experience chronic infections, although spontaneous recoveries have been reported.[4,25] Asymptomatic infections in immunocompromised individuals[26-28] have probably been underestimated (as is strongly indicated in the immunocompetent population).

B. CLINICAL SIGNS IN IMMUNOCOMPETENT HUMANS

Symptomatic immunocompetent patients usually present with mild to profuse, watery diarrhea, with or without mucous, rarely with blood or leukocytes. Frequent and voluminous bowel movements can contribute to rapid weight loss and dehydration. Other frequently reported symptoms include abdominal cramping, mild fever (<39°C), nausea and vomiting, neuralgia (headache), fatigue, and anorexia.[4,25] Cryptosporidiosis-associated symptoms for 586 individuals in 36 large-scale surveys were summarized.[11] The most commonly reported symptoms, ranked by frequency, were diarrhea (92%), nausea and vomiting (51%), abdominal pain (45%), and mild fever (36%). A later review of case reports[3] ranked symptoms in the following order: diarrhea, abdominal pain, anorexia/nausea/vomiting, mild fever, malaise/fatigue/weakness, and, finally, respiratory problems. As might be expected, many individuals experienced multiple symptoms. The relationship between oocyst excretion (an indication of parasite load in the intestinal tract) and severity of symptoms is not entirely clear. Significantly higher numbers of oocysts were excreted by *C. parvum*-infected individuals with diarrhea than by those with enteric symptoms in the absence of diarrhea.[18] However, oocyst excretion in infected but asymptomatic individuals was not significantly lower than the oocyst excretion observed in the infected with diarrhea group. This seemingly unusual observation may, in part, be a consequence of the small sample sizes in the study.

C. CLINICAL SIGNS IN IMMUNOCOMPROMISED HUMANS

Cryptosporidiosis in immunocompromised individuals (especially patients with AIDS) is associated with diarrhea that is often more severe than in immunocompetent patients and is considered a contributing factor leading to death. Symptomatic diarrhea may be marked by stool volumes that exceed 6 liters per day.[3,4] Intestinal infections may extend from the small intestine into the colon, stomach, and esophagus and may involve a variety of extraintestinal organs and tissues.

D. EXTRA-INTESTINAL INFECTIONS

Cryptosporidial colonization of extra-intestinal sites is more common in immunocompromised hosts. Infection of gall bladder and bile duct epithelium has resulted in acalculous cholecystitis and sclerosing cholangitis in several AIDS patients.[29-31] Reported symptoms include fever, nonradiating right upper quadrant pain, nausea, vomiting, and in some cases diarrhea and jaundice. Elevated serum bilirubin and liver enzyme levels accompany dilation and wall thickening of the gall bladder and bile ducts. Diagnosis has been confirmed by histologic examination of biopsy specimens or by identification of oocysts in bile (oocyst numbers may be below detectable limits in stool). Hepatobiliary infections were observed in SIV-infected Rhesus monkeys.[32]

Pancreatic cryptosporidiosis with colonization of the pancreatic ducts has been reported in several immunocompromised cases,[32-38] but only one case of pancreatic cryptosporidiosis has been reported in an immunologically normal individual, an adolescent farm girl. One week after cryptosporidial enteritis was diagnosed, high serum amylase levels implicated pancreatic involvement. Laboratory analyses subsequently revealed ascites associated with an enlarged pancreas. No other etiologic agent was identified.[36] AIDS patients with pancreatic cryptosporidial infections have also exhibited bile duct or gall bladder infections. A severe combined immunodeficient infant with cryptosporidial enteritis had cryptosporidia in the pancreatic duct epithelium at autopsy.[33]

Respiratory cryptosporidiosis has been increasingly reported, particularly in immunocompromised individuals.[28,33,39-51] Symptoms included cough, croup, wheezing, hoarseness, and shortness of breath. Few reports have demonstrated cryptosporidia in the bronchial mucosal epithelium.[33,39] Most cases have been diagnosed by oocysts in sputum, tracheal aspirates, bronchoalveolar lavage fluid, and alveolar exudate (obtained by lung biopsy). It is not clear whether oocysts detected in these specimens were due to aspiration from the gastrointestinal tract or colonization and replication in the respiratory epithelium. The species of *Cryptosporidium* present in the respiratory tract is probably *C. parvum,* as the avian-specific species *C. meleagridis* and *C. baileyi*, which can infect the respiratory epithelium of turkeys and other avians, have not been shown to infect mammals. One exception is a case report of *C. baileyi* infection in an immunocompromised (HIV-infected) human.[1] However, the DNA base sequence for ribosomal RNA from this isolate matched that found in *C. parvum* following polymerase chain reaction (PCR) analysis.[52] Attributing respiratory symptoms to cryptosporidiosis is also complicated by the observation that most patients diagnosed with oocysts in respiratory samples were concomitantly infected with other respiratory agents, particularly cytomegalovirus, *Pneumocystis carinii*, and *Mycobacterium*. Nevertheless, *Cryptosporidium* spp. were the only pathogens isolated from four symptomatic patients also infected with HIV.[46] Most immunocompromised patients with respiratory cryptosporidiosis did not recover. In spite of the many case reports of extra-intestinal cryptosporidiosis, no disseminated, systemic cryptosporidial infections have been reported.

III. DIAGNOSTIC METHODS FOR DEMONSTRATING OOCYSTS OR ANTIGENS IN CLINICAL SPECIMENS

Diagnosis of cryptosporidiosis in humans and animals has progressed from histologic staining of gut or other tissue biopsy specimens to simple and sensitive assays to detect oocyst or other antigens in stool samples. This shift in diagnostic methodologies has improved disease detection, since stool samples represent a sampling of the entire gut,[4] whereas the focal distribution of cryptosporidia in epithelial surfaces can lead to negative biopsy specimens in infected individuals. In addition to stool specimens, oocysts have also been detected in bile and in sputum.

A. SPECIMEN COLLECTION AND PREPARATION FOR DIAGNOSTIC ASSAYS

Stool specimens may be submitted fresh, preserved in 10% buffered formalin, or suspended in a storage medium composed of aqueous potassium dichromate (2.5% w/v, final concentration). Oocysts remain infectious in storage media for extended periods and should be considered a biological hazard. It is therefore recommended that stool specimens be preserved in 10% buffered formalin or sodium acetate-acetic acid-formalin (SAF) to render oocysts nonviable (contact time with formalin necessary to kill oocysts is not clear, but should be considered in excess of 18 to 24 hours).[53-56] The use of mercuric chloride-containing preservatives (e.g., polyvinyl alcohol, PVA) is not recommended due to incompatibilities with some methodologies[57] and the environmental hazards posed by the disposal of mercury-containing compounds.

Symptomatic cryptosporidiosis is usually associated with substantial oocyst numbers in stool samples, obviating the need for stool concentration before application of diagnostic methods.[4] Oocyst numbers can be quite variable, even in liquid stools; therefore, multiple stool samples should be tested before a negative diagnostic interpretation is reported. Oocyst excretion was intermittent in symptomatic and asymptomatic individuals experimentally infected with *C. parvum*.[18] Single-stool examination identified only 50% of infected individuals involved in an outbreak of waterborne cryptosporidiosis.[58]

In asymptomatic infections or in epidemiologic studies, stool concentration is of greater importance. Sheather's sucrose flotation has been popular in research laboratories, but formalin-ether and formalin-ethyl acetate sedimentation is reportedly more popular in clinical laboratories.[4] Other concentration methods include zinc sulfate flotation and saturated sodium chloride flotation.[4,59] While not advocated for diagnostic applications, additional concentration and purification methods include discontinuous sucrose, isopycnic or discontinuous Percoll®, or cesium chloride gradient centrifugation.[60,61]

Two potential shortcomings regarding oocyst concentration techniques should be considered. Sedimentation methods are generally performed using low speed centrifugation. Given their small size and mass, cryptosporidial oocysts may become trapped in the ether or ethyl acetate plug and fail to properly sediment. Increased centrifugation speed or time (10 minutes, 500 *g*) may be warranted when attempting to recover cryptosporidial oocysts.[57] Resolution of cryptosporidial infections is accompanied by increasing numbers of nonacid-fast, oocyst "ghosts".[62] Such oocysts may not float or sediment as expected, giving rise to false negative results. See Appendix for a formalin-ethyl acetate sedimentation method.

B. STAINING METHODS FOR MICROSCOPICAL OOCYST DETECTION

Microscopical methods for detecting *Cryptosporidium* are summarized in Table 1. Identification of cryptosporidia was initially performed by histologic staining of infected tissues (see color plate), mucosal scrapings, or gut contents.[63-66] Subsequently, oocysts could be demonstrated in feces by noninvasive Giemsa staining techniques,[67,68] a major improvement over biopsy sampling but one which could not differentially stain oocysts from similarly sized fecal yeasts and other debris. In 1981, the Ziehl-Neelsen acid-fast staining technique finally provided clinical and research laboratories with a simple and effective method for identifying oocysts in stool samples: bright red oocysts against a background of blue-green fecal debris and yeasts (see color plate).*[69]

* Color plate appears after page 54.

The acid-fast staining technique has been modified and improved upon in the intervening years. Hot[70] and cold[71] modified acid-fast stains were described in 1983. Further modifications included the incorporation of dimethyl sulfoxide (DMSO) into the acid-fast stain[72,73] and the incorporation of the detergent tergitol into the modified cold Kinyoun acid-fast method.[74] An advantage of the acid-fast staining methods is that other parasites can be detected in the fecal smears that would go unidentified if specific immunofluorescence or enzyme immunoassays are used (e.g., *Isospora*, *Cyclospora*).

Alternatives to the brightfield microscopic acid-fast methods include negative stains,[75-78] the hot safranin-methylene blue stain,[79] modified Kohn's stain,[80] modified Koster stain,[81] and aniline-carbol-methyl violet and tartrazine.[82] Fluorescent stains include auramine O,[83] auramine-rhodamine,[70,71] auramine-carbol-fuchsin,[84] acridine orange,[70,71] mepacrine,[85] and 4′,6-diamidino-2-phenylindole (DAPI) and propidium iodide staining[86] (see color plate). The fluorescent dye methods exhibit potentially higher sensitivities (lower signal to noise ratio) compared to the brightfield stains, but, like all nonspecific chemical staining methods, they yield false positives and may leave some oocysts unstained. These methods may be useful for screening samples, but identification should be confirmed with more specific assays (immunofluorescence assays, or IFA; enzyme immunoassays, or EIA).

Quick screening of fresh fecal samples using an iodine wet mount has also been recommended, but it is usually combined with a more sensitive confirmatory stain or assay.[4] Some investigators find direct observation of fresh or concentrated fecal specimens using phase contrast or differential interference contrast (or Nomarski) microscopy (see color plate) to be a valuable screening step. Wet mount observation will be most valuable with specimens containing moderate to high numbers of oocysts.

Morphometric characteristics of *Cryptosporidium* sp. oocysts are summarized in Table 2. Named (accepted) species include *C. baileyi*, *C. meleagridis*, *C. muris*, *C. nasorum*, *C. parvum*, and *C. serpentis*.[87-91] Unnamed isolates include oocysts described in lizards, ostriches, and quail.[92-94] While oocysts from guinea pigs were not described (or even found) in early reports,[95,96] morphometric data for oocysts have not been described in more recent studies of *C. wrairi*.[97,98]

C. IMMUNOFLUORESCENCE MICROSCOPY FOR OOCYST DETECTION

Immunologic techniques for the detection of cryptosporidia in stool specimens were introduced in 1985 and 1986. Indirect immunofluorescent assays were described for the detection of oocysts employing convalescent human serum[99] and oocyst-immunized rabbit antiserum.[100] Immunofluorescent assays employing oocyst-reactive monoclonal antibodies were also introduced[101-104] (see color plate). The immunofluorescent methodologies showed significantly increased sensitivities and specificities compared to conventional staining techniques[20,102,104-106] and have found widespread application in research and clinical laboratories, as well as for monitoring oocyst presence in environmental samples.[60,100,101,104,105,107-123] The assays generally work well with fresh or preserved stools (formalin, potassium dichromate), but some fixatives can cause problems (e.g., MIF). See Table 3 for a comparison of specificity and sensitivity between commercial IFA and EIA kits. New diagnostic kits not evaluated and reported in the scientific literature are not included in this table.

Immunofluorescence assays demonstrating cryptosporidial life-cycle stages (e.g., oocysts) in infected tissues or biopsy specimens have been reported[104,124] and can be performed using reagents available in commercial diagnostic kits.[102,105] These immunohistological assays are primarily of research value, given the broad availability of stool-based diagnostic assays and the ready identification of *Cryptosporidium* in tissue sections (cryptosporidia are uniquely found on the lumenal surface of epithelial cells and are apparent in specimens stained with hematoxylin and eosin or other routine histology stains).

D. ENZYME IMMUNOASSAYS FOR CRYPTOSPORIDIAL ANTIGEN DETECTION

Fecal enzyme immunoassays have been reported utilizing oocyst-reactive monoclonal antibodies.[125-127] The monoclonal antibodies were adapted for antigen detection in an antigen-capture EIA. One assay positively identified samples confirmed by immunofluorescent microscopy, but also produced false positives.[125] Attempts to verify the microscopically negative samples as positive using other methodologies were not successful. Another assay was reportedly more sensitive than microscopic examination of acid-fast stained samples, but less sensitive than the immunofluorescent procedure.[126] Yet another assay yielded a 96% agreement between the EIA and a fecal flotation method for detecting oocysts.[127] Nevertheless, the EIA was less sensitive than the modified Ziehl-Neelson stain or IFA (especially when oocyst numbers were small). The authors predicted that all stools with $\geq 10^6$ oocysts per ml would yield

Table 1 Diagnostic Procedures for *Cryptosporidium* spp. Oocyst Detection by Microscopy

Method	Fixation[a]	Concentration[b]	Oocyst Appearance	Yeast/Background Appearance	Advantages and Disadvantages	Ref.
Direct microscopy (no staining)	N	S	Refractile, slightly pink	NR	Good concentration	71
Brightfield stains						
Giemsa	N	N	Blue/blue green	Blue/blue green	Poor differentiation	204
Gram	NR	N	Pale blue	NR	Poor differentiation	205
	NR	NR	Unstained or pale red	Purple	Poor differentiation	206
Modified Kohn's	PD	S	Blue to blue-gray	Gray	Good morphology	80
Methylene blue/borax	NR	NR	Light blue or purple	Dark blue	Fast	206
Aniline-carbol-methyl violet	N	N	Blue to blue-violet	Unstained	High contrast	82
Safranin/methylene blue	N	N	Orange-pink	Blue	Fast, simple, but requires heat	207,208
Modified Kinyoun acid-fast	N	N or FES	Red	Green	Good differentiation	71
Modified Ziehl-Neelsen acid-fast	NR	NR	Red	Green	Good differentiation	69
Modified acid-fast	BF, PD	N or FS	Red	Blue	Good differentiation	70
DMSO modified acid-fast	N	N	Pink	Green	Fast, simple	72
	F,BF	N, FES	Red	Green	Fast, simple	73
Negative stains						
Carbol fuchsin	N	N	Unstained	Red	Fast, simple	75
Iodine	N	N or FES	Unstained	Brown	Fast, simple	71
	BF, PD	N or FS	Unstained	Brown	Insensitive, poor morphology	70
Light green	N, BF	N, FES	Unstained	Green	Fast, simple	78
Merbromine	N, BF	N, FES	Unstained	Orange	Fast, simple	78
Methanamine-silver	NR	NR	Unstained	Black	NR	209
	BF, PD	N or FS	NR	NR	Insensitive, poor	70
Nigrosin	BF	FES	Unstained	Unstained	Fast, simple	76
Periodic acid-Schiff	BF, PD	N or FS	Unstained	Red	Insensitive, poor	70
Phosphotungstic acid	G	N	Light brown	Black	Electron microscopy	79

Fluorochrome stains						
Acridine orange	N	N or FES	Orange/green	Orange	Poor	71
	BF, PD	N or FS	NR	NR	Insensitive, poor	70
Auramine O	N, BF	N, FES	NR	NR	Simple, sensitive	83
Auramine-rhodamine	N	N or FES	Orange (fluorescent)	Nonfluorescent	Sensitive	71
	BF, PD	N or FS	Orange (fluorescent)	NR	Fair to good morphology	70
Auramine-carbol fuchsin	F	N	Orange	Red	Includes negative stain, fast, sensitive	84
DAPI/propidium iodide	M	NR	Blue nuclei, red sporozoite cytoplasm, blue-green residual body	NR	Rapid scanning, reliable identification based on nuclei and sporozoite appearance	86
Mepacrine	BF	N	Fluorescent (color not described)	Nonfluorescent	Stain inexpensive, fast, sensitive	85
Immunofluorescence						
Monoclonal antibody	N, BF, PD	N, FS, FES, FEAS	Green	Unlabeled or eriochrome black counterstain	Sensitive and specific	101,102, 104,105
Monoclonal antibody	F	FES	Green	Unlabeled or Evans blue counterstain	Sensitive and specific	103
Polyclonal antibody	F	FES	Green	Unlabeled	Specific	99
	F, PD	N	Green	Unlabeled	Sensitive and specific	100

[a] N = none (fresh); F = formalin in water; BF = buffered formalin; PD = potassium dichromate; G = glutaraldehyde; M = methanol; NR = not reported.

[b] N = none; FEAS = formalin/ethyl acetate sedimentation; S = sucrose flotation; Z = zinc sulfate flotation; FS = formalin sediment (simple centrifugation), FES = formalin/ether sedimentation, NR = not reported.

Table 2 Morphometric Features of Oocysts from *Cryptosporidium* Species

	Host[a]	Host Class	Oocyst Length (μm)	Oocyst Width (μm)	Oocyst Shape Index (length/width)	Ref.
Named Species						
C. baileyi	Chicken	Avian	5.6–6.3 (6.2)	4.5–4.8 (4.6)	1.2–1.4 (1.4)	87
			6.0–7.5 (6.6)	4.8–5.7 (5.0)	1.0–1.8 (1.3)	88
C. meleagridis	Turkey	Avian	4.5–6.0 (5.2)	4.2–5.3 (4.6)	1.0–1.3 (1.1)	88
C. muris	Calf (bovine)	Mammal	6.6–7.9 (7.4)	5.3–6.5 (5.6)	1.1–1.5 (1.3)	89
C. nasorum[b]	Cichlid fish	Fish	3.5–4.7 (4.3)	2.5–4.0 (3.3)	0.9–1.9 (1.3)	90
C. parvum	Calf (bovine)	Mammal	4.5–5.4 (5.0)	4.2–5.0 (4.5)	1.0–1.3 (1.1)	89
C. serpentis	Snake	Reptile	5.6–6.6 (6.2)	4.8–5.6 (5.3)	1.0–1.3 (1.2)	91
Unnamed Species						
C. species	Lizard	Reptile	5.3–5.7 (5.7)	4.2–5.7 (4.7)	0.9–1.4 (1.2)	92
C. species	Ostrich	Avian	3.0–6.1 (4.6)	3.3–5.0 (4.0)	1.0–1.4 (1.2)	93
C. species	Quail	Avian	4.5–6.0 (5.0)	3.6–5.6 (4.3)	1.0–1.6 (1.2)	94

[a] Oocyst source for morphometric calculations.

[b] Calculations based on electron micrograph measurements (authors noted shrinkage without correction).

positive results. An indirect, double-antibody EIA using polyclonal antisera has also been developed.[128] Again, the test was not as sensitive as the immunofluorescent procedure. Specific detection of 18- and 20-kDa *C. parvum* coproantigens reportedly constitutes the foundation for a diagnostic stool EIA.[129] Batch processing of clinical fecal samples and large-scale epidemiological studies would be facilitated by specific and sensitive EIA methodologies.

At least four commercial EIA tests (Alexon, Inc.; Dako Corp.; Seradyn, Inc.; LMB Laboratories, Inc.) have been introduced for the detection of cryptosporidial antigen in stool samples.[130-137] These kits are reportedly superior to conventional microscopic examination (especially acid-fast staining) and show good correlation with the monoclonal antibody-based immunofluorescence assays. Kit sensitivities and specificities ranged from 66.3 to 100% and 93 to 100%, respectively (Table 3). Diagnostic kits not evaluated and reported in peer-reviewed journals are not included in this table.

E. OTHER IMMUNOASSAYS FOR CRYPTOSPORIDIAL ANTIGEN DETECTION

Oocyst detection in stool samples employing latex beads coated with antisera from oocyst-immunized rabbits has been reported.[138] This agglutination assay was applied to homogenized stool and gut content samples from *C. parvum*-infected mice. The assay was rapid and simple to perform but lacked specificity (false positives were observed).

A flow cytometry-based assay was reported for quantitation of oocysts in feces from experimentally infected laboratory mice.[139] The assay used an oocyst-specific monoclonal antibody (IgG, OW50) previously employed to detect oocysts in clinical specimens by immunofluorescence microscopy.[105] The performance of the assay probably benefits from the relatively consistent composition of mouse stool (standardized diet) and exhibited a useful detection limit of approximately 1000 oocysts per fecal pellet or 2500 oocysts per ml fecal suspension. Compared to immunofluorescence microscopy, the flow cytometry assay was approximately 10 to 15 times more sensitive. It is not yet clear whether this methodology will prove useful with human or veterinary clinical specimens.

F. RECOMMENDED METHODS FOR OOCYST OR ANTIGEN DETECTION IN CLINICAL SPECIMENS

The scientific literature is replete with diagnostic methods for *Cryptosporidium* detection. Since most investigators and clinicians prefer the method(s) that work best in their own laboratories, a universally appropriate diagnostic method is not apparent. Acid-fast staining methods, with or without stool concentration, are most frequently reported in clinical laboratories (see Appendix for stool concentration and acid-fast staining methods used at the CDC Diagnostic Parasitology Laboratory). For greatest sensitivity and specificity, immunofluorescent microscopy is the method of choice (followed closely by EIA) for clinical samples. Either assay can be used to confirm results obtained from nonspecific staining techniques. Several commercial IFA products are presently available. MeriFluor™ *Cryptosporidium/Giardia* is an immunofluorescent assay system based on an oocyst-specific monoclonal antibody (IgG,

Table 3 Summary of Selected Commercially Available Diagnostic Kits for Detection of *Cryptosporidium* spp. (Primarily in Clinical Specimens)

Kit Name (Clinical Specimens)	Manufacturer/Distributor	Type of Test[a]	Sensitivity[b]	Specificity[2]	Comparison Test	Ref.
ProSpecT/*Cryptosporidium*	Alexon, Inc.	EIA-plate	97	98	Acid-fast stain, IIF[c]	130
			98	98	Acid-fast stain	131
			96	99.5	M-DIF[d]	137
			94 (97)	99 (98)	M-DIF[d], acid-fast, Color Vue	136
IDEIA *Cryptosporidium*	Dako Corp.	EIA-plate	100 (93.1)	100 (98.7)	Auramine stain, N-DIF[e]	132
MeriFluor™ *Cryptosporidium/Giardia*	Meridian Diagnostics, Inc.	DFA, IgG	100	100	Acid-fast stain	105
			96 (100)	100 (100)	Acid-fast, ProSpecT, Color Vue	136
Color Vue *Cryptosporidium*	Seradyn, Inc.	EIA-plate	93	93	IIF[c]	133
			76	100	M-DIF[d]	137
			94 (92)	100 (100)	M-DIF[d], acid-fast, ProSpecT	136
Cryptosporidium Antigen Detection	LMB Laboratories	EIA-plate	66.3	99.8	Acid-fast, auramine	134
Microwell ELISA			93	99	IIF[c]	135

[a] EIA = enzyme immunoassay; DFA = direct immunofluorescence assay, IIF = indirect immunofluorescence assay, NA = not available.
[b] Percent specificity or specificity compared to conventional methods; numbers in parentheses indicate values reported by the manufacturer.
[c] IIF = indirect immunofluorescence (MeriFluor *Cryptosporidium/Giardia* assay).
[d] M-DIF = direct immunofluorescence (MeriFluor *Cryptosporidium/Giardia* assay).
[e] N-DIF = direct immunofluorescence (DetectIF *Cryptosporidium*, Shield Diagnostics, Ltd.).

OW50) produced by Arrowood and Sterling (Meridian Diagnostics, Inc.; Cincinnati, OH 45244).[102,104,105] Detect IF *Cryptosporidium* is an immunofluorescent assay system based on an oocyst-specific monoclonal antibody (IgM, C1) produced by McLauchlin and associates[103] (Shield Diagnostics, Ltd.; Dundee DD1 1 SW, Scotland, U.K.). Both assays incorporate counterstains (eriochrome black and Evans blue, respectively). The latter assay is available for research only in the U.S. Crypto IF Kit is an immunofluorescent assay system based on an oocyst-specific monoclonal antibody[119] that incorporates Evans blue as a counterstain (TechLab; Blacksburg, VA 24060). All of these reagents exhibit broad reactivity with *C. parvum* and other *Cryptosporidium* species, so they should be applicable to human and veterinary specimens. EIA kits were described above. Commercially available diagnostic assays for detecting *Cryptosporidium* in clinical samples are presented in Table 3.

IV. SERODIAGNOSTIC METHODS FOR CRYPTOSPORIDIOSIS

Oocyst excretion in most infected individuals is transient, affording investigators a limited window of opportunity to confirm infection by parasite or antigen detection. Serologic methods may identify recently infected individuals by demonstrating seroconversion, a significant increase in anticryptosporidial antibody titers, or by the identification of antigen-specific antibody responses to target antigens. The self-limiting and generally short-term nature of cryptosporidiosis in the immunocompetent host limits the opportunity to detect antibody responses in serologic assay (the earliest detectable responses appear as the infection is being resolved, i.e., approximately 2 weeks post-infection). This limitation may not be true of infections in immunocompromised individuals where cryptosporidiosis may be protracted or chronic. Data regarding serologic responses in the immunocompromised host are insufficient to predict whether a serologic assay will prove useful in diagnosing cryptosporidial infections within this population.

Humoral responses to cryptosporidial infections were first assayed by indirect immunofluorescence (IIF) against infected histologic tissue sections.[140,141] Anticryptosporidial immunoglobulins were demonstrated in the sera of 10 mammal species.[140] Immunocompetent humans that recovered from infection were shown to develop titers between 1:40 and 1:640 that persisted for at least 1 year.[141] In the same study, hypogammaglobulinemic individuals infected with *Cryptosporidium* did not develop demonstrable anti-cryptosporidial immunoglobulins. Five AIDS patients concurrently infected with *Cryptosporidium* all exhibited titers of at least 1:40.[141] Immunofluorescent assays have also been developed using target antigens comprised of purified oocysts[101,142-145] and mixtures of oocysts and sporozoites.[146-148] With the exception of the monoclonal antibody-based immunofluorescent assay for oocyst detection, IIF assays for humoral responses largely have been restricted to research laboratories.

Microtiter enzyme immunoassays for the detection of antibody responses to crude cryptosporidial antigens have been developed in several laboratories employing antigens prepared from sonicated oocysts;[27,149-152] whole, frozen oocysts;[145,153] zirconium bead disrupted and freeze-thawed oocysts;[154] and oocysts prepared in an undefined manner.[155] One study reported EIA methodologies to be more sensitive than immunoblotting techniques,[156] while other investigators have observed a poor correlation between microtiter EIA and immunoblotting results.[150] While several investigators have demonstrated a strong correlation between infection and EIA titers, reservations exist regarding the specificity of these assays.[4] Cross-reactivity to cryptosporidial antigens by sera from individuals infected with other parasites is a potential drawback to all serodiagnostic assays. In this regard, monoclonal antibodies raised against *Plasmodium chabaudi* cross react with *C. parvum* sporozoite antigens, and monoclonal antibodies raised against *C. parvum* cross react with *P. chabaudi* antigens in indirect immunofluorescent assays (Mead, unpublished data). Antigen cross-reactivity was reported between *C. parvum* and *Eimeria* spp. antigens in an IIF assay employing sera from *C. parvum*-immunized rabbit and lambs or *Eimeria* spp.-infected lambs.[157] Anti-*Eimeria* polyclonal sera (lamb) reacted with *C. parvum* antigens (29–30 and 66–69 kDa) in an immunoblot assay. The cross-reacting antigens did not include lower molecular weight *C. parvum* antigens (15, 17, and 27 kDa). No cross-reactivity was observed when anti-*Cryptosporidium* rabbit or lamb sera were tested against *Toxoplasma* and *Sarcocystis* spp. zoites by IFA. Microtiter EIA demonstrated *Cryptosporidium*-specific human serologic responses that were not inhibited (adsorbed) by *Giardia lamblia*, *Trichomonas vaginalis*, *Campylobacter jejuni*, *Escherichia coli*, or *Candida albicans*.[158] Further study will determine the significance of cross-reactivity in the sera of individuals exhibiting multiple parasitic infections.

Western blot methods have been developed for the detection of antigen-specific antibody responses to *Cryptosporidium* spp. using antigens prepared from sonicated oocysts;[150,156,159-162] purified, detergent-solubilized sporozoites;[146] boiled oocysts;[151,163] zirconium bead disrupted and freeze-thawed oocysts;[154]

and freeze-thawed oocysts.[147,154,164] Antigens with molecular weights ranging from 3 to over 200 kDa have reacted with immune sera from experimentally and naturally infected animals and humans. A 27-kDa antigen (also identified as a 20- or 23-kDa antigen) has been noted by most of these investigators to be universally identified by immune sera. Originally identified as an antigen from sonicated oocysts,[156] it was subsequently localized to the surface of sporozoites in a study employing purified sporozoites as antigen.[146] Immunoelectron microscopy[165] and surface labeling studies[146,166] confirmed the surface location of this antigen. Immunoblot reactivity to the 27-kDa antigen could be demonstrated as soon as 10 days post-infection (PI), peaked in intensity by 3 to 4 weeks PI, and was essentially absent within 4 to 5 months, using longitudinally collected serum samples from animals and humans.[146] Immunoblot profiles for calves and lambs have been described by several research groups.[146,147,150,151,154,159,163] The onset, peak, and diminution of immunoblot reactivity to *C. parvum* antigens were especially notable to 15-, 17- and 27-kDa antigens. The naïve status of the calves and lambs before natural or experimental infection probably contributed to the characteristic immunoblot patterns observed in the acute, convalescent, and postconvalescent sera.

Given the ubiquitous distribution of *C. parvum*, it is not surprising that human sera should present more complex immunoblot profiles compared to neonatal or juvenile livestock. Human pre-infection sera (when available) or random samples demonstrate significant background reactivity to cryptosporidial antigens: apparent evidence of prior infection and serologic responses to the parasite. Recently, immunoblot analyses were performed on sera collected from crew members of a Coast Guard cutter involved in a waterborne outbreak of cryptosporidiosis.[161] The well defined exposure period coupled with the opportunity to collect acute, convalescent and postconvalescent sera provided an unrivaled opportunity to evaluate the immunoblot methodology. The immunoblot profiles, including immunoglobulin response kinetics and signal amplitude (blot band intensity), to the 15-, 17-, and 27-kDa antigen groups were characteristic for crew members with parasitologically confirmed cryptosporidiosis and in crew members with suspected infections. The immunoblot methodology appeared to be twofold more sensitive in detecting cryptosporidial infections than parasitologic examination of single-stool specimens. Additional controlled studies (including cross-reactivity studies) are needed to define and validate the immunoblot methodology. While existing data do not support the application of serological assays for routine diagnosis of acute infections, these data suggest that immunoblot reactivity to these antigens may be useful as an epidemiologic marker of recent infection. There are currently no commercially available serologic assays for the detection of *Cryptosporidium*-specific antibodies.

V. BIOCHEMICAL- AND NUCLEIC ACID-BASED TECHNIQUES FOR *CRYPTOSPORIDIUM* DETECTION

The application of molecular biologic techniques has increased significantly in the last few years. Preliminary studies addressed strain differences among *C. parvum* isolates, species differences between *C. parvum* and *C. baileyi*, and the phylogenetic relationship between the genus *Cryptosporidium* and related organisms. DNA probes specific for *C. parvum* have been developed, and nucleic acid-based assays have been developed to detect parasites in stool specimens and environmental samples.

Karyotypic analyses of the chromosome-sized DNA of *C. parvum* and *C. baileyi* isolates were initiated using pulsed field gel electrophoresis.[167] Geographically distinct *C. parvum* isolates were collected from various mammalian hosts (a human isolate from Mexico; bovine isolates from Alabama, Arizona, and Iowa; and an equine isolate from Louisiana). An avian isolate (*C. baileyi* from chickens) was included for species differentiation. At least five chromosomal bands, separated using field inversion gel electrophoresis, were estimated to range in size from 1500 to over 3300 kilobase pairs (KB). The DNA electrophoretic patterns of the *C. parvum* isolates were essentially identical, while the *C. baileyi* pattern was readily distinguished by the separation of at least six bands in the same size range observed for the mammalian isolates. Later studies employing orthogonal pulsed field gel electrophoresis yielded more accurate size estimates for the chromosomes of *C. parvum* (900 to 1400 KB).[168,169] The conservative chromosomal DNA patterns observed among *C. parvum* isolates appear in contrast to the highly variable chromosomal DNA karyotypes observed among other protozoan parasites including species of *Plasmodium, Trypanosoma,* and *Leishmania*.[170-172]

Variation among five *C. parvum* isolates was investigated using two-dimensional gel electrophoretic analyses of sporozoite proteins.[173] Iowa (bovine), Peru (human), and Mexico (human) isolates were readily differentiated from one another. Alabama (bovine) and Louisiana (equine) isolates were indistinguishable but were differentiated from the Peru (human) and Iowa (bovine) isolates. An electrophoretic analysis of oocyst wall proteins from three *Cryptosporidium* species (*C. parvum*, *C. baileyi*, and *C. muris*) identified common and electrophoretically unique bands for each species.[164]

An isoenzyme study compared *C. parvum*, *C. baileyi*, and *C. muris* isolates.[174] Glucose phosphate isomerase (GPI) and phosphoglucomutase (PGM) zymograms differentiated the three species. Zymograms of GPI for five isolates of *C. parvum* were indistinguishable, but PGM zymograms differentiated a human isolate from three isolates obtained from calves, deer, and sheep (which showed the same pattern).[174] A later study showed that among 23 *C. parvum* isolates of various host origin, all had identical GPI zymograms.[175] Zymograms of PGM grouped nine animal isolates as distinct from eight of nine human isolates (one human isolate matched the nine animal isolates). Hexokinase (HK) zymograms grouped six animal isolates as distinct from four human isolates. The authors hypothesized that human and animal *C. parvum* cycles of transmission are largely independent. A disadvantage of the isoenzyme assay is the relatively large number of oocysts necessary to yield detectable bands in the starch gel electropherograms (approximately 10^7 oocysts per sample for PGM or HK and 10^8 oocysts for GPI).[174,175] Isoenzyme electrophoresis is probably impractical for environmental samples but may be useful for clinical samples where oocysts are available in larger quantities.

Phylogenetic relationships between *Cryptosporidium* and related apicomplexans have been investigated by the partial sequencing of *C. parvum* ribosomal RNA.[176] Subunit ribosomal RNA (srRNA) sequence comparisons revealed no close relationships. The srRNA sequences most closely related to those of *Cryptosporidium* were from *Tetrahymena* and *Plasmodium*. The 18S rRNA nucleotide sequence was determined for isolates of *C. parvum* and *C. muris* and showed greater than 99% sequence identity.[177] These data contrast the phylogenetic analysis using parsimony based on ultrastructural comparisons.[178] The latter study linked *Cryptosporidium* with the eimeriorins (*Sarcocystis* and *Eimeria*) and not with *Plasmodium*. The apparent srRNA sequence divergence from the apicomplexans may help to explain the failure of existing antiCoccidial agents to control infections by *Cryptosporidium* species. Clearly, further study is necessary to resolve these issues.

Several *Cryptosporidium* DNA and RNA regions have been sequenced and reported to be valuable as targets for parasite detection and determination of viability. After a *C. parvum* genomic DNA sequence was randomly selected for analysis, a 400-base sequence was specifically amplified using appropriate oligonucleotide primers and yielded a unique pattern upon restriction endonuclease digestion.[179] The assay was reportedly sensitive and the authors predicted it would detect "small" numbers of oocysts in clinical or environmental samples. No data were presented regarding cross-reactivity with other species of *Cryptosporidium*. This DNA sequence (probe) was the basis of a PCR-based viability assay reported to function by detecting DNA in sporozoites released during excystation, relying on the digestion of free DNA before excystation and assuming that sporozoite DNA detected following excystation treatment was evidence of viability (release of sporozoites).[180] A more recent PCR-based viability assay used a truncated DNA sequence derived from a *C. parvum* oocyst protein sequence to detect free sporozoites following excystation.[181] While detecting 100 sporozoites in a 30-cycle PCR reaction, it was speculated that this technique could be optimized further. A caveat in all of these studies has been the use of purified, viable oocysts. These methods should be carefully evaluated using oocysts subjected to various environmental insults to validate that nonviable sporozoite DNA is not released from oocysts and that the methods will work with environmentally isolated oocysts.

A PCR-based assay targeting a small (18S) subunit rRNA sequence was reported to be as sensitive as immunofluorescence for detecting oocysts in water, including wastewater.[182,183] The assay detected as few as 1 to 10 purified oocysts, but was up to 1000-fold less sensitive in the presence of environmental contaminants. Flow cytometry purification of environmental samples largely restored the sensitivity.

The 18S rRNA sequences of *C. parvum* and *C. muris* were the basis of another PCR assay.[184] Coupling the PCR product with restriction endonuclease digestion yielded a method to differentiate *Cryptosporidium* isolates at the level of species (*C. parvum*, *C. baileyi*, *C. muris*). The authors proposed the technique may be useful in detecting small numbers of oocysts in environmental samples.

A *C. parvum* 5S rRNA sequence has been reported which appears to be *Cryptosporidium* specific (other species not tested) and may be useful as a probe for sensitive detection assays.[185] The 18S rRNA sequence of an isolate of *C. parvum* was compared to two other published sequences for geographically distinct isolates of *C. parvum*.[176,177] The sequence homology was 95% and 91.6%, respectively. Despite questions about the quality of a reverse transcription sequence previously reported,[176] the authors concluded that *C.* parvum 18S rRNA sequences from diverse geographical areas exhibit demonstrable sequence variation. For more information on biochemistry and molecular biology, see Chapters 7 and 10.

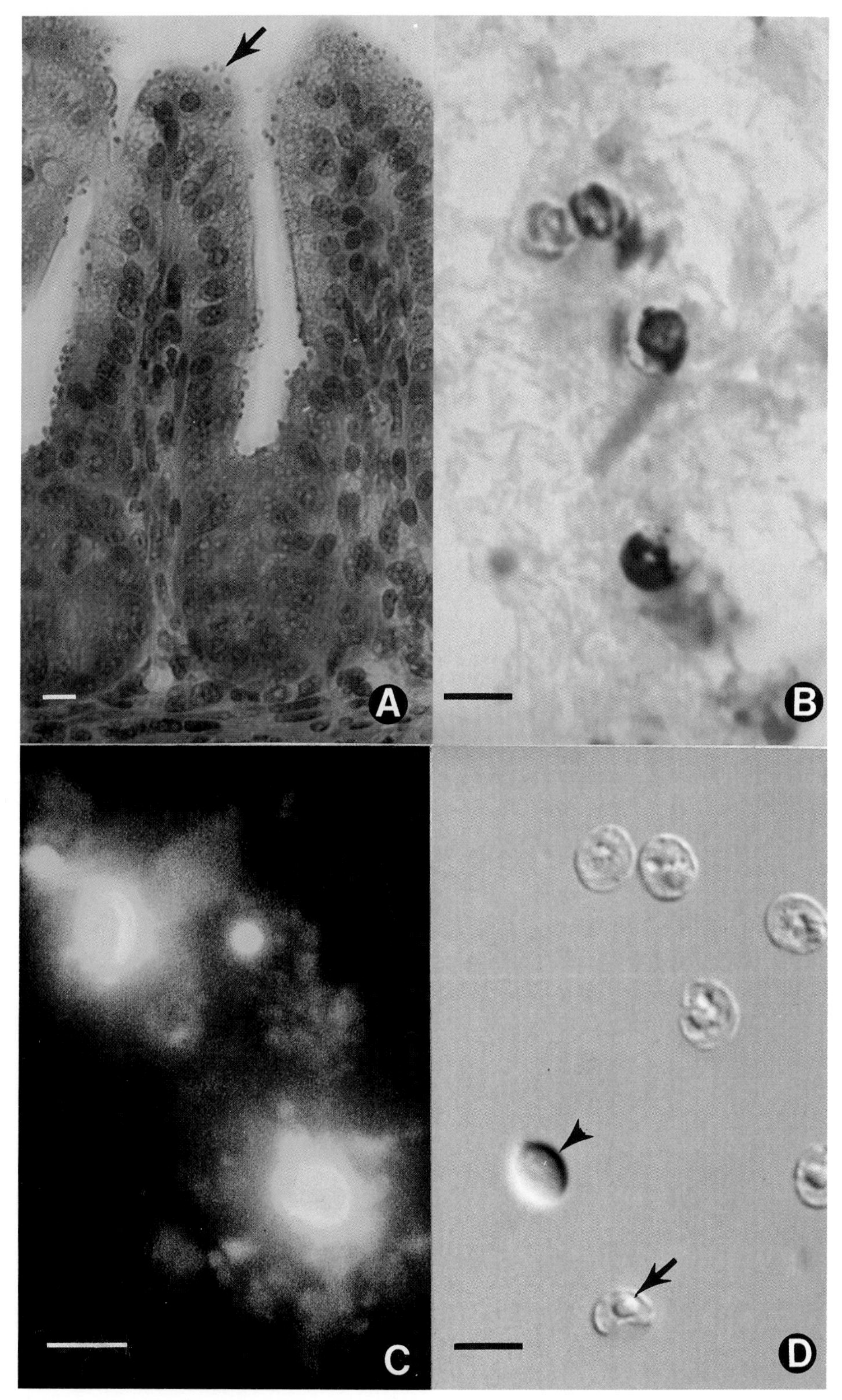

Chapter 2, Plate 1. Photomicrographs of *C. parvum*. **(A)** Hematoxylin- and eosin-stained neonatal BALB/c mouse ileum demonstrating numerous cryptosporidia (arrow) on epithelial cells. Bar = 25 μm. **(B)** Acid-fast-stained fecal smear demonstrating oocysts filled with an amorphous red mass or distinct crescent-shaped sporozoites. Bar = 5 μm. **(C)** Oocysts labeled by immunofluorescence (fluorescein) using a monoclonal antibody-based assay. Bar = 5 μm. **(D)** Differential interference contrast image of oocysts recovered from infected calf stool using Sheather's flotation. Note residual body (arrow) in the oocyst with the indented wall ("kidney shaped") and yeast cell (arrowhead). Bar = 5 μm.

VI. DETECTION METHODS FOR ENVIRONMENTAL SAMPLES

(See Chapter 4 for a brief description of the cartridge filter/immunofluorescence method for detecting *Cryptosporidium* oocysts in environmental samples.) The immunofluorescence method is applied following sample collection using a polypropylene cartridge filter or a flat, membrane filter.[107,109,110,113,121,122,186-192] The former filter method is used to collect relatively large sample volumes (100 to 400+ gallons), while the membrane filter method is used to collect smaller samples (approximately 10 liters). The proponents of the low volume membrane method contend that the higher efficiency of capture and recovery (elution) yield a performance equal to or better than that resulting from the purportedly inefficient and variable high-volume cartridge filter method. Both methods yield water concentrates that, unless modest in volume, are subjected to purification methods before analysis by microscopy. The quantity and composition of concentrate matrix impacts the performance of the immunofluorescence assay, i.e., high contaminant levels, especially algae, negatively impact the discrimination of cryptosporidial oocysts and their internal features. Calcium carbonate flocculation of small samples (10 liters) has been reported as an alternative to filtration,[193] but this method does not appear to have been widely adopted and has been reported to damage oocysts (loss of viability) before subsequent analyses.[194,195] Flow cytometry methods have been introduced for application to environmental samples.[119,120] These methods do not quantitate oocysts directly (unlike the stool assay describe above[139]), but rather the flow cytometer is used as a sample purification technique before subsequent immunofluorescent microscopy analyses.

The following provides additional detail regarding the various commercial assays and reagents used to identify and discriminate cryptosporidial oocysts. Commercial assays applicable to environmental samples follow. HydroFluor™ Combo is an immunofluorescent assay system based on an oocyst-specific monoclonal antibody (IgM, OW3) produced by Arrowood and Sterling[102,104] (EnSys, Inc.; Research Triangle Park, NC 27709). Detect IF *Cryptosporidium* is an immunofluorescent assay system based on an oocyst-specific monoclonal antibody (IgM, C1) produced by McLauchlin and colleagues[103] (Shield Diagnostics, Ltd.; Dundee DD1 1 SW, Scotland, U.K.). Crypto IF Kit is an immunofluorescent assay system based on an oocyst-specific monoclonal antibody[119] (TechLab; Blacksburg, VA 24060). Specificity differences between the first two products have been reported. HydroFluor™ Combo reacts with *C. parvum*, *C. meleagridis*, *C. wrairi*, *C. serpentis*, *Cryptosporidium* sp. (lizard and turtle), and some *C. muris* isolates but not *C. baileyi* (References 98, 104, 113, 196; Arrowood, unpublished data). No cross reactivity with other organisms was noted when tested against a wide variety of bacteria, yeasts, protozoa, and helminth ova.[105] Apparent cross-reactivity with algae in this indirect immunofluorescent assay was essentially eliminated by incorporating a blocking reagent (normal goat serum).[123] This observation indicates that the apparent cross-reactivity was contributed by the secondary antibody (goat anti-mouse-FITC conjugate) rather than the monoclonal antibody. Detect IF *Cryptosporidium* reacts with *C. parvum*, *C. baileyi*, *C. meleagridis*, *C. serpentis*, and *C. muris*.[103,113,196] Commercially available diagnostic assays for detecting *Cryptosporidium* in environmental samples are presented in Table 4.

Unlike bacteriological samples, which in many cases are essentially "grab samples", parasitological samples require relatively large volumes, lengthy collection times, and equally time-consuming processing to extract the captured particulate matter. In contrast to most bacteriological and virological assays, parasitological samples do not incorporate an enrichment step based on *in vitro* cultivation (amplification) of the captured organisms. Despite the promise of improved *in vitro* cultivation assays for *Cryptosporidium parvum*,[8,9,197-201] asexual development has not been reported to "recycle" (amplify) to any notable extent, nor has meaningful evidence of oocyst maturation been reported, especially of "thin-walled" oocysts thought responsible for life cycle recycling. The prospect for significant *in vitro* amplification of viable oocysts is discouraging (even with fresh, "viable" oocysts as inocula, *in vitro* development is inefficient, generating a "peak" number of developing stages approximately equal to the original number of oocysts in the inoculum). Nevertheless, *in vitro* culture improvements, especially coupled with highly sensitive nucleic acid-based assays, may eventually yield working viability assays that actually measure oocyst (sporozoite) infectivity. See Chapter 8 for more information on *in vitro* cultivation.

VII. CONCLUSIONS

Diagnostic methods for *Cryptosporidium* spp. have improved dramatically over the last 10 years. The transition from histological analyses of biopsy specimens to the development of fecal staining techniques set the stage for an avalanche of case reports and surveys demonstrating cryptosporidial infections are

Table 4 Summary of Selected Commercially Available Diagnostic Kits for Detection of *Cryptosporidium* spp. (Primarily in Environmental Samples)

Kit Name (Environmental Specimens)	Manufacturer/ Distributor	Type of Test[a]	*Cryptosporidium* Species	Detection	Ref.
HydroFluor Combo	EnSys, Inc.	IIF, IgM	*C. parvum*	+	102,104,113
			C. baileyi	–	
			C. meleagridis	+	
			C. muris	±	
			C. serpentis	+	
			C. wrairi	+	
DetectIF *Cryptosporidium*	Shield Diagnostics, Ltd.	DFA, IgM	*C. parvum*	+	103,113
			C. baileyi	+	
			C. meleagridis	NT	
			C. muris	+	
			C. serpentis	NT	
			C. wrairi	+	
Crypto IF Kit	TechLab	DFA, IgM	*C. parvum*	+	119
			C. baileyi	+	
			C. meleagridis	NT	
			C. muris	+	
			C. serpentis	NT	
			C. wrairi	NT	

[a] DFA = direct immunofluorescence assay, IIF = indirect immunofluorescence assay, NT = not tested.

not restricted to the rare immunocompromised host; rather, infection is commonplace, especially in neonatal livestock and human children (few wildlife studies have been conducted). The ubiquitous distribution of *Cryptosporidium* spp. led to the development of methods for detecting the parasite in clinical specimens and in environmental samples. Methods for demonstrating oocysts or parasite antigens in stool are reasonably sensitive and useful. Nevertheless, variable excretion of oocysts and other antigens justify efforts to develop new methodologies with improved sensitivity and specificity. Methods applicable to detecting the parasite in environmental samples are particularly in need of improvement. The inability to sensitively detect and differentiate *Cryptosporidium* at the level of species or subspecies (strain) constrains our understanding of the natural history, epidemiology, and zoonotic potential of *Cryptosporidium* isolates and confounds the assessment of the public health risk posed by oocyst contamination of water or foods. Given recent efforts to develop biochemical and nucleic acid-based detection methods, the next few years should yield tools that will effectively address many of the challenges presented by these parasites.

REFERENCES

1. Ditrich, O., Palkovic, L., Sterba, J., Prokopic, J., Loudova, J., and Giboda, M., The first finding of *Cryptosporidium baileyi* in man, *Parasitol. Res.*, 77, 44, 1991.
2. Wilson, M. and Arrowood, M. J., Diagnostic parasitology: direct detection methods and serodiagnosis, *Lab. Med.*, 24, 145, 1993.
3. Ungar, B. L. P., Cryptosporidiosis in humans (*Homo sapiens*), in *Cryptosporidiosis of Man and Animals*, Dubey, J. P., Speer, C. A., and Fayer, R., Eds., CRC Press, Boca Raton, FL, 1990, 59.
4. Crawford, F. G. and Vermund, S. H., Human cryptosporidiosis, *CRC Crit. Rev. Microbiol.*, 16, 113, 1988.
5. Dupont, H. L., Chappell, C. L., Sterling, C. R., Okhuysen, P. C., Rose, J. B., and Jakubowski, W., The infectivity of *Cryptosporidium parvum* in healthy volunteers, *N. Engl. J. Med.*, 332, 855, 1995.
6. Current, W. L. and Haynes, T. B., Complete development of *Cryptosporidium* in cell culture, *Science*, 224, 603, 1984.
7. McDonald, V., Stables, R., Warhurst, D. C., Barer, M. R., Blewett, D. A., Chapman, H. D., Connolly, G. M., Chiodini, P. L., and McAdam, K. P. W. J., *In vitro* cultivation of *Cryptosporidium parvum* and screening for anticryptosporidial drugs, *Antimicrob. Agents Chemother.*, 34, 1498, 1990.
8. Gut, J., Petersen, C., Nelson, R., and Leech, J., *Cryptosporidium parvum*: *in vitro* cultivation in Madin-Darby canine kidney cells, *J. Protozool.*, 38, S72, 1991.
9. Arrowood, M. J., Xie, L.-T., and Hurd, M. R., *In vitro* assays of maduramicin activity against *Cryptosporidium parvum*, *J. Euk. Microbiol.*, 41, 23S, 1994.
10. Current, W. L. and Reese, N. C., A comparison of endogenous development of three isolates of *Cryptosporidium* in suckling mice, *J. Protozool.*, 33, 98, 1986.

11. Fayer, R. and Ungar, B. L. P., *Cryptosporidium* spp. and cryptosporidiosis, *Microbiol. Rev.*, 50, 458, 1986.
12. Shepherd, R. C., Reed, C. L., and Sinha, G. P., Shedding of oocysts of *Cryptosporidium* in immunocompetent patients, *J. Clin. Pathol.*, 41, 1104, 1988.
13. Baxby, D. and Hart, C. A., Cryptosporidiosis, *Br. Med. J.*, 289, 1148, 1984.
14. Current, W. L., Reese, N. C., Ernst, J. V., Bailey, W. S., Heyman, M. B., and Weinstein, W. M., Human cryptosporidiosis in immunocompetent and immunodeficient persons. Studies of an outbreak and experimental transmission, *N. Engl. J. Med.*, 308, 1252, 1983.
15. Hart, C. A., Baxby, D., and Blundell, N., Gastro-enteritis due to *Cryptosporidium*: a prospective survey in a children's hospital, *J. Infect.*, 9, 264, 1984.
16. Hunt, D. A., Shannon, R., Palmer, S. R., and Jephcott, A. E., Cryptosporidiosis in an urban community, *Br. Med. J.*, 289, 814, 1984.
17. Baxby, D., Hart, C. A., and Blundell, N., Shedding of oocysts by immunocompetent individuals with cryptosporidiosis, *J. Hyg. (Camb.)*, 95, 703, 1985.
18. Chappell, C. L., Okhuysen, P. C., Sterling, C. R., and DuPont, H. L., *Cryptosporidium parvum*: intensity of infection and oocyst excretion patterns in healthy volunteers, *J. Infect. Dis.*, 173, 232, 1996.
19. Roberts, W. G., Green, P. H. R., Ma, J., Carr, M., and Ginsberg, A. M., Prevalence of cryptosporidiosis in patients undergoing endoscopy — evidence for an asymptomatic carrier state, *Am. J. Med.*, 87, 537, 1989.
20. Crawford, F. G., Vermund, S. H., Ma, J. Y., and Deckelbaum, R. J., Asymptomatic cryptosporidiosis in a New York City day care center, *Pediatr. Infect. Dis. J.*, 7, 806, 1988.
21. Lacroix, C., Berthier, M., Agius, G., Bonneau, D., Pallu, B., and Jacquemin, J. L., *Cryptosporidium* oocysts in immunocompetent children: epidemiologic investigations in the day care centers of Poitiers, France, *Eur. J. Epidemiol.*, 3, 381, 1987.
22. Garcia-Rodriguez, J. A., Martin-Sanchez, A. M., Canut Blasco, A., and Garcia Luis, E. J., The prevalence of *Cryptosporidium* species in children in day care centres and primary schools in Salamanca (Spain): an epidemiological study, *Eur. J. Epidemiol.*, 6, 432, 1990.
23. Tangermann, R. H., Gordon, S., Wiesner, P., and Kreckman, L., An outbreak of cryptosporidiosis in a day-care center in Georgia, *Am. J. Epidemiol.*, 133, 471, 1991.
24. Goodgame, R. W., Genta, R. M., White, A. C., and Chappell, C. L., Intensity of infection in AIDS-associated cryptosporidiosis, *J. Infect. Dis.*, 167, 704, 1993.
25. Ungar, B. L. P., Ward, D. J., Fayer, R., and Quinn, C. A., Cessation of *Cryptosporidium*-associated diarrhea in an acquired immunodeficiency syndrome patient after treatment with hyperimmune bovine colostrum, *Gastroenterology*, 98, 486, 1990.
26. Zar, F., Geiseler, P. J., and Brown, V. A., Asymptomatic carriage of *Cryptosporidium* in the stool of a patient with acquired immunodeficiency syndrome, *J. Infect. Dis.*, 151, 195, 1985.
27. Janoff, E. N., Limas, C., Gebhard, R. L., and Penley, K. A., Cryptosporidial carriage without symptoms in the acquired immunodeficiency syndrome (AIDS), *Ann. Intern. Med.*, 112, 75, 1990.
28. Gari-Toussaint, M., Marty, P., Le Fichoux, Y., Pesce, A., and Saint-Paul, M. C., Intestinal and respiratory cryptosporidiosis temporarily symptomless in an HIV seropositive man, *Medecine et Maladies Infectieuses*, 11, 843, 1988.
29. Pitlik, S. D., Fainstein, V., Rios, A., Guarda, L., Mansell, P. W. A., and Hersh, E. M., Cryptosporidial cholecystitis, *N. Engl. J. Med.*, 308, 967, 1983.
30. Blumberg, R. S., Kelsey, P., Perrone, T., Dickersin, R., LaQuaglia, M., and Ferruci, J., Cytomegalovirus- and *Cryptosporidium*-associated acalculous gangrenous cholecystitis, *Am. J. Med.*, 76, 1118, 1984.
31. Hinnant, K., Schwartz, A., Rotterdam, H., and Rudski, C., Cytomegaloviral and cryptosporidial cholecystitis in two patients with AIDS, *Am. J. Surg. Pathol.*, 13, 57, 1989.
32. Kaup, F. J., Kuhn, E. M., Makoschey, B., and Hunsmann, G., Cryptosporidiosis of liver and pancreas in rhesus monkeys with experimental SIV infection, *J. Med. Primatol.*, 23, 304, 1994.
33. Kocoshis, S. A., Cibull, M. L., Davis, T. E., Hinton, J. T., Seip, M., and Banwell, J. G., Intestinal and pulmonary cryptosporidiosis in an infant with severe combined immune deficiency, *J. Pediatr. Gastroenterol. Nutr.*, 3, 149, 1984.
34. Gross, T. L., Wheat, J., Bartlett, M., and O'Connor, K. W., AIDS and multiple system involvement with *Cryptosporidium, Am. J. Gastroenterol.*, 81, 456, 1986.
35. Alonso, J. F., Becerra, E. A., Heras, M. M. J., Rodriquez, J. L. V., Rodriquez, M. A. J., and Leal, J. A. L., Intestinal cryptosporidiosis with affection of bilio-pancreatic tract and bronchial tree in a child with acquired immunodeficiency syndrome, *Med. Clin. (Barc.)*, 89, 335, 1987.
36. Hawkins, S. P., Thomas, R. P., and Teasdale, C., Acute pancreatitis: an new finding in *Cyptosporidium* enteritis, *Br. Med. J.*, 294, 483, 1987.
37. Godwin, T. A., Cryptosporidiosis in the acquired immunodeficiency syndrome: a study of 15 autopsy cases, *Hum. Pathol.*, 22, 1215, 1991.
38. Chui, D. W. and Owen, R. L., AIDS and the gut [review], *J. Gastroenterol. Hepatol.*, 9, 291, 1994.
39. Forgacs, P., Tarshis, A., Ma, P., Federman, M., Mele, L., Silverman, M. L., and Shea, J. A., Intestinal and bronchial cryptosporidiosis in an immunodeficient homosexual man, *Ann. Intern. Med.*, 99, 793, 1983.
40. Brady, E. M., Margolis, M. L., and Korzeniowski, O. M., Pulmonary cryptosporidiosis in acquired immune deficiency syndrome, *J. Am. Med. Assoc.*, 252, 89, 1984.

41. Ma, P., Villanueva, T. G., Kaufman, D., and Gillooley, J. F., Respiratory cryptosporidiosis in the acquired immune deficiency syndrome, *J. Am. Med. Assoc.*, 252, 1298, 1984.
42. Miller, R. A., Wasserheit, J. N., Kirihara, J., and Coyle, M. B., Detection of *Cryptosporidium* oocysts in sputum during screening for *Mycobacteria, J. Clin. Microbiol.*, 20, 1192, 1984.
43. Manivel, C., Filipovich, A., and Snover, D. C., Cryptosporidiosis as a cause of diarrhea following bone marrow transplantation, *Dis. Colon Rectum*, 28, 741, 1985.
44. Harari, M. D., West, B., and Dwyer, B., *Cryptosporidium* as a cause of laryngotracheitis in an infant, *Lancet*, 1, 1207, 1986.
45. Kibbler, C. C., Smith, A., Hamilton-Dutoit, S. J., Milburn, H., Pattinson, J. K., and Prentice, H. G., Pulmonary cryptosporidiosis occurring in a bone marrow transplant patient, *Scand. J. Infect. Dis.*, 19, 581, 1987.
46. Hojlyng, N. and Jensen, B. N., Respiratory cryptosporidiosis in HIV-positive patients, *Lancet*, 1, 590, 1988.
47. Goodstein, R. S., Colombo, C. S., Illfelder, M. A., and Skaggs, R. E., Bronchial and gastrointestinal cryptosporidiosis in AIDS, *J. Am. Osteo. Assoc.*, 89, 195, 1989.
48. Travis, W. D., Schmidt, K., Maclowry, J. D., Masur, H., Condron, K. S., and Fojo, A. T., Respiratory cryptosporidiosis in a patient with malignant lymphoma — report of a case and review of the literature, *Arch. Pathol. Lab. Med.*, 114, 519, 1990.
49. Dautzenberg, B., Truffot, C., Legris, S., Meyohas, M. C., Berlie, H. C., Mercat, A., Chevret, S., and Grosset, J., Activity of clarithromycin against *Mycobacterium avium* infection in patients with acquired immune deficiency syndrome. A controlled clinical trial, *Am. Rev. Resp. Dis.*, 144, 564, 1991.
50. Rodriguez Perez, R., Fernandez Perez, B., Dominguez Alvarez, L. M., and Naval Calvino, G., Respiratory cryptosporidiosis in HIV patients. A report of 2 cases, *Rev. Clin. Espan.*, 190, 210, 1992.
51. Mifsud, A. J., Bell, D., and Shafi, M. S., Respiratory cryptosporidiosis as a presenting feature of AIDS, *J. Infect.*, 28, 227, 1994.
52. Slemenda, S., Paper presented at the Microsporidiosis and Cryptosporidiosis in Immunodeficient Patients Conf., Ceské Budejovice, Czech Republic, 1993.
53. Angus, K. W., Sherwood, D., Hutchison, G., and Campbell, I., Evaluation of the effect of two aldehyde-based disinfectants on the infectivity of faecal cryptosporidia for mice, *Res. Vet. Sci.*, 33, 379, 1982.
54. Campbell, I., Tzipori, S., Hutchison, G., and Angus, K. W., Effect of disinfectants on survival of *Cryptosporidium* oocysts, *Vet. Rec.*, 111, 414, 1982.
55. Pavlasek, I., Effect of disinfectants in infectiousness of oocysts of *Cryptosporidium* sp., *Cs. Epidem.*, 33, 97, 1984.
56. Blewett, D. A., Disinfection and oocysts, in *Cryptosporidiosis: Proc. 1st Int. Workshop*, Angus, K. W. and Blewett, D. A., Eds., The Animal Diseases Research Association, Edinburgh, Scotland, 1989, 107.
57. Current, W. L. and Garcia, L. S., Cryptosporidiosis, *Clin. Microbiol. Rev.*, 4, 325, 1991.
58. Hayes, E. B., Matte, T. D., O'Brien, T. R., McKinley, T. W., Logdson, G. S., Rose, J. B., Ungar, B. L. P., Word, D. M., Pinsky, P. F., Cummings, M. L., Wilson, M. A., Long, E. G., Hurwitz, W. S., and Juranek, D. D., Large community outbreak of cryptosporidiosis due to contamination of a filtered public water supply, *N. Engl. J. Med.*, 320, 1372, 1989.
59. Soravia, E., Martini, G., and Zasloff, M., Antimicrobial properties of peptides from *Xenopus* granular gland secretions, *FEBS Lett.*, 228, 337, 1988.
60. Arrowood, M. J. and Sterling, C. R., Isolation of *Cryptosporidium* oocysts and sporozoites using discontinuous sucrose and isopycnic Percoll gradients, *J. Parasitol.*, 73, 314, 1987.
61. Kilani, R. T. and Sekla, L., Purification of *Cryptosporidium* oocysts and sporozoites by cesium chloride and Percoll gradients, *Am. J. Trop. Med. Hyg.*, 36, 505, 1987.
62. Bogaerts, J., Lepage, P., Rouvroy, D., and Vandepitte, J., *Cryptosporidium* spp., a frequent cause of diarrhea in central Africa, *J. Clin. Microbiol.*, 20, 874, 1984.
63. Tyzzer, E. E., An extracellular coccidium, *Cryptosporidium muris* (gen. et sp. nov.), of the gastric glands of the common mouse, *J. Med. Res.*, 23, 487, 1910.
64. Tyzzer, E. E., *Cryptosporidium parvum* (sp. nov.), a coccidium found in the small intestine of the common mouse, *Arch. Protistenkd.*, 26, 394, 1912.
65. Slavin, D., *Cryptosporidium meleagridis* (sp. nov.), *J. Comp. Pathol.*, 65, 262, 1955.
66. Panciera, R. J., Thomassen, R. W., and Garner, F. M., Cryptosporidial infection in a calf, *Vet. Pathol.*, 8, 479, 1971.
67. Pohlenz, J., Bemrick, W. J., Moon, H. W., and Cheville, N. F., Bovine cryptosporidiosis: a transmission and scanning electron microscopic study of some stages in the life cycle and of the host-parasite relationship, *Vet. Pathol.*, 15, 417, 1978.
68. Tzipori, S., Angus, K. W., Campbell, I., and Gray, E. W., *Cryptosporidium*: evidence for a single-species genus, *Infect. Immun.*, 30, 884, 1980.
69. Henriksen, S. A. and Pohlenz, J. F. L., Staining of cryptosporidia by a modified Ziehl-Neelsen technique, *Acta Vet. Scand.*, 22, 594, 1981.
70. Garcia, L. S., Brewer, T. C., Bruckner, D. A., and Shimizu, R. Y., Techniques for the recovery and identification of *Cryptosporidium* oocysts from stool specimens, *J. Clin. Microbiol.*, 18, 185, 1983.
71. Ma, P. and Soave, R., Three-step stool examination for cryptosporidiosis in 10 homosexual men with protracted watery diarrhea, *J. Infect. Dis.*, 147, 824, 1983.
72. Bronsdon, M. A., Rapid dimethyl sulfoxide-modified acid-fast stain of *Cryptosporidium* oocysts in stool specimens, *J. Clin. Microbiol.*, 19, 952, 1984.

73. Pohjola, S., Jokipii, L., and Jokipii, A. M. M., Dimethylsulphoxide-Ziehl-Neelsen staining technique for detection of cryptosporidial oocysts, *Vet. Rec.*, 115, 442, 1985.
74. Ma, P., *Cryptosporidium* — biology and diagnosis, *Adv. Exp. Med. Biol.*, 202, 135, 1986.
75. Heine, J., An easy technique for the demonstration of cryptosporidia in faeces, *Zbl. Vet. Med. B.*, 29, 324, 1982.
76. Pohjola, S., Negative staining method with nigrosin for the detection of cryptosporidial oocysts: a comparative study, *Res. Vet. Sci.*, 36, 217, 1984.
77. Current, W. L., Human cryptosporidiosis, *N. Engl. J. Med.*, 309, 1326, 1983.
78. Chichino, G., Bruno, A., Cevini, C., Atzori, C., Gatti, S., and Scaglia, M., New rapid staining methods of *Cryptosporidium* oocysts in stools, *J. Protozool.*, 38, S212, 1991.
79. Baxby, C., Getty, B., Blundell, N., and Ratcliffe, S., Recognition of whole *Cryptosporidium* oocysts in feces by negative staining and electron microscopy, *J. Clin. Microbiol.*, 19, 566, 1984.
80. Asahi, H., Kumada, M., Kato, K., and Koyama, T., A simple staining method for cryptosporidian oocysts and sporozoites, *Jpn. J. Med. Sci. Biol.*, 41, 117, 1988.
81. Kageruka, P., Brandt, J. R. A., Taelman, H., and Jonas, C., Modified koster staining method for the diagnosis of cryptosporidiosis, *Ann. Soc. Belge Med. Trop.*, 64, 171, 1984.
82. Milacek, P. and Vitovec, J., Differential staining of cryptosporidia by aniline-carbol-methyl violet and tartrazine in smears from faeces and scrapings of intestinal mucosa, *Folia Parasit. (Prague)*, 32, 50, 1985.
83. Payne, P., Lancaster, L. A., Heinzman, M., and McCutchan, J. A., Identification of *Cryptosporidium* in patients with the acquired immunodeficiency syndrome, *N. Engl. J. Med.*, 309, 613, 1983.
84. Casemore, D. P., Armstrong, M., and Jackson, B., Screening for *Cryptosporidium* in stools, *Lancet*, 1, 734, 1984.
85. Ungureanu, E. M. and Dontu, G. E., A new staining technique for the identification of *Cryptosporidium* oocysts in faecal smears, *Trans. Roy. Soc. Trop. Med. Hyg.*, 86, 638, 1992.
86. Kawamoto, F., Mizuno, S., Fujioka, H., Kumada, N., Sugiyama, E., Takeuchi, T., Kobayashi, S., Iseki, M., Yamada, M., Matsumoto, Y., Tegoshi, T., and Yoshida, Y., Simple and rapid staining for detection of *Entamoeba* cysts and other protozoans with fluorochromes, *Jpn. J. Med. Sci. Biol.*, 40, 35, 1987.
87. Current, W. L., Upton, S. J., and Haynes, T. B., The life cycle of *Cryptosporidium baileyi* n. sp. (Apicomplexa: Cryptosporidiidae) infecting chickens, *J. Protozool.*, 33, 289, 1986.
88. Lindsay, D. S., Blagburn, B. L., and Sundermann, C. A., Morphometric comparison of the oocysts of *Cryptosporidium meleagridis* and *Cryptosporidium baileyi* from birds, *Proc. Helminthol. Soc. Wash.*, 56, 91, 1989.
89. Upton, S. J. and Current, W. L., The species of *Cryptosporidium* (Apicomplexa: Cryptosporidiidae) infecting mammals, *J. Parasitol.*, 71, 625, 1985.
90. Landsberg, J. H. and Paperna, I., Ultrastructural study of the coccidian *Cryptosporidium* sp. from stomachs of juvenile cichlid fish, *Dis. Aquat. Org.*, 2, 13, 1986.
91. Tilley, M., Upton, S. J., and Freed, P. S., A comparative study on the biology of *Cryptosporidium serpentis* and *Cryptosporidium parvum* (Apicomplexa: Cryptosporidiidae), *J. Zoo Wildlife Med.*, 21, 463, 1990.
92. Ostrovska, K. and Paperna, I., *Cryptosporidium* sp. of the starred lizard *Agama stellio*: ultrastructure and life cycle, *Parasitol. Res.*, 76, 712, 1990.
93. Gajadhar, A. A., Host specificity studies and oocyst description of a *Cryptosporidium* sp. isolated from ostriches, *Parasitol. Res.*, 80, 316, 1994.
94. Lindsay, D. S. and Blagburn, B. L., Cryptosporidiosis in birds, in *Cryptosporidiosis of Man and Animals,* Dubey, J. P., Speer, C. A., and Fayer, R., Eds., CRC Press, Boca Raton, FL, 1990, 59.
95. Jervis, H. R., Merrill, T. G., and Sprinz, H., Coccidiosis in the guinea pig small intestine due to a *Cryptosporidium*, *Am. J. Vet. Res.*, 27, 408, 1966.
96. Vetterling, J. M., Jervis, H. R., Merrill, T. G., and Sprinz, H., *Cryptosporidium wrairi* sp. n. from the guinea pig *Cavia porcellus*, with an emendation of the genus, *J. Protozool.*, 18, 243, 1971.
97. Chrisp, C. E., Reid, W. C., Rush, H. G., Suckow, M. A., Bush, A., and Thomann, M. J., Cryptosporidiosis in guinea pigs: an animal model, *Infect. Immun.*, 58, 674, 1990.
98. Chrisp, C. E., Suckow, M. A., Fayer, R., Arrowood, M. J., Healey, M. C., and Sterling, C. R., Comparison of the host ranges and antigenicity of *Cryptosporidium parvum* and *Cryptosporidium wrairi* from guinea pigs, *J. Protozool.*, 39, 406, 1992.
99. Casemore, D. P., Armstrong, M., and Sands, R. L., Laboratory diagnosis of cryptosporidiosis, *J. Clin. Pathol.*, 38, 1337, 1985.
100. Stibbs, H. H. and Ongerth, J. E., Immunofluorescence detection of Cryptosporidium oocysts in fecal smears, *J. Clin. Microbiol.*, 24, 517, 1986.
101. Sterling, C. R. and Arrowood, M. J., Detection of *Cryptosporidium* sp. infections using a direct immunofluorescent assay, *Pediatr. Infect. Dis.*, 5, s139, 1986.
102. Garcia, L. S., Brewer, T. C., and Bruckner, D. A., Fluorescent detection of *Cryptosporidium* oocysts in human fecal specimens by using monoclonal antibodies, *J. Clin. Microbiol.*, 25, 119, 1987.
103. McLauchlin, J., Casemore, D. P., Harrison, T. G., Gerson, P. J., Samuel, D., and Taylor, A. G., Identification of *Cryptosporidium* oocysts by monoclonal antibody, *Lancet*, 1, 51, 1987.
104. Arrowood, M. J. and Sterling, C. R., Comparison of conventional staining methods and monoclonal antibody-based methods for *Cryptosporidium* oocyst detection, *J. Clin. Microbiol.*, 27, 1490, 1989.

105. Garcia, L. S., Shum, A. C., and Bruckner, D. A., Evaluation of a new monoclonal antibody combination reagent for the direct fluorescent detection of *Giardia* cysts and *Cryptosporidium* oocysts in human fecal specimens, *J. Clin. Microbiol.*, 30, 3255, 1992.
106. Grigoriew, G. A., Walmsley, S., Law, L., Chee, S. L., Yang, J., Keystone, J., and Krajden, M., Evaluation of the MeriFluor immunofluorescent assay for the detection of *Cryptosporidium* and *Giardia* in sodium acetate formalin-fixed stools, *Diagn. Microbiol. Infect. Dis.*, 19, 89, 1994.
107. Musial, C. E., Arrowood, M. J., Sterling, C. R., and Gerba, C. P., Detection of *Cryptosporidium* in water by using polypropylene cartridge filters, *Appl. Environ. Microbiol.*, 53, 687, 1987.
108. Madore, M. S., Rose, J. B., Gerba, C. P., Arrowood, M. J., and Sterling, C. R., Occurrence of *Cryptosporidium* oocysts in sewage effluents and selected surface waters, *J. Parasitol.*, 73, 702, 1987.
109. Ongerth, J. E. and Stibbs, H. H., Identification of *Cryptosporidium* oocysts in river water, *Appl. Environ. Microbiol.*, 53, 672, 1987.
110. Rose, J. B., Kayed, D., Madore, M. S., Gerba, C. P., Arrowood, M. J., Sterling, C. R., and Riggs, J. L., Methods for the recovery of *Giardia* and *Cryptosporidium* from environmental waters and their comparative occurrence, in *Advances in* Giardia *Research,* Wallis, P. M. and Hammond, B. R., Eds., University of Calgary Press, Calgary, 1988, 205.
111. Stetzenbach, L. D., Arrowood, M. J., Marshall, M. M., and Sterling, C. R., Monoclonal antibody based immunofluorescent assay for *Giardia* and *Cryptosporidium* detection in water samples, *Water Sci. Technol.*, 20, 193, 1988.
112. Loose, J. H., Sedergran, D. J., and Cooper, H. S., Identification of *Cryptosporidium* in paraffin-embedded tissue sections with the use of a monoclonal antibody, *Am. J. Clin. Pathol.*, 91, 206, 1989.
113. Rose, J. B., Landeen, L. K., Riley, K. R., and Gerba, C. P., Evaluation of immunofluorescence techniques for detection of *Cryptosporidium* oocysts and *Giardia* cysts from environmental samples, *Appl. Environ. Microbiol.*, 55, 3189, 1989.
114. Smith, H. V., McDiarmid, A., Smith, A. L., Hinson, A. R., and Gilmour, R. A., An analysis of staining methods for the detection of *Cryptosporidium* spp. oocysts in water-related samples, *Parasitology*, 99, 323, 1989.
115. Smith, A. L. and Smith, H. V., A comparison of fluorescein diacetate and propidium iodide staining and *in vitro* excystation for determining *Giardia intestinalis* cyst viability, *Parasitology*, 99, 329, 1989.
116. LeChevallier, M. W., Norton, W. D., and Lee, R. G., *Giardia* and *Cryptosporidium* spp. in filtered drinking water supplies, *Appl. Environ. Microbiol.*, 57, 2617, 1991.
117. LeChevallier, M. W., Norton, W. D., and Lee, R. G., Occurrence of *Giardia* and *Cryptosporidium* spp. in surface water supplies, *Appl. Environ. Microbiol.*, 57, 2610, 1991.
118. Roach, P. D., Olson, M. E., Whitley, G., and Wallis, P. M., Waterborne *Giardia* cysts and *Cryptosporidium* oocysts in the Yukon, Canada, *Appl. Environ. Microbiol.*, 59, 67, 1993.
119. Vesey, G., Slade, J. S., Byrne, M., Shepherd, K., Dennis, P. J., and Fricker, C. R., Routine monitoring of *Cryptosporidium* oocysts in water using flow cytometry, *J. Appl. Bacteriol.*, 75, 87, 1993.
120. Vesey, G., Hutton, P., Champion, A., Ashbolt, N., Williams, K. L., Warton, A., and Veal, D., Application of flow cytometric methods for the routine detection of *Cryptosporidium* and *Giardia* in water, *Cytometry*, 16, 1, 1994.
121. LeChevallier, M. W., Norton, W. D., Siegel, J. E., and Abbaszadegan, M., Evaluation of the immunofluorescence procedure for detection of *Giardia* cysts and *Cryptosporidium* oocysts in water, *Appl. Environ. Microbiol.*, 61, 690, 1995.
122. Nieminski, E. C., Schaefer, F. W., III, and Ongerth, J. E., Comparison of two methods for detection of *Giardia* cysts and *Cryptosporidium* oocysts in water, *Appl. Environ. Microbiol.*, 61, 1714, 1995.
123. Rodgers, M. R., Flanigan, D. J., and Jakubowski, W., Identification of algae which interfere with the detection of *Giardia* cysts and *Cryptosporidium* oocysts and a method for alleviating this interference, *Appl. Environ. Microbiol.*, 61, 3759, 1995.
124. Bonnin, A., Petrella, T., Dubremetz, J. F., Michiels, J. F., Puygauthier-Toubas, D., and Camerlynck, P., Histopathological method for diagnosis of cryptosporidiosis using monoclonal antibodies, *Eur. J. Clin. Microbiol. Infect. Dis.*, 9, 664, 1990.
125. Chapman, P. A., Rush, B. A., and McLauchlin, J., An enzyme immunoassay for detecting *Cryptosporidium* in faecal and environmental samples, *J. Med. Microbiol.*, 32, 233, 1990.
126. Anusz, K. Z., Mason, P. H., Riggs, M. W., and Perryman, L. E., Detection of *Cryptosporidium parvum* oocysts in bovine feces by monoclonal antibody capture enzyme-linked immunosorbent assay, *J. Clin. Microbiol.*, 28, 2770, 1990.
127. Robert, B., Ginter, A., Antoine, H., Collard, A., and Coppe, P., Diagnosis of bovine cryptosporidiosis by an enzyme-linked immunosorbent assay, *Vet. Parasitol.*, 37, 1, 1990.
128. Ungar, B. L. P., Enzyme-linked immunoassay for detection of *Cryptosporidium* antigens in fecal specimens, *J. Clin. Microbiol.*, 28, 2491, 1990.
129. El-Shewy, K., Kilani, R. T., Hegazi, M. M., Makhlouf, L. M., and Wenman, W. M., Identification of low-molecular-mass coproantigens of *Cryptosporidium parvum*, *J. Infect. Dis.*, 169, 460, 1994.
130. Xia, Z., Sonnad, S., Turner, S., and Marasigan, M., Evaluation of a microtiter assay for detection of *Cryptosporidium* antigen in stool, 92nd Annual Meeting of the American Society for Microbiology, New Orleans, LA, 1992, 106.

131. Dagan, R., Fraser, D., El-On, J., Kassis, I., Deckelbaum, R., and Turner, S., Evaluation of an enzyme immunoassay for the detection of *Cryptosporidium* spp. in stool specimens from infants and young children in field studies, *Am. J. Trop. Med. Hyg.*, 52, 134, 1995.
132. Siddons, C. A., Chapman, P. A., and Rush, B. A., Evaluation of an enzyme immunoassay kit for detecting *Cryptosporidium* in faeces and environmental samples, *J. Clin. Pathol.*, 45, 479, 1992.
133. Sloan, L. M. and Rosenblatt, J. E., Evaluation of an immunoassay for the detection of *Cryptosporidium* stool specimens, 91st Annual Meeting of the American Society for Microbiology, Dallas, TX, 1991, 22.
134. Newman, R. D., Jaeger, K. L., Wuhib, T., Lima, A. A., Guerrant, R. L., and Sears, C. L., Evaluation of an antigen capture enzyme-linked immunosorbent assay for detection of *Cryptosporidium* oocysts, *J. Clin. Microbiol.*, 31, 2080, 1993.
135. Rosenblatt, J. E. and Sloan, L. M., Evaluation of an enzyme-linked immunosorbent assay for detection of *Cryptosporidium* spp. in stool specimens, *J. Clin. Microbiol.*, 31, 1468, 1993.
136. Kehl, K. S. C., Cicirello, H., and Havens, P. L., Comparison of four different methods for detection of *Cryptosporidium* species, *J. Clin. Microbiol.*, 33, 416, 1995.
137. Aarnaes, S. L., Blanding, J., Speier, S., Forthal, D., de la Maza, L. M., and Peterson, E. M., Comparison of the ProSpecT and Color Vue enzyme-linked immunoassays for the detection of *Cryptosporidium* in stool specimens, *Diagn. Microbiol. Infect. Dis.*, 19, 221, 1994.
138. Pohjola, S., Neuvonen, A., Niskanen, A., and Ranatanna, A., Rapid immunoassay for detection of *Cryptosporidium* oocysts, *Acta Vet. Scand.*, 27, 71, 1986.
139. Arrowood, M. J., Hurd, M. R., and Mead, J. R., A new method for evaluating experimental cryptosporidial parasite loads using immunofluorescent flow cytometry, *J. Parasitol.*, 81, 404, 1995.
140. Tzipori, S. and Campbell, I., Prevalence of *Cryptosporidium* antibodies in 10 animal species, *J. Clin. Microbiol.*, 14, 455, 1981.
141. Campbell, P. N. and Current, W. L., Demonstration of serum antibodies to *Cryptosporidium* sp. in normal and immunodeficient humans with confirmed infections, *J. Clin. Microbiol.*, 18, 165, 1983.
142. Koch, K. L., Phillips, D. J., Aber, R. C., and Current, W. L., Cryptosporidiosis in hospital personnel: evidence for person-to-person transmission, *Ann. Intern. Med.*, 102, 593, 1985.
143. D'Antonio, R. G., Winn, R. E., Taylor, J. P., Gustafson, T. L., Current, W. L., Rhodes, M. M., Gary, G. W. J., and Zajac, R. A., A waterborne outbreak of cryptosporidiosis in normal hosts, *Ann. Intern. Med.*, 103, 886, 1985.
144. Casemore, D. P., The antibody response to *Cryptosporidium*: development of a serological test and its use in a study of immunologically normal persons, *J. Infect.*, 14, 125, 1987.
145. Current, W. L. and Snyder, D. B., Development of and serologic evaluation of acquired immunity to *Cryptosporidium baileyi* by broiler chickens, *Poultry Sci.*, 67, 720, 1988.
146. Mead, J. R., Arrowood, M. J., and Sterling, C. R., Antigens of *Cryptosporidium* sporozoites recognized by immune sera of infected animals and humans, *J. Parasitol.*, 74, 135, 1988.
147. Hill, B. D., Blewett, D. A., Dawson, A. M., and Wright, S., Analysis of the kinetics, isotype and specificity of serum and coproantibody in lambs infected with *Cryptosporidium parvum*, *Res. Vet. Sci.*, 48, 76, 1990.
148. Tsaihong, J. C. and Ma, P., Comparison of an indirect fluorescent antibody test and stool examination for the diagnosis of cryptosporidiosis, *Eur. J. Clin. Microbiol. Infect. Dis.*, 9, 770, 1990.
149. Ungar, B. L. P., Soave, R., Fayer, R., and Nash, T. E., Enzyme immunoassay detection of immunoglobulin M and G antibodies to *Cryptosporidium* in immunocompetent and immunocompromised persons, *J. Infect. Dis.*, 153, 570, 1986.
150. Mosier, D. A., Kuhls, T. L., Simons, K. R., and Oberst, R. D., Bovine humoral immune response to *Cryptosporidium parvum*, *J. Clin. Microbiol.*, 30, 3277, 1992.
151. Peeters, J. E., Villacorta, I., Vanopdenbosch, E., Vandergheynst, D., Naciri, M., Ares-Mazas, E., and Yvore, P., *Cryptosporidium parvum* in calves: kinetics and immunoblot analysis of specific serum and local antibody responses (immunoglobulin A [IgA], IgG, and IgM) after natural and experimental infections, *Infect. Immun.*, 60, 2309, 1992.
152. Ortega-Mora, L. M. and Wright, S. E., Age-related resistance in ovine cryptosporidiosis: patterns of infection and humoral immune response, *Infect. Immun.*, 62, 5003, 1994.
153. Laxer, M. A., Alcantara, A. K., Javato-Laxer, M., Menorca, D. M., Fernando, M. T., and Ranoa, C. P., Immune response to cryptosporidiosis in Philippine children, *Am. J. Trop. Med. Hyg.*, 42, 131, 1990.
154. Whitmire, W. M. and Harp, J. A., Characterization of bovine cellular and serum antibody responses during infection by *Cryptosporidium parvum*, *Infect. Immun.*, 59, 990, 1991.
155. Williams, R. O., Measurement of class specific antibody against *Cryptosporidium* in serum and faeces from experimentally infected calves, *Res. Vet. Sci.*, 43, 264, 1987.
156. Ungar, B. L. P. and Nash, T. E., Quantification of specific antibody response to *Cryptosporidium* antigens by laser densitometry, *Infect. Immun.*, 53, 124, 1986.
157. Ortega-Mora, L. M., Troncoso, J. M., Rojo-Vazquez, F. A., and Gomez-Bautista, M., Cross-reactivity of polyclonal serum antibodies generated against *Cryptosporidium parvum* oocysts, *Infect. Immun.*, 60, 3442, 1992.
158. Janoff, E. N., Mead, P. S., Mead, J. R., Echeverria, P., Bodhidatta, L., Bhaibulaya, M., Sterling, C. R., and Taylor, D. N., Endemic *Cryptosporidium* and *Giardia lamblia* infections in a Thai orphanage, *Am. J. Trop. Med. Hyg.*, 43, 248, 1990.

159. Lazo, A., Barriga, O. O., Redman, D. R., and Beih-Nielsen, S., Identification by transfer blot of antigens reactive in the enzyme-linked immunosorbent assay (ELISA) in rabbits immunized and a calf infected with *Cryptosporidium* sp., *Vet. Parasitol.*, 21, 151, 1986.
160. Luft, B. J., Payne, D., Woodmansee, D., and Kim, C. W., Characterization of the *Cryptosporidium* antigens from sporulated oocysts of *Cryptosporidium parvum*, *Infect. Immun.*, 55, 2436, 1987.
161. Moss, D. M., Bennett, S. N., Arrowood, M. J., Hurd, M. R., Lammie, P. J., Wahlquist, S. P., and Addiss, D. G., Kinetic and isotypic analysis of specific immunoglobulins from crew members with cryptosporidiosis on a U.S. Coast Guard cutter, *J. Euk. Microbiol.*, 41, 52S, 1994.
162. Ortega-Mora, L. M., Troncoso, J. M., Rojo-Vazquez, F. A., and Gomez-Bautista, M., Identification of *Cryptosporidium parvum* oocyst/sporozoite antigens recognized by infected and hyperimmune lambs, *Vet. Parasitol.*, 53, 159, 1994.
163. Lumb, R., Lanser, J. A., and O'Donoghue, P. J., Electrophoretic and immunoblot analysis of *Cryptosporidium* oocysts, *Immunol. Cell Biol.*, 66, 369, 1988.
164. Tilley, M. and Upton, S. J., Electrophoretic characterization of *Cryptosporidium parvum* (KSU-1 isolate) (Apicomplexa: Cryptosporidiidae), *Can. J. Zool.*, 68, 1513, 1990.
165. Lumb, R., Smith, P. S., Davies, R., Odonoghue, P. J., Atkinson, H. M., and Lanser, J. A., Localization of a 23000 MW antigen of *Cryptosporidium* by immunoelectron microscopy, *Immunol. Cell Biol.*, 67, 267, 1989.
166. Tilley, M., Upton, S. J., Blagburn, B. L., and Anderson, B. C., Identification of outer oocyst wall proteins of 3 *Cryptosporidium* (Apicomplexa: Cryptosporidiidae) species by I-125 surface labeling, *Infect. Immun.*, 58, 252, 1990.
167. Mead, J. R., Arrowood, M. J., Current, W. L., and Sterling, C. R., Field inversion gel electrophoretic separation of *Cryptosporidium* spp. chromosome-sized DNA, *J. Parasitol.*, 74, 366, 1988.
168. Nelson, R. G., Kim, K., Gooze, L., Petersen, C., and Gut, J., Identification and isolation of *Cryptosporidium parvum* genes encoding microtubule and microfilament proteins, *J. Protozool.*, 38, S52, 1991.
169. Kim, K., Gooze, L., Petersen, C., Gut, J., and Nelson, R. G., Isolation, sequence and molecular karyotype analysis of the actin gene of *Cryptosporidium parvum*, *Mol. Biochem. Parasitol.*, 50, 105, 1992.
170. Kemp, D. J., Corcoran, L. M., Coppel, R. L., Stahl, H. D., Bianco, A. E., Brown, G. V., and Anders, R. F., Size variation in chromosomes from independant cultured isolates of *Plasmodium falciparum*, *Nature*, 315, 347, 1985.
171. Scholler, J. K., Reed, S. G., and Stuart, K., Molecular karyotype of species and subspecies of *Leishmania*, *Mol. Biochem. Parasitol.*, 20, 279, 1987.
172. Aymerich, S. and Goldenberg, S., The karyotype of *Trypanosoma cruzi* dm-28c — comparison with other *T. cruzi* strains and trypanosomatids, *Exp. Parasitol.*, 69, 107, 1989.
173. Mead, J. R., Humphreys, R. C., Sammons, D. W., and Sterling, C. R., Identification of isolate-specific sporozoite proteins of *Cryptosporidium parvum* by 2-dimensional gel electrophoresis, *Infect. Immun.*, 58, 2071, 1990.
174. Ogunkolade, B. W., Robinson, H. A., McDonald, V., Webster, K., and Evans, D. A., Isoenzyme variation within the genus *Cryptosporidium*, *Parasitol. Res.*, 79, 385, 1993.
175. Awad-el-Kariem F, M., Robinson H, A., Dyson D, A., Evans, D., Wright, S., Fox M, T., and Mcdonald, V., Differentiation between human and animal strains of *Cryptosporidium parvum* using isoenzyme typing, *Parasitology*, 110, 129, 1995.
176. Johnson, A. M., Fielke, R., Lumb, R., and Baverstock, P. R., Phylogenetic relationships of *Cryptosporidium* determined by ribosomal RNA sequence comparison, *Int. J. Parasitol.*, 20, 141, 1990.
177. Cai, J., Collins, M. D., McDonald, V., and Thompson, D. E., PCR cloning and nucleotide sequence determination of the 18S rRNA genes and internal transcribed spacer 1 of the protozoan parasites *Cryptosporidium parvum* and *Cryptosporidium muris*, *Biochim. Biophys. Acta*, 1131, 317, 1992.
178. Barta, J. R., Phylogenetic analysis of the class Sporozea (phylum Apicomplexa Levine 1970): evidence for the independant evolution of the heteroxenous life cycles, *J. Parasitol.*, 75, 195, 1989.
179. Laxer, M. A., Timblin, B. K., and Patel, R. J., DNA sequences for the specific detection of *Cryptosporidium parvum* by the polymerase chain reaction, *Am. J. Trop. Med. Hyg.*, 45, 688, 1991.
180. Filkorn, R., Wiedenmann, A., and Botzenhart, K., Selective detection of viable *Cryptosporidium* oocysts by PCR, *Zentralblatt fur Hygiene und Umweltmedizin*, 195, 489, 1994.
181. Wagner-Wiening, C. and Kimmig, P., Detection of viable *Cryptosporidium* parvum oocysts by PCR, *Appl. Environ. Microbiol.*, 61, 4514, 1995.
182. Johnson, D. W., Pieniazek, N. J., and Rose, J. B., DNA probe hybridization and PCR detection of *Cryptosporidium* compared to immunofluorescence assay, *Water Sci. Technol.*, 27, 77, 1993.
183. Johnson, D. W., Pieniazek, N. J., Griffin, D. W., Misener, L., and Rose, J. B., Development of a PCR protocol for sensitive detection of *Cryptosporidium* oocysts in water samples, *Appl. Environ. Microbiol.*, 61, 3849, 1995.
184. Awad-el-Kariem, F. M., Warhurst, D. C., and McDonald, V., Detection and species identification of *Cryptosporidium* oocysts using a system based on PCR and endonuclease restriction, *Parasitology*, 109, 19, 1994.
185. Taghi-Kilani, R., Remacha-Moreno, M., and Wenman, W. M., Three tandemly repeated 5s ribosomal RNA-encoding genes identified, cloned and characterized from *Cryptosporidium parvum*, *Gene*, 142, 253, 1994.
186. LeChevallier, M. W., Trok, T. M., Burns, M. O., and Lee, R. G., Comparison of the zinc sulfate and immunofluorescence techniques for detecting *Giardia* and *Cryptosporidium*, *J. Am. Water Works Assoc.*, 82, 75, 1990.

187. Anon., D-19 Proposal P 229, Proposed test method for *Giardia* cysts and *Cryptosporidium* oocysts in low-turbidity water by a fluorescent antibody procedure, in *Annual Book of the ASTM Standards,* 11.02, Water (11), American Society for Testing and Materials (ASTM), Philadelphia, PA, 1993, 899.
188. APHA, AWWA, and WEF, Section 9711, Pathogenic protozoa, proposed methods for *Giardia* and *Cryptosporidium* spp., in *Standard Methods for the Examination of Water and Wastewater,* 19th ed., APHA, American Water Works Association, WEF, 1995, 9/110.
189. U.S. EPA, Monitoring requirements for public drinking water supplies: proposed rule, *Fed. Reg.*, 59, 6332, 1994.
190. Hansen, J. S. and Ongerth, J. E., Effects of time and watershed characteristics on the concentration of *Cryptosporidium* oocysts in river water, *Appl. Environ. Microbiol.*, 57, 2790, 1991.
191. Dawson, D. J., Maddocks, M., Roberts, J., and Vidler, J. S., Evaluation of recovery of *Cryptosporidium parvum* oocysts using membrane filtration, *Lett. Appl. Microbiol.*, 17, 276, 1993.
192. Aldom, J. E. and Chagla, A. H., Recovery of *Cryptosporidium* oocysts from water by a membrane filter dissolution method, *Lett. Appl. Microbiol.*, 20, 186, 1995.
193. Vesey, G., Slade, J. S., Byrne, M., Shepherd, K., and Fricker, C. R., A new method for the concentration of *Cryptosporidium* oocysts from water, *J. Appl. Bacteriol.*, 75, 82, 1993.
194. Fricker, C. R., Viability of *Cryptosporidium parvum* oocysts concentrated by calcium carbonate flocculation, *J. Appl. Bacteriol.*, 77, 120, 1994.
195. Campbell, A. T., Robertson, L. J., Smith, H. V., and Girdwood, R., Viability of *Cryptosporidium parvum* oocysts concentrated by calcium carbonate flocculation, *J. Appl. Bacteriol.*, 76, 638, 1994.
196. Smith, H. V. and Rose, J. B., Waterborne cryptosporidiosis, *Parasitol. Today*, 6, 8, 1990.
197. Buraud, M., Forget, E., Favennec, L., Bizet, J., Gobert, J.-G., and Deluol, A.-M., Sexual stage development of cryptosporidia in the Caco-2 cell line, *Infect. Immun.*, 59, 4610, 1991.
198. Upton, S. J., Tilley, M., Nesterenko, M. V., and Brillhart, D. B., A simple and reliable method of producing *in vitro* infections of *Cryptosporidium parvum* (Apicomplexa), *FEMS Microbiol. Lett.*, 118, 45, 1994.
199. Upton, S. J., Tilley, M., and Brillhart, D. B., Effects of select medium supplements on *in vitro* development of *Cryptosporidium* parvum in HCT-8 cells, *J. Clin. Microbiol.*, 33, 371, 1995.
200. Woods, K. M., Nesterenko, M. V., and Upton, S. J., Development of a microtitre ELISA to quantify development of *Cryptosporidium parvum in vitro*, *FEMS Microbiol. Lett.*, 128, 89, 1995.
201. You, X., Arrowood, M. J., Lejkowski, M., Xie, L., Schinazi, R. F., and Mead, J. R., A chemiluminescence immunoassay for evaluation of *Cryptosporidium parvum* growth *in vitro*, *FEMS Microbiol. Lett.*, 1996.
202. Melvin, D. M. and Brooke, M. M., Laboratory procedures for the diagnosis of intestinal parasites, 3rd ed., U.S. Department of Health and Human Services pub. no. (CDC)82-8282, Atlanta, GA, 1982.
203. Ash, L. R. and Orihel, T. C., *Parasites: A Guide to Laboratory Procedures and Identification,* American Society of Clinical Pathologists Press, Chicago, 1987.
204. Pohlenz, J., Moon, H. W., Cheville, N. F., and Bemrick, W. J., Cryptosporidiosis as a probable factor in neonatal diarrhea of calves, *J. Am. Vet. Med. Assoc.*, 172, 452, 1978.
205. Garza, D., Hopfer, R. L., Eichelberger, C., Eisenbach, S., and Fainstein, V., Fecal staining methods for screening *Cryptosporidium* oocysts, *J. Med. Technol.*, 1, 560, 1984.
206. Cross, R. F. and Moorhead, P. D., A rapid staining technic for cryptosporidia, *Mod. Vet. Pract.*, 65, 307, 1984.
207. Baxby, D. and Blundell, N., Sensitive, rapid, simple methods for detecting *Cryptosporidium* in faeces, *Lancet*, 11, 1149, 1983.
208. Baxby, D., Blundell, N., and Hart, C. A., The development and performance of a simple, sensitive method for the detection of *Cryptosporidium* oocysts in faeces, *J. Hyg. (Camb.)*, 92, 317, 1984.
209. Angus, K. W., Campbell, I., Gray, E. W., and Sherwood, D., Staining of faecal yeasts and *Cryptosporidium* oocysts, *Vet. Rec.*, 108, 173, 1981.

APPENDIX

A. FORMALIN-ETHYL ACETATE SEDIMENTATION (CONCENTRATION)

(Method adapted from Melvin and Brooke[202] by Suzanne Wahlquist, M.S., Biology and Diagnostics Branch, Division of Parasitic Diseases, National Center for Infectious Diseases, Centers for Disease Control and Prevention; Atlanta, GA.)

1. Thoroughly mix the formalinized fecal specimen.
2. Strain 5 ml of fecal suspension through wet gauze placed in a paper funnel into a 15-ml conical centrifuge tube.
3. Add distilled water to 15 ml, mix, and centrifuge at 2000 to 2500 rpm (500 *g*) for 10 minutes.
4. Decant supernatant. Add 10 ml of 10% formalin to the sediment and mix thoroughly with applicator sticks.

5. Add 4 ml of ethyl acetate, stopper the tube, and shake vigorously in an inverted position for 30 seconds. Carefully remove stopper.
6. Centrifuge at 2000 to 2500 rpm (500 *g*) for 10 minutes.
7. Free the plug of debris from the top of the tube by ringing the sides with an applicator stick. Decant top layers of supernatant. Use a cotton swab to remove adhering debris from the sides of the tube.
8. Add several drops of 10% formalin to the sediment and mix well; stopper the tube and save for examination.

B. CARBOL-FUCHSIN STAIN FOR CRYPTOSPORIDIUM SPP. OOCYSTS

(Method adapted from Ash and Orihel[203] by Suzanne Wahlquist, M.S., Biology and Diagnostics Branch, Division of Parasitic Diseases, National Center for Infectious Diseases, Centers for Disease Control and Prevention; Atlanta, GA.)

1. Preparation of Reagents

- Carbol-fuchsin stain: May be purchased commercially.
- Acid alcohol: 10 ml sulfuric acid + 90 ml absolute ethanol.
- Malachite green: 3 g malachite green + 100 ml distilled water.

2. Preparation of Fecal Smear

Smear 1 to 2 drops of 10% formalin-fixed specimen on a slide. Do not make the smear too thick; you should be able to see through the wet material before it dries. Heat fix on slide warmer at 60°C until dry (about 5 minutes).

3. Staining Procedure

1. Fix (immerse) in absolute methanol — 30 seconds.
2. Stain in carbol-fuchsin stain — 1 minute.
3. Rinse with distilled water and drain.
4. Decolorize in 10% sulfuric acid alcohol — 2 minutes.
5. Rinse with distilled water and drain.
6. Counterstain in 3% malachite green — 2 minutes.
7. Rinse *very briefly* with distilled water and drain.
8. Dry on slide warmer at 60°C until dry (about 5 minutes).
9. Mount with coverslip (no. 1 thickness) using Permount™ (Fisher Scientific; Philadelphia, PA) or comparable mounting medium.

Stained smears are examined at high power magnification (400×) and oil immersion (1000×). *Cryptosporidium* oocysts stain bright red to pink against a green background. Oocysts may appear filled with an amorphous red mass or distinct crescent-shaped sporozoites (see color plate). Empty oocysts ("ghosts") may also be observed, especially as infections are being resolved, but these are generally poorly stained or appear unstained. See Table 2 for morphometric (size) characteristics of oocysts from various *Cryptosporidium* species.

Chapter 3

Cryptosporidiosis — Human and Animal Epidemiology

David P. Casemore, S. E. Wright, and Robert L. Coop

CONTENTS

0-8493-7695-5/97/$0.00+$.50

I. INTRODUCTION

A. GENERAL INTRODUCTION

Cryptosporidium is an important and widely distributed enteric pathogen of young livestock and humans and is common in other hosts, in which it is often asymptomatic. Many early reports of human infection were in immunocompromised subjects, particularly those with the then newly emergent AIDS, and the immunosuppressed. Even when a source was not apparent, the infection was generally regarded as an opportunist zoonosis. Surveys in the early 1980s indicated the importance of the parasite as a cause of acute, sporadic gastroenteritis in otherwise healthy subjects, particularly children. No attempt will be made in this chapter to fully reference earlier reports, which have been extensively reviewed elsewhere.[1-10]

Cryptosporidium causes acute, self-limiting, gastroenteritis in normal humans and persistent and potentially fatal infection in the immunocompromised worldwide. Millions of cases are estimated to occur every year, in both developed and underdeveloped countries.[1,5,8,11-13] It is a cause of endemic sporadic infection, travelers' diarrhea, and outbreaks, some major.[14-19] In many laboratories it is among the most commonly reported human enteric pathogens. Patterns of prevalence by age, season, and geographic location and of routes of transmission have been described. In livestock, the infection causes morbidity and sometimes mortality and is thus of clinical veterinary and economic concern.[1,6,7,10]

Cryptosporidiosis represents a threat to AIDS patients and other immunocompromised persons, with infection rates from <1 to >50% worldwide.[20-23] The epidemiological picture for immunocompetent and immunocompromised subjects differs little except where indicated below.

There is increasing documentation of outbreaks, associated with public water supplies (see Chapter 4), hospitals, childcare centers, and recreational farm visits, indicating the need for control measures which must be based on an understanding of the natural history of the organism and the epidemiology and dynamics of the infection.[17-19]

B. NATURAL HISTORY

Most clinical infections in humans and mammalian livestock are due to *C. parvum*, although infection in a severely immunocompromised patient with a *C. baileyi*-like organism has been reported.[24] *C. muris* infection, reported in animals other than mice, has not been substantiated in humans and will not be discussed in this chapter. Large (8 to 10 μm) acid-fast *Cryptosporidium*-like bodies, also described as algal-like bodies and detected by methods used for *Cryptosporidium*, are a newly recognized cyclosporan species, *Cyclospora cayetanensis*.[25] Details described below and the term *Cryptosporidium* refer to infection with *C. parvum*, although this may include isolates which, though morphologically indistinguishable, differ in some molecular and biological characteristics. General epidemiological features of cryptosporidiosis are determined by the natural history of this protozoan parasite (see Chapter 1)[1,2,6,9,10,19] and are summarized in Table 1 and in Figure 1.

Table 1 Biological Features of *Cryptosporidium* which Affect its Epidemiology

- Small (4 to 6 μm), environmentally resistant oocyst stage, excreted fully sporulated (capable of transmission directly from host to host)
- Large livestock animal and human reservoir
- Ubiquitous, homo(mono)xenous, and able to cross-infect multiple host species
- Probable low infective dose (≤10 to 100 oocysts)
- Ability to multiply to large numbers (>10^{10}) in a single host animal
- Resistance to disinfection
- Resistance to specific therapy

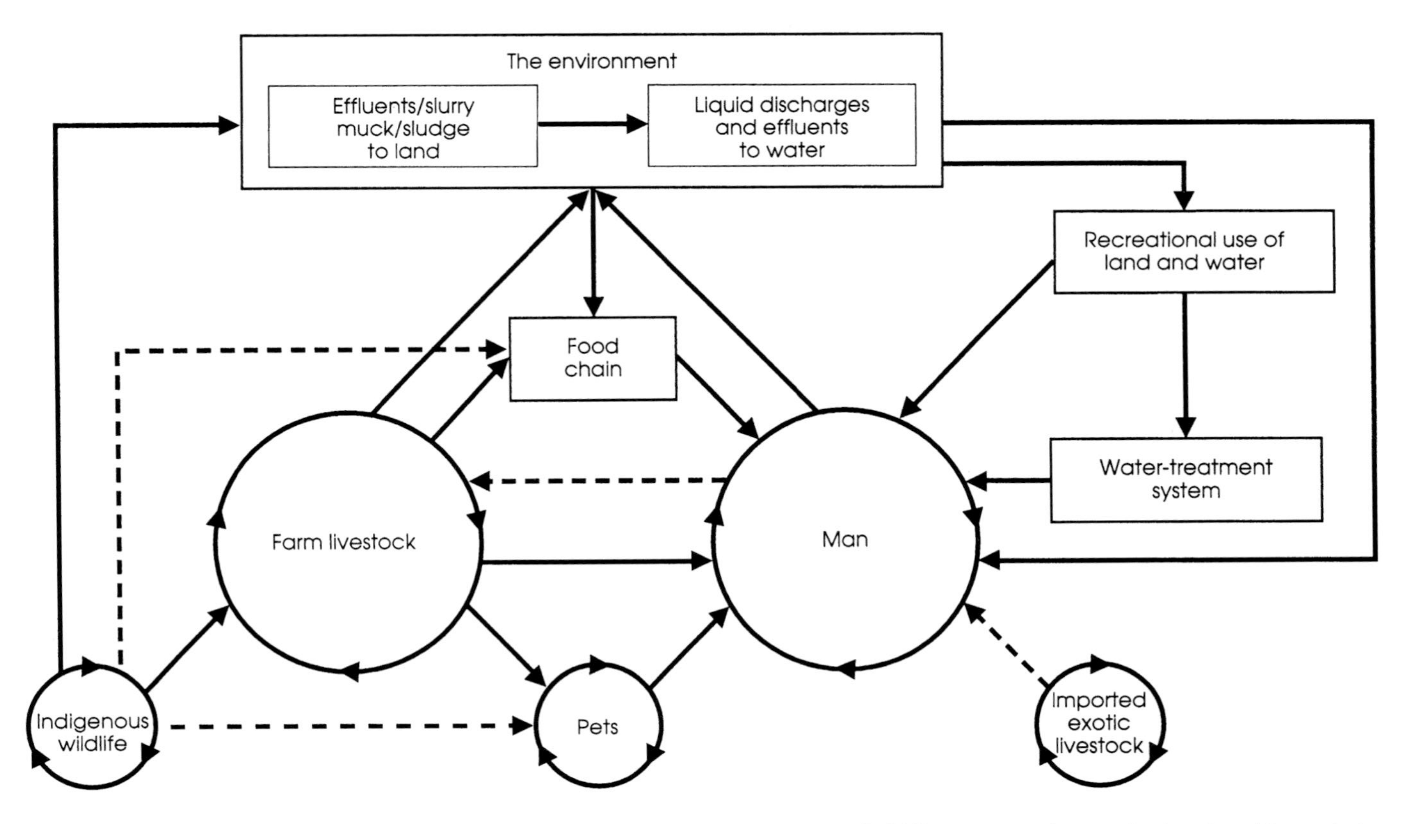

Figure 1 Reservoirs of infection and routes of transmission of *Cryptosporidium parvum*. Solid lines represent recognized routes of transmission; broken lines represent probable routes of transmission. (Modified from Casemore, D. P., *Epidemiol. Infect.*, 104, 1, 1990.)

The dynamics and epidemiology of cryptosporidiosis are also influenced by the molecular characteristics of the parasite (see below and Chapter 10) and by host immunological factors, including herd immunity (see Chapter 6). Nonspecific factors, such as climate, social factors reflected in diet, animal contact, travel exposure, etc., must be considered when studying the epidemiology of this infection.

II. EPIDEMIOLOGY IN ANIMALS

A. INTRODUCTION

Epidemiological data for cryptosporidial infection in animals are largely confined to economically important livestock, particularly ruminants. The parasite has been recorded in about 80 mammalian species worldwide.[1,2,10] Clinical infections are confined largely to neonates, and age-related susceptibility to infection has been demonstrated experimentally for several species.[26-28] Ingestion of normal colostrum does not seem to reduce susceptibility to infection,[29] though failure to achieve adequate passive immunity may reduce the animal's ability to thrive.[30] Colostrum with high antibody titers against cryptosporidial antigens (hyperimmune colostrum) has been tested as a therapeutic in animals and humans with varying degrees of success (see Chapter 6).[31-34] Increasing evidence of infection in adult animals[35-37] has raised the possibility of transmission from dam to offspring, making control difficult in the absence of effective chemotherapeutic agents to break the cycle.

B. FERAL AND WILD ANIMALS

Epidemiological data for such animals are sparse, often confined to chance observations of small groups or populations or wild animals transferred to zoological parks. They may be important potential reservoirs of infection for livestock, either by direct contact or environmental contamination. Since first identified in mice,[38] *C. parvum* has been recorded in populations of other rodents (see Chapter 7).[39-43] Infections appear asymptomatic, with small numbers of oocysts excreted intermittently in feces of mature animals and possibly over a protracted period. The widespread presence of rodents in all agricultural enterprises provides a potential reservoir of infection for agriculturally important animals — as the infectivity of *C. parvum* from mice for calves has been verified.[40]

A cottontail rabbit[44] and grey squirrel[45] have been recorded with *C. parvum* infections, but no prevalence data exist. Examination of dead or dying grey foxes over 18 years in the southern U.S. revealed *Cryptosporidium* infection in only 3 of 157 foxes.[46] All had canine distemper. Asymptomatic juvenile raccoons were found infected.[47-49] Several species of deer, both farmed and wild caught,[50-52] may be reservoirs of infection. Mortality rates were significant among farmed offspring of wild caught deer,[51] although the relatively crowded farm conditions may facilitate the spread and enhance the severity of infection.

Though feral, farm, and domestic cats have asymptomatic infections, prevalence data is sparse.[53-55] Of 57 farm cats around Glasgow, Scotland, 8.8% excreted oocysts, compared to 12% of feral cats and 5% of domestic cats within the city. The latter compares with 3.8% of urban cats in Japan.[56] Symptomatic infections in cats[57] occasionally have been associated with feline immunodeficiency virus (FIV).[58-59]

C. DOMESTICATED ANIMALS

1. Cattle

Most data on the incidence of *C. parvum* infection in domesticated animals relates to cattle. Infections have been recorded worldwide.[1,10,60] Clinical infections seem largely confined to neonates. The prepatent period of 5 to 12 days is followed by oocyst shedding which is usually coincident with onset of diarrhea; patency lasts 3 to 12 days when the number of oocysts shed may exceed 10^{10}, with peaks of 10^6 to 10^7 oocysts per gram of feces.[61-63] Clinical signs usually subside before shedding stops.

Much survey data from the U.S. indicates an enormous variation in incidence. A survey of healthy beef calves on 22 Montana farms revealed no cryptosporidial infection,[64] whereas 41 of 73 Idaho farms had infected calves,[65] and 9 of 12 Maryland dairy farms had infected calves (with prevalence rates from 8.3 to 75%).[66] Of eight geographically well dispersed veal units involving 460 calves, *C. parvum* was the most commonly identified gastrointestinal pathogen.[30] In a nationwide U.S. survey, 59% of 1103 farms and 22% of 7369 calves were infected.[67] Almost half the 7- to 21-day-old calves were infected.

Cryptosporidiosis is recognized as a common gastrointestinal problem in calves in the U.K. The U.K. Veterinary Investigation Service diagnosed 2177 episodes of infection in calves in 1994 vs. 216 in 1984. This is not the total number of animals affected but reflects samples from sites where one or more calves were infected at a specific time, although repeat samples from the same source are not included in the

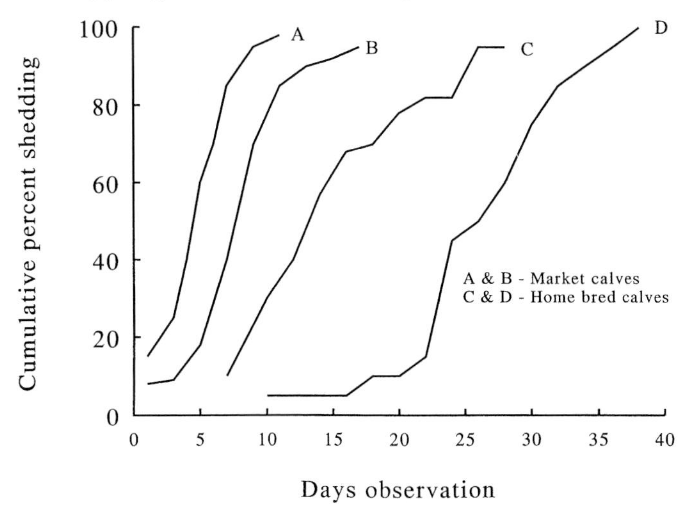

Figure 2 Cumulative percentage of calves infected with *Cryptosporidium parvum* in a survey of homebred and market-bought calves.

data. It is an underestimate of infection, since it does not include farms with infections from previous years or farms with undiagnosed or unreported cases.

In New Zealand, 37% of 550 samples from scouring calves in the spring of 1984–86 were positive for oocysts, calves 4 to 7 days old predominating.[68] In Trinidad and Tobago, 8.7% of calves under 24 weeks of age were infected.[69] Samples were taken from both intensive and extensive rearing systems, and diarrheic and nondiarrheic animals were included in the sampling regime (there were no statistically significant differences in incidence between the rearing systems).

Seasonal production of livestock often results in large groups of neonates in close proximity. Because they have little or no collective (herd) immunity, they are equally susceptible to infection, which can spread rapidly through direct contact with affected individuals or contamination of the environment. Under such conditions, *C. parvum* can spread through the exposed population and produce a very high prevalence rate (see Figure 2). In five surveys of naturally infected housed calves in the U.K. and the Netherlands, calves were sampled on alternate days over 3 to 4 weeks, and fecal smears were examined for oocysts.[63] Prevalence rates varied from 60 to 100%, and market-purchased calves became infected more rapidly than homebred calves (see Figure 2).

Asymptomatic infection has been reported in adult cattle.[35,36,70,71] Cattle had 25 to 1.8×10^4 oocysts per gram of feces (mean 900 oocysts per gram), the highest number being in summer, and there was no apparent correlation between calving date and oocyst excretion.[35] The number of oocysts shed by adults varied greatly, possibly reflecting the difficulty of detecting few oocysts in voluminous stools and problems with nonspecific stains. One group had to concentrate 10 g of cattle feces to detect oocysts,[35] while another detected numerous oocysts in direct fecal smears.[36]

A high prevalence of *Cryptosporidium* antibody (63% by immunofluorescence, or IFAT, and 51.4% by enzyme-linked immunoabsorbent assay, or ELISA) found in adult cattle[36] compared well with the limited sero-prevalence data published by others.[72] However, because significant cross-reaction was found between serum from lambs infected with *Eimeria* species and cryptosporidial antigens in an IFAT[73] and because ruminants are known to have high exposure rates to eimerian species, such data must be interpreted with caution. Zebu cattle (*Bos indicus*) and water buffalo (*Bubalus bubalis*) calves in India,[74] Italy, and Egypt[75,76] become infected, but prevalence data for these species are lacking.

2. Sheep

While there is less information on infected sheep than cattle, distribution appears to be worldwide. Symptomatic infection, confined to lambs, varies in severity. Of 40 artificially reared lambs infected with *Cryptosporidium* but no other enteric pathogens, 16 died.[77] Of naturally reared lambs on the same farm, the earliest lambs appeared healthy, but scouring commenced midway through lambing and of 60 animals affected, 12 died.[78] Of 532 lambs born later on that farm, 200 scoured and 58 died; no other pathogen was found. By contrast, naturally reared lambs from a small Idaho flock had "uncomplicated" diarrhea, suffered no inappetence, and recovered quickly, while losses in a group of orphan lambs were attributed to multiple infections including *Cryptosporidium*, *Escherichia coli*, corona virus-like particles, and bovine virus diarrhea-like particles.[75]

Of 196 neonatal deaths on five sheep farms in Chile, oocysts were found in 7.7%;[80] two thirds of the lambs were older than 7 days. Some evidence of seasonality in prevalence was observed in lambs on farms in northern Spain, where the prevalence was greater in spring (90% of farms) than in autumn (40% of farms).[81] During spring, 40% of lambs were infected, while in autumn, only 8%. In an Ohio sheep flock where 100% of lambs suffered diarrhea, 18 of 23 lambs 2 to 3 weeks old and 4 of 23 clinically normal ewes were shedding oocysts.[37] Further observations led investigators to suggest a periparturient rise in *C. parvum* shedding among ewes.[82] Other workers have identified asymptomatic infection in adult sheep which could infect offspring.[83-85]

3. Goats

The situation for goats seems much the same as for sheep. Nagy et al.[86] found *Cryptosporidium* to be the most frequently isolated pathogen in goat kids with gastrointestinal disease. In Denmark, an outbreak of cryptosporidiosis in the kids from a dairy goat flock of 250 achieved a morbidity rate approaching 100% and mortalities of 20 to 40%, in spite of good husbandry.[87] Kids experienced diarrhea during the second week of life. In northern Spain, up to 70% of kids sampled were infected, and all the farms visited had cryptosporidiosis in their flocks.[81]

4. Pigs

Infection in pigs has been reported sporadically worldwide, but the prevalence remains unclear as most studies have involved only small numbers of animals. Unlike experimentally induced infections, which can produce clinical symptoms, [88,89] natural infections seem to be largely asymptomatic in pigs. In Canada, from 1981 to 1985, 5.3% of 3491 pigs submitted from 133 farms for histological diagnosis were positive. Of these infected animals 26% had diarrhea, but most harbored at least one other primary diarrheic agent in addition to *Cryptosporidium*. In a California auction yard, over a 3-month period it was found that 5% of 200 pigs were infected with *Cryptosporidium*; 7 appeared healthy, and 3 had diarrhea.[91] No other bacterial or viral screening was performed on these animals, so mixed infections cannot be ruled out. In Trinidad and Tobago, 19.6% (54 of 275) of piglets sampled were infected with *Cryptosporidium*,[69] approximately half of which animals experienced diarrhea. On two pig farms in Ohio,[92] light infections were confined to nursing and weanling pigs. *C. parvum* infection in pigs appears common and has potential for environmental contamination, but infections may be unnoticed due to their asymptomatic nature.

5. Horses

First described in immunocompromised Arabian foals,[93] infection also has been reported in normal foals, with and without diarrhea. Infection rates in nondiarrheic animals of 2.4 to 60% have been recorded,[94] and it appears that infection may be widespread, occasionally producing clinical symptoms. Clinical infection in older animals has been recorded also.[95,96]

Studies in France reported asymptomatic infection with oocyst shedding in yearlings and infection rates in mares as high as 80% with possibile transmission from mares to foals.[95,96] Neither mares nor yearlings appeared infected on farms where 17 to 31% of the foals were infected.[97] This latter study used an immunofluorescence assay, rather than the less specific acid-fast staining utilized by previous surveys. These conflicting results make the role of the mare in the transmission of *Cryptosporidium* to her foal unclear. On premises where foals and mares were housed individually until the foal was 9 days old and then allowed to mix with other mares and foals, oocyst excretion began when the foals reached 4 weeks old.[97] This would seem to imply that infection came from mixing with other foals.

6. Dogs

Prevalence data for infection in domestic dogs are sparse. Individual cases in young dogs have been reported, often with concurrent infections such as canine distemper virus.[98-100] Fecal analysis of 57 adults in Finland,[101] 200 in the Federal Republic of Germany,[102] and 101 in the Edinburgh area of Scotland[103] failed to detect oocysts, while in California, only 4 of 200 (2%) impounded stray dogs were infected.[104] In more recent coprological surveys, 1 of 100 canine stools from public parks in Scotland contained oocysts,[105] while 21 of 107 (19.6%) fecal samples taken from public parks in Australia proved to be positive.[106] This latter survey also indicated that 29 of 190 stray dogs (15%), 3 of 44 (6.8%) breeding kennel dogs, and 1 of 152 (0.7%) dogs at veterinary clinics were shedding oocysts.

7. Adult Animals

A problem associated with detection of oocysts in adult animals is the difficulty of accurately identifying oocysts in the often fibrous feces of such animals. Because there are too many objects of appropriate size in the feces of adult animals which absorb acid-fast stains but are not *Cryptosporidium* oocysts, specific methods such as immunofluorescence using monoclonal antibodies (MAbs) and measurement are essential for accurate assessment.

8. Serology

Similarly, serological surveys in animals suggest excessively high prevalence rates, particularly in adults, given the potential for cross reaction of sera from animals infected with eimerian species with cryptosporidial antigens in the IFAT.[73] In a limited seroprevalence survey of ten species, very high rates of apparent exposure to *Cryptosporidium* were found.[72] Cross reaction with gastrointestinal bacteria and viruses was checked for, but no reference was made to other protozoa, probably a more relevant test under the circumstances.

9. Prepatent Period

The prepatent period can vary considerably in *C. parvum* infections reflecting the host species, strain of parasite, age of host at infection, or size and condition of the inoculum. Age-related resistance to infection has been demonstrated in several species,[26-28] and the age of the animal when it is exposed to the parasite can radically alter the severity of the infection and extent of oocyst shedding. The number of oocysts ingested can also considerably influence the length of the prepatent period,[107] fewer oocysts extending the prepatent period several days. As little as one oocyst can initiate infection in a gnotobiotic lamb.[107]

III. EPIDEMIOLOGY IN HUMANS

A. DISTRIBUTION OF HUMAN CRYPTOSPORIDIOSIS

1. Frequency of Occurrence

Population surveys began in the early 1980s, often of selected populations based on diagnostic specimens submitted routinely to the laboratory, especially from children.[1,4,5,8,9,108] Age- and sex-specific data often were not recorded; a few surveys were period-prevalence studies and included cases in which it was not possible to ascribe a causative role. A number of studies were of short duration, and few were adequately controlled. Laboratory-based surveys are subject to a variety of biases; ascertainment will inevitably vary according to specimen selection criteria (clinical and laboratory), diagnostic test method, availability of facilities, and reporting systems.[109-116] Accurate identification is essential but cannot be assumed for all reported studies, and some methods are significantly more sensitive than others.[109,114,116] It is difficult, therefore, to gain an accurate picture of incidence rates or of prevalence, or even to demonstrate unequivocally an etiological role. The data do, however, indicate the ubiquity of the parasite and there is now little doubt about its contribution to the considerable worldwide burden of gastrointestinal disease.[11-13]

2. Sero-Epidemiology

Several attempts have been made to study the immune response to infection or to determine sero-prevalence using IFAT or ELISA and more recently by molecular methods.[8,9,20,117-130] Sero-prevalence rates, generally 25 to 35% among industrialized population groups,[8,9,20,119] were generally higher in underdeveloped countries; volunteer Peace Corps workers visiting West Africa, with a 32% initial positivity, showed up to 13.6% sero-conversions.[121] Evidence of nosocomial transmission (see below) was obtained by serological means.[117] Sero-prevalence by ELISA was 44% among dairy farmers in Wisconsin, compared with 24.0% in other subjects (relative risk [RR] = 1.9);[125] in the lower prevalence group, positivity was also linked with exposure to cattle. In Oklahoma, sero-positivity increased with age, from 13% at <5 years to 58% of adolescents, and with attendance at daycare facilities.[127] Up to 64% of subjects were positive in Latin America, depending on antibody isotype.[8] In China and Brazil, prevalence rates were 42.3 to 57.5%; they increased with age in subjects over 1 year old.[128] In the area of Brazil studied, almost all children were positive by 2 years of age, but only 16.9% of children and young adults from Virginia were positive; the findings reflected the expected risk factors. In northeast Brazil, 191/202 (94.6%) of children had antibodies, with evidence of a secondary transmission rate of 19%.[126] Most children (32/35) tested in a small point prevalence survey among orphan infants in Thailand were positive.[123]

Studies on cryptosporidial antigens have indicated that they are complex but have some characteristics which are of considerable epidemiological significance (see below and Chapters 6 and 10).[130] There is some evidence of pre-existing immunity in asymptomatic close family contacts of some cases, possibly associated with previous exposure, and of prior exposure preventing outbreaks in populations, perhaps through a herd immunity mechanism.[9,119,131] However, an epidemiological study of an outbreak associated with consumption of fresh-pressed apple cider suggested that presumed previous exposure did not correlate with a demonstrable protective effect against symptomatic infection.[132]

As might be expected, seroprevalence rates tend to be higher than the frequency of oocyst detection.[22] In addition to giving some indication of prevalence of infection in particular populations, some of these studies have also provided evidence of recent infection, especially for cases occurring in outbreaks.[9,119,129,130] Although some studies have shown persistence of IgA and IgM, some subjects become seronegative, particularly when opportunity for re-exposure is limited. Persistence of IgM and IgA, together with apparently poor IgG production in some cases, implies the need for careful interpretation of serological findings.[119,121,126] Detailed examination of the immune response of the crew of a Coast Guard cutter using sensitive molecular methods indicated that certain small molecular weight antigens were diagnostic of recent infection.[130] Further studies are required to elucidate some of the questions raised by sero-epidemiological studies, especially the importance of herd immunity in relation to outbreaks and to molecular antigenic studies to identify the origin of outbreak strains.[17,18,131,133]

3. Relative Frequency of Occurrence

Cryptosporidium was one of the most commonly identified enteric pathogens, especially in developing countries where it may commonly be associated with mixed infections.[1,4-6,8,11,134-141] Such findings reflect the demographic setting, the age of subjects sampled, and seasonal variation in incidence of the different agents involved. In the U.K., *Campylobacter* is the most commonly identified enteric pathogen, but, during certain periods of the year, *Cryptosporidium* is the most common finding in 1- to 5-year-old children, occurring twice as often as salmonella and several times more commonly than shigellae.[135] Molbak and others found this to be the most prevalent parasite significantly associated with diarrhea (odds ratio [OR] = 2.79, $p = 0.0006$) and to be associated with an epidemic at the onset of the rainy season.[136]

4. Age Distribution

Early reports of human cryptosporidiosis were often in adults, reflecting a high proportion of immunocompromised subjects. Attention was drawn in the early 1980s to sporadic infection in the community, especially in otherwise healthy children.[1-3,108] Subsequent studies have confirmed a peak incidence in children aged 1 to 5 years in most areas, but generally at the lower end of that range in developing countries (see below). A secondary peak in laboratory-proven incidence was seen in young adults age 20 to 40 years, often resulting from family contact with children or occupational exposure. Screening for *Cryptosporidium* is often restricted to specimens from children, but the role of the parasite in diarrheal disease in young adults in developed countries should not be underestimated. Clinical infection is uncommon in the U.K. over the age of 45,[135,142] but in susceptible adults, it may be particularly unpleasant and occasionally leads to severe disease.[143] A significant association ($p = 0.004$) between severity and increasing age has been reported.[135,143-44] Asymptomatic infection has been found in adults who are close family contacts of cases but concentration methods may be required for its detection; there is little evidence generally of increased incidence in the elderly which may reflect frequency of exposure and hence immunity.

Infection in some Scandinavian countries appears mainly in adults, most of whom are believed to have acquired infection abroad, especially Russia, or by occupational exposure.[145-146] The preponderance of adult cases and relative infrequency of symptomatic cases among children has not been adequately explained, although association with travel would be expected to involve predominantly adults. The low incidence in children may imply that autochthonous infection is uncommon there except in rural subjects in whom it is often asymptomatic, presumably again because of past exposure.

In a 2-year U.K. survey involving over 60,000 patients, and over 12 years in Wales, 2/3 of infections were in the 1- to 5-years age group.[9,135,142] Surveys included patients of all ages, and examination of denominator data indicates that the peak is not the result of age sampling bias. Infection was less common in adults over 45 and children under 1 year and was infrequent in those under 6 months of age. Similar reports come from the U.K., the U.S., Central and South America, parts of Africa, and India.[1,5,9,11,141,146-151] However, in other surveys in the U.K., in Eire, and in some developing countries, infection was common in children under 1 year.[9] The reasons for these differences are not clear but may relate to levels of passively acquired maternal immunity, breast-feeding practices, and frequency of exposure. In rural settings generally, clinical cryptosporidiosis in adults is uncommon. Infection in rural infants often appears when they are becoming mobile, are teething, and are changing diet. This pattern probably reflects the frequency of and opportunity for exposure and declining levels of passively acquired immunity. Repeated or recurrent infection has been noted occasionally and may be associated with reduced severity.[9,13,143] It is not yet possible to distinguish between continued low-level or intermittent excretion, recrudescent infection, or reinfection. A protective effect in infants from breastfeeding has previously been noted in both developed and some developing countries (see below).[9] Children with diarrhea in Mexico had prevalences of 9.4% in a rural household study population, but 29.6% in a hospital-based urban population, of which 38 and 41%, respectively, were <6 months of age.[148] These differences may partly reflect increased levels of malnutrition in urban children, but the study was probably confounded by differing selection criteria for the two cohorts studied, as urban children were selected clinically, whereas rural subjects were obtained on a door-to-door basis.

5. Distribution of Cases by Sex

The distribution of cases by sex appears to be generally unremarkable. However, experience with campylobacter infections, for example, shows that detailed examination of rates of infection may reveal differences in both age- and sex-specific distribution from those shown by analysis of simple incidence figures. In a 2-year study of hospitalized children in India,[149] there was no significant sex-specific difference, although in Delhi infection was predominant in males.[150] An increased risk (OR = 1.9, 95% confidence interval [CI] 1.0 to 3.4) was found for boys in Guinea-Bissau,[147] while an increased incidence was found for females in Nigeria.[151] In Scotland,[152] there was a preponderance of cryptosporidiosis in males over females in young children, while in young adults there were more females than males. This difference may be explained by the observation of a slight bias towards males in rates for infectious diseases generally among children, while the increased rate for young adult females may be explained by the tendency for adult females to be more commonly responsible for childcare, including diaper changing and toiletting.

6. Geographic Distribution

Several published reviews have listed extensive bibliographies indicating that *Cryptosporidium* has a worldwide distribution and confirming its ubiquity as a human enteric pathogen.[1,4-6,8,11] Rates varied from less than 1% to more than 30%. Some variation may be attributed to genuine geographic variation and some to demographic or temporal factors, especially for surveys of short duration. In some instances, methodological factors may be important.

Reports of laboratory-confirmed incidence generally yield figures in developed countries of approximately 1 to 2% overall and about 4 to 5% in children, with a maximum incidence in the 1- to 5-year-old age group.[1,4-6,8,9,11] A few surveys showing low prevalence (<1%) in certain areas may reflect low risk, but some call into question study duration and sometimes methodology. Generally, surveys in developed countries are based more on diagnostic specimens in a healthcare setting and thus are more likely to show association with symptoms, asymptomatic individuals being less likely to be sampled. Surveys among daycare attenders have indicated high, and in some cases increasing, rates of prevalence and suggest that cryptosporidiosis is endemic in many such units.[1,5,6,8,9,153-59] In the U.K. study described above, 12 outbreaks over a 2-year period were associated with childcare centers or schools.[135]

B. RISK FACTORS FOR ACQUISITION OF CRYPTOSPORIDIOSIS IN DEVELOPING COUNTRIES

Changing patterns of infection have been noted in recent years with the appearance of a number of newly recognized pathogens, including *Cryptosporidium*. Some of this relates to improved diagnostic techniques but some appears related to a real increase in incidence. Thus, in addition to the traditional exposure factors, such as farming, etc., there have been changes in social and behavioral patterns and hence increased exposure.[9,159-160] Risk factors include:[1,2,5,9]

- Deficient immunity (AIDS and other acquired or congenital immunodepression, immunosuppression, malnutrition, etc.)
- Zoonotic contact through leisure activity (camping, backpacking, farm visits, etc.)
- Occupational exposure (veterinary, agricultural, nursing or medical, laboratory, child daycare, etc.)
- Poor hygienic and sanitary conditions (including drinking water and food, deficient sewerage or waste disposal, poor fly control)
- Exposure to untreated surface or recreational water (leisure activities) or inadequately treated water supply (inadequate treatment or a breakdown in treatment)
- Consumption of raw foods such as unpasteurized milk, raw meat, etc. (has shown paradoxical negative association, i.e., a protective effect with regular consumption in some case-control studies).
- Travel, especially from developed to underdeveloped or from urban to rural areas
- Younger age (including weaning and teething, pica and finger sucking, wearing diapers; poor infection control such as separation of areas and staff for diapering and food preparation)
- Contact with a case of diarrhea (including daycare attendance, household contact, parenting, etc.).

1. Indigenous (Autochthonous) Infection in Developing Countries

Reported rates are generally higher in developing countries (range approximately 2.5 to >30%, mean 8.5%) especially in younger children (<2 years), in whom it may be asymptomatic or significantly associated with diarrhea and sometimes with persistence, enteropathy, severe morbidity, and some mortality.[1,8,9,11,123-124,126-128,136-139,141,147-151,161-174] High rates of asymptomatic infection (in a few reports) probably reflect hyperendemicity and recurrent re-infection in immune subjects. In such settings, it is more difficult to define discrete episodes; because an individual appears to be asymptomatic when sampled does not mean that they were so when they acquired their primary infection. Association with malnutrition has been noted.[5,123-124,134,136,138-139,163-166,168,173,175] This, in turn, may further increase the malnourishment and impair immune function, thus increasing susceptibility. Mixed, intercurrent, and co-infection and persistence are common in such settings.

2. Risk Factors for Infection in Developing Countries

The peak age for incidence tends to be earlier than in developed countries, generally less than 3 years old, and infection has been noted to occur frequently in those less than 1 year old. The reported age-specific rates of infection differ in different areas of the world. The reasons are not clear, but some may be methodological, reflecting variation in sampling and detection methods. Such differences in incidence between urban and rural populations in the U.K. and in some developing countries were attributed to differences in breastfeeding practices.[1,8,9,136,147] A positive association with bottle-feeding and with overcrowding in urban slums was noted in Liberia,[161] while in India association with bottle-feeding in rural children was positive.[162] In Guinea-Bissau, looking analytically at risk factors which emphasized the importance of breastfeeding and also the keeping of pigs and dogs, prolonged food storage, and gender (OR for boys = 1.9, CI 1.0 to 3.4), an association was noted with young age (<18 months) and onset of the rainy season.[136,147] There is conflicting evidence for the protective role of breastfeeding and of colostral antibody in resistance to infection, although differences in isotype between human and animal (particularly herbivore) colostrum and also in the transplacental transfer of antibody might be significant. Increased opportunity for exposure in bottle-fed infants and in children who are being weaned may be as important as the immunological properties of breast milk for the differing rates of infection.

Close association with livestock from an early age may result in increased immunity and increased prevalence of asymptomatic oocyst excreters.[70,85,141,147,166,176] Infection among families of livestock handlers presumably has been introduced into the household from their animals, followed by secondary propagation (person-to-person transmission). It has been suggested that early infection, probably by person-to-person transmission, is followed by immunity to clinical infection. On the other hand, immunity impaired by malnutrition may delay clearing and hence prolong excretion. Significantly different levels of malnutrition between urban (40%) and rural (22%) children might account for the increased incidence of cryptosporidiosis in the former.[148] Severe measles seen in developing countries may exacerbate concurrent (or intercurrent) cryptosporidiosis,[129,143] and depressed immunity associated with viral infections may in turn increase susceptibility.

3. Ethnic Differences

Differences in incidence in particular ethnic groups within the same study area (for example, lower rates among muslims) may reflect dietary differences or differing exposure to livestock or to toiletting practices.[9,147]

4. Temporal Distribution and Seasonality

Seasonal or temporal trends in increased incidence vary from country to country, including summer in Australia; rainy season in Central America, South Africa, Guinea Bissau, and India; spring or late summer in North America; and late summer in Germany.[1,3,5,8,9,11,134,136,146-147,177-178] Conversely, in the Sudan detection decreased when the weather changed from hot and dry to colder and wetter.[169] Particularly in the U.K.,[9,135,142] peaks of incidence in the spring and in late autumn or early winter appear to be associated with factors including farming activity and peak rainfall. An association with ingestion of surface waters or contaminated potable water supplies reflect the same factors (see Chapter 4).[1,9,14,16,178-184] These trends in urban and rural areas may reflect direct zoonotic contact and indirect effects of rainfall, farming events such as lambing and calving, and environmental pollution with farm waste (Figure 1). In the U.K., outbreaks or temporal clusters of apparently unconnected sporadic cases occur in different parts of the country, often about the same time each year but not necessarily in the same locality. Whether this relates to local herd immunity levels affecting both primary incidence and secondary person-to-person spread or to other dynamic factors (see below) is not known. Detailed sero-epidemiological studies might help in elucidating this.

C. ROUTES OF TRANSMISSION FOR HUMAN CRYPTOSPORIDIOSIS

Major reservoirs and routes of transmission outlined in Figure 1 are zoonotic, environmental (indirect), and non-zoonotic (person-to-person), although distinctions may be somewhat artificial, simply reflecting the potential host range rather than a necessary source or route of transmission.

1. ZOONOTIC TRANSMISSION

Early reports drew attention to the association of human infection with exposure to infected calves.[1,2,5,7] *C. parvum* has been identified in about 80 species of mammals, and cross-transmission has been demonstrated between a variety of host species (see Chapter 1).[1,2,8,10] There is, therefore, a potentially large zoonotic reservoir for animals and humans. Many human infections have been derived from young cattle and sheep, either directly or indirectly. School children and students have become exposed during visits to working farms and livestock markets during the lambing and calving seasons or when lambs are taken to urban schools and nurseries for educational purposes.[2,9,160,185-186] Educational visits to farms and livestock markets may also lead to exposure to a variety of other infectious agents.[160] A number of cases of human cryptosporidiosis have been investigated in the U.K. for whom the only identified risk factor was recent attendance at riding stables or recent delivery of horse manure for garden fertilizer in which small numbers of oocysts were identified.[9] There has been secondary spread within households or play-groups following such zoonotic exposure.[9,85,176] Zoonotic exposure of urban subjects also occurs when camping on farmland, but in such circumstances exposure to more than one confounding risk factor (livestock animals, their excreta, raw milk, and water, etc.) is common, often combined with lower standards of hygiene. Backpackers' giardiasis resulting from consumption of untreated, presumed pristine, upland surface waters is well recognized; such exposure may also lead to cryptosporidiosis and campylobacteriosis (unpublished observations). Small rodents may also be capable of acting as reservoirs of human infection, as described above.

a. Association with Companion Animals

Companion animals, particularly cats and dogs, shown to be infected have been implicated in human disease occasionally (see above). Small mammals and exotic pets have been found infected and might be a risk to those cleaning cages and caring for the pets. However, despite the frequency with which pets are present in households of infected patients, rarely have they been implicated as a source of infection. The finding of oocysts in the stools of pets does not confirm the source of human infection because of the possibility that the pet and owners were exposed to a common source of infection or that the pet was infected by human contact. Oocysts in small numbers in fecal pellets of small animals such as rodents tend to die rapidly, probably through desiccation (unpublished observations). Wild rodents in the home may potentially transmit infection to pets directly or by contamination of food.

b. Occupational Transmission

Human infection (especially among veterinary students) has resulted from occupational exposure to infected animals.[2,9,13] A negative association has also been noted in some instances between bovine exposure and symptomatic human infection, particularly with rural adults who are occupationally exposed.[9] Occupational exposure probably leads to repeated mild or asymptomatic infection and high levels of immunity.

c. Indirect Zoonotic Transmission

Occupationally exposed individuals may inadvertently carry infected material home, exposing family members, especially the young, sometimes with further secondary spread.[9,176] Indirect transmission may have been the route for some cases in the U.K., including severe maternal periparturient infection.[143] Person-to-person transmission occurs among household contacts of calf and sheep handlers and in household, daycare, and school contacts of cases resulting from educational and recreational farm visits.[9,70,85,160,174,176] In elementary school children in Jordan,[174] 4% were positive for *Cryptosporidium*, for which the main risk factors appeared to be residence in a rural village with potential for zoonotic exposure and unsatisfactory water supplies.

2. ENVIRONMENTAL SOURCES

Recognition of the importance of transmission through the environment has increased, especially with the occurrence of large waterborne outbreaks. Co-infection with *Giardia* and campylobacter[9] suggests a common mode of transmission, possibly involving contaminated water or fruit and vegetables as sources of infection, as well as by person-to-person transmission, especially through poor hygienic practices.

a. Waterborne Transmission

(See also Chapter 4.) Waterborne cryptosporidiosis associated with water supplies results from either human or animal pollution of the environment. The relative contribution of animals or human excreta is, however, currently unknown.[9,14-19] Temporal or seasonal peaks in incidence of cryptosporidiosis have coincided with periods of maximal rainfall. The widespread practice of disposal of both animal and human excreta to land, e.g., by muck and slurry spreading on pasture, may lead to infection directly by aerosol spread or indirectly by contamination of water courses and reservoir feeder streams. Surface waters polluted naturally or by these practices may lead to contamination of water supplies or of food crops during irrigation. The impact of an outbreak may be considerable, not only on laboratory, epidemiological, and health-care services, but also on the food and catering industry, etc.[17-19,187-190]

Low-level intermittent contamination may account for some unexplained sporadic cases or small clusters of cases.[17,18,191] The source, viability, and pathogenicity of oocysts found in water or other environmental samples cannot yet be reliably determined. The significance of small numbers of oocysts is often, therefore, uncertain although their presence should be taken as presumptive evidence for careful and detailed epidemiological study and enhanced control measures, including optimizing water treatment. There is a need for routine epidemiological surveillance and clinical and environmental monitoring to give early warning of problems (see below).[18,19,111,113,115,131,192-193]

The outbreak in Milwaukee, WI, in 1993 led to a number of detailed studies within the main outbreak investigation. For example, a U.S. Coast Guard cutter filled its water tanks during the outbreak and crew members subsequently became ill.[130] Sera were obtained from 50 crew members, of whom 62% complained of symptoms; 20% were positive for oocysts in their stools. The study produced the most detailed information to date of the serological response to infection in humans, provided evidence of additional cases, and demonstrated diagnostically useful antigens. Recurrence of symptoms during the course of cryptosporidiosis and the risk of secondary transmission were studied.[194] Cases generally seemed to be more severe than in other reported series. Of confirmed cases, 39% had a recurrence of symptoms following apparent recovery and 21% among clinically diagnosed unconfirmed cases. Recurrence of symptoms following initial recovery has also been noted in the U.K. (Casemore, unpublished observations). The significance of these findings for transmissibility is unknown. Evidence of secondary transmission was found in only 5% of cases when the index case was an adult. An exploratory case-control study of risk factors for post-outbreak transmission found person-to-person transmission that declined rapidly within 2 months.[195] Immunosuppression and having a child less than 5 years, but not drinking Milwaukee water, were independently associated with infection. The effectiveness of point-of-use water filters in preventing cryptosporidiosis during the outbreak was also monitored.[196] Of 155 inquiry responders, those with submicron filters or reverse osmosis units had lower attack rates and lower household transmission rates, suggesting infection could be reduced substantially if drinking water away from home was avoided. A study of children during the Milwaukee outbreak showed that of 209 who submitted stool samples to the laboratory, 57 (27%) were positive for *Cryptosporidium*.[197] They were more likely to have had more than one sample submitted, to live in the area supplied with contaminated water, to be tested later in their illness, to have evidence of underlying disease, and to be older than 1 year. Clinical illness was more prolonged for this group and associated with weight loss and abdominal cramps. Many tests were falsely negative early in the illness, suggesting either delayed or intermittent excretion or failure of tests to detect oocysts; confirmed positive cases more likely had more than one specimen examined. A study of factors affecting *Cryptosporidium* testing in Connecticut showed that testing practice was inadequate to allow proper assessment of impact or detection of outbreaks.[115] Outbreaks associated with swimming pools and other recreational waters have been reported in the U.S., Canada, and the U.K.[115,198-205] Outbreaks have followed defecation incidents or have resulted from defects in pool filtration or sewage disposal resulting in pollution of the pool water or associated areas. The main controlling factors of fecal contamination of a pool are dilution and efficient filtration. In well managed pools this infection will not normally be a major problem, although those sharing a pool with an infected child who accidentally defecates will stand a high probability of infection. Fluid stool may rapidly disperse in a pool; diapered children should wear additional waterproof protection and should not enter the pool if they are unwell.

b. Foodborne Transmission

Cryptosporidiosis has sometimes been associated with consumption of certain foods.[1,9,132,135,206-210] Direct identification of food in the transmission of *Cryptosporidium* is hampered by the lack of the equivalent of bacteriological enrichment culture for recovery of small numbers of oocysts and for confirmation of viability, although reports show this is possible using molecular probes.[211] In Guinea-Bissau there was an increased risk of infection with consumption of stored cooked food.[147]

Based on epidemiological evidence from the U.K.,[9,135] consumption of certain foods, especially raw fresh sausages, offal, and raw milk, appears to be a risk factor for infection with *Cryptosporidium*. The association with raw milk consumption in the U.K. has been found to occur particularly in those who only occasionally consume it or have recently commenced doing so, presumably because those who regularly consume raw milk will maintain sufficient levels of immunity to *Cryptosporidium*. Doorstep delivery of bottled pasteurized milk in the U.K. has been associated with *Campylobacter* infection as a result of corvid birds pecking the foil caps after contact with cow dung in adjacent pasture.[212] *Cryptosporidium* clearly could be transmitted by the same route. Several outbreak studies in the U.K. have revealed evidence of a negative association (protective effect) with the regular consumption of raw vegetables and may also reflect increased exposure and immunity (Public Health Laboratory Service, Communicable Disease Surveillance Centre (PHLS CDSC), unpublished outbreak reports).

In the first fully documented report of foodborne transmission of *Cryptosporidium*,[132] approximately 26% of attendees at a country fair in Maine suffered illness after drinking fresh pressed apple cider — 160 primary cases (33 laboratory confirmed, 127 clinically defined) and 53 secondary cases (17 laboratory confirmed) were identified; 54% of those exposed to the cider were ill, compared with 2% not known to have been exposed. On average, symptoms began 6 days later (range <1 to 13 days), with a median duration of 6 days (range 1 to 16 days); 84% had diarrhea and 82% had vomiting; unusually, 16% of primary cases reported vomiting without diarrhea. Oocysts were recovered from some retained cider, the cider press, and a stool specimen from a calf on the farm which had supplied the apples. Incubation periods were estimates based on known exposure, but the earliest were unusually short. Because there was no evidence of other infecting agents, the short incubation periods and high rates for vomiting might be explained by ingestion of very high doses or by dietary effects, or both. This would be borne out by the differing pattern with lower rates of vomiting (60%) in secondary cases who had a median incubation period of 8 days (range 1 to 24 days).

The fermented alcoholic beverage sold as cider in parts of Europe is usually stored for some months and may be filtered. Laboratory studies have shown that viability of oocysts soon declines on storage in this product (Casemore, unpublished data).

3. NON-ZOONOTIC TRANSMISSION

a. Urban Transmission

Person-to-person transmission, particularly involving children,[1,3,5,8,9,13] is common, confirming the hypothesis that cryptosporidiosis is not necessarily a zoonosis. However, the distinction is somewhat artificial, and urban-propagated transmission may be initiated directly or indirectly from zoonotic sources. In western urban daycare centers, rates for cryptosporidiosis of over 60% have been noted.[1,3] Sporadic community cases and daycare center outbreak cases are often associated with secondary cases among families or a history of recent diarrhea among contacts who had not been investigated and also with asymptomatic excreters among contacts.[9,13,126,153-159] A case-controlled study in Switzerland showed a 4.6% rate of infection in children needing medical care; the highest risk factor was contact with another person with diarrhea, followed by a history of foreign travel.[158,213] In a survey of centers in Georgia, the prevalence of *Cryptosporidium* had increased significantly in recent years and appeared to be endemic.[156] A comprehensive review of cryptosporidiosis in a child-care setting showed the high prevalence in such centers.[159]

b. Infection in the Immunocompromised

Cryptosporidiosis continues to represent a serious threat to AIDS sufferers and others who are severely immunocompromised, with rates varying from <1 to >50% in different parts of the world. Presentation and severity of infection vary from resolving, intermittent, chronic, or profuse and may even resemble classical cholera.[214-224]

Other immunocompromised states may lead to increased incidence, for example, a high rate of infection (17%) in Turkish patients with neoplasia and with diarrhea compared with controls.[223] In general, the epidemiological picture differs little between immunocompetent and immunocompromised subjects except for the so-called gay bowel syndrome of sexually transmitted enteric infection.[22,143,225]

Respiratory symptoms are common in cryptosporidiosis in immunocompetent as well as immunocompromised patients. *Cryptosporidium* has been found in sputum, particularly in AIDS patients and in vomit, and these may provide additional routes for transmission.[136,142,158,169,224] Transmission may occur via fomites, although survival of oocysts on surfaces is likely to be limited by the susceptibility of oocysts to desiccation.[226]

c. Nosocomial Infection

Cross infection from patient to patient (immunocompromised and immunocompetent) and between patients and staff in hospitals is further evidence of person-to-person transmission.[117,224,227-234] It is of considerable potential importance for immunocompromised patients. Given the known resistance of oocysts to disinfectants and the apparently low infective dose, it is surprising that nosocomially acquired infection has not been documented more frequently.

In the U.S., *Cryptosporidium* was transmitted to hospital staff caring for infected immunocompromised patients.[117] Serological studies suggested increased exposure among the staff caring for an AIDS patient with cryptosporidiosis. A nurse acquired infection from a bone-marrow transplant recipient who had developed cryptosporidiosis.[228] The nurse was believed to have had minimal unprotected exposure, suggesting that the minimum infective dose was small. In a bone-marrow transplant unit in Italy,[229] five patients developed cryptosporidiosis after admission to the unit of a sixth patient with the infection. Contamination of the ward environment was demonstrated. In a renal unit in Argentina, 11 of 14 patients with diarrhea and a member of the nursing staff and her husband were found to have cryptosporidiosis; a number of asymptomatic patients also had evidence of infection.[230] Transmission was thought to have resulted from the sharing of toilet facilities. In the U.K., three of six leukemic children with cryptosporidiosis, two of whom died, were thought to have acquired their infection while in the hospital.[231] Investigation of an outbreak among pediatric patients in Mexico suggested that poor hand-washing practice and naso-gastric feeding tubes were important factors in transmission.[232] In Denmark,[233] 18 HIV-positive patients in a ward developed cryptosporidiosis, as did the departmental secretary and a visiting relative. There was a high mortality rate among the patients. Evidence suggested a low infective dose with transmission via an ice machine contaminated by the hands of an incontinent patient who was psychotic as a result of AIDS CNS involvement. In a north Wales infectious diseases unit, five nursing staff were confirmed to have contracted infection from a terminally ill AIDS patient who was admitted with cryptosporidiosis.[224] Cryptosporidia were demonstrated in his feces and vomit, raising the possibility of fecal-oral and aerosol transmission. His dementia, fecal incontinence, and intractable vomiting necessitated close contact, making strict adherence to infection control protocols difficult.

Two hospital-associated outbreaks have involved child daycare centers within hospitals. In the U.S.,[234] eight children and also daycare staff and family members were affected. In Portugal,[235] 28 children and one member of staff were affected. Such centers may provide a source of infection for hospital staff and may then be transmitted, especially to immunocompromised patients with potentially fatal consequences.

Transmission among AIDS patients in a residential care home would be subject to some of the same risk factors as hospitalized AIDS patients, although residents may be less ill.[236] Such centers, and HIV daycare centers, sometimes ask about the advisability of permitting companion animals; the attendant risks have to be set against the psycho-social benefits to the centers' attenders.

A study of renal transplant and AIDS patients in a hospital, following a widespread waterborne outbreak in the U.K., suggested that renal patients had no more risk than normal subjects.[179,237] AIDS patients, in contrast, were more prone to acquire cryptosporidiosis and to suffer more severe infection. However, for AIDS patients, epidemiological evidence showed no increase in risk of cryptosporidiosis associated with water consumption in another area of study.[225]

d. Travelers' Diarrhea

Cryptosporidiosis has emerged as an important cause of travelers' diarrhea,[1,5,8-10] often as mixed infections, especially with *Giardia,* suggesting a common epidemiology involving contaminated water or food and poor hygienic practices. Increased exposure to livestock and the rural environment may be important in some areas. Direct and indirect person-to-person transmission is important where poor hygienic conditions prevail in many underdeveloped countries, including many holiday and business travel destinations. Flies and cockroaches have been implicated in transmission in developing countries.[238-239]

Travelers' diarrhea is usually associated with travel to less developed countries. However, urban subjects from developed countries have also been shown to have acquired infection when vacationing in more rural areas of the same or other developed country.

IV. EPIDEMIOLOGY — SOME GENERAL CONSIDERATIONS

Outbreaks of water-associated and waterborne cryptosporidiosis have been increasingly recognized in both the U.K. and North America (see above and Chapter 4). Similar data are not generally available for other developed countries, but this probably represents under-detection and under-reporting.

Enteric protozoal infection may often be community-propagated outbreaks initiated by water. The contaminated water may no longer be present in the supply by the time an increased incidence is noted, leading to difficulties in laboratory confirmation of the source and complicating decisionmaking for control. Justifying a boil advisory notice affecting thousands of people and setting criteria for removal of the notice are made difficult.[18,189-190] Trigger or action levels based on routine monitoring data, rather than on (undetermined) peak concentrations, are of doubtful validity.

Because infection rates determined by the laboratory are selective data, they are essentially artificial, although trends can be deduced. Recognition of an outbreak, unless unusually large, will be determined by ascertainment, based on clinical criteria for submission and laboratory criteria for the examination. Outbreaks can thus be artefacts created by changes in ascertainment practice. Pseudo-outbreaks can be created by failure of good laboratory practice and misidentification of artefacts. This may lead to increased sampling and thus an apparent increase in the incidence of diarrhea.[17-19,109-110,116]

A. DYNAMIC FACTORS AND EPIDEMIOLOGICAL INVESTIGATION

The size of a waterborne or water-associated outbreak will be determined not only by the efficiency of water treatment following a challenge, but also by both host and parasite factors, described as the dynamics of transmission. This involves the fluctuations or flow of changes in parasite numbers and characteristics, of host numbers and their susceptibility, and host exposure factors.

1. Herd Immunity Factors

The term herd describes any defined species population of potential hosts, animal or human. Such factors are complex, operating indirectly and by specific mechanisms.[133] Immunity results from an immunological antigenic (e.g., parasite) challenge when the host's immune system recognizes the antigen again (although the host is not necessarily refractory to reinfection). A fully susceptible individual may require a smaller infective dose than a partially immune subject. Recurrent exposure may lead to limited establishment of the parasite and asymptomatic infection. Such individuals generally have more limited ability to transmit their infection. Oocysts excreted late in an infection may be less viable than those excreted earlier (unpublished data), raising the possibility that such immune infected subjects may limit the infectivity of the oocysts excreted through poorly understood immune mechanisms. Dynamically speaking, an individual can be placed in one of several categories (the so-called compartmental model):

- Susceptible (variable, depending on their immune status)
- Infected but with limited infectiousness (incubation or pre-patent period)
- Acutely ill and fully infectious
- "Immune" but infected (e.g., low-level asymptomatic carrier)
- Immune and fully recovered (but increasingly re-infectable with time)

Each of these stages may merge into the next. One implication is that the minimum infective dose (MID) may vary according to the position of the exposed person in relation to this model.

If herd immunity is low to moderate, in an urban area in a developed country, then routes of transmission to susceptible adults are limited. It would seem then that the water route would potentially be relatively important in transmission to urban adults. The first indication of a waterborne outbreak may thus be an increase in numbers of infected adults, many of whom will be household primary cases. This has been a finding of a number of outbreak investigations.[15,18,180,183-184] In several outbreaks in the U.K., early primary cases have a markedly higher mean age (>15) and no history of contact with another case of diarrhea; mean age decreases with time for later cases and many more are identifiable as secondary cases. Thus, it is essential that laboratories screen stools for *Cryptosporidium* from adult patients as well as children. In one outbreak in southern England, the median age for the outbreak was low but included some elderly residents of local nursing homes, most of whom had retired to the area from elsewhere, and some adult holiday visitors. Epidemiological study strongly suggested an association with drinking water, part of which came from a river with unacceptably high turbidity suggesting contamination (PHLS CDSC, unpublished outbreak report).[18] The age distribution of local cases suggested a high degree of herd immunity. A further outbreak, resulting in over 500 confirmed cases, occurred in the same area in the summer of 1995. This was shown epidemiologically to be linked to the same water treatment works (Harrison et al., unpublished report). Waterborne transmission was suspected to have initiated clusters of propagated cases with a low median age in an area of southeast England which previously had several such episodes (PHLS CDSC, unpublished outbreak reports). This emphasizes the way in which so-called waterborne cryptosporidiosis may often be more accurately described as a community outbreak initiated by water and that predicting the outcome of contamination involves consideration of complex dynamics outlined in Figure 3.

Given an understanding of the complexity of the dynamics of transmission it is likely that low-level intermittent contamination, especially in an area of raised herd immunity, results in low incidence of apparently sporadic cases with little secondary spread. Indeed, low-level intermittent contamination probably helps maintain herd immunity. Such mechanisms probably underlie the observation of an outbreak in a town in Oregon associated with the temporary use of water supplied from another town.[132]

2. Parasite Dynamic Factors

It must be assumed that environmental isolates will vary in viability, in potential infectivity for man, and in virulence, although little is known of this aspect. If strains of *C. parvum* are animal host adapted, the MID may differ from human strains in primary and secondary transmission. In well adapted strains in gnotobiotic hosts, the MID may be as low as one oocyst.[107] In normal, colostrum-fed lambs the same isolate still showed an MID in single figures but infection was somewhat slower to develop and was clinically less severe (S.E. Wright and D. P. Casemore, unpublished data). Recent human feeding trials using a calf isolate suggested an MID of ≤30 and a median dose of 132 oocysts.[240] Other isolates might have a different MID. An ID_{50} is a standardized dose concept in scientific experiments, but because water is almost universally consumed and a particular supply may be distributed to large numbers of people, then an ID ≤ 1% might expose significant numbers of consumers who may then transmit infection to others, thus amplifying the outbreak.

3. Parasite Environmental Dispersion and Host Consumption Factors

Because parasites in water are not homogeneously dispersed and consumption varies, the dose ingested may vary considerably for any given count on a sample (which itself is subject to wide statistical, sampling, and analytical variation). The distribution "curve" for oocysts in water moves both temporally and geographically within a source collection, treatment, and distribution system. Taking relatively small samples (compared with total reservoir or river flow volume) at infrequent intervals (weekly or monthly) is thus unlikely to detect significantly increased contamination, except by chance. If positive, it is then not possible to predict which part of a distribution "curve" it represents. One might expect that a very low dose in the water would not be sufficient to infect consumers; however, nonhomogeneous dispersal (resulting, for example, from a bolus of contaminated water) may lead to clusters of oocysts even when

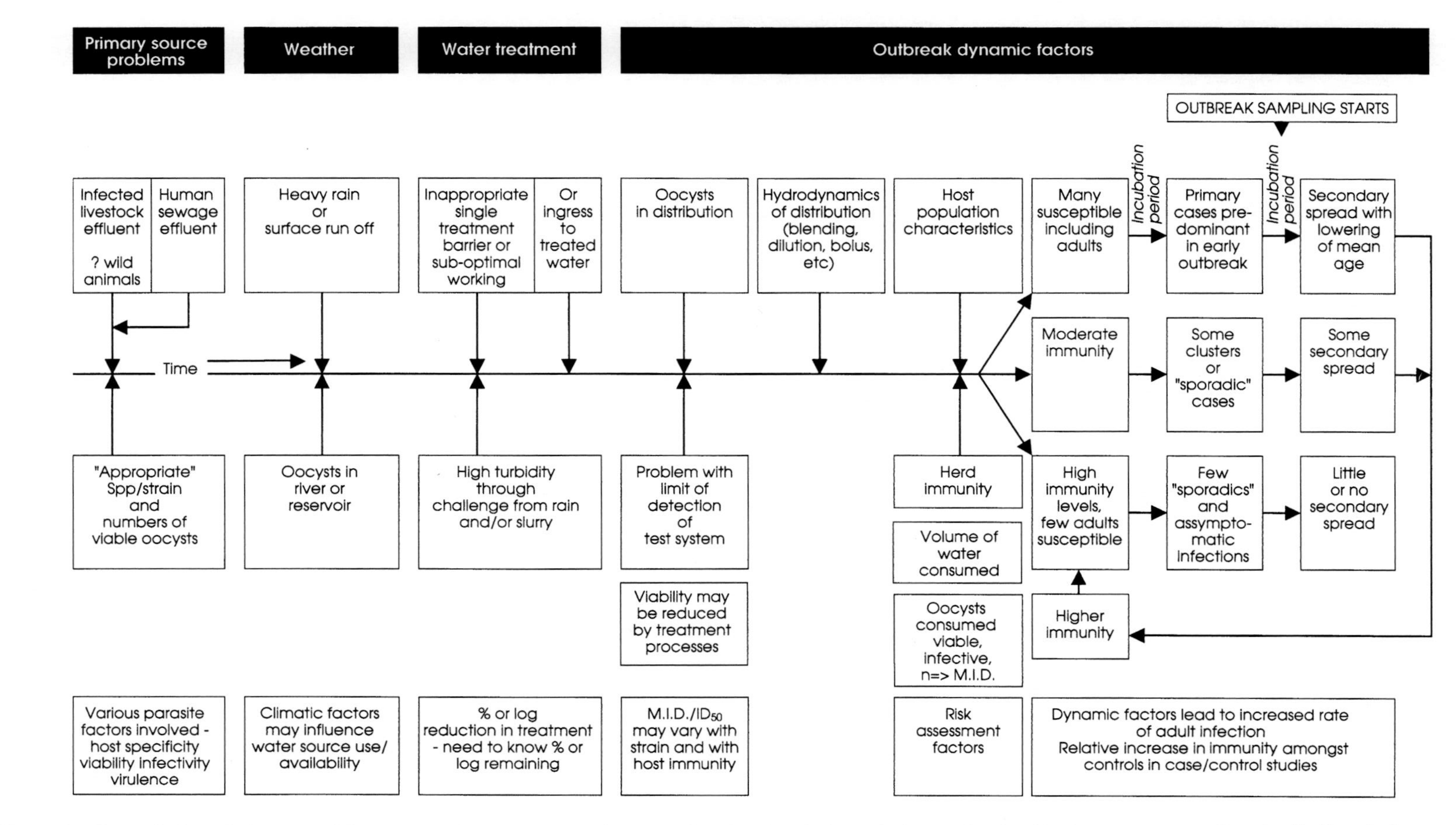

Figure 3 Transmission of *Cryptosporidium parvum* in water: the environmental and outbreak dynamic factors involved in the events controlling the likelihood of an outbreak.

only small numbers of oocysts are present. This can result in a susceptible consumer, especially a severely immunocompromised one, becoming infected. If one or two cases result, they are unlikely to be detected even with good epidemiological surveillance. If such a case is found by chance, the role of water is unlikely to be suspected, could not be confirmed epidemiologically, and thus would add to the background of apparently unexplained sporadic cases. Given the paucity and variability of ascertainment, it is only when large numbers of cases occur in an area within a finite period of time that an outbreak is likely to be recognized. This does not mean that transmission is not occurring.

Because water is supplied to so many consumers within an urban distribution area of a developed country, if oocysts are present the possibility of a MID being ingested by a susceptible individual is higher than for other vehicles. The greater the number of oocysts, the greater the probability that a significant number of consumers will be exposed, including those with partial immunity susceptible to lower MID. However, with the extreme variability and unpredictability of these events, one must assume for regulatory purposes that the worst combination of factors will apply.[241-242]

4. Descriptive and Analytical Epidemiological Studies

An epidemiologist looks for clusters of cases in place and time (descriptive data) and then attempts to quantify risk associations by statistical evaluation of data from groups of cases and appropriately defined controls (analytical study). The findings should show characteristics which include temporal, geographic, and biological consistency and plausibility and strength of association. It has been possible in such investigations to show not only significant association but also a dose response effect with volume of water ingested.

A number of difficulties arise in the investigation of community-wide outbreaks of cryptosporidiosis.[17-19,112] If an outbreak of cryptosporidiosis results from contaminated potable water supplies, the time, place, and extent (except in general terms) of exposure is uncertain. One can define a household primary case, but this individual may have become infected from the primary vehicle (water) at home or elsewhere or from a different source. This will tend to introduce a bias which will reduce the level of association found with the suspected vehicle of transmission. Individuals also vary in their pattern of water consumption at home and away. It is important to establish normal patterns of water consumption, as volumes consumed are likely to have increased as a result of infection.

If an analytical (case-control) study is set up, controls are sought from within that community who are likely to receive the same water supply and would seem, superficially, to be ideal as controls. However, if potential controls had symptoms they will be excluded because they may have been undetected cases and thus would have been susceptible at the putative date of exposure. That being so, those who were asymptomatic and "acceptable" as controls must be more likely to have been immune and would not have become ill even though exposed to the same degree. This introduces another bias reducing the apparent level of significance of association.

The possibility of an outbreak in an area thus depends on a combination of factors (see Figure 3):

- A sufficient number of parasites and their viability, infectivity, and virulence
- A sufficient number of susceptible individuals for ingestion to lead to primary infection
- A sufficient number of susceptible individuals for whom contact with a primary case may lead to secondary transmission, thus amplifying the incidence
- Adequate laboratory and epidemiological surveillance and reporting

To this add one or more water-related factors, which may include:

- Abnormal weather conditions (e.g., heavy rainfall) leading to agricultural wash off and pollution of water sources
- High turbidity surface water, streaming in a reservoir, etc., leading to an increased challenge to treatment
- Breakdown in integrity of an aquifer (ground waters), possibly following heavy rain after a prolonged drought
- Suboptimal working of water treatment
- Distribution pipework or service reservoir fault (ingress incident)

V. MOLECULAR EPIDEMIOLOGY

Typing of infective agents has traditionally depended on methods for culture-enriched organisms, based on both phenotypic and genotypic characteristics, e.g., serology, biochemical typing, antibiograms, etc., but has become increasingly reliant on so-called molecular epidemiology. Given the obligate parasitic nature of *Cryptosporidium*, the typing and subtyping of isolates below species level must depend largely upon such molecular methods. Antigen and antibody analysis at the molecular level has begun to yield epidemiologically useful information, including differences between isolates from different host species (see Chapter 10).[243-249] Isoenzyme (zymodeme) profiling suggests there are different subtypes or strains, although this technique requires considerable numbers of oocysts ($\geq 10^8$). Other techniques, such as single and two-dimensional gel electrophoresis of antigens, western blotting, etc., confirm this diversity and have suggested that there may be distinct human and animal strains and geographic variation. Although there are several possible explanations for these findings, some currently conflict with the epidemiological evidence and need to be clarified. The extraction and purification of nucleic acids and other molecular components has led the way to the production of gene probes and to other more sensitive means of detection and identification, the development of which is essential for epidemiology. The differences between these laboratory findings and epidemiological evidence need to be resolved by further study.[249]

VI. DETECTION AND CONTROL

The primary purpose of epidemiology is to identify sources of infection, to control outbreaks, and to learn lessons that might prevent future outbreaks. What is the relationship between the prevalence of infection and incidence of disease and oocysts numbers in water? Where prevalence is very high, oocysts may be commonly present, but the risks to the community may be low. Conversely, if incidence is low, then oocysts are probably not often present but may represent an increased risk when they are. At present, one cannot safely define "permissible" oocyst numbers or trigger (action) values or assess risk to communities from the numbers detected in water with any reliability. What can be done is to estimate risk based on "worse case scenarios" using certain assumptions.[241-242] From this it follows that the regulators and the public need to understand the concept of an acceptable level of risk — a political and educational problem. When risk is defined in terms of both the probability of harm and the severity of harm produced, then it can be seen that a dual standard might be required. For the average, healthy subject, occasional enteric infection is not especially serious, although unpleasant and disruptive and collectively costly to society. To AIDS patients and other seriously immunocompromised patients, it may be life threatening, even though the probability of ingesting a given number of oocysts may be no greater for an immunocompromised person (although the MID might be lower). The consequences of oocysts in the supply are thus more serious and the risk of serious outcome may be unacceptable. The immunocompromised should avoid exposure as much as possible, for example, by washing salads and fruit in boiled water and avoiding consumption of unboiled water, as travelers might do while visiting underdeveloped countries. They should also limit their contact with livestock and diapered children.

An essential requirement of epidemiological outbreak detection and control is adequate, reliable screening and careful interpretation of data. It is known that the infectious dose may be low and oocysts may no longer be present in a supply when an outbreak is recognized, which underlies the problem of evaluating risk associated with potable water based on routine monitoring data. When an incident occurs, data need to be assessed carefully by a local multidisciplinary group.

Point-of-use filters may protect supplies in some circumstances, although this carries with it other risks such as bacterial colonization and concentration of oocysts in the event of contamination of the supply. Chemical tablets for the treatment of water have little effect on *Cryptosporidium*. Oocysts are susceptible to moderate heat treatment and water need only be raised to the boil. The most direct means of control, as with other enteric infectious agents, is personal hygiene, including attention to hand washing, especially in child daycare and hospital settings. Terminal AIDS patients, suffering from a variety of infections including cryptosporidiosis, impose severe demands on nursing staff. Those with CNS involvement present an increased risk because of behavioral difficulties. Hospitals serving AIDS and other seriously immunocompromised patients must be particularly vigilant in management of patients with cryptosporidiosis. Even minor gastrointestinal symptoms should be investigated to minimize risk of spread to patients or to other staff. Fomites, suggested as vehicles of transmission in nosocomial cases and in the daycare setting, must have limited duration of infectivity given the susceptibility of oocysts to desiccation. Vomiting and the respiratory route also have been suggested as routes of transmission. Despite the resistance of oocysts to disinfectants, physical cleaning to reduce numbers and reduce the protective effect of body fluids should reduce risk of transmission. Water remains a significant vehicle for transmission that cannot easily be controlled, especially if water treatment systems fail, except by the introduction of a boil water advisory notice, which itself carries identifiable risks such as scald injuries and a potential decrease in personal hygiene.

REFERENCES

1. Fayer, R. and Ungar, B. L. .P., *Cryptosporidium* spp. and cryptosporidiosis, *Microbiol. Rev.*, 50, 458, 1986.
2. Current, W. L., *Cryptosporidium*: its biology and potential for environmental transmission, *CRC Crit. Rev. Environ. Contr.*, 17, 21, 1986.
3. Soave, R. and Armstrong, D., *Cryptosporidium* and cryptosporidiosis, *Rev. Infect. Dis.*, 8, 1012, 1986.
4. Janoff, E. N. and Reller, L. B., *Cryptosporidium* species, a protean protozoan, *J. Clin. Microbiol.*, 25, 967, 1987.
5. Crawford, F. G. and Vermund, S. H., Human cryptosporidiosis, *CRC Crit. Rev. Microbiol.*, 16, 113, 1988.
6. Tzipori, S., Cryptosporidiosis in perspective, in *Advances in Parasitology*, Vol. 27, Baker, J. R. and Muller, R., Eds., Academic Press, New York, 1988, 63.
7. Chermette, R. and Boufassa-Ouzrout, S., *Cryptosporidiosis — A Cosmopolitan Disease in Animals and in Man*, Office International des Epizooties, Paris, 1988.
8. Ungar, B. L. P., Cryptosporidiosis in humans (*Homo sapiens*), in *Cryptosporidiosis of Man and Animals*, Dubey, J. P., Speer, C. A., and Fayer, R., Eds., CRC Press, Boca Raton, FL, 1990, 1.
9. Casemore, D. P., Epidemiological aspects of human cryptosporidiosis, *Epidemiol. Infect.*, 104, 1, 1990.
10. O'Donoghue, P. J., *Cryptosporidium* and cryptosporidiosis in man and animals, *Int. J. Parasitol.*, 25, 139, 1995.
11. Reinthaler, F. F., Epidemiology of cryptosporidiosis in children in tropical countries, *J. Hyg. Epidemiol. Microbiol. Immunol.*, 33, 505, 1989.
12. Anon., WHO/PAHO Informal Consultation on Intestinal Protozoal Infections, CDS/IPI/92.2, Mexico, 1991, World Health Organization, Geneva, 1992.
13. Current, W. L., Cryptosporidium parvum: household transmission, *Ann. Intern. Med.*, 120, 518, 1994.
14. Smith, H. V. and Rose, J. B., Waterborne cryptosporidiosis, *Parasitol. Today*, 6, 8, 1990.
15. Anon., *Cryptosporidium* in water supplies, Report of the Group of Experts, Chairman, Sir John Badenoch, Department of the Environment/Department of Health, Her Majesty's Stationery Office, London, 1990.
16. MacKenzie, W. R., Neil, M. D., Hoxie, N. J., Mary, M. S., Proctor, M. E., Gradus, M. S., Blair, K. A., Peterson, D. E., Kazmierczak, J. J., Addiss, D. G., Fox, K. R., Rose, J. B., and Davis, J. P., A massive outbreak in Milwaukee of *Cryptosporidium* infection transmitted through the public water supply, *N. Engl. J. Med.*, 331, 161, 1994.
17. Casemore, D. P., Enteric protozoa and the water route of transmission — epidemiology and dynamics, in *Water and Public Health*, Golding, A. M. B., Noah, N., and Stanwell-Smith, R., Eds., Smith-Gordon, Great Britain, 1994, 123.
18. Casemore, D. P., The problem with parasites, in *Protozoan Parasites and Water* (proceedings of a conference, University of York, 1994), Betts, W. B., Casemore, D. P., Fricker, C., Smith, H., and Watkins, J., Eds., Royal Society of Chemistry, London, 1995.
19. Meinhardt, P. L., Casemore, D. P., Miller, K. B., Epidemiological aspects of human cryptosporidiosis and the role of waterborne transmission, *Epidemiol. Rev.*, in press.
20. Ungar B. L. P., Cryptosporidium in infectious diseases and their etiologic agents, in *Principles and Practice of Infectious Diseases*, Vol. 2, Part III, Mandell, G. L., Bennett, J. E., Dolin, R., Eds., Churchill Livingstone, New York, 1995, 2500.
21. Petersen, C., Cryptosporidiosis in patients infected with the human immunodeficiency virus, *Clin. Infect. Dis.*, 15, 903, 1992.
22. Gellin, B. G., Soave, R., Coccidian infections in AIDS, in *The Medical Clinics of North America*, White, D. A. and Gold, J. W. M., Eds., 76, 205, 1992.
23. Mannheimer, S. B. and Soave, R., Protozoal infections in patients with AIDS: cryptosporidiosis, isosporiasis, cyclosporiasis and microsporidiosis, in *Infectious Disease Clinics of North America*, Vol. 8, Cornell University Medical College, New York, 1994, 483.
24. Ditrich, O., Palkovic, L., Sterba, J., Prokopic, J., Loudova, J., and Giboda, M., The first finding of *Cryptosporidium baileyi* in man, *Parasitol. Res.*, 77, 44, 1991.
25. Ortega, Y. R., Sterling, C. R., Gilman, R. H., Cama, V. A., and Diaz, F., Cyclospora species — a new protozoan pathogen of humans, *N. Engl. J. Med.*, 328, 1308, 1993.
26. Sherwood, D., Angus, K. W., Snodgrass, D. R., and Tzipori, S., Experimental cryptosporidiosis in laboratory mice, *Infect. Immun.*, 38, 471, 1982.
27. Harp, J. A., Woodmansee, Moon, D. B., and Moon, H. W., Resistance to calves to *Cryptosporidium parvum*: effects of age and previous exposure, *Infect. Immun.*, 58, 2237, 1990.
28. Ortega-Mora, L. M. and Wright, S. E., Age-related resistance in ovine cryptosporidiosis: patterns of infection and humoral immune response, *Infect. Immun.*, 62, 5003, 1994.
29. Harp, J. A., Woodmansee, D. B., and Moon, H. W., Effects of colostral antibody on susceptibility of calves to *Cryptosporidium parvum* infection, *Am. J. Vet. Res.*, 50, 2117, 1989.
30. McDonough, J. P., Stull, C. L., and Osburn, B. I., Enteric pathogens in intensively reared veal calves, *Am. J. Vet. Res.*, 55, 1516, 1994.
31. Fayer, R., Guidry, A., and Blagburn, B. L., Immunotherapeutic efficacy of bovine colostral immunoglobulin from a hyperimmunised cow against cryptosporidiosis in neonatal mice, *Infect. Immun.*, 52, 2962, 1990.
32. Doyle, D. S., Crabb, J., and Petersen, C., Anti-*Cryptosporidium parvum* antibodies inhibit infectivity *in vitro* and *in vivo*, *Infect. Immun.*, 61, 4079, 1993.

33. Tzipori, S., Robertson, D., and Chapman, C., Remission of diarrhea due to cryptosporidiosis in an immunodeficient child treated with hyperimmune bovine colostrum, *Brit. Med. J.*, 293, 1276, 1986.
34. Ungar, B. P. L., Ward, D. J., Fayer, R., and Quinn, C. A., Cessation of *Cryptosporidium*-associated diarrhea in an acquired immunodeficiency syndrome patient after treatment with hyperimmune bovine colostrum, *Gastroenterology*, 98, 486, 1990.
35. Scott, C. A., Smith, H. V., and Gibbs, H. A., Excretion of *Cryptosporidium parvum* oocysts by a herd of beef suckler cows, *Vet. Rec.*, 134, 172, 1994.
36. Lorenzo-Lorenzo, M. J., Ares-Mazas, E., and Villacorta, I., Detection of oocysts and IgG antibodies to *Cryptosporidium parvum* in asymptomatic adult cattle, *Vet. Parasitol.*, 47, 9, 1993.
37. Xiao, L., Herd, R. P., and Rings, D. M., Diagnosis of *Cryptosporidium* on a sheep farm with neonatal diarrhea by immunofluorescence assays, *Vet. Parasitol.*, 47, 17, 1993.
38. Tyzzer, E. E., *Cryptosporidium parvum*, a coccidium found in the small intestine of the common mouse, *Arch. Protistenkd.*, 26, 394, 1912.
39. Iseki, M., Two species of *Cryptosporidium* naturally infecting house rats, *Rattus norvegicus*, *Jpn. J. Parasitol.*, 35, 521, 1986.
40. Klesius, P. H., Haynes, T. B., and Malo, L. K., Infectivity of *Cryptosporidium* sp. isolated from wild mice for calves and mice, *J. Am. Vet. Med. Assoc.*, 189, 192, 1986.
41. Sinski, E., Hlebowicz, E., and Bednarska, M., Occurrence of *Cryptosporidium parvum* infection in wild small mammals in District of Mazury Lake (Poland), *Acta Parasitol.*, 38, 59, 1993.
42. Chalmers, R. M., Sturdee, A. P., Casemore, D. P., Curry, A., Miller, A., Parker, N. D., and Richmond, T. M., *Cryptosporidium muris* in wild house mice (*Mus musculus*): first report in the U.K., *Eur. J. Protist.*, 30, 151, 1994.
43. Chalmers, R. M., Sturdee, A., Miller, A., and Bull, S., Animal Reservoirs for *Cryptosporidium*, in *Protozoan Parasites and Water* (proceedings of a conference, University of York, 1994), Betts, W. B., Casemore, D. P., Fricker, C., Smith, H., Watkins, J., Eds., Royal Society of Chemistry, London, 1995.
44. Ryan, M. J., Sundberg, J. P., Sauerschell, R. J., and Todd, K. S., *Cryptosporidium* in a wild cottontail rabbit (*Sylvilagus floridanus*), *J. Wildl. Dis.*, 22, 267, 1986.
45. Sundberg, J. P., Hill, D., and Ryan, M. J., Cryptosporidiosis in a grey squirrel, *J. Am. Vet. Med. Assoc.*, 181, 1420, 1982.
46. Davidson, W. R., Nettles, V. F., Hayes, L. E., Howerth, E. W., and Couvillion, C. E., Diseases diagnosed in grey foxes (*Urocyon cinereoargentus*) from the southeastern United States, *J. Wildl. Dis.*, 28, 28, 1992.
47. Carlson, B. L. and Neilsen, S. W., Cryptosporidiosis in a raccoon, *J. Am. Vet. Med. Assoc.*, 181, 1405, 1982.
48. Snyder, D. E., Indirect immunofluorescent detection of oocysts of *Cryptosporidium parvum* in the feces of naturally infected raccoons (*Procyon lotor*), *J. Parasitol.*, 74, 1050, 1988.
49. Martin, H. D. and Zeidner, N. S., Noncomitant cryptosporidia, coronavirus and parvovirus infection in a raccoon (*Procyon lotor*), *J. Wildl. Dis.*, 28, 113, 1992.
50. Tzipori, S., Angus, K. W., Campbell, I., and Sherwood, D., Diarrhea in young red deer associated with infection with *Cryptosporidium*, *J. Infect. Dis.*, 144, 170, 1981.
51. Simpson, V. R., Cryptosporidiosis in newborn red deer (*Cervus elaphus*), *Vet. Rec.*, 130, 116, 1992.
52. Heuschele, W. P., Oosterhuis, J., Janseen, D., Robinson, P. T., Ensley, P. K., Meier, E., Olson, T., Anderson, M. P., and Benirschke, K., Cryptosporidial infections in captive wild animals, *J. Wildl. Dis.*, 22, 493, 1986.
53. Iseki, M., *Cryptosporidium felis* sp. n. (Protozoa: Eimeriorina) from the domestic cat, *Jpn. J. Parasitol.*, 28, 285, 1979.
54. Mtambo, M. M. A., Nash, A. S., Blewett, D. A., Smith, H. V., and Wright, S., *Cryptosporidium* infection in cats: prevalence of infection in domestic and feral cats in the Glasgow area, *Vet. Rec.*, 129, 502, 1991.
55. Nash, A. S., Mtambo, M. M. A., and Gibbs, H. A., *Cryptosporidium* infection in farm cats in the Glasgow area, *Vet. Rec.*, 133, 576, 1993.
56. Uga, S., Matsumura, T., Ishibashi, K., Yoda, Y., Yatomi, and K., Katoaoka, N., Cryptosporidiosis in dogs and cats in Hyogo prefecture, Japan, *Jpn. J. Parasitol.*, 38, 139, 1989.
57. Bennett, M., Baxby, D., Blundell, N., Gaskell, C. J., Hart, C. A., and Kelly, D. F., Cryptosporidiosis in the domestic cat, *Vet. Rec.*, 116, 73, 1985.
58. Poonacha, K. B. and Pippin, C., Intestinal cryptosporidiosis in a cat, *Vet. Pathol.*, 19, 708, 1982.
59. Monticello, T. M., Levy, M. G., Bunch, S. E., and Fairly, R. A., Cryptosporidiosis in a feline leukaemia virus-positive cat, *J. Am. Vet. Med. Assoc.*, 191, 705, 1987.
60. Panciera, R. J., Thomassen, R. W., and Garner, F. M., Cryptosporidial infection in a calf, *Vet. Pathol.*, 8, 479, 1971.
61. Anderson, B. C., Patterns of shedding of cryptosporidial oocysts in Idaho calves, *J. Am. Vet. Med. Assoc.*, 178, 982, 1981.
62. Henriksen, S. A., Epidemiology of cryptosporidiosis in calves, cryptosporisiosis, in *Proceedings of the First International Workshop*, Angus, K. W. and Blewett, B. A., Eds., Moredun Research Institute, Edinburgh, 1989, 79.
63. Blewett, D. A., Quantitative techniques in *Cryptosporidium* research, cryptosporidiosis, in *Proceedings of the First International Workshop*, Angus, K. W. and Blewett, D. A., Eds., Moredun Research Institute, Edinburgh, 1989, 85.
64. Myers, L. L., Firehammer, B. D., Border, M. M., and Schoop, D. S., Prevalence of enteric pathogens in the feces of healthy beef calves, *Am. J. Vet, Res.*, 45, 1544, 1984.
65. Anderson, B. C. and Hall, R. F., Cryptosporidial infection in Idaho dairy calves, *J. Am. Vet. Med. Assoc.*, 181, 484, 1982.

66. Leek, R. G. and Fayer, R., Prevalence of *Cryptosporidium* infections, and their relation to diarrhea in calves on 12 dairy farms in Maryland, *Proc. Helminthol. Soc. Wash.,* 51, 360, 1984.
67. Garber, L. P., Salman, M., Hurd, H. S., Keefe, T., and Schlater, J. L., Potential risk factors for *Cryptosporidium* infections in dairy calves, *J. Am. Vet. Med. Assoc.*, 205, 86, 1994.
68. Townsend, K. G. and Lance, D. M., Cryptosporidiosis in calves, *N. Z. Vet. J.*, 35, 216, 1987.
69. Kaminjolo, J. S., Adesiyun, A. A., Loregnard, R., and Kitson-Pigott, W., Prevalence of *Cryptosporidium* oocysts in livestock in Trinidad and Tobago, *Vet. Parasitol.*, 45, 209, 1993.
70. Nouri, M. and Toroghi, R., Asymptomatic cryptosporidiosis in cattle and humans in Iran, *Vet. Rec.*, 128, 358, 1991.
71. Kemp, J. S., Wright, S. E., and Bukhari, Z., On farm detection of *Cryptosporidium parvum* in cattle, calves and environmental samples, in *Protozoan Parasites and Water* (proceedings of a conference, University of York, 1994), Betts, W. B., Casemore, D., Fricker, C., Smith, H., and Watkins, J., Eds., Royal Society of Chemistry, London, 1995.
72. Tzipori, S., and Campbell, I., Prevalence of *Cryptosporidium* antibodies in 10 animal species, *J. Clin. Microbiol.*, 14, 455, 1981.
73. Ortega-Mora, L. M., Troncoso, J. M., Rojo-Vasquez, F. A., and Gomez-Bautista, M., Cross-reactivity of polyclonal serum antibodies generated against *Cryptosporidium parvum* oocysts, *Infect. Immun.*, 60, 3442, 1992.
74. Dubey, J. P., Fayer, R., and Rao, J. R., Cryptosporidial oocysts in feces of water buffalo and Zebu calves in India, *J. Vet. Parasitol.*, 6, 55, 1992.
75. Canestri-Trotti, G. and Quesada, A., Primo reperto de *Cryptosporidium* sp. in bufali italini (*Bubalus bubalis*), *Atti. Soc. Ital. Buitaria.*, 16, 443, 1984.
76. Iskander, A. R., Tawfeek, A., and Farid, A. F., Cryptosporidial infection among buffalo calves in Egypt, *Ind. J. Anim. Sci.*, 57, 1057, 1987.
77. Tzipori, S., Angus, K. W., Campbell, I., and Clerihew, L. W., Diarrhea due to *Cryptosporidium* infection in artificially reared lambs, *J. Clin. Microbiol.*, 14, 100, 1981.
78. Angus, K. W., Appleyard, W. T., Menzies, J. D., Campbell, I., and Sherwood, D., An outbreak of diarrhea associated with cryptosporidiosis in naturally reared lambs, *Vet. Rec.*, 110, 129, 1982.
79. Anderson, B. C., Cryptosporidiosis in Idaho lambs: natural and experimental infections, *J. Am. Vet. Med. Assoc.*, 181, 151, 1982.
80. Valenzuela, G., Sci, M. V., Grandon, W., Quintana, M. V. I., and Tadich, N., Prevalencia de *Cryptosporidium* spp. en corderos muertos en la provincia de Valdivia, Chile, *Arch. Med. Vet.*, 23, 81, 1991.
81. Matos-Fernandez, M. J., Ortega-Mora, L. M., Periera Bueno, J., Gonzalez-Paniello, R. M., Reguera de Castro, E. N., Reyero Fernandez, F., Alvarez-Pacios, C., and Rojo-Vazquez, F. A., Epidemiologia de la Cryptosporidiosis en el ganado ovino y caprino de la montana de Leon, *Med. Vet.,* 11, 147, 1994.
82. Xiao, L., Herd, R. P., and McClure, K. E., Periparturient rise in the excretion of *Giardia* sp. cysts and *Cryptosporidium parvum* oocysts as a source of infection for lambs, *J. Parasitol*, 80, 55, 1994.
83. Chermette, R., Polack, B., Boufassa, S., Bariaud, F., Tarnau, C., Couderc, O., and Creuzon, F., Cryptosporidies chez les animaux adultes en France: role epidemiologique, *Commun. Congr. Natl. Soc. France Parasitol.*, Rennes, Frances, 1984.
84. Papadopoulou, C., Xylouri, E., Mantizios, A., Spyropoulos, and G., Stoforos, S., Cryptosporidiosis in farm animals in Greece, in *Proceedings of the First International Workshop,* Angus, K. W. and Blewett, D. A., Eds., Animal Diseases Research Association, Edinburgh, 1989, 123.
85. Nouri, M. and Karami, M., Asymptomatic cryptosporidiosis in nomadic shepherds and their sheep, *J. Infect.*, 23, 331, 1991.
86. Nagy, B., Bozso, M., Palfi, V., Nagy, Gy., and Sahiby, M. A., Studies on cryptosporidial infections of goat kids, Les Maladies de la Chevre, Oct. 9–11, 1984, *Les Colloques de l'INRA,* Niort, France, 1984, 443.
87. Thamsborg, S. M., Cryptosporidiosis in kids of dairy goats, *Vet. Rec.*, 127, 627, 1990.
88. Tzipori, S., McCartney, E., Lawson, G. H. K., and Rowland, A. C., Experimental infection of piglets with *Cryptosporidium, Res. Vet. Sci.*, 31, 358, 1981.
89. Moon, H. W. and Bemrick, W. J., Fecal transmission of calf cryptosporidia between calves and pigs, *Vet. Pathol.*, 18, 248, 1981.
90. Sanford, S. E., Enteric cryptosporidial infection in pigs: 184 cases (1981–1985) *J. Am. Vet. Med. Assoc.*, 190, 695, 1987.
91. Tacal, J. V., Sobieh, M., and El-Ahraf, A., *Cryptosporidium* in market pigs in southern California, U.S.A., *Vet. Rec.*, 120, 615, 1987.
92. Xiao, L., Herd, R. P., and Bowman, G. L., Prevalence of *Cryptosporidium* and *Giardia* infections on two Ohio pig farms with different management systems, *Vet. Parasitol.*, 52, 331, 1994.
93. Snyder, S. P., England, J. J., and McChesney, A. E., Cryptosporidiosis in immunodeficient Arabian foals, *Vet. Pathol.*, 15, 12, 1978.
94. Legronne, D., Regnier, G., Veau, P., Chermette, R., Boufassa, S., Soule, C., Cryptosporidiose chez des poulaines diarrheiques, *Point Vet.*, 17, 528, 1985.
95. Chermette, R., Boufassa, S., Soule, C., Tarnau, C., Couderc, O., and Legronne, D., La cryptosporidiose equine: une parasitose meconnue, in *CEREOPA Etudes et Realisations Pedagogigues sur le Cheval*, 13eme Journee d'etude Paris, *CEREOPA 81*, 1987.
96. Xiao, L. and Herd, R. P., Review of equine *Cryptosporidium* infection, *Equine Vet. J.*, 26, 9, 1994.
97. Xiao, L. and Herd, R. P., Epidemiology of equine *Cryptosporidium* and *Giardia* infections, *Equine Vet. J.,* 26, 9, 1994.

98. Fukushima, K. and Helman, R. G., Cryptosporidiosis in a pup with distemper, *Vet. Pathol.,* 21, 247, 1984.
99. Turnwald, G. H., Barta, O., Taylor, H. W., Kreeger, J., Coleman, S. U., and Pourciau, S. S., Cryptosporidiosis associated with immunosuppression attributable to distemper in a pup, *J. Am. Vet. Med. Assoc.*, 192, 79, 1988.
100. Sisk, D. B., Gosser, H. S., and Styer, E. L., Intestinal cryptosporidiosis in two pups, *J. Am. Vet. Med. Assoc.*, 184, 835, 1984.
101. Pojhola, S., Survey of cryptosporidiosis in feces of normal healthy dogs, *Nord. Vet. Med.*, 36, 189, 1984.
102. Augustin-Bichl. G., Boch, J., and Henkel, G., Kryptosporidien — Infektionen bei Hund und Katze, *Berl. Muench. Tieraerztl. Wochenschr.*, 97, 179, 1984.
103. Simpson, J. W., Burnie, A. G., Miles, R. S., Scott, J. L., and Lindsay, D. I., Prevalence of *Giardia* and *Cryptosporidium* infection in dogs in Edinburgh, *Vet. Rec.*, 123, 445, 1988.
104. El-Ahraf, A., Tacal, J. V., Sobih, M., Amin, M., Lawrence, W., and Wilcke, B. W., Prevalence of cryptosporidiosis in dogs and human beings in San Bernardino County, California, *J. Am. Vet. Med. Assoc.*, 198, 631, 1991.
105. Grimason, A. M., Smith, H. V., Parker, J. F. W., Jackson, M. H., Smith, P. G., and Girdwood, R. W. A., Occurrence of *Giardia* sp. cysts and *Cryptosporidium* sp. oocysts in feces from public parks in the west of Scotland, *Epidemiol. Infect.*, 110, 641, 1993.
106. Johnston, J., and Gasser, R. B., Copro-parasitological survey of dogs in southern Victoria, *Austral. Vet. Pract.*, 23, 127, 1993.
107. Blewett, D. A., Wright, S. E., Casemore, D. P., Booth, N. E., and Jones, C. E., Infective dose size studies on *Cryptosporidium parvum* using gnotobiotic lambs, *Water Sci. Technol.,* 27, 61, 1993.
108. Casemore, D. P., Sands, R. A., and Curry, A., *Cryptosporidium* species: a "new" human pathogen, *J. Clin. Pathol.*, 38, 1321, 1985.
109. Casemore, D. P., Laboratory methods for diagnosing cryptosporidiosis, *J. Clin. Pathol.*, 44, 445, 1991.
110. Casemore, D. P., Roberts, C., Guidelines for screening for *Cryptosporidium* in stools: report of a joint working group, *J. Clin. Pathol.*, 46, 2, 1993.
111. Breck, K., and Semple, C., The role of a clinical laboratory in an outbreak of cryptosporidiosis associated with a municipal swimming pool, *Can. J. Med. Technol.*, 55, 170, 1993.
112. Casemore, D. P., Problems associated with sampling and examination for *Cryptosporidium*, in *Proceedings of Workshop on Cryptosporidium in Water Supplies,* Dawson, A. and Lloyd, A., Eds., Her Majesty's Stationery Office, London, 1990, 11.
113. Mac Kenzie, W. R., Addiss, D. G., Davis, J. P., *Cryptosporidium* and the public water supply (letter), *N. Engl. J. Med.*, 331, 161, 1994.
114. Alles, A. J., Waldron, M. A., Sierra, L. S., and Mattia, A. R., Prospective comparison of direct immunofluorescence and conventional staining methods for detection of *Giardia* and *Cryptosporidium* spp. in human fecal specimens, *J. Clin. Microbiol.*, 33, 1632, 1995.
115. Roberts, C. L., Morin, C., Mshar, P., Addiss, D., Cartter, M., and Hadler, J., Do laboratories in Connecticut screen for *Cryptosporidium*?, in *Proceedings of 44th Annual Conference Epidemic Intelligence Service,* Centers for Disease Control and Prevention, Atlanta, GA, 1995.
116. Casemore, D. P., A pseudo-outbreak of cryptosporidiosis, Public Health Laboratory Service, *Communicable Dis. Rep.,* Outbreak Forum — IV, 2, R66, 1992.
117. Koch, K. L., Phillips, D. J., Aber, R. C., and Current, W. L., Cryptosporidiosis in hospital personnel, evidence for person-to-person transmission, *Ann. Intern. Med.*, 102, 593, 1985.
118. Ungar, B. L. P., Soave, R., Fayer, R., and Nash, T. E., Enzyme immunoassay detection of immunoglobulin M and G antibodies to *Cryptosporidium* in immunocompetent and immunocompromised persons, *J. Infect. Dis.*, 153, 570, 1986.
119. Casemore, D. P., The antibody response to *Cryptosporidium*: development of a serological test and its use in a study of immunologically normal persons, *J. Infect.*, 14, 125, 1987.
120. Ungar, B. L. P., Gilman, R. H., Lanata, C. F., and Perez-Schael, I., Seroepidemiology of *Cryptosporidium* infection in two Latin American populations, *J. Infect. Dis.*, 157, 551, 1988.
121. Ungar, L. P., Mulligan, M., and Nutman, T. B., Serologic evidence of *Cryptosporidium* infection in U.S. volunteers before and during peace corps service in Africa, *Arch. Intern. Med.*, 149, 894, 1989.
122. Garcia-Rodriguez, J. A., Sanchez, A. M. M., Canut, A., and Luis, G., The seroepidemiology of *Cryptosporidium* species in different population groups in Spain, *Serodiagn. Immunother. J. Infect. Dis.*, 3, 367, 1989.
123. Janoff, E. N., Mead, P. S., Mead, J. R., Echeverria, P., Bodhidatta, L., Bhaibulaya, M., Sterling, C. R., and Taylor, D. C., Endemic *Cryptosporidium* and *Giardia lamblia* infections in a Thai orphanage, *Am. J. Trop. Med. Hyg*., 43, 248, 1990.
124. Laxer, M. A., Alcantara, A. K., Javato-Laxer, M., Menorca, D. M., Fernando, M. T., and Ranoa, C. P., Immune response to cryptosporidiosis in Philippine children, *Am. J. Trop. Med. Hyg.*, 42, 131, 1990.
125. Lengerich E. J., Addiss, D. G., Marx, J. J., Ungar, B. L. P., and Juranek, D. D., Increased exposure to cryptosporidia among dairy farmers in Wisconsin, *J. Infect. Dis.*, 167, 1252, 1993.
126. Newman, R. D., Shu-Xian, Z., Wuhib, T., Lima, A. A. M., Guerrant, R. L., and Sears, C. L., Household epidemiology of *Cryptosporidium parvum* infection in an urban community in Northeast Brazil, *Ann. Intern. Med.*, 120, 500, 1994.
127. Kuhls, T. L., Mosier, D. A., Crawford, D. L., and Griffis, J., Seroprevalence of cryptosporidial antibodies during infancy, childhood, and adolescence, *Clin. Infect. Dis.*, 18, 731, 1994.

128. Zu, S-X., Jin-Fen, L., Barrett, L. J., Fayer, R., Shu-Yu, S., McAuliffe, J. F., Roche, J. K., and Guerrant, R. L., Seroepidemiologic study of *Cryptosporidium* infection in children from rural communities of Anhui, China, and Fortaleza, Brazil, *Am. J. Trop. Med. Hyg.*, 51, 1, 1994.
129. Groves, V. J., Lehmann. D., and Gilbert, G. L., Seroepidemiology of cryptosporidiosis in children in Papua New Guinea and Australia, *Epidemiol. Infect.*, 113, 491, 1994.
130. Moss, D. M., Bennett, S. N., Arrowood, M. J., Hurd, M. R., Lammie, P. J., Wahlquist, S. P., and Addiss, D. G., Kinetic and isotopic analysis of specific immunoglobulins from crew members with cryptosporidiosis on a U.S. Coast Guard cutter, *J. Euk. Microbiol.*, 41, 52S, 1994.
131. McAnulty, J., Keene, W., and Fleming, D. W., A water system in one town causing an outbreak of cryptosporidiosis in another town, *33rd Intersci. Conf. Antimicrob. Agents Chemother.*, 33, 387, 1993.
132. Millard, P. S., Gensheimer, K. F., Addiss, D. G., Sosin, D. M., Beckett, G. A., Houck-Jankoski, A., and Hudson, A., An outbreak of cryptosporidiosis from fresh-pressed apple cider, *J. Am. Med. Assoc.*, 272, 1592, 1994.
133. Fine, P. E. M., Herd immunity history, theory, practice, *Epidemiol. Rev.*, 15, 265, 1993.
134. Steele, A. D., Gove, E., and Meewes, P. J., Cryptosporidiosis in white patients in South Africa, *J. Infect.*, 19, 281, 1989.
135. Palmer, S. R. and Biffin, A. H., Cryptosporidiosis in England and Wales: prevalence and clinical and epidemiological features, *Brit. Med. J.*, 300, 774, 1990.
136. Molbak, K., Hojlyng, N., Inghold, L., Jose Da Silva, A. P., Jepsen, S., and Aaby, P., An epidemic outbreak of cryptosporidiosis: a prospective community study from Guinea Bissau, *Pediatr. Infect. Dis. J.*, 9, 566, 1990.
137. Ghosh, A. R., Nair, G. B., Dutta, P., Pal, S. C., and Sen, D., Acute diarrheal diseases in infants aged below six months in hospital in Calcutta, India: an aetiological study, *Trans. Roy. Soc. Trop. Med. Hyg.*, 85, 796, 1991.
138. Lima, A. A. M., Gang, G., Schorling, J. B., de Albuquerque, L., McAuliffe, J. F., Mota, S., Leite, R., and Guerrant, R. L., Persistent diarrhea in Northeast Brazil: etiologies and interactions with malnutrition, *Acta Pediatr. Suppl.*, 381, 39, 1992.
139. Arthur, J. D., Bodhidatta, L., Echeverria, P., Phuphaisan, S., and Paul, S., Diarrheal disease in Cambodian children at a camp in Thailand, *Am. J. Epidemiol.*, 135, 541, 1992.
140. Gunzburg, S., Gracey, M., Burke, V., and Chang, B., Epidemiology and microbiology of diarrhea in young Aboriginal children in the Kimberley region of Western Australia, *Epidemiol. Infect.*, 108, 67, 1992.
141. Chacin-Bonilla, L., Mejia De Young, M., Cano, G., Guanipa, N., Estevez, J., and Bonilla, E., Cryptosporidium infections in a suburban community in Maracaibo, Venezuela, *Am. J. Trop. Med. Hyg.*, 49, 63, 1993.
142. Casemore, D. P., Cryptosporidium — detection and control, *J. Sterile Serv. Manag.*, 3, 14, 1992.
143. Casemore, D. P., Human cryptosporidiosis, in *Recent Advances in Infection*, Reeves, D. S. and Geddes, A. M., Eds., Churchill Livingstone, New York, 1989.
144. Bannister, P. and Mountford, R. A., *Cryptosporidium* in the elderly: a cause of life threatening diarrhea, *Am. J. Med.*, 86, 507, 1989.
145. Pohjola, S., Jokipii, A. M. M., Jokipii, L., Sporadic cryptosporidiosis in a rural population is asymptomatic and associated with contact to cattle, *Acta Vet. Scand.*, 27, 91, 1986.
146. Vuorio, A. F., Jokipii, A. M. M., and Jokipii, L., *Cryptosporidium* in asymptomatic children, *Rev. Infect. Dis.*, 13, 261, 1991.
147. Molbak, K., Aaby, P., Hojlyng, N., and Jose de Silva, A. P., Risk factors for *Cryptosporidium* diarrhea in early childhood: a case-control study from Guinea-Bissau, West Africa, *Am. J. Epidemiol.*, 139, 734, 1994.
148. Miller, K., Duran-Pinales, C., Cruz-Lopez, A., Morales-Lechuga, L., Taren, D., and Enriquez, F. J., *Cryptosporidium parvum* in children with diarrhea in Mexico, *Am. J. Trop. Med. Hyg.*, 51, 322, 1994.
149. Das, P., Sengupta, P. D., Bhattacharya, M. K., Pal, S. C., and Bhattacharya, S. K., Significance of *Cryptosporidium* as an aetiologic agent of acute diarrhea in Calcutta: a hospital based study, *Am. J. Trop. Med. Hyg.*, 96, 124, 1993.
150. Mahajan, M., Mathur, M., Talwar, C., Cryptosporidiosis in east Delhi children, *J. Commun. Dis.*, 24, 133, 1992.
151. Okafor, J. I., Okunji, P. O., Cryptosporidiosis in patients with diarrhea in five hospitals in Nigeria, *J. Commun. Dis.*, 26, 75, 1994.
152. Marshall, R., Lothian cases of cryptosporidiosis 1989: a surveillance scheme, *Communic. Dis. Environ. Health Scot. Wkly. Rep.*, 25, 5, 1991.
153. Diers, J. and McCallister, G. L., Occurrence of *Cryptosporidium* in home daycare centers in west-central Colorado, *J. Parasitol.*, 75, 637, 1989.
154. Lazar, L. and Radulescu, S., Cryptosporidiosis in children and adults: parasitological and clinico-epidemiological features, *Arch. Roum. Path. Exp. Microbiol.*, 48, 357, 1989.
155. Garcia-Rodriguez, J. A., Martin-Sanchez, A. M., Canut Blasco, A. C., and Garcia Luis, E. J., The prevalence of *Cryptosporidium* species in children in day care centers and primary schools in Salamanca (Spain): an epidemiological study, *Eur. J. Epidemiol.*, 6, 432, 1990.
156. Addiss, D. G., Stewart, J. M., Finton, R. J., Wahlquist, S. P., Williams, R. M., Dickerson, J. W., Spencer, H. C., and Juranek, D. D., *Giardia lamblia* and *Cryptosporidium* infections in child day-care centers in Fulton County, Georgia, *Pediatr. Infect. Dis. J.*, 10, 907, 1991.
157. Tangermann, R. H., Gordon, S., Wiesner, P., and Kreckman, L., An outbreak of cryptosporidiosis in a day-care center in Georgia, *Am. J. Epidemiol.*, 133, 471, 1991.
158. Mäusezahl, D., Egger, M., Odermatt, P., Tanner, M., Klinik und epidemiologie der kryptosporidiose bei immunkompetenten kindern, *Schweiz. Rundschau. Med. (PRAXIS)*, 80, 936, 1991.

159. Cordell, R. L., Addiss, D. G., Cryptosporidiosis in child care settings: a review of the literature and recommendations for prevention and control, in *Pediatr. Infect. Dis. J.*, 13, 310, 1994.
160. Dawson, A., Griffin, R., Fleetwood, A., Barrett, N. J., Farm visits and zoonoses, Public Health Laboratory Service, *Communic. Dis. Rep. (Rev.),* 5, R81, 1995.
161. Hojlyng, N., Molbak, K., and Jepsen, S., *Cryptosporidium* spp., a frequent cause of diarrhea in Liberian children, *J. Clin. Microbiol.*, 23, 1109, 1986.
162. Malla, N., Sehgal, R., Ganguly, N. K., and Mahajan, R. C., Cryptosporidiosis in children in Chandrigarh, *Ind. J. Med. Res.*, 86, 722, 1987.
163. Soave, R., Ruiz, J., Garcia-Saucedo, V., Garrocho, C., and Kean, B. H., Cryptosporidiosis in a rural community of central Mexico, *J. Infect. Dis.*, 159, 1160, 1989.
164. Subramanyam, V. R., Broadhead, R. L., Pal, B. B., Pati, J. B., and Mohanty, G., Cryptosporidiosis in children of eastern India, *Ann. Trop. Pediatr.*, 9, 122, 1989.
165. Sarabia-Arce, S., Salazar-Lindo, E., Gilman, R. H., Naranjo, J., and Miranda, E., Case-control study of *Cryptosporidium parvum* infection in Peruvian children hospitalized for diarrhea: possible association with malnutrition and nosocomial infection, *Pediatr. Infect. Dis. J.*, 9, 627, 1990.
166. Rahman, M., Shahid, N. S., Rahman, H., Sack, D. A., Rahman, N., and Hossain S., Cryptosporidiosis: a cause of diarrhea in Bangladesh, *Am. J. Trop. Med. Hyg.*, 42, 127, 1990.
167. Perera, J., and Lucas, G. N., Cryptosporidiosis — oocyst shedding and infection in household contacts, *Ceylon Med. J.*, 35, 11, 1990.
168. Vidal, T. P., Gamboa, C, C., Henriquez, I. M. B., and Biolley, A. M., Cryptosporidiosis in 22 infants in a nutritional center of Temuco, Chile, *Rev. Med. Chile,* 119, 1136, 1991.
169. Sallon, S., el Showwa, R., el Masri, M., Khalil, M., Blundell, N., and Hart, C. A., Cryptosporidiosis in children in Gaza, *Ann. Trop. Pediatr.*, 11, 277, 1991.
170. Zu, S-X., Li, J-F., Barrett, L. J., Fayer, R., Shu, S-Y., McAuliffe, J. F., Roche, J. K., and Guerrant, R. L., Seroepidemiologic study of *Cryptosporidium* infection in children from rural communities of Anhui, China and Fort Aleza, Brazil, *Am. J. Trop. Med. Hyg.,* 51, 1, 1994.
171. You-Gui, C., Fu-Bao, Y., Hai-Si, L., Wen-Sheng, S., Mei-Xin, D., and Ming, L., *Cryptosporidium* infection and diarrhea in rural and urban areas of Jiangsu, People's Republic of China, *J. Clin. Microbiol.*, 30, 492, 1992.
172. Zu S-X., Shu-Yu, Z., and Jin-Fen, L., Human cryptosporidiosis in China, *Trans. Roy. Soc. Trop. Med. Hyg.*, 86, 639, 1992.
173. Jirapinyo, P., Ruangsiri, K., Tesjaroen, S., Limsathayourat, N., Sripiangjan, J., Yoolek, A., and Junnoo, V., High prevalence of *Cryptosporidium* in young children with prolonged diarrhea, *S. E. Asian J. Trop. Med. Public Health,* 24, 730, 1993.
174. Nimri, L. F. and Batchoun, R., Prevalence of *Cryptosporidium* species in elementary school children, *J. Clin. Microbiol.*, 32, 1040, 1994.
175. Zu S-X, Z., G-D, F., Fayer, R., and Guerrant, R. L., Cryptosporidiosis: pathogenesis and immunology, *Parasitol. Today,* 8, 24, 1992.
176. Miron, D., Kenes, J., and Dagan, R., Calves as a source of an outbreak of cryptosporidiosis among young children in an agricultural closed community, *Pediatr. Infect. Dis. J.*, 10, 438, 1991.
177. Leeuwen, P. V., Lawrence, A., and Hansman, D., An outbreak of cryptosporidial infection among children in Adelaide, *Med. J. Austral.*, 154, 708, 1991.
178. Wuhib, T. and Silva, T. et al., *Cryptosporidium* infection in HIV positive patients in north eastern Brazil, *Am. J. Trop. Med. Hyg.*, 47, 166, 1992.
179. Richardson, A. J., Frankenberg, R. A., Buck, A. C., Selkon, J. B., Colbourne, J. S., Parsons, J. W., and Mayone-White, R. T., An outbreak of waterborne cryptosporidiosis in Swindon and Oxfordshire, *Epidemiol. Infect.*, 107, 485, 1991.
180. Joseph, C., Hamilton, G., O'Connor, M., Nicholas, S., Marshall, R., Stanwell-Smith, R., Sims, R., Ndawula, E., Casemore, D. P., Gallagher, P., and Harnett, P., Cryptosporidiosis in the Isle of Thanet, an outbreak associated with local drinking water, *Epidemiol. Infect.*, 107, 509, 1991.
181. Weinstein, P., Macaitis, M., Walker, C., and Cameron, S., Cryptosporidial diarrhea in South Australia: an exploratory case-control study of risk factors for transmission, *Med. J. Austral.*, 158, 117, 1993.
182. Anon., A large outbreak of cryptosporidiosis in Jackson County, Office of Epidemiology and Health Statistics, *CD Summary,* 41, 1992.
183. Morgan, D., Allaby, M., Crook, S., Casemore, D. P., Healing, T. D., Soltanpoor, N., Hill, S., and Hooper, W., Waterborne cryptosporidiosis associated with a borehole supply, Public Health Laboratory Service, *Communic. Dis. Rep. (Rev.),* 5(7), R93, 1995.
184. Atherton, F., Newman, C. P. S., and Casemore, D. P., An outbreak of waterborne cryptosporidiosis associated with a public water supply in the U.K., *Epidemiol. Infect.*, 123–131, 1995.
185. Shield, J., Baumer, J. H., Dawson, J. A., and Wilkinson, P. J., Cryptosporidiosis — an educational experience, *J. Infect.*, 21, 297, 1990.
186. Coop, R. L., Wright, S. E., and Casemore, D. P., Cryptosporidiosis, in *Textbook on Zoonoses Control*, Palmer, S. R., Soulsby, L., and Simpson, D., Eds., Oxford University Press, in press.

187. Hegmann, K. B., Brummit, C. F., Hegmann, K. T., and Anderson, A. J., The impact of a *Cryptosporidium* epidemic on a hospital, *Am. J. Epidemiol. 27th Ann. Mtg. Soc. Epidemiol. Res.,* 139, 539, 1994.
188. Gradus, M. S., Singh, A., and Sedmak, G. V., The Milwaukee *Cryptosporidium* outbreak: its impact on drinking water standards, laboratory diagnosis and public health surveillance, *Clin. Microbiol. Newslett.*, 16, 57, 1994.
189. Dawson, A. and Lloyd, A., Eds., Proceedings of Workshop on *Cryptosporidium* in Water Supplies, Department of the Environment/Welsh Office/Department of Health, Her Majesty's Stationery Office, London, 1994.
190. Anon., *Cryptosporidium* in Water Supplies, Second Report of the Group of Experts (Chairman, Sir John Badenoch), Department of the Environment/Department of Health, Her Majesty's Stationery Office, London, 1995.
191. Goldstein, S., Hightower, A., Ravenholdt, O., Reich, R., Martin, D., Mesnik, J., Griffiths, S., Bryant, A., and Juranek, D., An outbreak of cryptosporidiosis associated with municipal drinking water, in *Proceedings of the 44th Annual Epidemic Intelligence Service Conference*, Centers for Disease Control, Atlanta, GA, 1995, 15.
192. Klaucke, D. N., Buehler, J. W., Thacker, S. B., Parrish, R. G., Trowbridge, F. L., Berkelman, R. L., and the Surveillance Coordination Group, Guidelines for evaluating surveillance systems, *Morbid. Mortal. Wkly. Rep.*, 37 (Suppl. 5), 1, 1988.
193. Nazareth, B., Stanwell-Smith, R. E., Rowland, M. G. M., and O'Mahony, M. C., Surveillance of waterborne disease in England and Wales, Public Health Laboratory Service, *Comm. Dis. Rep. (Rev.),* 4(8), R93, 1994.
194. Mac Kenzie, W. R., Schell, W. L., Blair, K. A., Addiss, D. G., Peterson, D. E., Hoxie, N. J., Kazmierczak, J. J., and Davis, J. P., Massive waterborne outbreak of *Cryptosporidium* infection Milwaukee, Wisconsin: Recurrence of illness and risk of secondary transmission, *Clin. Inf. Dis.,* 21, 57, 1995.
195. Osewe, P., Addiss, D., Blair, K., Hightower, A., Kamb, M., and Davis, J. P., Cryptosporidiosis in Wisconsin: an exploratory case-control study of risk factors for post-outbreak transmission, in press.
196. Addiss, D. G., Pond, R. S., Remshak, M., Juranek, D. D., Stokes, S., and Davis, J. P., Reduction of risk of watery diarrhea with point-of-use water filters during a massive outbreak of waterborne *Cryptosporidium* infection in Milwaukee, 1993, *J. Appl. Environ. Microbiol.*, in press.
197. Cicirello, H. G., Kehl, K. S., Addiss, D. G., Chusid, M. J., Glass, R. I., Davis, J. P., and Havens, P. L., Cryptosporidiosis in children during a massive waterborne outbreak, Milwaukee, Wisconsin: clinical, laboratory, and epidemiology findings, in press.
198. Gallaher, M. M., Herndon, J. L., Nims, L. J., Sterling, C. R., Grabowski, D. J., and Hull, H. F., Cryptosporidiosis and surface water, *Am. J. Public. Health*, 79, 39, 1989.
199. Kramer, M. H., Sorhage, F., Dalley, E., Wahlquist, S., Goldstein, S., and Herwaldt, B., Outbreak of cryptosporidiosis associated with recreational exposure to lake water, New Jersey, in *Proceedings of the 44th Annual Epidemic Intelligence Service Conference*, Centers for Disease Control, Atlanta, GA, 1995, 16.
200. Sorvillo, F. J., Fujioka, K., Kebabjian, R. S., Tolushige, W., Mascola, L., Schweid, S., Hillario, M., Waterman, S. H., Swimming-associated cryptosporidiosis — Los Angeles County, *Morbid. Mortal. Wkly. Rep.*, 39, 343, 1990.
201. Joce, R. E., Bruce, J., Kiely, D., Noah, N. D., Dempster, W. B., Stalker, R., Gumsley, P., Chapman, P. A., Norman, P., Watkins, J., Smith, H. V., Price, T. J., and Watts, D., An outbreak of cryptosporidiosis associated with a swimming pool, *Epidemiol. Infect.*, 107, 497, 1991.
202. Bell, A., Guasparini, R., Meeds, D., Mathias, R. G., and Farley, J. D., A swimming pool-associated outbreak of cryptosporidiosis in British Columbia, *Can. J. Public Health*, 84, 334, 1993.
203. Hunt, D. A., Sebugwawo, S., Edmondson, S. G., and Casemore, D. P., Cryptosporidiosis associated with a swimming pool complex, *Comm. Dis. Rep.*, 4(2), R20, 1994.
204. McAnulty, J. M., Fleming, D. W., and Gonzalez, A. H., A community-wide outbreak of cryptosporidiosis associated with swimming at a wave pool, *J. Am. Med. Assoc.*, 272, 1597, 1994.
205. Bongard, J., Savage, R., Dern, R., Bostrum, H., Kazmierczak, J., Keifer, S., Anderson, H., Davis, J. P., *Cryptosporidium* infections associated with swimming pools — Dane County, Wisconsin, 1993, *J. Am. Med. Assoc.*, 272, 914, 1994.
206. Freidank, H. and Kist, M., Cryptosporidia in immunocompetent patients with gastroenteritis, *Eur. J. Clin. Microbiol.*, 6, 56, 1986.
207. Elsser, K. A., Moricz, M., Proctor, E. M., *Cryptosporidium* infections: a laboratory survey, *Can. Med. Assoc. J.,* 135, 211, 1986.
208. Casemore, D. P., Foodborne illness — foodborne protozoal infection, *Lancet*, 336, 1427, 1990.
209. Smith, J. L., *Cryptosporidium* and *Giardia* as agents of foodborne disease, *J. Food Protect.*, 56, 451, 1993.
210. Fayer, R., Diseases caused by viruses, parasites, and fungi, in *Foodborne Disease Handbook*, Vol. 2, Hui, Y. H., Gorham, J. R., Murrell, K. D., and Cliver, D. O., Eds., Marcel Dekker, New York, 1994, 331.
211. Bankes, P., The detection of *Cryptosporidium* oocysts in milk and beverages, in *Protozoan Parasites and Water* (proceedings of a conference, University of York, 1994), Betts, W. B., Casemore, D., Fricker, C., Smith, H., and Watkins, J., Eds., Royal Society of Chemistry, London, 1994.
212. Palmer, S. R. and McGuirk, S. M., Bird attacks on milk bottles and campylobacter infection, *Lancet*, 345, 326, 1995.
213. Egger, M., Mausezahl, D., Odermatt, P., Marti, H. P., and Tanner, M., Symptoms and transmission of intestinal cryptosporidiosis, *Arch. Dis. Childhood*, 65, 445, 1990.
214. Bretagne, S., Liance, M., Breuil, J., Bougnoux, M. E., and Hoin, R., Prevalence de la cryptosporidiose dans differentes populations: importance de cette parasitose en milieu pediatrique, *Bull. Soc. Franc. Parasitol.*, 7, 185, 1989.

215. Faucherre, V., Jarry, D., Brunel, M., and Rioux, J. A., Enquete realisee en 1988 aupres des laboratoires de parasitologie sur l'incidence des examens lies a l'infection VIH, *Med. Mal. Infect.*, 21, 738, 1991.
216. Keusch, G. T., Thea, D. M., Kamenga, M., Kakanda, K., Mbala, M., Brown, C., and Davachi, F., Persistent diarrhea associated with AIDS, *Acta Pediatr. Suppl.*, 381, 45, 1992.
217. Petersen, C., Cryptosporidiosis in patients infected with the Human Immunodeficiency Virus, *Clin. Infect. Dis.*, 15, 903, 1992.
218. Brandonisio, O., Maggi, P., Panaro, M. A., Bramante, L. A., Di Coste, A., and Angarano, G., Prevalence of cryptosporidiosis in HIV-infected patients with diarrheal illness, *Eur. J. Epidemiol.*, 9, 190, 1993.
219. Cotte, L., Rabodonirina, M., Piens, M. A., Perreard, M., Mojon, M., and Trepo, C., Prevalence of intestinal protozoans in french patients infected with HIV, *J. Acquired Immun. Defic. Syndr.*, 6, 1024, 1993.
220. Sorvillo, F. J., Lieb, L. E., Kerndt, P. R., and Ash, L. R., Epidemiology of cryptosporidiosis among persons with acquired immunodeficiency syndrome in Los Angeles County, *Am. J. Trop. Med. Hyg.*, 51, 326, 1994.
221. Mengesha, B., Cryptosporidiosis among medical patients with the acquired immunodeficiency syndrome in Tikur Anbessa Teaching Hospital, Ethiopia, *E. Africa Med. J.*, 71, 376, 1994.
222. Sanchez-Mejorada, G. and Ponce-de-Leon, S., Clinical patterns of diarrhea in AIDS: etiology and prognosis, *Rev. Invest.Clin.*, 46, 187, 1994.
223. Tanyuksel, M., Gun, H., and Doganci, L., Prevalence of *Cryptosporidium* sp. in patients with neoplasia and diarrhea, *Scand. J. Infect. Dis.*, 27, 69, 1995.
224. Casemore, D. P., Gardner, C. A., O'Mahony, C. O., Cryptosporidial infection, with special reference to nosocomial transmission of *Cryptosporidium parvum*: a review, *Folia Parasitol.*, 41, 17, 1994.
225. Sorvillo, F., Lieb, L. E., Nahlen, B., Miller, J., Mascola, L., and Ash, L. R., Municipal drinking water and cryptosporidiosis among persons with AIDS in Los Angeles County, *Epidemiol. Infect.*, 113, 313, 1994.
226. Robertson, L. J., Campbell, A. T., and Smith, H. V., Survival of *Cryptosporidium parvum* oocysts under various environmental pressures, *Appl. Environ. Microbiol.*, 58, 3494, 1992.
227. Araujo, V., Fang, G., Guerrant, R. L., Nosocomial gastrontestinal infections, *Current Opin. Infect. Dis.*, 4, 549, 1991.
228. Dryjanski, J., Gold, J. W. M., Ritchie, M. T., Kurtz, R. C., Lim, S. M., Armstrong, D., Cryptosporidiosis: case report in a health team worker, *Am. J. Med.*, 80, 751, 1986.
229. Martino, P., Gentile, G., Caprioli, A., Baldassarri, L., Donelli, G., Arcese, W., Fenu, S., Micozzi, A., Venditti, M., Mandelli, F., Hospital acquired cryptosporidiosis in a bone marrow transplantation unit, *J. Infect. Dis.*, 158, 647, 1988.
230. Roncoroni, A. J., Gomez, M. A., Mera, J., Cagnoni, P., Michel, M. D., *Cryptosporidium* infection in renal transplant patients, *J. Infect. Dis.*, 160, 559, 1989.
231. Foot, A. B. M., Oakhill, A., Mott, M. G., Cryptosporidiosis and acute leukaemia, *Arch. Dis. Childhood*, 65, 236, 1990.
232. Navarrete, S., Stetler, H. C., Avila, C., Aranda, J. A. G., Santos-Preciado, J. I., An outbreak of *Cryptosporidium* diarrhea in a pediatric hospital, *Pediatr. Infect. Dis.*, 10, 248, 1991.
233. Ravn, P., Lundgren, J. D., Kjaeldgaard, P., Holten-Andersen, W., Hojlyng, N., Nielsen, J. O., Gaub, J., Nosocomial outbreak of cryptosporidiosis in AIDS patients, *Brit. Med. J.*, 302, 277, 1991.
234. Combee, D. L., Collinge, M. L, Britt, E. M., Cryptosporidiosis in a hospital-associated day care center, *Pediatr. Infect. Dis.*, 5, 528,1986.
235. Melo, Cristino, J. A. G., Carvalho, M. I. P., Salgado, M. J., An outbreak of cryptosporidiosis in a hospital day-care center, *Epidemiol. Infect.*, 101, 355, 1988.
236. Heald, A. E., and Bartlett, J. A., *Cryptosporidium* spread in a group residential home, *Ann. Int. Med.*, 121, 467, 1994.
237. Clifford, C. P., Crook, D. W. M., Conlon, C. P., Fraise, A. P., Day, D. G., Peto, T. E. A., Impact of waterborne outbreak of cryptosporidiosis on AIDS and renal transplant patients, *Lancet*, 335, 1455, 1990.
238. Sterling, C. R., Miranda, E., and Gilman, R. H., The potential role of flies (*Musca domestica*) in the mechanical transmission of *Giardia* and *Cryptosporidium* in a Pueblo Joven community of Lima, Peru, *Am. J. Trop. Med. Hyg.*, 349, 233, 1987.
239. Zerpa, R. and Huicho, L., Childhood cryptosporidial diarrhea associated with identification of *Cryptosporidium* sp. in the cockroach *Periplaneta americana*, *Pediatr. Infect. Dis.*, 13, 546, 1994.
240. DuPont, H. L., Chappell, C. L., Sterling, C. R., Okhuysen, P. C., Rose, J. B., and Jakubowski, W., The infectivity of *Cryptosporidium parvum* in healthy volunteers, *N. Engl. J. Med.*, 332, 855, 1995.
241. Rose, J. B., Lisle, J. T., and Haas, C. N., Risk assessment methods for *Cryptosporidium* and *Giardia* in contaminated water, in *Protozoan Parasites and Water* (proceedings of a conference, University of York, 1994), Betts, W. B., Casemore, D., Fricker, C., Smith, H., and Watkins, J., Eds., Royal Society of Chemistry, London, 1995.
242. Medema, G. J., Teunis, P. F. M., Gornik. V., Havelaar, A. H., and Exner, M., The use of water quality data to estimate the risk of infection with *Cryptosporidium* via drinking water, in *Protozoan Parasites and Water* (proceedings of a conference, University of York, 1994), Betts, W. B., Casemore, D. P., Fricker, C., Smith, H., and Watkins, J., Eds., Royal Society of Chemistry, London, 1994.
243. Mead, J. R., Humphreys, R. C., Sammons, D. W., and Sterling, C. R., Identification of isolate-specific sporozoite proteins of *Cryptosporidium parvum* by two-dimensional gel electrophoresis, *Infect. Immun.*, 58, 2071, 1990.
244. Nichols, G. L., McLauchlin, J., and Samuel, D., Workshop on pneumocystis, *Cryptosporidium* and microsporidia — a technique for typing *Cryptosporidium* isolates, *J. Protozool.*, 38, 237S, 1991.

245. Nina, J. M. S., McDonald, V., Deer, R. M. A., Wright, S. E., Dyson, D. A., Chiodini, P. L., and McAdam, K. P. W. J., Comparative study of the antigenic composition of oocyst isolates of *Cryptosporidium parvum* from different hosts, *Parasite Immunol.*, 14, 227, 1992.
246. Nina, J. M. S., McDonald, V., Dyson, D. A., Catchpole, J., Uni, S., Iseki, M., Chiodini, P. L., and McAdam, K. P. W. J., Analysis of oocysts wall and sporozoite antigens from three *Cryptosporidium* species, *Infect. Immun.*, 60, 1509, 1992.
247. Peterson, C., Cellular biology of *Cryptosporidium parvum*, *Parasitol. Today*, 9, 87, 1993.
248. Awad-el-Kariem, F. M., Robinson, H. A., Dyson, D. A., Evans, D., Wright, S., Fox, M. T., and McDonald V., Differentiation between human and animal strains of *Cryptosporidium parvum* using isoenzyme typing, *Parasitology*, 110, 129, 1995.
249. Casemore, D. P., Is human cryptosporidiosis a zoonotic disease?, *Lancet*, 342, 312, 1993.
250. Centers for Disease Control and Prevention, Assessing the public health threat associated with waterborne cryptosporidiosis: report of a workshop, *Morbid. Mortal. Wkly. Rep.*, 44(RR-6), 1–19, 1995.

Chapter 4

Waterborne Cryptosporidiosis: Incidence, Outbreaks, and Treatment Strategies

Joan B. Rose, John T. Lisle, and Mark LeChevallier

CONTENTS

I. INTRODUCTION

Cryptosporidium parvum has become the most important newly recognized contaminant in drinking water in the U.S. The exogenous oocyst stage of this protozoan parasite is much more resistant to conventional water treatment-disinfection processes than waterborne bacteria or viruses. The organism is ubiquitous and infectious for most mammals and causes diarrhea in humans. Fecal contamination of waterways has led to massive outbreaks. In Milwaukee, WI, an outbreak in 1993 was estimated to cost the community millions of dollars, and 1 year later immunocompromised patients were still affected. The U.S. Environmental Protection Agency (USEPA) has issued new rules to control this biological contaminant in drinking water, and much has been learned in the past 5 years about the potential risk of waterborne cryptosporidiosis.

II. OVERVIEW OF WATERBORNE OUTBREAKS OF CRYPTOSPORIDIOSIS

A. OUTBREAKS IN DRINKING WATER SYSTEMS

The five well-documented outbreaks of cryptosporidiosis in drinking water in the U.S. (Table 1) affected from 500 to 400,000 persons. Rivers, lakes, springs, and ground water have all been implicated as sources, and contaminated water was linked to evidence of suboptimal treatment. Suboptimal coagulation, flocculation, filtration, and/or disinfection were partially responsible for epidemic levels of cryptosporidiosis in three communities in Georgia, Wisconsin, and Oregon. Despite operational deficiencies and high turbidity spikes, treated drinking water met the existing USEPA requirements for turbidity (average <1.0 NTU) and coliforms (<1 CFU/100 ml). Disinfection appeared sufficient at the time of the outbreaks

0-8493-7695-5/97/$0.00+$.50

Table 1 *Cryptosporidium* Outbreaks in Drinking Water Supplies in the U.S.

Year	Location	Population Exposed	Population Infected	Source Water	Treatment	Suspected Cause	Ref.
1984	Braun Station, TX	5900	2006	Ground water	Chlorination	Sewage-contaminated well	1
1987	Carrollton, GA	32,400	12,960	River	Conventional[a]	Treatment deficiencies[b]	5
1991	Pennsylvania	NA	551	Ground water	Chlorination	Treatment deficiencies[b]	3
1992	Jackson Co., OR	160,000	15,000	Spring/river	Chlorination/package filtration plant	Treatment deficiencies[b]	2
1993	Milwaukee, WI	1,600,000	403,000	Lake	Conventional[a]	Treatment deficiencies[b]	4
1993	Washington state	Unknown	7	Well	—	Untreated	6,6a
1993	Minnesota	Unknown	27	Lake	Filtered, chlorinated	Unknown	6,6a
1993	Las Vegas, NV	Unknown	103	Lake	Conventional	Unknown	6,6a
1994	College Place, WA	Unknown	104	Well	—	Sewage contamination	6,6a

Note: NA = information not available.

[a] Conventional treatment includes coagulation, flocculation, filtration, and disinfection.

[b] Treatment deficiencies include suboptimal performance of one or more steps in conventional treatment.

but oocysts are extremely resistant to chlorination, and conventional disinfection cannot completely protect the water supply once it has been contaminated.

All five outbreaks were associated with unfiltered water supplies (Table 1), two from groundwater and one each from a spring, river, or lake.[1-5] All source waters were considered pristine, requiring only chlorination to satisfy regulatory criteria. Wastewater was the suspected contaminant in the Texas and Pennsylvania outbreaks; however, the sources of contamination in Georgia, Oregon, and Wisconsin remain unknown.

The Milwaukee, WI, outbreak was the largest in the U.S. and exemplifies the factors found to contribute to *Cryptosporidium* outbreaks in drinking water supplies.[4] The source water comes from Lake Michigan and is treated in a northern or southern facility.

During the week of April 5, 1993, the health department noted increased cases of gastrointestinal illness. Subsequently, *Cryptosporidium* oocysts were identified in stools from persons in south Milwaukee. A rapid and widespread increase in cryptosporidiosis in this section of the city suggested a common source and implicated the southern treatment facility. A boil water advisory was announced. An epidemiological survey examined operational records from both treatment facilities. Turbidity in finished water from the southern plant increased on March 21 from 0.25 to 1.70 NTU; the facility was temporarily closed on April 9. Turbidity in finished water in the northern treatment facility never exceeded 0.45 NTU, all drinking water samples were negative for coliforms, and all operational parameters were within regulations from February to April.

The southern facility treated water by adding chlorine and polyaluminum chloride, followed by rapid mixing, mechanical flocculation, sedimentation, and rapid sand filtration. Filters were cleaned by backwashing with treated water. Backwash was recycled through the treatment process. The coagulant, polyaluminum chloride, was replaced with alum on April 2. On April 5, the turbidity in the finished water increased to 1.5 NTU. Coagulant dosage was not adjusted because the streaming-current monitor was improperly installed and turbidimeters for monitoring the filtered water were not in use. Turbidity was measured every 8 hours from clear well grab samples.

The two rivers that discharge into Lake Michigan were swelled by unusually heavy rains and snow runoff, impacting the area where the southern facility draws its water. The watersheds of these rivers contain abattoirs, human sewage discharges, cattle grazing ranges, and environments with numerous species of wild and domesticated mammals. Neither the source nor time of entrance of oocysts into the southern treatment facility could be determined.

The Milwaukee,[4] Carrollton,[5] and Jackson County[2] outbreaks share one or more of the following treatment deficiencies:

1. Monitoring equipment (e.g., turbidimeter) to optimize filtration during periods of rapid change in source water was improperly installed, poorly maintained, turned off, or ignored. Other equipment (flocculators, chemical injectors, filters) was temporarily inoperable.
2. Treatment plant personnel did not respond to faulty or inoperable monitoring equipment by increasing the frequency or types (i.e., jar tests) of testing to compensate for deficiencies.
3. Filter backwash was returned to the head of the treatment process, possibly concentrating oocysts in the feed water. This process may be accelerated by high oocyst numbers in the source water.
4. Sources of high concentrations of oocysts were near the treatment facility intake with no mitigative barrier (e.g., retention ponds) to prevent or reduce introduction of oocysts into receiving streams or groundwater sources during periods of high surface runoff.
5. Sources and concentrations of *Cryptosporidium* were unknown in watersheds prior to the outbreak event.
6. Natural events (i.e., heavy rain, snow melts) may have flushed areas of high oocyst concentrations into receiving waters upstream of the treatment facility intakes.
7. During periods of high turbidity plus increased flushing of oocysts from land sources, filtration processes were altered or suboptimal, and increased levels of turbidity or spikes were seen in finished water.

Four other outbreaks of cryptosporidiosis were recently investigated.[6,6a] In Washington state, a small groundwater well was apparently contaminated with irrigation water. Another small outbreak was reported in Minnesota with 27 cases reported. An outbreak in Las Vegas, NV, was of great concern because no treatment deficiency could be found.[6] The treatment plant was well equipped with on-line turbidimeters and particle counters, and the quality of filtered water was consistently high (<0.5 NTU). Neither the source nor the level of contamination was determined. While there were only 103 cases, the mortality among AIDS patients exposed was 53%.

Although *Cryptosporidium* oocysts have been found in most surveyed source waters,[38,39] outbreaks or epidemics of cryptosporidiosis can be minimized if treatment plant personnel are conscientious and well trained, if equipment is properly maintained, and if potential sources of oocysts in the watershed have been identified and avoided or reduced. Furthermore, knowledge of impending natural events such as heavy rainfall can necessitate modifications in treatment and should forewarn personnel, providing time to optimize the filtration process.

B. RECREATIONAL WATERBORNE OUTBREAKS

The first outbreak associated with recreational water was in 1988,[7] when 44 persons became ill as a result of a fecal accident and inoperable filters in a swimming pool in Los Angeles. The attack rate was 73%. Four additional recreational water outbreaks (Table 2) have been associated with swimming pools.[8-11] Fecal accidents and, in one case, a cross-connection to the sewage system were responsible for the contamination. Attack rates ranged from 6 to 75%, with children identified as the primary cases in most outbreaks and children's pools identified as a risk factor.

Table 2 *Cryptosporidium* Infections Associated with Swimming Pools

Location	Outbreak Dates	Attack Rates	Notes	Ref.
Los Angeles, CA	7/13/88–8/14/88	73%	1–3 diatomaceous earth filters inoperative	7
London, U.K.	6/20/89–10/24/89	32 primary cases	30 cm of sewage covering drain in children's pools; also found sewage in main pool	8
British Columbia, Canada	11/1/90–1/91	6–75%	Fecal accident; median age with illness: 8 years of age	9
Madison, WI	8/1/93–9/11/93	Pool A: 47%, Pool B: 55%	Median age with illness: 4 years of age	10
Lane County, OR	6/92–10/92	37% primary cases	Median age with illness: 5 years of age; oocysts found in filter backwash water	11

In a pool in the U.K., oocysts were detected at 50 to 80 per liter; after cleaning, 0 to 6 oocysts per liter were detected.[8] Approximately 1 oocyst per liter was detected in backwash water from the filter of a wave pool.[11] As in drinking water outbreaks, disinfection of pools and maintenance of chlorine residuals did not inactivate the oocysts. The release of potentially large numbers of oocysts from a single fecal accident places individuals using the pool at risk even if small volumes of pool water are ingested. Although filtration is an important barrier for drinking water, there is very little information regarding the efficacy of sand or diatomaceous earth pool filters for the removal of oocysts. To remedy the contamination, some pools were closed; some were super-chlorinated (80 mg/l for 90 minutes) and swept, filters cleaned or changed; and some pools were drained.[7-11]

III. ANALYTICAL METHODS

Detection of infectious *C. parvum* oocysts in environmental samples is key to identifying a waterborne outbreak, determining source water densities, and evaluating treatment plant performance. It is vital, therefore, that specific, rapid, and highly sensitive methods be available to public health workers and environmental specialists. Existing immunofluorescence methodology does not fulfill these requirements (see Chapter 2).

A. IMMUNOFLUORESCENCE METHOD

Immunofluorescence assays for detection of environmental oocysts were essentially modifications of methods for *Giardia*.[12] Commercial availability of monoclonal antibodies for *Giardia* and *Cryptosporidium* facilitated development of these methods.[13-15] One method filtered water through a polypropylene cartridge, eluted filtrate from the filter using a mechanical shaker, concentrated the filtrate by centrifugation at 1200× *g*, removed extraneous debris by flotation over Sheather's sucrose solution (1.29 specific gravity g/ml), labeled the supernate with a fluorescein conjugated antibody against *Cryptosporidium*, and examined it by epifluorescence microscopy.[16] Others determined that oocysts could be effectively recovered using a 1.10-specific gravity Percoll-sucrose gradient.[17] High recoveries from Percoll-sucrose gradients were consistent with *Giardia* cyst recoveries by various flotation media.[18] A combined method for recovery of both *Giardia* and *Cryptosporidium*[17] was found superior to the previous standard method for *Giardia*[19] and is described in detail.[20,21] An overview is presented in Figure 1.

The immunofluorescence method can be used to examine large volumes of water, is relatively specific for *Giardia* and *Cryptosporidium* spp., and permits examination by both epifluorescence and phase-contrast or differential interference contrast microscopy. The method has utility for determining the presence and density of protozoan contamination within a watershed, provides design criteria for calculating the size and complexity of the water treatment process, and can be used to evaluate the effectiveness of cyst reduction through various stages of treatment. Disadvantages of the method include low recovery efficiencies (5 to 25%), long processing times (at best 1 to 2 days, but typically 1 to 2 weeks), the need for a highly trained analyst, high cost (approximately $300 per sample), inability to discriminate viable or virulent strains, and cross-reactivity with several species of *Cryptosporidium*. Difficulties in performing the assay were reflected in an evaluation of 16 commercial laboratories that found recoveries for *Giardia* and *Cryptosporidium* averaged 9.1 and 2.8%, respectively.[22] Four laboratories misidentified an algal cell (*Oocystis minuta*) as a *Cryptosporidium* oocyst, four others failed to recover *Giardia*, and six others failed to detect *Cryptosporidium*, although samples contained approximately 740 cysts and oocysts.

A USEPA-sponsored evaluation of ten expert laboratories yielded better results. Recoveries of *Giardia* cysts from spiked filters averaged 44.1% (range 1.3 to 139%), whereas *Cryptosporidium* recoveries averaged 25.2% (range 0 to 106%). Negative controls were correctly processed by all the laboratories, but many could not detect a tenfold difference in oocysts levels, a level deemed necessary to differentiate pristine from polluted sources.

Such limitations in analytical procedures call to question the significance of current tests for cysts or oocysts in drinking water. A negative result does not necessarily mean the water is not contaminated with pathogenic protozoa, and a positive result does not necessarily indicate a public health threat. A recent workshop sponsored by the Centers for Disease Control (Atlanta, GA) concluded that protozoa monitoring results must be viewed within the total context of source water quality and treatment plant integrity, and that a positive immunofluorescence result alone should not trigger a public health notice.[23]

New analytical methods have been developed to overcome these limitations. One method aims to determine oocyst viability based on exclusion or inclusion of two fluorogenic vital dyes, 4′,6-diamidino-2-phenylindole (DAPI) and propidium iodide (PI).[24] A significant correlation (r = 0.997) was found between the reaction of the vital stains and *in vitro* excystation; however, excystation may not always relate directly with viability or infectivity. Others have suggested that recoveries of cysts and oocysts could be improved with cotton instead of polypropylene filters, washing filter material twice, limiting the number of centrifugations, increasing centrifugation speed, and using a Percoll-sucrose gradient of

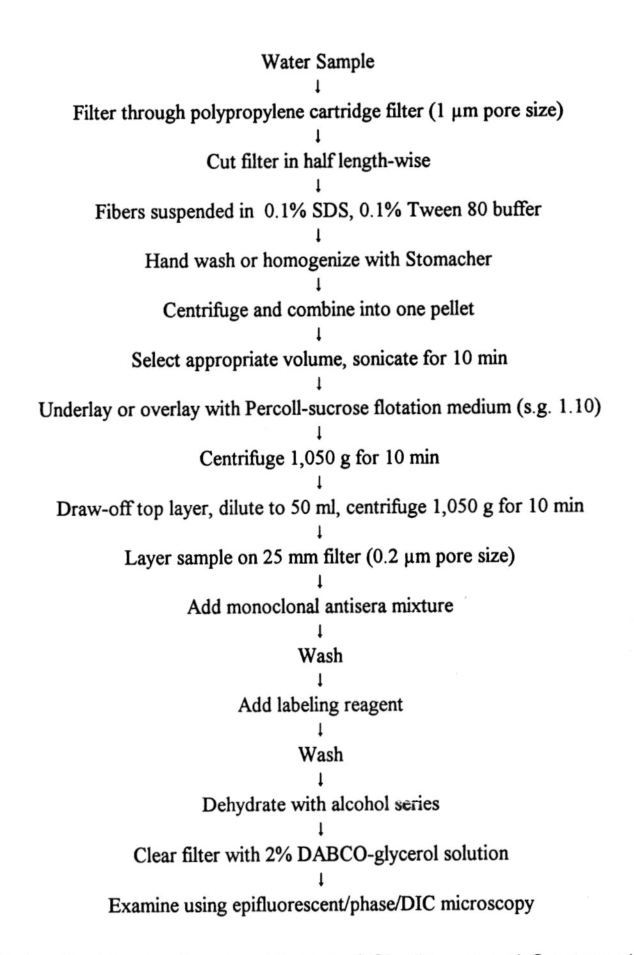

Figure 1 Flow chart of method for simultaneous detection of *Giardia* cysts and *Cryptosporidium* oocysts in water.

1.15 specific gravity.[25] Data to determine if these procedures increase overall recovery efficiencies and reduce sample variation from a variety of environmental waters are limited.

B. OTHER METHODS

The importance of an accurate, sensitive, simple, and inexpensive method to detect viable oocysts in environmental samples has led to a variety of approaches, including flow cytometry,[26] gene probes and polymerase chain reaction (PCR) methods,[27,28] immunomagnetic separation, electrorotation,[29] ELISA,[30] colorimetric methods, and cultural methods.[31] None of these techniques has been sufficiently developed nor thoroughly evaluated enough to replace the immunofluorescence procedure.

IV. PRESENCE OF *CRYPTOSPORIDIUM* IN WATER

A. SURVEYS OF SURFACE WATERS

Immunofluorescent assay (IFA) procedures have been used to detect oocysts in waters throughout the U.S. (Table 3). Of seven studies conducted in the U.S. and one in Canada, six focused on local water sources.[32-37] Data from these studies dispelled the presumption that *Cryptosporidium* was present infrequently and was geographically isolated. The average percent positive samples ranged from 9.1 to 100% with oocyst concentrations ranging from 0 to 240/l. The stream described by Madore et al.[36] received water from domestic and agricultural wastes and had levels as high as 5800 oocysts/l (not included in the oocyst concentration range previously stated). These studies and outbreaks of cryptosporidiosis from drinking water in Braun Station, TX,[1] and Carrollton, GA,[5] helped regulatory agencies and health authorities recognize the public health threat of waterborne *Cryptosporidium*. Data on the national prevalence of *Cryptosporidium* in surface and ground waters were not yet available.

Table 3 Studies on the Prevalence of *Cryptosporidium* in Surface and Groundwaters in North America

Water Source	Number of Samples	Average Positive (%)	Range	Geometric Average (oocysts/l)	Ref.
Stream/river[a]	6	100	0.80–5,800	1920[b]	36
Stream	19	73.7	0–240	1.09	34
Stream/river	58	77.6	0.04–18	0.94	32
Stream/river	38	73.7	<0.001–44	0.66	38
River[a]	11	100	2–112	25[c]	33
River/lake	85	87.1	0.07–484	2.70	39
River	22	31.8	0.01–75.7	0.58	39a
River/lake[a]	18	NA	7.1–28.5	17.8	35
Lake	20	70.7	0–22	0.58	34
Lake/reservoir	32	75.0	1.1–8.9	0.91	32
Lake	24	58.3	<0.001–3.8	1.03	38
Lake	44	27.3	0.11–251.7	4.74	39a
Pristine river	3	NA	NA	0.08	35
Pristine river	59	32.2	NA	0.29	38
Pristine lake	34	52.9	NA	0.093	38
Pristine spring	7	28.6	<0.003–0.13	0.04	38
Pristine lake	11	9.1	0–0.003	0.003	37
Well	18	5.6	NA	0.003[d]	38

Note: NA = information not available.

[a] Impacted by domestic and/or agricultural waste.

[b] Arithmetic mean.

[c] Data adjusted for recovery efficiencies.

[d] Single sample value.

Two studies addressed this deficit and focused on sources of drinking water from diverse geographical regions in the U.S. and Canada.[38-39] They were designed to provide information on what types of drinking water supplies may require more treatment than others to meet current and future regulatory criteria for removal/inactivation of *Cryptosporidium* oocysts. One study examined surface and groundwater in 17 states.[38] Single samples were collected from rivers, streams, lakes, springs, and groundwater wells used as sources of drinking water. Samples were also collected from water used to backwash rapid sand filters. Samples were categorized as pristine or nonpristine. They were pristine if there was little or no human activity in the watershed or water, restricted access, no agricultural activity within the watershed, and no sewage treatment facility discharges impacting the water upstream from the sampling site. Sample volumes were maintained at approximately 400 l for surface water and 1000 l for groundwater. Equivalent sample volumes examined for *Cryptosporidium* were maintained at 100 l for surface waters and 400 l for ground water. Maximum densities of oocysts recovered from nonpristine watersheds ranged from <0.001 to 44 oocysts/l. Average oocyst levels were 2.3 times greater in nonpristine rivers and 11.0 times greater in nonpristine lakes than in similar pristine sources. Oocysts were detected in 39.0% of all pristine waters sampled. Ground water was positive for oocysts in 1 of 18 samples analyzed at a density of 0.003 oocysts per liter. Rapid sand filter backwash samples had a geometric average of 2.2 oocysts per liter.[38]

In a similar study, LeChevallier et al.[39] examined the source water supplies of 66 surface water treatment facilities in 14 states and one Canadian province for *Cryptosporidium*. Surface water sources included lakes, reservoirs, and streams. Sample locations were categorized as protected from or impacted by industrial pollution. For a site to be impacted, it had to have agricultural, sewage, and/or urban runoff within the watershed and upstream of the sample site. Sample volumes were maintained at approximately 378 l. Equivalent volumes ranged from 0.04 to 242 l and averaged 1.37 l.[39] Multiple samples were taken from each site (Table 3); 87.1% of the samples were positive for *Cryptosporidium* oocysts with densities ranging from 0.07 to 484 oocysts per liter.[39] The geometric mean was 2.70 oocysts per liter. The maximum oocyst densities were found in three large rivers — the Mississippi, Ohio, and Missouri — and ranged from 10 to 484 oocysts per liter. These values are comparable to those discussed previously.[38] A tenfold difference in oocyst densities was found between urbanized and pristine watersheds, as in the previous study.[38] In contrast, LeChevallier et al.[39] found a significant correlation between oocyst density and water

Table 4 Studies on the Prevalence of *Cryptosporidium* in Drinking Water

Study Site	Filtration	Number of Samples	Positive (%)	Range (oocysts/l)	Geometric Average (oocysts/l)	Ref.
United States	Yes	28	14.3	0.005–0.007	0.001	38
United States	Yes	82	26.8	0.001–0.48	0.015	40
United States	No	6	33.3	0.001–0.017	0.002	38
Canada	No	42	3.8	0.002–0.005	NA	37
Scotland	NA	142	40.1	0.007–0.72	NA	40a

Note: NA = information not available.

turbidity, perhaps a reflection of the overall quality of waters being analyzed in the two studies. Categorizing waters as pristine or nonpristine was qualitative and possibly subjective. Rose et al.[38] examined more samples from the western U.S., where surface water is generally of higher quality than in the eastern states, where LeChevallier et al.[39] conducted their study. The two studies indicated that surface water receiving runoff from pristine watersheds can, on occasion, have concentrations of oocysts as high as those considered polluted.[38,39] Watershed protection may minimize the prevalence and concentration of oocysts discharged into receiving waters. Several municipalities are using watershed protection programs, presuming these efforts will maintain pristine water in that area and prevent discharge of oocysts into receiving waters and eventually drinking-water source supplies. If future filtration requirements are based on source-water quality, these efforts could have regulatory significance.

B. SURVEYS OF DRINKING WATER

The presence of oocysts in treated drinking water may indicate a breakdown in treatment barriers when accompanied by other indicators of treatment failure (e.g., turbidity, particle counts, mechanical defects). Three studies have examined the presence of oocysts in finished drinking water in North America (Table 4). Of the 158 samples collected, 3.8 to 33.3% were positive for *Cryptosporidium* in the range of 0.001 to 0.48 oocysts per liter.

In one study collection volumes ranged from 400 to 1000 l, and 34 samples were examined (Table 4).[38] Oocysts were detected in nonfiltered drinking water (2 of 6 samples, or 33.3%) at a frequency 2.3 times greater than filtered waters (4 of 28 samples, or 14.3%). In another study, oocysts were found in 26.8% of 82 samples from filtered drinking water supplies (Table 4).[40] Oocyst concentrations ranged from 0.001 to 0.48 oocysts per liter, with a geometric mean of 0.015; however, the oocyst concentrations were fifteen times greater than those found in the prior study.[38] Treatment plants with greater concentrations of oocysts in source water frequently had oocysts in their finished product. Interestingly, each treatment plant was using some type of filtration (sand, dual media, mixed media, or granular activated carbon) and had no record of suboptimal operation. A third study, in Canada, reflected the prevalence of oocysts in nonfiltered pristine systems (Table 4).[37] Only 3.8% of 42 samples were positive, but at concentrations similar to those reported by others.[38] The aforementioned surveys indicate *Cryptosporidium* oocysts are ubiquitous in North American surface waters. Filtration was shown, in one study,[40] to effectively remove oocysts on average by 99.5% to a level that apparently minimizes the probability of an epidemic during normal source water quality treatment conditions. However, even in properly operated filtration systems, low levels of oocysts can be found in treated drinking water. A contributing factor may be the ability of filters to retain and concentrate oocysts. In rapid sand filter backwash waters which had been valved to waste, 2.2 oocysts per liter were found.[38] The presence of oocysts at this level in backwash waters increases the probability of filtered water contamination if filtration plants are not operated properly to prevent breakthrough and/or backwash waters are recycled to the head of the treatment facility without intermediate treatment to remove oocysts.

C. *CRYPTOSPORIDIUM* SURVIVAL IN THE ENVIRONMENT

The assessment of oocyst viability is important in examining survival in the environment, treatment processes, and public health risks. Attempts have been made to determine viability by infectivity, excystation, and dye inclusion tests.

The exclusion/inclusion of vital fluorogenic stains (i.e., DAPI and PI) were used to assess the influence of environmental conditions on oocyst viability (Table 5).[41] When modeling water treatment processes, the efficiency of killing oocysts with lime, ferric sulfate, and alum was directly related to the pH of the reaction solution. When pH and contact times were adjusted to model treatment used in central Scotland

Table 5 Effects of Environmental Conditions on the Viability of *Cryptosporidium* Oocysts

Environmental Conditions	Dead Oocysts %	Experimental Notes
Tap water	96	176 days in a laboratory flow-through system at room temperature
Water treatment processes	0	Conventional treatment processes and contact times; laboratory scale at room temperature and 4°C
River water	94	176 days submerged in river at ambient temperatures
Cow feces	66	176 days submerged in semisolid feces at ambient temperatures
Human feces	78	178 days at 4°C
Sea water	38	35 days at 4°C in static laboratory conditions
Freezing	>99	Immersion in liquid nitrogen
Freezing	67	21 hours in –22°C
Desiccation	>99	4 hours at room temperature

Source: Robertson, L. et al., *Appl. Environ. Microbiol.,* 58, 3494, 1992. With permission.

(approximately 1 hour at pH 6.0), no significant die-off was observed. Similar testing indicated that oocysts survived when suspended in a flow-through chamber connected to the laboratory drinking water tap or river water. Respective die-off rates were 96 and 94% after 176 days. In most distribution systems, 2 days is a more realistic contact time and only 37% of the oocysts had become nonviable by this time. This indicates that even after the initial contamination event has been recognized and remedial actions have been taken, the distribution system can remain a source of viable oocysts for days.

Oocysts survive in human and cattle feces, supporting the suggestion in several outbreaks that sewage discharges and/or runoff from lands contaminated with feces from wild or domesticated animals could be significant sources of viable oocysts. Fecal material possibly protects the oocyst from desiccation, prolonging viability, in the environment. Once in the receiving body of water (i.e., river water), the die-off rates increase dramatically. However, in marine waters it appears the oocysts are significantly more resistant to becoming nonviable. This has important implications to coastal recreational waters contaminated with sewage, where the probability of accidental ingestion of oocysts would be significantly increased. However, the static laboratory conditions used in the marine studies may have resulted in an overestimation of survival.

High temperatures render *C. parvum* oocysts in water noninfectious.[42] Infectivity was verified in neonatal mouse pups. After heating 10^6 oocysts per ml in water to 72.4°C or higher for 1 minute or when the temperature was held at 64.2°C or higher for 5 minutes, infectivity was lost. These findings suggest that *C. parvum* oocysts can be rendered noninfectious in a relatively short time at temperatures far below boiling. However, these results were obtained in pure water, and the extension of these results to water that contains high turbidity and/or dissolved solids or other complex solutions may not be valid.

D. MODELS FOR RISK ASSESSMENT

As technological advancements in microbial detection and knowledge of microbial interactions in drinking water increases, directors of municipalities and health officials must ensure that the public is protected from epidemic and endemic waterborne disease. Through risk assessment, health officials can communicate with the water utility, interpreting water quality surveys and defining the adequacy of drinking-water treatment at acceptable public health risks.

Dose-response data have been used to develop risk estimates for microbiological infections, including viruses and *Giardia*.[43-47] These models have been used to set treatment goals in the USEPA's Surface Water Treatment Rule (SWTR) for *Giardia* and established a goal of 10^{-4} annual risk for determining the microbiological quality of drinking water in the U.S.[58] To meet this goal, various treatment configurations which may include new treatment processes may be required, depending on source water quality.

In a human dose-response study on infectivity and morbidity resulting from oral ingestion of *Cryptosporidium* oocysts, exposure to as few as 30 oocysts initiated infection in one of six persons.[48] Using data from this study, an exponential dose response model was developed[49] and used to estimate the most likely level of contamination present during the Milwaukee outbreak.[4,49] The following assumptions were used to complete the risk assessment analysis: (1) the morbidity ratio (illness:infection) was 100%, (2) the daily water ingestion rate was 1948 ml, (3) the contamination event lasted 21 days, (4) each day during the outbreak represents an independent identical exposure event. The risk equation may be described as follows, in terms of the probability of infection (P_i):

$$P_i = 1 - \exp(-rN)$$

where r is the fraction of microorganisms that are ingested which survive to initiate infection (which is organism specific) and N is the daily exposure, assumed to be 2 liters of drinking water per day. For *Cryptosporidium*, r = 0.00467 (95% C.I., 0.00195 to 0.0097).

Oocysts per liter were calculated for 1-, 10-, and 30-day exposures, using the risk model with an attack rate of 0.14 to 0.52 f or P_i. The 1-, 10-, and 30-day exposure values predicted from the risk model were 1.6, 0.16, and 0.54 oocysts per liter, for a 0.14 attack rate and 79, 7.9, and 2.6 oocysts per liter for the 0.52 attack rate. Based on environmental monitoring on duplicate ice samples corrected for loss of oocysts due to freezing, thawing, and method recovery deficiencies, oocyst exposures ranged from 0.6 to 1.3 oocysts per liter.

The *Cryptosporidium* risk assessment model was also used to analyze historical data[49] from a survey of finished drinking water.[4] In this study, 26.8% of the finished drinking water samples were positive for oocysts in the range of 0.001 to 0.48 oocysts per liter. The analysis assumed that all oocysts were viable, that conventional chlorine disinfection did not affect oocyst infectivity, and that 2 l of water were consumed. The calculated risk estimate was 1.4×10^{-4} with a range of 1.2×10^{-3} to 44×10^{-4}. This suggests that, throughout the year, risks may range between 1/100,000 and 4/1000, respectively. This is within the endemic levels of the disease. Because cryptosporidiosis is a nonreportable disease in the U.S., these risks levels would be difficult to ascertain. The minimum concentration of viable oocysts that could be present in drinking water and still meet the annual risk safety level of 10^{-4} was found to be ≤1 oocyst per 34,000 l. In terms of monitoring criteria, if a utility monitored the finished water daily, the presence of 1 viable oocyst per 340 l for not more than 8 to 10 days (2 to 2.8% of the time) would still be within the annual safety goal of 10^{-4}. Current monitoring techniques do not permit such an evaluation.

V. DRINKING WATER TREATMENT

A. SOURCE-WATER CONTROL

The multiple barrier concept for prevention of waterborne disease in drinking water includes protection of source waters from contamination. Protected watersheds generally have lower oocyst levels than sites receiving agricultural, sewage, or urban runoff (Table 2). Limiting these activities in a watershed might help reduce the burden on the water treatment process,[50] but storm events that wash fecal material into receiving streams, animal migration, or epizootic infections may create peaks in the oocyst densities. Because no studies have quantified the relative contribution of various sources of contamination, the watershed manager is left to guessing at the significant sources of contamination. Source-water protection is clearly an area where additional research is warranted.

Impoundment of water in a reservoir can reduce variations in water quality. Small size and low density of an oocyst results in a low sedimentation velocity (0.5 µm/sec) requiring more than a year for the oocyst to settle to the bottom of a 20-m deep reservoir.[51] Although a reservoir may not reduce the number of oocysts through sedimentation, oocyst viability would be expected to decline over this period.

B. COAGULATION, SEDIMENTATION, AND FILTRATION

Physical removal of oocysts through coagulation, sedimentation, and filtration is the primary barrier against waterborne cryptosporidiosis. Most waterborne outbreaks have been associated with problems with one of these processes (Table 1). When deficiencies were corrected, rates of cryptosporidiosis decreased,[2] indicating that properly operated filtration can control waterborne disease in a community.

Proper chemical conditioning of raw water through appropriate selection of coagulant chemicals is important for optimum removal of oocysts. Effective coagulation of oocysts has been achieved using alum, ferric chloride, and polyaluminum chloride.[52-54] Addition of a polymer along with the metal salt generally improves oocyst removal.[55,56] Enhanced coagulation for optimum removal of total organic carbon also improved removal of oocysts compared to conventional baseline conditions.[57] These results are encouraging, because the USEPA will require enhanced coagulation for reduction of trihalomethane precursors.[58]

Properly operated conventional treatment (coagulation, sedimentation, filtration) can remove 99% or more of oocysts.[52,55,56,59] Microfiltration and ultrafiltration membrane processes can remove all oocysts.[60] In addition, treatment processes such as direct filtration (with chemical pretreatment), high-rate filtration, dissolved-air flotation, diatomaceous earth, and slow-sand filtration can also effectively remove oocysts.[51-53, 55, 56] One of the critical times when oocysts can breach the filtration barrier is following backwash of the filter. Optimization of the backwash procedure, including addition of coagulant to the final portion of the backwash water and slowly ramping the filter rate, or filtering to waste can minimize the passage of oocysts.[55,61] Other important operational parameters are flowrate variations or surges in rates due to changes in plant production, cycling service pumps, on/off plant operation, negative head development in filters with resulting air-binding, and nonoptimal design or functioning of filtration rate control systems.[62,63]

In general, treatment of *Cryptosporidium* is more difficult than *Giardia*, possibly because of (1) its smaller size, (2) its lower sedimentation rate, and (3) its increased disinfection resistance. Of 71 surface water treatment plants examined on multiple occasions (94% of the sites were examined five or more times), oocysts were detected in 39 (54.9%) treatment plant effluents; 15 of the systems were effluent positive on multiple occasions.[64] By comparison, *Giardia* cysts were detected in 20 of 71 (28%) treated effluents; six systems were cyst positive on multiple occasions. The authors conclude that controlling coagulation and filtration procedures for removal of *Cryptosporidium* will be more difficult than control of *Giardia*.

Coagulation procedures can be monitored using jar tests, zeta potential, or streaming current detectors.[53] Improper installation of the streaming current detector in Milwaukee was partially responsible for the inadequate treatment and resulting outbreak. Filtration can be monitored using on-line turbidimeters or particle counters. Correlations between oocysts, particle count, and turbidity removal have been documented.[65] Although filtered particle count values are related to raw water levels, in general, filtered turbidity of 0.1 NTU or particle counts (>3 μm) less than 50/ml are indicative of high quality drinking water.

C. WASTE STREAM RECYCLE

Efforts to conserve water in arid areas, to minimize impact on receiving streams, and to reduce waste have led to the recycling of water used to wash treatment filters. These waters can contain high levels of oocysts;[66,67] however, if properly handled, the microbial impact of washwater recycle can be minimized.[55,67] Many systems, however, are not designed or operated for this process. Often times, the recycle pumps are designed to quickly drain the washwater holding tanks, increasing the proportion of recycled water to 10% or more of the total flow. To reduce the number of oocysts it has been recommended that recycled water be constantly returned at low rates and that washwater be treated with polymers.[67] Others concurred, indicating that recycled turbidities should be less than 5 NTU or have a suspended solids content of less than 10 mg/l.[55] In the evaluation of the Waterloo (Ontario, Canada) treatment plant, other options for handling backwash water were deemed to pose potentially greater risks than recycling the water through the treatment plant.[61]

D. DISINFECTION

Disinfection has always been the major barrier for control of microbial contaminants in water. For control of the waterborne protozoan a combination of filtration and disinfection will be needed. Using particle counting as a surrogate for oocyst removal by sedimentation and filtration the disinfection requirements to meet acceptable risk levels was calculated (Figure 1).[64] The results showed that the average plant will need to provide 2.0 $\log_{10}$ (99%) disinfection (range 0.5 to 2.6 $\log_{10}$), along with effective particle removal to meet the 10^{-4} annual risk of *Cryptosporidium* infection goal.

Chlorine-based disinfectants generally have had a low level of effectiveness for oocyst inactivation (Table 6). Early investigations found *Cryptosporidium* oocysts resistant to a variety of hospital disinfectants, including bleach.[68] Because as much as 80 mg/l of free chlorine or monochloramine required 90 minutes to produce 90% oocyst inactivation, the authors concluded that conventional disinfection practices would do little to inactivate waterborne *Cryptosporidium*.[69] Chlorine disinfection alone has not been effective for eliminating cryptosporidiosis in drinking-water and recreational-water outbreaks (Tables 1 and 2). A noticeable exception is shown in Table 4, where the disinfection effectiveness of several biocides is based on their oxidation/reduction potential (ORP).[70] The authors claimed that it was necessary to maintain an effective ORP (approximately 800 mV) for an appropriate contact time to achieve oocyst inactivation. Chlorine dioxide has demonstrated some efficacy as a disinfectant for

Table 6 Summary of Free Chlorine, Monochloramine, and Chlorine Dioxide Disinfection Results for *Cryptosporidium*

Compound	Chlorine Residual (mg/l)	Contact Time (min)	CT product (mg·min/l)	Temperature (°C)	pH	Percent Inactivation	Analytical Method	Ref.
Chlorine	80	90	7,200	25	7	>99	Mouse infectivity	69
Chlorine	15	240	3,600	22	8	47	Mouse infectivity	77
Chlorine	968	1440	1,393,970	10	7	85	Excystation	72
Chlorine	17	30	510	NR	NR	99	Excystation	70
Monochloramine	80	90	7200	25	7	99	Mouse infectivity	69
Monochloramine	15	240	3600	22	8	99.6	Mouse infectivity	77
Monochloramine	3.75	1440	5414	10	7	80.5	Excystation	72
Chlorine dioxide	1.3	60	78	25	7	90	Mouse infectivity	69
Chlorine dioxide	0.07	16	1.12	Room	NR	97	Mouse infectivity	71
Chlorine dioxide	0.22	30	6.6	Room	NR	94	Mouse infectivity	71
Chlorine dioxide	4.03	15	60	10	7	95.8	Excystation	72

Note: NR = not reported.

Table 7 Summary of Ozone Disinfection Results for *Cryptosporidium*

Chlorine Residual (mg/l)	Contact Time (min)	CT Product (mg·min/l)	Temperature (°C)	Percent Inactivation	Analytical Method	Ref.
1	5	5	25	90–99	Mouse infectivity	69
1	10	10	25	>99		
0.77	6	4.6	Room	>99	Mouse infectivity	71
0.51	8	4.1	Room	>99		
0.16–1.3	5–15	7	7	99	Mouse infectivity	73
0.17–1.9	5–15	3.5	22	99		
2.4 (avg.)	2.3	5.5	22–25	99	Mouse infectivity	72c
1.25	15	18.75	10	98.6	Excystation	72
4 (approx.)	2	8	Room	>90	Excystation	72b
1–5	10	10–50	5	18–39	Stain	72a
1–5	10	10–50	20	70–>99		
0.7–1.5	14–25	9.8–27	8–10	42–84	Stain	55

Cryptosporidium inactivation (Table 6). Thirty minutes of exposure to 0.22 mg chlorine dioxide per liter significantly reduced oocyst infectivity, although some oocysts remained viable.[71] In contrast, others found that a disinfectant concentration multiplied by a contact time (CT) product of 60 to 80 mg·min/l was necessary to produce 90 to 96% inactivation.[69,72]

Ozone has shown promise for effective disinfection of *Cryptosporidium* (Table 7, Figure 2). Noticeable in the data contained in Table 7 is the impact of the analytical method on CT values. Generally, these excystation and vital stain methods are more conservative measures of oocyst inactivation than animal infectivity. Reliance on these excystation and vital stain methods alone could greatly overestimate disinfection requirements for *Cryptosporidium.* On average, 4.5 mg·min/l CT was required for 99% oocyst inactivation (measured by mouse infectivity) by ozone at 20 to 25°C (Table 7). However, the conventional method of determining CT by using the final concentration of reactants at the end of the contact time overestimates the CT needed for disinfection and unduly increases treatment costs.[73] The Holm disinfection model that integrates the disinfectant concentration and time throughout the reactor might provide a more accurate model for estimating CT.[73] Using the Holm model, CTs for *Cryptosporidium* inactivation were 6.9 mg·min/l at 7°C, and 2.4 mg·min/l at 22°C.

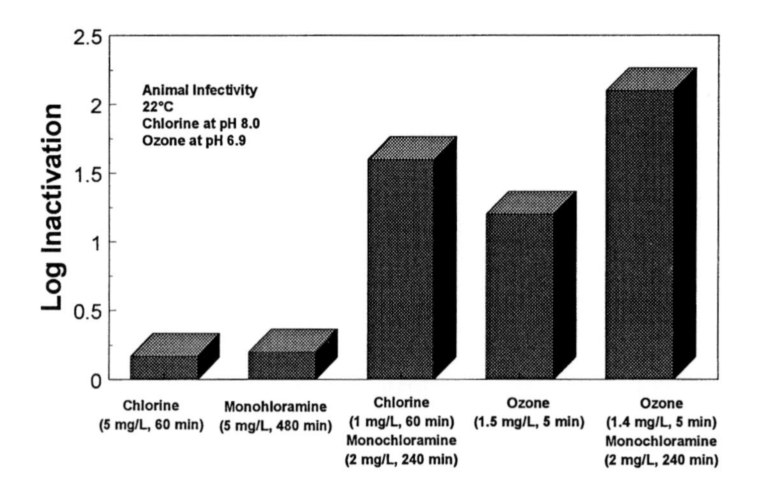

Figure 2 Impact of single and multiple disinfectants on infectivity of *Cryptosporidium*. (Adapted from Finch, G. R. et al., in *Proceedings of the American Water Works Association Water Quality Technology Conference* (San Francisco, CA, November 6–10, 1994), American Water Works Association, Denver, CO, 1994, 1303.)

Ultraviolet (UV) light has only limited application for *Cryptosporidium* inactivation. Complete inactivation of 2500 oocysts per milliliter required exposure for 150 minutes in thin-layer suspensions (10 ml suspension in 64 cm^2 petri dish) to UV light (15,000 mW/sec at a distance of 22 cm).[74] Others concurred, indicating that the irradiation dose for *Cryptosporidium* (120 mW sec/cm^2) was 7.5 times higher than similar values for *Giardia lamblia* and 4 times the level recommended by the U.S. Public Health Service.[72]

E. SYNERGISTIC EFFECTS

Despite the apparent resistance of *Cryptosporidium* to conventional disinfection, more than 90% of oocysts detected in chlorinated drinking water appeared empty (e.g., lacking observable sporozoites) and were probably not viable.[40] Oocyst viability (measured using DAPI and PI) decreased 50, 99.7, and 100% when oocysts were shaken with sand for 5, 90, and 120 minutes, respectively.[75] Viability decreased an additional 25% when oocysts were shaken with sand for 5 minutes, then exposed to free chlorine (1 mg/l at 20°C) for 5 minutes. Others reported similar results.[76] The synergism of multiple disinfectants was found when free chlorine was followed by monochloramine producing oocyst inactivation greater than the sum of both disinfectants examined separately (Figure 3).[77] The combination of free chlorine (1 mg/l for 60 minutes) and chloramines (2 mg/l for 240 minutes) approach the 99% inactivation levels suggested in Figure 1. Synergism for ozone and chloramines was also observed.

That multiple stresses can act synergistically to reduce *Cryptosporidium* viability may impact control strategies. Oocyst exposure to sewage treatment, followed by environmental stressors in the receiving streams, rechlorination, pH adjustments, sedimentation, filtration, and abrasion in conventional drinking water filter media might provide the combination of factors needed to limit the infectivity of oocysts in treated water supplies. One of the most significant factors in each identified *Cryptosporidium* waterborne outbreak to date has been the introduction of fresh fecal material relatively close to the potable water intake. Contamination close to the intake limits exposure of oocysts to environmental stress and places the burden for disinfection on conventional treatment barriers that sometimes fail. The proximity of contamination to the intake may also explain why there have been no recognized outbreaks of cryptosporidiosis related to unfiltered water systems, where watershed protection measures limit introduction of organisms near water intakes and environmental stressors inactivate upstream organisms. Clearly, examination of combinations of disinfectants and environmental effects is an important area that requires further investigation to better understand inactivation mechanisms and disinfection theory.

VI. *CRYPTOSPORIDIUM* REGULATIONS

No drinking water regulations in the U.S. specifically address *Cryptosporidium* in potable supplies, although a number of related regulations provide some protection. The Clean Water Act regulates point and nonpoint discharges into receiving waters. Limiting discharges, especially to areas where the receiving waters are sources for drinking water, provides some benefit. The problem with this approach

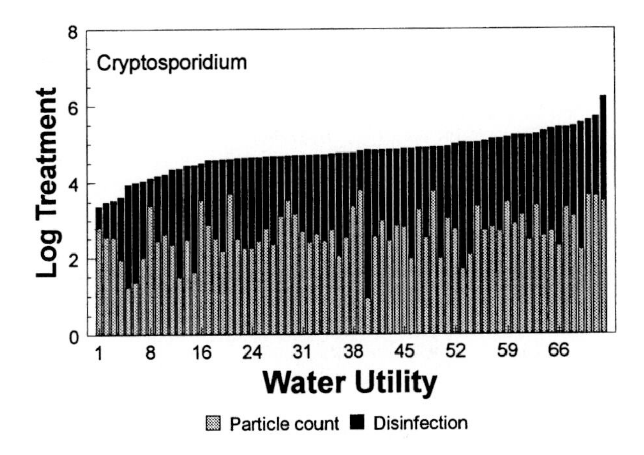

Figure 3 Evaluation of particle count removals (>3 μm) and disinfection requirements to meet *Cryptosporidium* treatment levels.

is that only coliform bacteria are regulated, and because coliform bacteria are much more susceptible to disinfection than oocysts, effluent discharges many contain low levels of bacteria but high levels of *Cryptosporidium*.

The USEPA's Surface Water Treatment Rule required filtration and disinfection of all surface water supplies and ground water directly impacted by surface water.[79] The rule specified a series of treatment requirements for surface and ground water under the influence of surface water. Requirements included a minimum treatment level of 3 $\log_{10}$ for *Giardia* and 4 $\log_{10}$ for viruses, although treatment levels could be increased for poor source water quality. The rule also lowered the acceptable limit for turbidity in finished drinking water from a monthly average of 1.0 NTU to a level not to exceed 0.5 NTU in 95% of 4-hour measurements. These requirements went into effect on June 29, 1993, just months after the outbreak in Milwaukee. Although the target organism for the rule was *Giardia*, many of the treatment requirements (filtration, lower turbidity levels, sanitary survey, etc.) also provided some protection against *Cryptosporidium*. However, upon examination of *Giardia* and *Cryptosporidium* levels in source waters for 68 surface water treatment facilities, it was determined that all the utilities would require more than the minimum 3 $\log_{10}$ treatment specified by the SWTR.

The USEPA recognized the possibility that regulations developed to limit the levels of disinfectants and disinfectant byproducts (D/DBP) could inadvertently increase the risk from microbial agents.[58] Utilizing the disinfection byproducts regulatory analysis model (DBPRAM), the health and economic implications of various approaches to DBP regulation were examined.[79] For several DBP control scenarios, the predicted increases in *Giardia* infection were orders of magnitude higher than decreases in cancer rates. To ensure that implementation of the D/DBP rule did not increase microbial risk, the USEPA considered it necessary to review and revise the SWTR. The revised rule, which includes regulation of *Cryptosporidium*, is called the Enhanced Surface Water Treatment Rule (ESWTR).[80]

The USEPA intends to use the microbial data generated by the Information Collection Rule[81] to help formulate the final draft of the ESWTR. The four principle treatment options outlined in the draft ESWTR are summarized in Table 8.

In March 1995, the USEPA announced a voluntary program to optimize treatment processes, principally for the control of *Cryptosporidium*. The program, called "Partnership for Safe Water" and endorsed by national water organizations and state regulators, consisted of a series of self-assessments, optimizations, and independent review. Within the first month of the program, nearly 300 water utilities serving over 80 million people enlisted in the partnership. Better water quality is the real key to providing public health protection from *Cryptosporidium*.

VII. CONCLUSIONS

Water is recognized as an important potential source of *Cryptosporidium* in communities. That *Cryptosporidium* oocysts are highly prevalent in surface and treated waters has been documented in more than 25 published studies using IFA technology. Oocysts have been found in 5.6 to 87.1% of the source waters at 0.003 to 4.74 oocysts per liter and in 3.8 to 40.1% of the drinking waters at 0.002 to

Table 8 Summary of the Proposed Enhanced Surface Water Treatment Rule

Parameter	Action
Definitions	
Groundwater under the influence	Includes *Cryptosporidium*
MCLG	
Cryptosporidium	MCLG = 0
Criteria to avoid filtration:	
Watershed control program for *Giardia, Cryptosporidium,* and viruses	Tightens requirements to avoid filtration
Analytical and monitoring	
Sanitary survey	1. All public water systems 2. Frequency 3–5 years 3. Conducted by state, agent, or system
Treatment requirements:	
Treatment based on *Giardia*	3–6 log treatment requirement based on the raw water levels
Treatment based on *Cryptosporidium*	3–6 log treatment requirement based on the raw water levels
Minimum treatment	Specify minimum treatment of 2 log removal of *Cryptosporidium* in addition to existing SWTR requirements for *Giardia* and viruses
No change	Do not modify existing SWTR levels for removal/inactivation

Note: MCLG = maximum contaminant level goal; SWTR = Surface Water Treatment Rule (USEPA).

0.015 oocysts per liter. Risk assessment models evaluate the risk of infection from low levels of oocysts in water; however, current methods are inadequate to address many of the important public health questions raised by detection of low levels of oocysts in water. When conventional water treatment practices are properly designed and operated, removal efficiencies for oocysts can approach 99.9%. Disinfection studies show *Cryptosporidium* to be very resistant to conventional disinfection; however, ongoing research indicates synergistic effects of multiple disinfectants and combined processes. The presence of *Cryptosporidium* in water challenges regulators to develop public policies that limit contamination of source waters, improve water treatment, and protect public health.

REFERENCES

1. D'Antonio, R. G., Winn, R. E., Taylor, J. P., Gustafson, T. L., Current, W. L., Rhodes, M. M., Gary, G. W., Zajac, R. A., A waterborne outbreak of cryptosporidiosis in normal hosts, *Ann. Intern. Med.*, 103, 886, 1985.
2. Leland, D., McAnulty, J., Keene, W., Stevens, G., A cryptosporidiosis outbreak in a filtered-water supply. *J. Am. Water Works Assoc.,* 85, 34, 1993.
3. Moore, A., Herwaldt, B., Craun, G., Calderon, R., Highsmith, A., Juranek, D., Surveillance for waterborne disease outbreaks — United States, 1991–1992, in *Morbid. Mortal. Wkly. Rep.,* SS-5, 1993.
4. MacKenzie, W., Neil, M., Hoxie, N., Proctor, M., Gradus, M., Blair, K., Peterson, D., Kazmierczak, J., Addiss, D., Fox, K., Rose, J., Davis, J., A massive outbreak in Milwaukee of *Cryptosporidium* infection transmitted through the public water supply, *N. Engl. J. Med.*, 331, 161, 1994.
5. Hayes, E. B., Matte, T. D., O'Brien, T. R., McKinley, T. W., Logsdon, G. S., Rose, J. B., Ungar, B. P., Word, D. M., Pinsky, P. F., Cummings, M. L., Wilson, M. A., Long, E. G., Hurwitz, E. S., Sauch, J. F., Use of immunofluorescence and phase-contrast microscopy for detection and identification of *Giardia* cysts in water samples, *Appl. Environ. Microbiol.*, 50, 1434, 1985.
6. Kramer, M. H., Herwaldt, B. L., Craun, G. F., Calderon, R. L., and Juranek, D. D., Waterborne disease: 1993–1994, *J. Am. Water Works Assoc.,* 88(3), 66–80, 1996.

6a. Goldstein, S. T., Juranek, D. D., Ravenholte, O., Hightower, A. W., Martin, D. G., Mesnik, J. L., Griffiths, S. D., Bryant, A. J., Reich, R. R., and Herwaldt, B. L., Cryptosporidiosis, an outbreak associated with drinking water despite state-of-the-art treatment, *Ann. Intern. Med.,* 124(5), 459–468, 1996.

7. Sorville, F.J., Swimming-associated cryptosporidiosis — Los Angeles County, *Morbid. Mortal. Wkly. Rep.*, 39, 343, 1990.
8. Joce, R. E., Bruce, J., Kiely, D., Noah, N. D., Dempster, W. B., Stalker, R., Gumsley, P., Chapman, P. A., Norman, P., Watkins, J., Smith, H. V., Price, T. J., Watts, D., An outbreak of cryptosporidiosis associated with a swimming pool, *Epidemiol. Infect.*, 107, 497, 1991.
9. Bongard, J., *Cryptosporidium* infections associated with swimming pools — Dane County, Wisconsin, 1993, *Morbid. Mortal. Wkly. Rep.,* 43, 562, 1994.
10. Bell, A., Guasparini, R., Meeds, D., Mathias, R. G., Farley, J. D., A swimming pool-associated outbreak of cryptosporidiosis in British Columbia, *Rev. Can. Sante Publique*, 84, 334, 1993.

11. McAnulty, J. M., Fleming, D. W., and Gonzalez, A. H., A community-wide outbreak of cryptosporidiosis associated with swimming at a wave pool, *J. Am. Med. Assoc.*, 272, 1597, 1994.
12. Sauch, J. F., Use of immunofluorescence and phase-contrast microscopy for detection and identification of *Giardia* cysts in water samples, *Appl. Environ. Microbiol.*, 50, 1434, 1985.
13. Riggs, J. L., Dupuis, K. W., Nakamura, K., and Spath, D. P., Detection of *Giardia lamblia* by immunofluorescence, *Appl. Environ. Microbiol.*, 45, 698, 1983.
14. Sterling, C. R., Arrowood, M. J., Marshall, M. M., and Stetzenbach, L. D., The detection of *Giardia* and *Cryptosporidium* from water sources using monoclonal antibodies, in *Proc. AWWA Water Quality Tech. Conf.*, American Water Works Association, Denver, CO, 1987, 271.
15. Sterling, C. R., Kutob, R. M., and Gizinski, M. J., Verastequi M, Stetzenbach L, Giardia *Detection Using Monoclonal Antibodies Recognizing Determinants of* In Vitro *Derived Cysts*, University of Calgary Press, Alberta, 1988, 219.
16. Musial, C. E., Arrowood, M. J., Sterling, C. R., and Gerba, C.P., Detection of *Cryptosporidium* in water by using polypropylene cartridge filters, *Appl. Environ. Microbiol.*, 53, 687, 1987.
17. LeChevallier, M. W., Norton, W. D., and Lee, R. G., Evaluation of a method to detect *Giardia* and *Cryptosporidium* in water, in *Monitoring Water in the 1990s: Meeting New Challenges,* ASTM STP 1102, Hall, J. R. Glysson, G. D., Ed., American Society for Testing and Materials, Philadelphia, 1992, 483.
18. Rose, J. B., Kayed, D., Madore, M. S., Gerba, C. P., Arrowood, M. J., Sterling, C. R., and Riggs, J. L., *Methods for the Recovery of* Giardia *and* Cryptosporidium *From Environmental Waters and their Comparative Occurrence,* U.S. Environmental Protection Agency, Cincinnati, OH, 1988.
19. LeChevallier, M. W., Trok, T. M., Burns, M. O., Lee, R. G., Comparison of the zinc sulfate and immunofluorescent techniques for detecting *Giardia* and *Cryptosporidium* in water, *J. Am. Water Works Assoc.,* 82(9), 75, 1990.
20. American Public Health Association, *Standard Methods for the Examination of Water and Wastewater* (suppl.), 18th ed., American Public Health Association, Washington, D.C., 1994.
21. ASTM, Proposed test method for *Giardia* cysts and *Cryptosporidium* oocysts in low-turbidity water by a fluorescent antibody procedure, *Ann. Book ASTM Stand.*, 11(01), 925, 1991.
22. Clancy, J. L., Gollnitz, W. D., and Tabib, Z., Commercial labs: how accurate are they?, *J Am. Water Works Assoc.,* 86(5), 89, 1994.
23. Addiss, D. G., Arrowood, M. J., Bartlett, M. E., Colley, D. G., Juranek, D. D., and Kaplan, J. E., Assessing the public health threat associated with waterborne cryptosporidiosis: report of a workshop, *Morbid. Mortal. Wkly. Rep.*, 44(RR-6), 1–19, 1995.
24. Campbell, A. T., Robertson, L. J., and Smith, H. V., Viability of *Cryptosporidium parvum* oocysts: correlation of *in vitro* excystation with inclusion or exclusion of fluorogenic vital dyes, *Appl. Environ. Microbiol.*, 58(11), 3488, 1992.
25. LeChevallier, M. W., Norton, W. D., Siegel, J. E., and Abbaszadegan M, Evaluation of the immunofluorescence procedure for detection of *Giardia* cysts and *Cryptosporidium* oocysts in water, *Appl. Environ. Microbiol.*, 61(2), 690, 1995.
26. Vesey, G., Slade, J. S., Byrne, M., Shepherd, K., Dennis, P. J., and Fricker, C. R., Routine monitoring of *Cryptosporidium* oocysts in water using flow cytometry. *J. Appl. Bacteriol.*, 75, 87, 1993.
27. Johnson, D. W., Pieniazek, N. J., and Rose, J. R., DNA probe hybridization and PCR detection of *Cryptosporidium* compared to immunofluorescence assay, *Water Sci. Technol.*, 27(3–4), 77, 1993.
28. Webster, K. A., Pow, J. D. E., Giles, N., Catchpole, J., and Woodward, M. J., Detection of *Cryptosporidium parvum* using a specific polymerase chain reaction, *Vet. Parasitol.*, 50(1–2), 35, 1993.
29. Beardsley, T., Putting a spin on parasites, *Sci. Am.*, 269, 87, 1993.
30. Chapman, P. A., Rush, B. A., and McLauchlin, J., An enzyme immunoassay for detecting *Cryptosporidium* in faecal and environmental samples, *J. Med. Microbiol.*, 32, 233, 1990.
31. Upton, S. J., Tilley, M., Nesterenko, M. V., and Brillhart, D. B., A simple and reliable method of producing *in vitro* infections of *Cryptosporidium parvum* (Apicomplexa), *FEMS Microbiol.*, 118, 45, 1994.
32. Rose, J. B., Occurrence and significance of *Cryptosporidium* in water, *J. Am. Water Works Assoc.,* 80, 53, 1988.
33. Ongerth, J. E. and Stibbs, H. H., Identification of *Cryptosporidium* oocysts in river water, *Appl. Environ. Microbiol.,* 53, 672, 1987.
34. Rose, J. B., Darbin, H., and Gerba, C. P,. Correlations of the protozoa, *Cryptosporidium* and *Giardia,* with water quality variables in a watershed, *Proc. Int. Conf. Water Wastewater Microbiol.* 2, 43.1, 1988.
35. Rose, J. B., Kayed, D., Madore, M. S., Gerba, C. P., Arrowood, M. J., and Sterling, C. R., Methods for the recovery of *Giardia* and *Cryptosporidium* from environmental waters and their comparative occurrence, in *Advances in* Giardia *Research*, Wallis, P. and Hammond, B., Eds., University of Calgary Press, Calgary, 1988.
36. Madore, M. S., Rose, J. B., Gerba, C. P., Arrowood, M. J., and Sterling, C. R., Occurrence of *Cryptosporidium* oocysts in sewage effluents and select surface waters, *J. Parasitol.*, 73, 702, 1987.
37. Roach, P., Olson, M., Whitely, G., and Wallis, P., Waterborne *Giardia* cysts and *Cryptosporidium* oocysts in the Yukon, Canada, *Appl. Environ. Microbiol.*, 59, 67, 1993.
38. Rose, J. B., Gerba, C. P., and Jakubowski, W., Survey of potable water supplies for *Cryptosporidium* and *Giardia*, *Environ. Sci. Technol.*, 25, 1393, 1991.
39. LeChevallier, M. W., Norton, W. D., and Lee, R. G., Occurrence of *Giardia* and *Cryptosporidium* spp. in surface water supplies, *Appl. Environ. Microbiol.*, 57, 2610, 1991.

39a. Stezenbach, L., Arrowood, M., Marshal, M., and Sterling, C., Monoclonal antibody based on immunofluorescent assay for *Giardia* and *Cryptosporidium* detection in water samples, *Water Sci. Technol.,* 20, 193, 1988.

40. LeChevallier, M. W., Norton, W. D., and Lee, R. G., *Giardia* and *Cryptosporidium* spp. in filtered drinking water supplies, *Appl. Environ. Microbiol.*, 57, 2617, 1991.

40a. Smith, H., Grimason, A., Benton, C., and Parker, J., The occurance of *Cryptosporidium* spp. oocysts in Scottish waters, and the development of a fluorogenic viability assay for individual *Cryptosporidium* spp. oocysts, *Water Sci. Technol.,* 24, 169, 1991.

41. Robertson, L., Campbell, A., and Smith, H., Survival of *Cryptosporidium parvum* oocysts under various environmental pressures, *Appl. Environ. Microbiol.*, 58, 3494, 1992.

42. Fayer, R., Effect of high temperature on infectivity of *Cryptosporidium parvum* oocysts in water, *Appl. Environ. Microbiol.*, 60, 2732, 1994.

43. Gerba, C. and Haas, C., Assessment of risks associated with enteric viruses in contaminated drinking water, *ASTM Spec. Tech. Publ.*, 679, 489, 1988.

44. Haas, C., Rose, J., Gerba, C., and Regli, S., Risk assessment of viruses in drinking water, *Risk Anal.*, 13, 545, 1993.

45. Regli, S., Rose, J., Haas, C., and Gerba, C., Modeling the risk from *Giardia* and viruses in drinking water, *J. Am. Water Works Assoc.,* 83, 76, 1991.

46. Rose, J., Haas, C., and Regli, S., Risk assessment and control of waterborne giardiasis, *Am. J. Public Health*, 81, 709, 1991.

47. Rose, J., Haas, C., and Gerba, C., Waterborne pathogens: assessing health risks, *Health Environ. Dig.*, 7, 1, 1993.

48. Dupont, H., Chappell, C., Sterling, C., Okhuysen, P., Rose, J., and Jakubowski, W., Infectivity of *Cryptosporidium parvum* in healthy volunteers, *N. Engl. J. Med.*, 332, 855, 1995.

49. Haas, C. and Rose, J., Reconciliation of microbial risk models and outbreak epidemiology: the case of the Milwaukee outbreak, in *Proceedings of the American Water Works Association's Annual Conference* (New York, June 5–9), American Water Works Association, Denver, CO, 1994.

50. Glicker, J. L., Engineering and water quality concerns surface water source protection, in *Methods for the Investigation and Prevention of Waterborne Disease Outbreaks*, U.S. Environmental Protection Agency, Washington, D.C., 1990, 157.

51. Badenoch, J., Bartlett, C. L. R., Benton, C., Casemore, D. P., Cawthorne, R., Earnshaw, F., Ives, K. J., Jeffery, J., Smith, H. V., Vaile, M. S.B., Warrell, D. A., and Wright, A. E., *Cryptosporidium* in water supplies, *Rep. Group Experts*, London, 1990.

52. Nieminski, E. C., *Giardia* and *Cryptosporidium* cysts removal through direct filtration and conventional treatment, in *Proceedings of the American Water Works Association's Annual Conference* (New York, June 19–23, 1994), American Water Works Association, Denver, CO, 1994, 463.

53. Lytle, D. L. and Fox, K. R., Particle counting and zeta potential measurements for optimizing filtration treatment performance, in *Proceedings of the American Water Works Association's Water Quality Technology Conference,* Part I (San Francisco, CA, November 6–10, 1994), American Water Works Association, Denver, CO, 1994, 833.

54. Wilczak, A., Gramith, K. M., Oppenheimer, J. A., Jacangelo, J. G., Riley, L. R., Gallagher, B., and Patania, N. L., Effect of treatment conditions on the removal of protozoan cysts and MS2 virus by clarification and protozoan cysts and MS2 virus by clarification and automatic backwash filtration, in *Proceedings of the American Water Works Association's Water Quality Technology Conference,* Part I (San Francisco, CA, November 6–10, 1994), American Water Works Association, Denver, CO, 751, 1994.

55. Hall, T., Pressdee, J., and Carrington, E., *Removal of* Cryptosporidium *Oocysts by Water Treatment Process,* Foundation for Water Research, Bucks, United Kingdom, 1994.

56. West, T., Daniel, P., Meyerhofer, P., DeGraca, A., Leonard, S., and Gerba, C., Evaluation of *Cryptosporidium* removal through high-rate filtration, in *Proceedings of the American Water Works Association's Annual Conference* (New York, June 19–23, 1994), American Water Works Association, Denver, CO, 1994, 493.

57. Ibrahim, E. A., Abbaszadegan, M., and LeChevallier, M. W., Enhanced coagulation for removal of dbp precursors, particulate and microbial contaminants, in *Proceedings of the American Water Works Association's Annual Conference* (New York, June 19–23, 1994), American Water Works Association, Denver, CO, 1995.

58. USEPA, National Primary Drinking Water Regulations; Disinfectants and Disinfection Byproducts; Proposed Rule, Federal Register, 59(145), 38668-28829, U.S. Environmental Protection Agency, Washington, D.C., 1994.

59. Nieminski, E. C., *Giardia* and *Cryptosporidium* — where do the cysts go?, in *Proceedings of the American Water Works Association's Water Quality Technology Conference,* Part I (Toronto, November 15–19, 1992), American Water Works Association, Denver, CO, 1992, 205.

60. Adham, S. S., Jacangelo, J. G., and Laine, J., Effect of membrane type on the removal of *Cryptosporidium parvum, Giardia muris*, and MS2 virus, in *Proceedings of the American Water Works Association's Annual Conference* (New York, June 19–23, 1994), American Water Works Association, Denver, CO, 1994, 477.

61. Welker, R., Porter, R., Pett, W. B., Provart, M. R., and Schwartz, M., Cryptosporidiosis outbreak in Kitchener — Waterloo: assessment and future prevention, in *Proceedings of the American Water Works Association's Annual Conference* (New York, June 19–23, 1994), American Water Works Association, Denver, CO, 1994, 104.

62. Cleasby, J. L., Filtration, in *Water Quality and Treatment*, 4th ed., Pontius, F. W., Ed., McGraw-Hill, New York, 1990, 455.

63. Bellamy, W. D., Cleasby, J. L., Logsdon, G. S., and Allen, M. J., Assessing treatment plant performance, *J. Am. Water Works Assoc.,* 85(12), 34, 1993.
64. LeChevallier, M. W. and Norton, W. D., Occurrence of *Giardia* and *Cryptosporidium* in raw and finished drinking water, *J. Am. Water Works Assoc.*, 87(9), 54–68, 1996.
65. LeChevallier, M. W. and Norton, W. D., Examining relationships between particle counts and *Giardia, Cryptosporidium*, and turbidity, *J. Am. Water Works Assoc.,* 84(12), 54, 1992.
66. Colbourne, J. S., Thames Utilities experience with *Cryptosporidium*, in *Proceedings of the American Water Works Association's Water Quality Technology Conference* (Philadelphia, PA, November 12–16, 1989), American Water Works Association, Denver, CO, 1989.
67. Cornwell, D. A. and Lee, R. G., *Recycle Stream Effects on Water Treatment,* American Water Works Association Research Foundation, Denver, CO, 1993.
68. Campbell, I., Tzipori, S., Hutchinson, G., and Angus, K. W., Effect of disinfectants on survival of *Cryptosporidium* oocysts, *Vet. Res.*, 111, 414, 1982.
69. Korich, D. G., Mead, J. R., Madore, M. S., Sinclair, N. A., and Sterling, C. R., Effects of ozone, chlorine dioxide, chlorine, and monochloramine on *Cryptosporidium parvum* oocyst viability, *Appl. Environ. Microbiol.*, 56, 1423, 1990.
70. Rasmussen, V., Nissen, B., Lund, E., and Strand, R. L., A comparison of *Cryptosporidium, Giardia* and virus inactivation using chlorine residual and redox monitoring, in *Proceeding of the American Water Works Association's Annual Conference* (New York, June 19–23, 1994), American Water Works Association, Denver, CO, 1994, 332.
71. Peeters, J. E., Mazas, E. A., Masschelein, W. J., Villacorta, M. I., and Debacker, E., Effect of disinfection of drinking water with ozone or chlorine dioxide on survival of *Cryptosporidium parvum* oocysts, *Appl. Environ. Microbiol.*, 55, 1519, 1989.
72. Ransome, M. E., Whitmore, T. N., and Carrington, E. G., Effect of disinfectants on the viability of *Cryptosporidium parvum* oocysts, *Water Supply*, 11(1), 103, 1993.
72a. Parker, J. F. W., Greaves, G. F., and Smith, H. V., The effect of ozone on the viability of *Cryptosporidium parvum* oocysts and a comparison of experimental methods, *Water Sci. Technol.,* 27, 93, 1993.
72b. Armstrong, D. C., Casemore, D. P., Couper, A. M., Martin, A. D., and Naylor, P. J., *Disinfection of* Cryptosporidium parvum *with Ozone at High Concentrations,* ICI Chemicals and Polymers, Ltd., Cheshire, U.K., 1994.
72c. Owens, J. H., Miltner, R. J., Schaefer, III, F. W., and Rice, E. W., Pilot-scale ozone inactivation of *Cryptosporidium* and *Giardia,* in *Proceedings of the American Water Works Association's Water Quality Technology Conference,* Part II (San Francisco, CA, November 6–10, 1994), American Water Works Association, Denver, CO, 1994, 1319.
73. Finch, G. R., Black, E. K., Gyurek, L., and Belosevic, M., Ozone inactivation of *Cryptosporidium parvum* in demand-free phosphate buffer determined by *in vitro* excystation and animal infectivity, *Appl. Environ. Microbiol,* 59(12), 4203, 1993.
74. Lorenzo-Lorenzo, M. J., Ares-Mazas, M. J., de Maturana, V. M., and Duran, D. D., Effect of ultraviolet disinfection of drinking water on the viability of *Cryptosporidium parvum* oocysts, *J. Parasitol.*, 79(1), 67, 1993.
75. Parker, J. F. W. and Smith, H. V., Destruction of oocysts of *Cryptosporidium parvum* by sand and chlorine, *Water Res.*, 27, 729, 1993.
76. Carrington, E. G. and Ransome, M. E., *Factors Influencing the Survival of* Cryptosporidium *Oocysts in the Environment,* Foundation for Water Research, Bucks United Kingdom, 1994.
77. Finch, G. R., Kathleen, B., and Gyurek, L. L., Ozone and chlorine inactivation of *Cryptosporidium*, in *Proceedings of the American Water Works Association's Water Quality Technology Conference* (San Francisco, CA, November 6–10, 1994), American Water Works Association, Denver, CO, 1994, 1303.
78. USEPA, National Primary Drinking Water Regulations; Filtration and Disinfection; Turbidity; *Giardia lamblia*, Viruses, *Legionella*, and Heterotrophic Bacteria, Federal Register 54(124), 27486-27541, U.S. Environmental Protection Agency, Washington, D.C., 1989.
79. Regli, S., Macler, B. A., Cromwell, J. E., Zang, X., Gelderloos, A. B., Grubbs, W. D., and Letkiewicz F., Framework for decision making: EPA perspective, in *Safety of Water Disinfection: Balancing Chemical and Microbial Risks*, Craun, G. F., Ed, ILSI Press, Washington, D.C., 1993, 487.
80. USEPA, National Primary Drinking Water Regulations: Enhanced Surface Water Treatment Requirements; Proposed Rule, Federal Register, 59(159), 38832-38858, U.S. Environmental Protection Agency, Washington, D.C., 1994.
81. USEPA, Monitoring Requirements for Public Drinking Water Supplies; Proposed Rule, Federal Register, 59(28), 6332-6444, U.S. Environmental Protection Agency, Washington, D.C., 1994.

Chapter 5

Prophylaxis and Chemotherapy: Human and Animal

Byron L. Blagburn and Rosemary Soave

CONTENTS

I. INTRODUCTION

Despite numerous years of active research on treatment and control of cryptosporidiosis, we still lack a consistently effective, approved cryptosporidicidal agent for use in either humans or animals. Major research initiatives by the U.S. National Institutes of Health and other sources of research funds, due principally to the AIDS epidemic, have engendered much interest in development of *in vivo* and *in vitro* methods for evaluation of cryptosporidicidal drugs (see Chapters 8 and 9). Although progress in identification and evaluation of promising lead compounds has been slow, research efforts have identified molecules with efficacy against *Cryptosporidium*, using target animals, laboratory animal models, or *in vitro* techniques. Some agents have undergone or are are undergoing evaluations in humans. In this chapter, we review progress in the evaluation of drugs with either chemotherapeutic or chemopreventive potential. We begin with drugs currently under evaluation in humans and proceed to drugs evaluated in animals, in cell cultures, or using other *in vitro* techniques.

II. TREATMENT AND PROPHYLAXIS IN HUMANS

A. CLINICAL MANIFESTATIONS

Human clinical cryptosporidial infection ranges from mild, self-limited diarrhea to fulminant, cholera-like enteritis complicated by extraintestinal disease. Asymptomatic infection occurs, but the frequency is unknown. Clinical manifestations of symptomatic cryptosporidial infection includes watery diarrhea, abdominal pain, nausea, vomiting, flatulence, bloating, urgency, incontinence, anorexia, and weight loss.

The severity and duration of human cryptosporidial disease varies with host immune competence.[1,2] When immune function is intact, cryptosporidiosis is usually explosive at onset, lasts approximately 10 to 14 days,

0-8493-7695-5/97/$0.00+$.50

and is followed by a complete clinical and parasitologic recovery. Oocyst shedding may lag behind clinical resolution of symptoms by as long as a few weeks. In the immunologically impaired host, onset of cryptosporidiosis is usually insidious — diarrheal symptoms precede detection of the organism in stool by weeks to months. In these individuals, the enteritis becomes chronic, and severity may wax and wane. In persons with the acquired immunodeficiency syndrome (AIDS), diarrhea frequency and volume often escalate unrelentingly as immune function becomes more deranged. However, cryptosporidiosis may also resolve spontaneously in patients anywhere along the spectrum of HIV infection, thus complicating the interpretation of uncontrolled treatment data. For the most part, cryptosporidiosis is a devastating complication for persons with AIDS — it contributes significantly to morbidity and hastens death.[3,4] In addition to AIDS, immunological deficiencies and other conditions associated with protracted cryptosporidiosis include congenital hypogammaglobulinemia, IgA deficiency, concurrent viral infections, malnutrition, and exogenous immunosuppression, as with corticosteroids.[5,6] If exogenous immunosuppression can be discontinued, cryptosporidiosis will resolve.

Biliary cryptosporidiosis has been well documented only in the immunocompromised host. Because definitive diagnosis requires invasive procedures that may not be justified in the absence of useful treatment options, the true incidence of this complication is not known.[7-10] Pancreatitis and reactive arthritis associated with cryptosporidiosis have also been described in immunocompetent as well as AIDS patients.[9-12] Respiratory cryptosporidiosis, well recognized in birds, has been reported in a few humans, but in these cases contamination of respiratory secretions from intestinal infection has not been satisfactorily ruled out.[4-6,13]

B. CHEMOTHERAPEUTIC AGENTS

There is no consistently efficacious therapy for cryptosporidiosis in humans. The limited availability and poor reproducibility of *in vivo* and *in vitro* methods for screening drugs and conducting preclinical studies have greatly hindered efforts to identify effective therapy. Progress has been made in developing animal models of the disease, but none are widely accepted. Furthermore, little is known of how cryptosporidial species and strain differences impact parasite virulence.

Despite the paucity of supportive preclinical data, the urgent need to identify effective therapy for this disease in persons with AIDS has led to the unprecedented administration of a vast array of chemotherapeutic, immunomodulatory, and palliative agents to this population. The largely anecdotal experience thus generated has resulted in a formidable list of approximately 100 ineffective compounds.[13,14] Over the past decade, controlled treatment trials have provided useful insights for designing effective studies of cryptosporidial therapy but have fallen short of identifying efficacious agents for this infection. The following is a compilation of chemotherapeutic and palliative attempts at treating cryptosporidiosis. Immunomodulatory therapy has been covered in the previous chapter.

1. Macrolides

a. Oral Spiramycin (Rhône-Poulenc Rorer Pharmaceuticals; Collegeville, PA)

Initially, interest centered on spiramycin, a macrolide discovered in 1957 and, though not licensed in the U.S., used worldwide. Spiramycin has broad spectrum antibacterial activity but interest in its potential as an anticryptosporidial agent derives from its activity against a related protozoan, *Toxoplasma gondii*. In the mid- to late-1980s, there were at least five anecdotal reports of success and failure with spiramycin for cryptosporidiosis.[15-19] These divergent results led to the first controlled cryptosporidiosis treatment trials, but they, too, provided divergent data. In a prospective study, 15 patients were randomized to either spiramycin or erythromycin; both antibiotics were associated with clinical response, but side effects limited their use.[20] In a placebo-controlled "n of 1" trial, no difference was shown between spiramycin and placebo.[21]

There have been two placebo-controlled trials of spiramycin for cryptosporidiosis in immunocompetent children and one in adults with AIDS. Of 39 immunocompetent infants and children in Durban, South Africa, randomized to 5 days of treatment with either spiramycin (75 mg/kg) or placebo,[22] no difference was found between active drug and placebo possibly because the dose was too low or the duration of therapy was too short. Of 44 immunocompetent infants in Costa Rica randomized to spiramycin (100 mg/kg/day) or matching placebo (which contained lactose) for 10 days, a statistically significant decrease in diarrhea and oocyst excretion was found in the spiramycin-treated group.[23] In a randomized, double-blind, placebo-controlled trial, 3.0 MIU (approximately 3 g) spiramycin 3 times daily for 3 weeks were no better than placebo in the treatment of cryptosporidial diarrhea in 73 patients with AIDS (Soave, unpublished). A food interaction study in normal hosts indicated that spiramycin absorption was significantly decreased in the presence of food, perhaps explaining the poor results obtained in the controlled trial in AIDS patients. Interestingly, less rigorously obtained data from the open-label compassionate use program that ran concomitantly with the placebo-controlled study revealed that 28 of 37 AIDS patients had a favorable clinical response, and 12 had parasitologic eradication, thus underscoring the importance of placebo-controlled clinical treatment trials.[19]

b. Intravenous Spiramycin

To circumvent absorption problems, an efficacy and safety study of intravenous spiramycin was conducted from 1989 to 1991. In this multi-center, single-blind, placebo-controlled, National Institutes of Allergy and Infectious Diseases (NIAID)-sponsored AIDS Clinical Trials Group (ACTG #113) trial, two doses of intravenous spiramycin (3.0 and 4.5 MIU) were examined. Of 31 AIDS patients enrolled, 5 (18%) had a favorable response to treatment, i.e., both clinical and parasitologic improvement, whereas 16 (57%) had partial benefit but did not meet the favorable response criteria. Analysis of the group revealed a statistically significant drop in oocyst count while receiving spiramycin as compared to placebo.[24] However, administration of intravenous spiramycin was associated with paresthesias, taste perversion, nausea, and vomiting, and at doses greater than 75 mg/kg/day there were cases of severe colitis due to intestinal injury.[25]

c. Azithromycin (Pfizer Labs Division; Parsippany, NJ)

Azithromycin is an azalide with a long half-life (6 to 8 hours) that achieves high tissue concentrations, particularly in the biliary tree and gallbladder. Experience with azithromycin for cryptosporidiosis in humans includes anecdotal data, a placebo-controlled study, and open-label protocols. There was complete resolution of cryptosporidial diarrhea with 40 mg/kg azithromycin in two children, immunocompromised after receiving chemotherapy.[26] In a multi-center, double-blind, crossover trial, 85 AIDS patients with cryptosporidial diarrhea were randomized to receive either 900 mg azithromycin, once daily for 3 weeks, or matching placebo.[27] Preliminary analysis of the data reveals no significant trend towards improvement in bowel movement frequency, stool oocyst shedding, and weight stabilization in the azithromycin treated group. Data from the first 33 patients revealed a highly significant decrease in stool oocyst counts at day 7 and day 21 in those subjects with the highest serum azithromycin levels ($p < .01$). In an ongoing compassionate-use program conducted by Pfizer Pharmaceuticals, the first 60 patients had clinical improvement but marginal reduction of stool oocyst counts.[28] A lactose-free form of azithromycin in 300 mg tablets was used in the placebo-controlled trial and is available through Pfizer's compassionate-use program.

To further investigate the effect of drug absorption in treatment of this disease, five very ill patients with AIDS and cryptosporidial diarrhea who had failed oral azithromycin were given intravenous azithromycin in a prospective, open-labeled, case-control study.[29] While bowel movement frequency and stool oocyst counts remained unchanged, serum alkaline phosphatase levels decreased in all five subjects who received intravenous azithromycin compared to 4/13 controls ($p < 0.05$). The data suggest a therapeutic effect on biliary tree infection but may be confounded by concomitant opportunistic infection with *Mycobacterium avium* complex or other organisms susceptible to azithromycin. Intravenous azithromycin at doses up to 2 g daily for as long as 2 weeks were well tolerated.

d. Clarithromycin and Other New Macrolides

Very little information is available on other macrolides including clarithromycin, roxithromycin,[30] and dirithromycin. A study to investigate the efficacy, absorption, and safety of two oral clarithromycin preparations, provided by Abbott Laboratories (liquid vs. tablet), has recently been initiated at two sites in New York City.

2. Benzeneacetonitrile Derivatives

a. Diclazuril (Janssen Research Foundation; Titusville, NJ)

Interest in this class of agents for treatment of cryptosporidiosis stems from their activity against the related coccidian, *Eimeria*. In addition to two anedoctal reports,[31,32] diclazuril sodium was tested in a randomized, double-blind, placebo-controlled, escalating-dose trial in AIDS patients at three centers in New York City.[33] Doses ranging from 50 to 800 mg were no more efficacious than placebo. However, the very small number of complete responders were the only subjects who had detectable serum diclazuril levels, suggesting that lack of efficacy may have been due to poor drug bioavailability.

b. Letrazuril (Janssen Research Foundation; Titusville, NJ)

Poor absorption of diclazuril led to synthesis of the *p*-fluor analog, letrazuril. This agent was studied in a randomized, double-blind, placebo-controlled, multi-center treatment trial with a pharmacokinetic arm (NIAID-ACTG 198). Letrazuril bioavailability was better than that of diclazuril, but patients experienced little clinical improvement. There was a highly significant parasitologic effect when stool specimens were examined by acid-fast staining, but specimens that were acid-fast negative were positive by ELISA antigen capture and indirect immunofluorescent assay, suggesting that letrazuril may interfere with acid-fast staining of oocysts.[34] Although data obtained in other open-label studies and anecdotal cases was more promising, in the majority, acid-fast staining of stool smears was used to assess drug parasitologic effect.[35-40] The benzeneacetonitrile derivatives are not currently available for use in human cryptosporidiosis.

3. Miscellaneous Agents

a. Paromomycin (Humatin, Parke-Davis; Morris Plains, NJ)

Paromomycin is synonymous with aminosidine, which is marketed outside the U.S. It is a poorly absorbed, oligosaccharide aminoglycoside, related to neomycin, kanamycin, and other aminoglycosides of the catenulin group, that achieves high concentrations in the colon.[41] The oral form has been available in the U.S. for many years for treatment of amebiasis, whereas the intravenous form is not available in the U.S. because of its toxicity.

Since 1990, when its use in the treatment of cryptosporidiosis in patients with AIDS was first reported,[42] there have been numerous reports of dramatic responses to this agent.[43-50] Nonetheless, many patients continue to shed oocysts and others relapse while on therapy, suggesting that the agent is static and not cidal for *Cryptosporidium* and/or that a nonabsorbable agent may not be able to induce a complete and long-lasting response. Two of three placebo-controlled trials of paromomycin, conducted to date, have provided promising results.[51,52] Perhaps most compelling, ten patients randomized to paromomycin or placebo showed a decrease in both bowel movement frequency and oocyst shedding with paromomycin. Problems with the study include the small sample size and the persistence of a median of 109 $\times 10^6$ oocysts in the stool of paromomycin-treated responders. Also worrisome is that four patients developed biliary tract disease and three required cholecystectomy during the followup period. Analysis of the NIAID-sponsored (ACTG 192), placebo-controlled trial of paromomycin in 35 AIDS patients is currently underway. Despite conflicting data, controlled and uncontrolled, paromomycin currently is used widely as the first-line agent against cryptosporidial infection in patients with AIDS. Most patients enter treatment trials after having failed paromomycin. Whether delay in seeking other therapy increases their chances of failure, possibly because of spread to the biliary tree, is not known but worth considering. The role of paromomycin in treatment of cryptosporidiosis continues to be problematic.

b. Nitazoxanide

This is a nitrothiazole benzamide compound first synthesized by Rossignol in 1976 with a wide spectrum of activity against protozoan, helminthic, and bacterial pathogens including flagellates, Coccidians, amoebae, nematodes, cestodes, and trematodes.[53,54] Preliminary data from studies in Mexico and Mali suggest activity in cryptosporidial infection of AIDS patients, and the agent appears to be well tolerated in humans.[55,56] A phase I/II open-label study of the pharmacokinetics, efficacy, and safety of four different doses of nitazoxanide is currently nearing completion in New York City.

c. Difluoromethylornithine (DFMO) (Eflornithine, Marion Merril Dow; Kansas City, MO)

This is an irreversible inhibitor of ornithine decarboxylase that interferes with the biosynthesis of putrescine and other polyamines. Despite two reports of some success with the use of this agent in AIDS patients with cryptosporidiosis, serious toxicity (bone marrow suppression, gastrointestinal intolerance, and hearing impairment) has impeded any further development.[57,58]

d. Atovaquone (Mepron®, Burroughs Wellcome; Research Triangle Park, NC)

This is an antimalarial hydroxy-naphthoquinone evaluated for activity against *Pneumocystis carinii*, *Toxoplasma gondii*, and *Cryptosporidium* in patients with AIDS. Results obtained in trials for cryptosporidosis are not readily available but reportedly are disappointing.[59]

There have also been anecdotal reports of improvement of cryptosporidial diarrhea in AIDS patients with other agents, including zidovudine (AZT; Burroughs Wellcome)[60,61] and recombinant interleukin-2.[62] These effects are most likely due to augmentation of immune function rather than a direct antiparasitic action.

C. SUPPORTIVE THERAPY

Nonspecific antidiarrheal agents including kaolin plus pectin (Kaopectate®), loperamide (Imodium®), diphenoxylate (Lomotil®), bismuth subsalicylate (Pepto-Bismol®), and opiates (tincture of opium, paregoric) are often helpful, but regimens must be individualized for each patient. The safety of nonspecific antidiarrheal therapy in patients with cryptosporidiosis is not known.

A synthetic cyclic octapeptide analog of somatostatin, Octreotide (Sandostatin®, Sandoz Pharmaceuticals; East Hanover, NJ) has shown efficacy in patients with severe refractory secretory diarrhea caused by pancreatic cholera syndrome and carcinoid syndrome. A number of anecdotal reports suggest successful use of this agent in treatment of AIDS-related diarrhea.[63-66] A prospective, multicenter trial of escalating doses of subcutaneously administered octreotide in 51 AIDS patients had a response rate of 41%; most responders had neither cryptosporidial infection nor other identifiable pathogens.[67] In a similar study of 34 AIDS patients in France, another somatostatin analog, vapreotide, was more likely to yield a response in patients with conditions other than cryptosporidiosis.[68] Experience with intravenous somatostatin analogs is limited.[30]

In the absence of consistently effective therapy for human cryptosporidiosis, careful management of fluid and electrolyte balance is of paramount importance. Patients not able to maintain stable hydration and electrolyte balance may need intravenous replenishment and possibly nutritional support. The use of total parenteral nutrition is controversial because it requires an invasive procedure and is very expensive. Psychological support is frequently needed to help patients overcome the tremendous anxiety they experience due to fear of incontinence and, hence, the inability to move around freely.

D. FUTURE DIRECTIONS

The search for efficacious therapy for cryptosporidiosis has been frustrating.[14,69,70] It is likely that the numerous failures of therapeutic intervention in patients with AIDS and cryptosporidiosis may reflect factors other than lack of drug efficacy such as profound immune dysfunction, multiple concomitant infections and therapies, and inadequate drug delivery to intracellular and perhaps biliary targets. Data obtained to date underscore the importance of placebo-controlled trials and raise numerous design issues. The latter include the need to stratify patients by immune function and to carefully monitor concomitant medication, particularly antiretrovirals that may correct immune dysfunction. Selection of endpoints and optimal, accurate, practical methods for quantitating diarrhea, as well as oocyst shedding, are of paramount importance.

In addition, to develop efficacious therapeutic modalities for cryptosporidiosis, we need to further develop *in vitro* and *in vivo* models that will help unravel the biologic secrets of this organism. The consensus of a workshop, convened by NIAID in 1991, was that discovery and development of therapeutic agents for cryptospidiosis depended on a concerted effort to enhance laboratory investigation and recruit young investigators and trained parasitologists into this area of study.[71]

III. TREATMENT AND PROPHYLAXIS IN ANIMALS

Numerous compounds have undergone efficacy evaluations against *Cryptosporidium* spp. in nonhuman hosts (Tables 1 to 3).[72-127] Results were obtained from studies that focused on treatment of naturally acquired infections, or treatment or prophylaxis of experimentally induced infections or disease. Although both therapeutic and prophylactic drug regimens have been used, most efficacy evaluations of candidate agents consisted of the latter.

Most studies evaluating potential anticryptosporidial agents were conducted in laboratory rodents (see Chapter 9),[72-104] including inbred and outbred strains of mice, as well as laboratory rats and hamsters. In many instances, laboratory species were rendered more susceptible to infection with *Cryptosporidium* by treatment with immunosuppressive agents such as hydrocortisone or dexamethasone or consisted of genetically immunodeficient stock, such as nude or SCID mice.

Many compounds show promise in laboratory rodents.[72-104] Among them are maduramicin, alborixin, lasalocid, several aromatic amidines, salinomycin, dehydroepiandosterone, paromomycin, L-arginine, glucanthine, clarithromycin, azithromycin, erythromycin, oleandomycin, spiramycin, pristinamycin, arprinocid, halofuginone, sinefungin, metronidazole, sulfadimethoxine, sulfamerazine, sulfameter, sulfamethazine, sulfaquinoxaline, sulfisoxazole, dinitolmide, mepacrine, diclazuril, glucanthine, norfloxacin, mefloquine, and pentamidine (Table 1). In some instances, efficacies exceeded 90% compared to nontreated or diluent-treated controls. Treatment with other compounds resulted in modest but demonstrable

Table 1 Efficacies of Potential Anticryptosporidial Drugs in Laboratory Rodents

Animal	Age	Drug(s)	Dosage and Method of Treatment[a]	Efficacy	Ref.
C57BL6N mice (immunosuppressed with dexamethasone)	Mature	Dehydroepiandrosterone	10 mg/kg/day (subcutaneously days 6–19)	+	72
C.B-17 SCID mice	Mature	Maduramicin alborixin	3 mg/kg/day	+	73
			Administereed in drinking water for 3 weeks	+	
C57BL/6N mice (treated with dexamethasone)	Mature	Paromomycin	0.25 g/kg/day	–	74
		Paromomycin	0.5 g/kg/day	–	
		Paromomycin	1 g/kg/day	+	
		Paromomycin	2 g/kg/day (orally for 10 consecutive days)	+	
BALB/c and SCID mice	Mature	Aminoguanidine	50 mg/kg by I.P. injection every 12 hours	–	75
Nude mice	Mature	L-arginine	4% in diet for 14 days	+	76
Athymic mice (intact and splenectomized)	Mature	Lasalocid	120 mg/kg/day for 3 days	+	77
		Lasalocid	20 mg/kg/day for 8 days	–	
		Sinefungin	10 mg/kg/day for 8 days	–	
		Dehydroepiandrosterone	120 mg/kg/day for 8 days	±	
BALB/c and SCID mice	5- to 6-day-old and 7-week-old	Clarithromycin 14-OH Clarithromycin	17.5–200 mg/kg	±	78
BALB/c mice	Neonate	Artemisinin	200 or 400 mg/kg	–	79
		β-Arteether	Administered subcutaneously	–	
		β-Artemether	or intrarectally	–	
BALB/c mice	Neonate	Azithromycin	100 or 200 mg/kg	+	80
		Clarithromycin	Orally prior to infection	±	
		Erythromycin	(0 hours) and 24, 48, and	±	
		Paromomycin	72 hours after infection	+	
SCID mice	Neonate	Atovaquone	≥100 mg/kg/day	–	81
Outbred mice (immunosuppressed with prednisolone)	Mature	Lasalocid	64 mg/kg/day	+	82
		Lasalocid	128 mg/kg/day	+	
		Azithromycin	400 mg/kg/day (for 3 days)	+	
Outbred mice	3-day-old	Maduramicin	1.0, 2.5 mg/kg	+	83
		Alborixin	1.5, 2.5 mg/kg	+	
		Enrofloxacin	1.0, 3.0 mg/kg	–	
		Aromatic amidines	2.8, 11.3 mg/kg (orally daily for 5 days)	±	
Outbred mice	7-day-old	Lasalocid	20–30 mg/kg (daily for 5–7 days)	+	84
Inbred and outbred mice	1- to 3-day-old	Clopidol	0.25 mg/mouse/day	–	85
		Clopidol × 10	2.5 mg/mouse/day	–	
		Methylbenzoquate	0.25 mg/mouse/day	–	
		Methylbenzoquate × 10	2.5 mg/mouse/day	–	
		Clopidol + Methylbenzoquate	0.1 mg/mouse/day	–	
		Clopidol + Methylbenzoquate × 10	1.0 mg/mouse/day	–	
		Robenidine HCl	0.03 mg/mouse/day	–	
		Robenidine HCl × 10	0.3 mg/mouse/day	–	
		Decoquinate	2.0 mg/mouse/day	–	
		Decoquinate × 2	4.0 mg/mouse/day	–	
		Furazolidone	0.4 mg/mouse/day	–	
		Furazolidone × 5	2.0 mg/mouse/day	–	
		Amprolium	0.25 mg/mouse/day	±	
		Arprinocid	0.06 mg/mouse/day	+	
		Nicarbazin	0.125 mg/mouse/day	–	

Table 1 (continued) Efficacies of Potential Anticryptosporidial Drugs in Laboratory Rodents

Animal	Age	Drug(s)	Dosage and Method of Treatment[a]	Efficacy	Ref.
Inbred and outbred mice (continued)		Dinitolmide	1.25 mg/mouse/day	±	
		Furaltedone	0.2 mg/mouse/day	–	
		Furaltedone × 10	2.0 mg/mouse/day	–	
		Sulphaquinoxaline	2.8 mg/mouse/day	±	
		Salinomycin	0.06 mg/mouse/day	+	
		Halofuginone	0.003 mg/mouse/day (treatment began 2 days before infection and continued for about 7 days)	Toxic	
C57 mice	Neonate	Ethopabate	28 mg/mouse/day	–	86
		Nicarbazin	0.1 mg/mouse/day	–	
		Sulfaquinoxaline	30 mg/mouse/day	–	
		Furaltadone	0.1 mg/mouse/day	–	
		Enterolyte-N	0.02 ml/mouse/day	–	
		Sulphamethazine	5 mg/mouse/day	–	
		Trinamide	3 mg/mouse/day	–	
		Amprolium	0.02, 8.0 mg/mouse/day	–	
		Phenamidine	0.01 ml/mouse/day	–	
		Zoaquin	0.5 mg/mouse/day	–	
		Halofuginone	6 mg/mouse/day	–	
		Salinomycin	0.02 mg/mouse/day	–	
		Monensin	0.04 mg/mouse/day	–	
		Emtryl	4 mg/mouse/day	–	
		Arprinocid	8 mg/mouse/day (daily for 8 days after infection) (dose doubled for last 4 days)	–	
BALB/c mice	Neonate	Thiabendazole	100, 200 mg/kg	–	87
		Parbendazole	30 mg/kg	–	
		Oxibendazole	5, 10 mg/kg	–	
		Mebendazole	8, 15 mg/kg	–	
		Albendazole	7.5, 10, 15 mg/kg (daily for 3 days [-1, 0, 1]; mice infected on day 0)	–	
Hamster	20–24 months	Dehydroepiandrosterone	120 μg/g/day subcutaneously for 7 days prior to infection	+	88
Hamster	5-day-old, 12-day-old	Arprinocid	0.5 or 1.0 mg/animal on days 3,7,11,14 or 6,10,14,18 after infection	±	89
Rat (immunosuppressed with dexamethasone)	Mature	Halofuginone	37.5 μg/kg/day	–	90
		Halofuginone	75 μg/kg/day	–	
		Halofuginone	150 μg/kg/day	+	
		Halofuginone	300 μg/kg/day	+	
		Halofuginone	600 μg/kg/day	+	
		Halofuginone	900 μg/kg/day (drug was added to feed for 11 days)	+	
Rat (immunosuppressed with dexamethasone)	Mature	Paromomycin	50 mg/kg/day	–	91
		Paromomycin	100 mg/kg/day	+	
		Paromomycin	200 mg/kg/day	+	
		Paromomycin	400 mg/kg/day	+	
Rat (immunosuppressed with hydrocortisone)	Mature	Paromomycin	10 mg/kg/day	–	92
			50 mg/kg/day	–	
			100 mg/kg/day (treated for 10 days)	+	

Table 1 (continued) Efficacies of Potential Anticryptosporidial Drugs in Laboratory Rodents

Animal	Age	Drug(s)	Dosage and Method of Treatment[a]	Efficacy	Ref.
Rat (Immunosuppressed with dexamethasone)	Mature	Paromomycin	200, 400 mg/kg/day	+	93
		Gentamicin	200 mg/kg/day	–	
		Neomycin	200 mg/kg/day	–	
		Kanamycin A	200 mg/kg/day	–	
		Polymixin B	200 mg/kg/day	–	
		Streptomycin	200 mg/kg/day	–	
		Azithromycin	400 mg/kg/day	+	
Rat (immunosuppressed with hydrocortisone)	Mature	Sulfadoxinepyrimethamine	3–60 mg/kg/day	–	94
		Quinicrine	?	–	
		Trimethoprimsulfamethoxazole	6–250 mg/kg/day	–	
		Bleomycin	0.05 mg/kg/day	–	
		Elliptinium	5 mg/kg/day	–	
		Daunorubicin	2 mg/kg/day	–	
		Pentamidine	60 mg/kg/day	–	
		α-difluoro-methylornithin	25, 100, 400 mg/kg/day	–	
		Diclazuril	4 mg/kg/day	–	
		N-methyglucamine	50 mg/kg/day	–	
		Vitamin A	10,000, 25,000 IU/day	±	
		Sinefungin	0.25, 2, 6, 10 mg/kg/day	±	
		Lasalocid	2,10 mg/kg/day	+	
		Metronidazole	25, 50 mg/kg/day	+	
		Sulfadimethoxine	10, 100 mg/kg/day	+	
Rat (immunosuppressed with hydrocortisone)	Mature	Sinefungin	0.01–10 mg/kg/day (curative)	+	95
		Sinefungin	0.01–10 mg/kg/day (preventive)	+	
Rat (immunosuppressed with dexamethasone)	Mature	Dehydroepiandrosterone	60, 120 mg/kg/day	+	96
Rat (immunosuppressed with hydrocortisone)	Mature		**Drugs given days 14 to 28 (curative)**		97
		Sulfadimethoxine	60 mg/kg/day	+	
		Bleomycin	0.05 mg/kg/day	–	
		Lasalocid	10 mg/kg/day	+	
		Metronidazole	25 mg/kg/day	+	
		Mepacrine	10 mg/kg/day	+	
		Elliptinium	5 mg/kg/day	–	
		Daunorubicin	2 mg/kg/day	–	
		Diclazuril	4 mg/kg/day	±	
		Glucanthine	50 mg/kg/day	±	
		α-Difluoro-methylornithine	400 mg/kg/day	–	
			Drugs given days 4 to 10 (preventive)		
		Spiramycin	100 mg/kg/day	–	
		Pristinamycin	70 mg/kg/day	±	
		Lincomycin	50 mg/kg/day	–	
		Clindamycin	100 mg/kg/day	–	
		Norfloxacin	10 mg/kg/day	±	
		Paromomycin	50 mg/kg/day	–	
		Mepacrine	10 mg/kg/day	–	
		Mefloquine	25 mg/kg/day	±	
		Pentamidine	2 mg/kg/day	+	
		Metronidazole	25 mg/kg/day	±	
		Lasalocid	2 mg/kg/day	+	
		Lasalocid	10 mg/kg/day	+	
		Sulfadimethoxine	10 mg/kg/day	±	
		Sulfadimethoxine	60 mg/kg/day	+	
		α-Difluoro-methylornithine	400 mg/kg/day	–	

Table 1 (continued) Efficacies of Potential Anticryptosporidial Drugs in Laboratory Rodents

Animal	Age	Drug(s)	Dosage and Method of Treatment[a]	Efficacy	Ref.
		Sulfaquinoxaline + Vitamin K	160 mg/kg/day	–	
		Sulfamethoxazole + Trimethoprim	250 mg/kg/day	–	
		Sulfadoxine + Pyrimethamine	50 mg/kg/day	–	
		Diclazuril	4 mg/kg/day	±	
		Diclazuril	0.2 mg/kg/day		
		Diclazuril	3 mg/kg/day		
Rat (immunosuppressed with dexamethasone)	Mature	Succinylsulfathiazole	360 mg/kg/day	–	98
		Sulfabenzamide	240 mg/kg/day	–	
		Sulfacetamide	120 mg/kg/day	–	
		Sulfachloropyridazine	360 mg/kg/day	–	
		Sulfadiazine	250 mg/kg/day	–	
		Sulfadimethoxine	120 mg/kg/day	+	
		Sulfadoxine	160 mg/kg/day	–	
		Sulfaguanidine	120 mg/kg/day	–	
		Sulfamerazine	200 mg/kg/day	+	
		Sulfameter	120 mg/kg/day	+	
		Sulfamethazine	175 mg/kg/day	+	
		Sulfamethizole	480 mg/kg/day	–	
		Sulfamethoxazole	320 mg/kg/day	–	
		Sulfamethoxypyridazine	160 mg/kg	–	
		Sulfanilamide	120 mg/kg	–	
		Sulfanilic acid	120 mg/kg/day	–	
		Sulfanitran	200 mg/kg/day	–	
		Sulfapyridine	240 mg/kg/day	–	
		Sulfasalazine	400 mg/kg/day	–	
		Sulfathiazole	120 mg/kg/day	–	
		Sulfisomidine	240 mg/kg/day	–	
		Sulfisoxizole	120 mg/kg/day (administered in food or water)	+	
Rat (immunosuppressed with dexamethasone)	Mature	Azithromycin	50, 100, 200 and 400 mg/kg/day in feed for 11 days	±	99
		Clarithromycin		±	
		Erythromycin		±	
		Oleandomycin		±	
		Spiramycin	200 mg/kg/day only	±	
Rat (immunosuppressed with dexamethasone)	Mature	Azithromycin	200 mg/kg/day (administered by gavage or in feed for 1 hour before infection through 11 days after infection)	+	100
		Spiramycin		–	
Rat (immunosuppressed with dexamethasone)	Mature	Arprinocid	50 mg/kg	+	101
		Arprinocid	25 mg/kg	+	
		Arprinocid	12.5 mg/kg (in feed for 11 days)	–	
Rat (immunosuppressed with dexamethasone)	Mature	Lasalocid	18 mg/kg/day	+	102
		Lasalocid	9.0 mg/kg/day	+	
		Lasalocid	4.5 mg/kg/day	+	
		Lasalocid	2.25 mg/kg/day	+	
		Lasalocid	1.12 mg/kg/day	–	
		Monensin	9.0 mg/kg/day	–	
		Salinomycin	9.0 mg/kg/day (both therapeutic and prophylactic regimens used)	–	
Rat (immunosuppressed with dexamethasone)	Mature	Sulfadimethoxine	120 mg/kg/day	+	103
		Sulfadimethoxine	80 mg/kg/day	+	
		Sulfadimethoxine	40 mg/kg/day	+	
		Sulfadimethoxine	20 mg/kg/day	+	
		Sulfadimethoxine	10 mg/kg/day	–	

Table 1 (continued) Efficacies of Potential Anticryptosporidial Drugs in Laboratory Rodents

Animal	Age	Drug(s)	Dosage and Method of Treatment[a]	Efficacy	Ref
Rat (immunosuppressed with dexamethasone	Mature		**Prophylactic** (daily for 11 days from 1 hour before inoculation)		104
		Diethyldithiocarbamate	900 mg/kg/day	+	
		Diethyldithiocarbamate	600 mg/kg/day	+	
		Diethyldithiocarbamate	300 mg/kg/day	+	
		Diethyldithiocarbamate	150 mg/kg/day	+	
		Diethyldithiocarbamate	75 mg/kg/day	+	
		Diethyldithiocarbamate	37.5 mg/kg/day	–	
		Amphotericin B	20 mg/kg/day	–	
		Eflornithine	3000 mg/kg/day	–	
		Ivermectin	0.4 mg/kg/day	–	
		Levamisole	4 mg/kg/day	–	
		Thiabendazole	50 mg/kg/day	–	
		Thalidomide	150 mg/kg/day	–	
			Therapeutic (daily from days 10–21)		
		Diethyldithiocarbamate	600 mg/kg/day (drugs given subcutaneously or in feed for all studies)	+	

Note: + = demonstrable activity against *C. parvum*; – = no demonstrable activity against *C. parvum*; ± = partial activity, activity at one or more dosages for a compound, or activity for one or more compounds in a group.

[a] Both therapeutic (curative) and prophylactic (preventive) regimens used in some studies.

reductions in parasite numbers compared to controls. Occasionally, infection relapsed when drugs were discontinued.

Table 2 summarizes data derived from evaluation of potential anticryptosporidial agents in ruminants. Among drugs tested, paromomycin, lasalocid, halofuginone, and sulfaquinoxaline possessed demonstrable or partial activity against *C. parvum* infections in ruminant species.[105-118] In many cases, these data support results obtained in laboratory animal models.

Table 3 summarizes drugs evaluated in miscellaneous animal species.[119-127] Though little information is available in animals other than rodents and ruminants, some success was achieved in treating cryptosporidiosis in felids and elaphid snakes.[120,126]

Attempts to identify effective cryptosporidicidal drugs against *Cryptosporidium* spp. in avian hosts have been uniformly unsuccessful. Unfortunately, drugs that possessed activity against *C. parvum* did not show positive results against *Cryptosporidium* spp. in avian hosts. It is unclear as to whether this lack of activity is due to differences in *Cryptosporidium* species, avian vs. mammalian hosts, or to failure to use effective doses or regimens.

Little information is available regarding treatment of cryptosporidial infections in reptilian hosts. Spiramycin did appear somewhat effective in the treatment of naturally acquired *Cryptosporidium* infections in elaphid snakes from a zoological park,[126] but was ineffective in subsequent studies.[126] Halofugione also was ineffective against *C. serpentis* in the latter report.[127]

In vitro efficacy evaluation of anticryptosporidial agents has confirmed the *in vivo* efficacies of agents such as maduramicin, paromomycin, and sinefungin and also has identified additional agents such as colchicine and vinblastine (Table 4; see Chapter 8).[128-133] Efficacies also were demonstrated for some small peptides, which function presumably by interacting with and lysing membranes of invasive sporozoites and merozoites.

Similar to the recommendations for humans, effective treatment of animals suffering from cryptosporidiosis may require oral or parenteral rehydration with fluids and electrolytes, in addition to antidiarrheals and attempted chemotherapy with putative anticryptosporidial drugs. Rehydration is particularly important in young animals, immunocompromised animals, or those suffering from intercurrent disease. The latter should receive appropriate antibiotic therapy if bacterial co-pathogens are involved.

Table 2 Efficacies of Potential Anticryptosporidial Drugs in Ruminants

Animal	Age	Drug(s)	Dosage and Method of Treatment*	Efficacy	Ref.
Calf	Neonate	Paromomycin	100 mg/kg	+	105
		Paromomycin	50 mg/kg	±	
		Paromomycin	25 mg/kg (twice daily in milk for 11 consecutive days)	±	
Calf	Neonate	Halofuginone lactate	30 μg/kg	–	106,107
		Halofuginone lactate	60 μg/kg	+	
		Halofuginone lactate	120 μg/kg (days 2–8 after infection)	+	
Calf	Neonate	Sulfadimethoxine	5-g bolus daily for 21 days	–	108
Calf	Neonate	Halofuginone lactate	30 μg/kg	–	109
		Halofuginone lactate	60 μg/kg	+	
		Halofuginone lactate	125 μg/kg	+	
		Halofuginone lactate	250 μg/kg	+	
		Halofuginone lactate	500 μg/kg (in milk daily for 3–14 days)	+	
Calf	4- to 5-day-old	Lasalocid-Na	15 mg/kg (administered daily for 3 days)	+	110
Calf	1–10 weeks	Sulphadimidine	Therapeutic (200 mg/kg for 3, 3-day courses)	–	111
		Sulphadimidine	Preventive (30 mg/kg for 2, 7-day courses)	–	
		Sulphadimidine	Preventive (40 mg/kg for 14 and 7 days at interval of 6 days	–	
Calf	Up to 14 days old	Amprolium	0.45 g/calf/day	–	112
		Sulphadimidine	5.0 g/calf/day	–	
		Trimethoprim +	0.2 g/calf/day	–	
		Sulphadiazine	1.0 g/calf/day		
		Dimetridazole	1.0 g/calf/day	–	
		Ipronidazole	1.0 g/calf/day	–	
		Quinacrine	0.5 g/calf/day	–	
		Monensin	0.2 g/calf/day	–	
		Lasalocid	0.3 g/calf/day	+ (toxic)	
		Lasalocid	0.03 g/calf/day (drugs administered twice daily throughout study)	–	
Calf	2 days to 3 months	Lasalocid	6–8 mg/kg (daily for 3–4 days)	+	113
Lamb	1-day-old	Halofuginone lactate	0.5 mg/kg/day, 3 or 5 days beginning 2 days after inoculation	+ (toxic?)	114
Goat	2- to 4-day-old	Paromomycin	100 mg/kg twice daily (total daily dose) from day before infection through day 10 after infection	+	115
Goat	?	Sulfaquinoxaline	100 mg/kg plus Vitamin K1 for 10 days	±	116
Goat	Neonate	Sulfadimethoxine	75 mg/kg daily for 5 days	–	117
Camel	4-year-old	Amprolium	Pellets fed orally for 10 days	–	118

Note: + = demonstrable activity against *Cryptosporidium* spp.; – = no demonstrable activity against *Cryptosporidium* spp.; ± = partial activity, activity at one or more dosages for a compound, or activity for one or more compounds in a group.

[a] Both therapeutic (curative) and prophylactic (preventive) regimens used in some studies.

Table 3 Efficacies of Potential Anticryptosporidial Drugs in Miscellaneous Animals

Animal	Age	Drug(s)	Dosage and Method of Treatment	Efficacy	Ref.
Pig	1-day-old	DL-α-difluoro-methylornithine	0.38–1.25/g/kg; fed in milk daily for 10 days after pigs were shedding oocysts	–	119
Cat	6-month-old	Paromomycin	165 mg/kg, (daily for 5 days)	+	120
Chicken (*C. baileyi*)	7-day-old	Lasalocid	Drugs added to feed at 1× or 2× recommended level, alone or in combination with the antioxidant duokvin	–	121
		Maduramicin		–	
		Monensin		–	
		Narasin		–	
		Salinomycin		–	
		Semduramicin		–	
Chicken (*C. baileyi*)	1 day	Halofuginone	3mg/kg feed	–	122
		Salinomycin	60 mg/kg feed	–	
		Lasalocid	75 mg/kg feed	–	
		Monensin	110 mg/kg feed (daily from 5 days before through 20 days after infection)	–	
Quail	4- to 6-day-old	Oxytetracycline	?	–	123
		Neomycin	?	–	
		Furazolidone	?	–	
Turkey	7-week-old	Amprolium	?	–	124
		Chlortetracycline	?	–	
Peacock	3- to 4-day-old	Oxytetracycline HCl	200 ppm administered in water	–	125
Black rat snake, Yellow rat snake, Corn snake	Mature	Spiramycin	80 mg/kg, 3 doses (1 dose every second day) mixed in baby food	±	126
Snakes (7 genera, 13 species)	?	Spiramycin	80 mg/kg given as 3 doses daily for 3 days	–	127
		Halofuginone	1, 0.5 mg/kg given daily or on alternate days	–	

Note: + = demonstrable activity against *C. parvum* or *C. baileyi*; – = no demonstrable activity against *C. parvum* or *C. baileyi*; ± = partial activity, activity at one or more dosages of a compound, or activity for one or more compounds in a group.

Prevention of cryptosporidiosis in animals, again similar to recommendations for humans, is best achieved by eliminating contact with viable oocysts. This is difficult, particularly on farms and in zoos, given the resistance of oocysts to disinfectants (see Chapter 1). Prevention is based largely on knowledge of the biology, life cycle, and modes of transmission of *Cryptosporidium* spp. Infected animals should be quarantined in facilities that can be cleaned and disinfected. Contaminated fomites should be cleaned thoroughly or discarded. Animal care personnel should wear clothing that can be cleaned regularly. Clean food and water must be provided. Rodents and wild mammals should be restricted from access to animal quarters. Neonatal mammals should receive adequate amounts of colostrum early in life. Mammals that do not suckle should be fed milk replacer and perhaps parenteral vitamins to increase their appetites.

In general, methods of control of cryptosporidiosis in animals remain the same as those published previously.[134,135] Additional evaluations of newer compounds that proved effective in laboratory animals and in cell cultures should be conducted in farm, zoo, and companion animals, when possible, to identify those with therapeutic potential. Subsequently, a combination of hygienic practices, effective chemotherapy, and supportive measures should result in effective control of most outbreaks of animal cryptosporidiosis.

Table 4 Efficacies of Potential Anticryptosporidial Drugs or Compounds *In Vitro*

Cell Line or Technique	Drug(s)	Dosage, Concentration, or Method of Treatment	Efficacy	Ref.
MDCK cell line (canine kidney)	Maduramicin	0.5 µg/ml (drug added to culture media for 24–168 hours)	+	128
CaCo-2 cell line (human adenocarcinoma)	Sinefungin	1 µg/ml	+	129
	Diminazene	10 µg/ml	–	
	Paromomycin	100 µg/ml (drugs added to culture media for study period of 3 days)	+	
HT29.74 cell line (human enterocyte)	Colchicine	10^{-4} *M*	+	130
	Vinblastine	10^{-4} *M* (oocysts incubated with cells and media containing drugs)	+	
HT29.74 cell line (human enterocyte)	Paromomycin	5 µg/ml	–	131
		50 µg/ml	±	
		500 µg/ml	±	
		5000 µg/ml (excysted sporozoites incubated in media containing drug for 24 hours)	+	
Not applicable	Lytic peptides			132
	Hecate-1	10 µM	+	
	Shiva-10	100 µM	+	
	Cecropin-b	100 µM (excysted sporozoites incubated with test article for 60 min.; viability determined by vital dye staining)	–	
L929 cell line (mouse fibroblast)	Amphotericin B, Amprolium, Arprinocid, Chloroquine, Cycloguanil, Diclazuril, Glycarbilamide, Halofantrine, Methyl benzoquate, Monensin, Nigericin, Oxytetracycline, Robenidine, Proguanil, Pyrimethamine, Spiramycin, Sulfaquinoxaline, Venturicidin, Zidovudine	.0064–20 µg/ml (excysted sporozoites incubated in media containing drug for 24 hours)	±	133

Note: + = demonstrable activity against *C. parvum*; – = no demonstrable activity against *C. parvum*; ± = partial activity, activity at one or more dosages of a compound, or activity for one or more compounds in a group.

REFERENCES

1. Soave, R., Danner, R. L., Honig, C. L., Ma, P., Hart, C. C., Nash, T., and Roberts, R. B., Cryptosporidiosis in homosexual men, *Ann. Intern. Med.*, 100, 504, 1984.
2. Wolfson, J. S., Richter, J. M., Waldron, M. A., Weber, D. J., McCarthy, D. M., and Hopkins, C. C., Cryptosporidiosis in immunocompetent patients, *N. Engl. J. Med.*, 312, 1278, 1985.
3. Petersen, C., Cryptosporidiosis in patients infected with the human immunodeficiency virus, *Clin. Infect. Dis.*, 15, 903, 1992.

4. Mannheimer, S. B. and Soave, R., Protozoal infections in patients with AIDS, *Inf. Dis. Clin. N. Am.*, 8, 483, 1994.
5. Ungar, B. L. P., *Cryptosporidium* and cryptosporidiosis, in *Textbook of AIDS Medicine*, Broder, S., Merigan, T. C., and Bolognesi, D., Eds., Williams & Wilkins, Baltimore, 1994, chap. 21.
6. Current, W. L. and Garcia, L. S., Cryptosporidiosis, *Clin. Microbiol. Rev.*, 4, 325, 1991.
7. Bouche, H., Housse, T. C., Dumont, J. L., Carnot, F., Menu, Y., Aveline, B., Belghiti, J., Boboc, B., Erlinger, S., Berthelot, Pl, and Pol, S., AIDS-related cholangitis: diagnostic features and course in 15 patients, *J. Hepatol.*, 17, 34, 1993.
8. Texidor, H. S., Godwin, T. S., and Ramirez, E. A., Cryptosporidiosis of the biliary tract in AIDS, *Radiology*, 180, 51, 1991.
9. Hinnant, K., Swartz, A., Rotterdam, H., Bell, E., and Tappen, M., Cytomegaloviral and cryptosporidial chlecystitis in two patients with AIDS, *Am. J. Surg. Pathol.*, 13, 57, 1989.
10. Gross, T. L., Wheat, J., Bartlett, M., and O'Connor, K. W., AIDS and multiple system involvement with *Cryptosporidium, Am. J. Gastroenterol.*, 81, 456, 1986.
11. Miller, T. L., Winter, H. S., Luginbuhl, L. M., Orav, E. J., and McIntosh, K., Pancreatitis in pediatric human immunodeficiency virus infection, *J. Pediatr.*, 120, 223, 1992.
12. Shepherd, R. C., Smail, P. J., and Sinha, G. P., Reactive arthritis complicating cryptosporidial infection, *Arch. Dis. Child.*, 64, 743, 1989.
13. Fayer, R. and Ungar, B. L., *Cryptosporidium* spp. and cryptosporidiosis, *Microbiol. Rev.*, 50, 458, 1986.
14. Soave, R., Treatment strategies for cryptosporidiosis, *Ann. N. Y. Acad. Sci.*, 3, 616, 442, 1990.
15. Anon., Update: treatment of cryptosporidiosis in patients with acquired immunodeficiency syndrome, *Morbid. Mortal. Wkly. Rep.*, 33, 9, 1984.
16. Portnoy, D., Whiteside, M. E., Buckley, E., and Macleod, C. L., Treatment of intestinal cryptosporidiosis with spiramycin, *Ann. Intern. Med.*, 101, 202, 1984.
17. Collier, A. C., Miller, R., and Meyers, J., Cryptosporidiosis after marrow transplantation: person-to-person transmission and treatment with spiramycin, *Ann. Intern. Med.*, 101, 2, 1984.
18. Fafard, J. and Lalonde, R., Long-standing symptomatic cryptosporidiosis in a normal man: clinical response to spiramycin, *J. Clin. Gastroenterol.*, 12, 2, 190, 1990.
19. Moskovitz, B. L., Stanton, T. L., and Kusmierek J. J. E., Spiramycin therapy for cryptosporidial diarrhea in immunocompromised patients, *J. Antimicrob. Chemo.*, 22(Suppl. B), 89, 1988.
20. Connolly, G. M., Dryden, M. S., Shanson, D. C., and Gazzard, B. G., Cryptosporidial diarrhea in AIDS and its treatment, *Gut*, 29, 593, 1988.
21. Woolf, G. M., Townsend, M., and Guyatt, G., Treatment of cryptosporidiosis with spiramycin in AIDS. An "N of 1" trial, *J. Clin. Gastroenterol.*, 9(6), 632, 1987.
22. Wittenberg, D. F., Miller, N. M., and van den Ende, J., Spiramycin is not effective in treating *Cryptosporidium* diarrhea in infants: results of a double-blind randomized trial, *J. Infect. Dis.*, 159, 131, 1989.
23. Saez-Llorens, X., Odio, C. M., Umana, M. A., and Morales, M. V., Spiramycin vs. placebo for treatment of acute diarrhea caused by *Cryptosporidium, Pediatr. Infect. Dis. J.*, 8, 136, 1989.
24. Soave, R., Petillo, J., Dieterich, D. T., Green, K., Flanigan, T., Weikel, C., Cheeseman, S., Bitar, R., Kasemkawat, C., and Moskovitz, B. L., Single-blind efficacy evaluation of intravenous spiramycin in patients with AIDS-related cryptosporidial diarrhea, (submitted).
25. Weikel, C., Lazenby, A., Belitsos, P., McDewit, M., Fleming, H., Jr., and Barbacci, M., Intestinal injury associated with spiramycin therapy of *Cryptosporidium* infection in AIDS, *J. Protozool.*, 38, 147S, 1991.
26. Vargas, S. L., Shenep, J. L., Flynn, P. M., Pui, C. H., Santana, V. M., and Hughes, W. T., Azithromycin for the treatment of severe *Cryptosporidium* diarrhea in two children with cancer, *J. Pediatr.*, 123, 154, 1993.
27. Soave, R., Havlir, D., Lancaster, D., Joseph, P., Leedom, J., Clough, W., Geisler, P., and Dunne, M., Azithromycin therapy of AIDS-related cryptosporidial diarrhea: a multi-center, placebo-controlled, double-blind study, *33rd Interscience Conference on Antimicrobial Agents and Chemotherapy*, Abstract 193, American Society for Microbiology, New Orleans, 1993.
28. Dunne, M. W., Williams, D. J., and Young, L. S., Azithromycin and the treatment of opportunistic infections, *Rev. Contemp. Pharmacother.*, 5, 373, 1994.
29. Friedman, C. R., Soave, R., and Bremer, S., Intravenous azithromycin for cryptosporidiosis in AIDS, *Lancet*, (in press).
30. Kreinik, G., Burstein, O., Landor, M., Bernstein, L., Weiss, L. M., and Wittner, M., Successsful management of intractable cryptosporidial diarrhea with intravenous octreotide, a somatostatin analogue, *AIDS*, 5, 765, 1991.
31. Connolly, G. M., Youle, M., and Gazzard, B. G., Diclazuril in the treatment of severe cryptosporidial diarrhea in AIDS patients, *AIDS*, 4, 700, 1990.
32. Menichetti, F., Moretti, M. V., Marroni, M., Papili, R., and DiCandilo, F., Diclazuril for cryptosporidiosis in AIDS (letter), *Am. J. Med.*, 90, 271–272, 1991.
33. Soave, R., Dieterich, D., Kotler, D., Gassyuk, E., Tierney, A., Lieber, L., and Legendre, R., Oral disclazuril for cryptosporidiosis, 6th International Conference on AIDS, Abstract Th.B, San Francisco, CA, 1990, 519.
34. Soave, R., Dieterich, D., Lew, E., and Poles, M., Letrazuril for AIDS-related cryptosporidial diarrhea: a multi-center, pharmacokinetic study and double-blind, placebo-controlled trial, (submitted).

35. Harris, M., Deutsch, G., Maclean, J. D., and Tsoukas, C. M., A phase I study of letrazuril in AIDS-related cryptosporidiosis, *AIDS*, 8, 1109, 1994.
36. Victor, G. H., Conway, B., Hawley-Foss, N. C., Manion, D., and Sahai, J., Letrazuril therapy for cryptosporidiosis: clinical response and pharmacokinetics (letter), *AIDS*, 7, 438, 1993.
37. Hamour, A. A., Bonington, A., Hawthorne, B., and Wilkins, E. G. L., Successful treatment of AIDS-related cryptosporidial sclerosing cholangitis, *AIDS*, 7, 1449, 1993.
38. Loeb, M., Walach, C., Phillips, J., Fong, I., Salit, I., and Rachlis, A., Walmsley, S., Treatment with letrazuril of refractory cryptosporidial diarrhea complicating AIDS, *J. AIDS Retrovirol.*, 10, 48, 1995.
39. Guillem, S., Gomez, M., Romeu, J., Raventos, A., Fernandez, A., Condom, M. J., and Clotet, B., Letrazuril for treatment of severe cryptosporidial diarrhea in AIDS, 8th International Conference on AIDS/III STD World Congress, Abstract PoB3257, Amsterdam, 1992.
40. Murdoch, D. A., Bloss, D. E., and Glover, S. C., Successful treatment of cryptosporidiosis in and AIDS patient with letrazuril, *AIDS,* 7, 1279, 1980.
41. Scaglia, M., Atzori, C, Marchetti. G., Orso, M., Maserati, R, Orani, A., Novati, S., and Olliaro, P., Effectiveness of aminosidine (paromomycin) sulfate in chronic *Cryptosporidium* diarrhea in AIDS patients: an open, uncontrolled, prospective clinical trial (letter), *J. Infect. Dis.*, 170, 1349, 1994.
42. Gathe, J., Jr., Piot, D., Hawkins, K., Bernal, A., Clemmons, J., and Stool, E., Treatment of gastrointestinal cryptosporidiosis with paromomycin (Abstract 2121), 6th International Conference on AIDS, June 20–23, San Francisco, CA, 6, 384, 1990.
43. Andreani, T., Modigliani, R., Le Charpentier, Y., Galian, A., Brouet, J.-C., Liance, M., Lechance, J., Messing, B., and Vernisse, B., Acquired immunodeficiency with intestinal cryptosporidiosis: possible transmission by Haitian whole blood, *Lancet*, 1, 1187, 1983.
44. Whiteside, M. E., Barkin, J. S., May, R. G., Weiss, S. D., Fischl, M. A., and Macleod, C. L., Enteric coccidiosis among patients with the acquired immunodeficinecy syndrome, *Am. J. Trop. Hyg.*, 33, 1065, 1984.
45. Clezy, K., Gold, J., Blaze, J., and Jones, P., Paromomycin for the treatment of cryptosporidial diarhea in AIDS patients (letter), *AIDS*, 5, 1146, 1991.
46. Armitage, K., Flanigan, T., Carey, J., Frank, I., MacGregor., R. R., Ross, P., Goodgame, R., and Turner, J., Treatment of cryptosporidiosis with paromomycin: a report of five cases, *Arch. Intern. Med.*, 152, 2497, 1992.
47. Fichtenbaum, C. J., Ritchie, D. J., and Powderly, W. G., Use of paromomycin for treatemnt of cryptosporidiosis in patients with AIDS, *Clin. Infect. Dis.*, 16, 298, 1993.
48. Danziger, L. H., Kanyok, T. P., and Novak, R. M., Treatment of cryptosporidial diarrhea in an AIDS patients with paromomycin, *Ann. Pharmacother.*, 27, 1460, 1993.
49. Wallace, M. R., Nguyen, M. T., and Newton, J. A., Use of paromomycin for the treatment of cryptosporidiosis in patients with AIDS, *Clin. Infect. Dis.,* 17, 1070, 1993.
50. Bissuel, F., Cotte, L., Rabodonirina, R. P., Piens, M. A., and Trepo, C., Paromomycin: an effective treatemnt for cryptosporidial diarrhea in patients with AIDS, *Clin. Infect. Dis.*, 18, 447, 1994.
51. Kanyok, T. P., Novak, R. M., and Danziger, L. H., Preliminary results of a randomized, blinded, control study of paromomycin vs. placebo for the tretament of cryptosporidial diarrhea in AIDS patients, 9th International Congress on AIDS, Abstract Po-B10-1508, Berlin, 1993.
52. White, A. C., Jr., Chappell, C. L., Hayat, C. S., Kimball, K. T., Flanigan, T. P., and Goodgame, R. W., Paromomycin for cryptosporidiosis in AIDS: a prospective, double-blind trial, *J. Infect. Dis.*, 170, 419, 1994.
53. Rossignol, J. F., Maisonneuve, H., and Cho, Y. W., Nitroimidazoles in the treatment of trichomoniasis, giardiasis and amoebiasis, *Int. J. Clin. Pharmacol. Ther. Toxicol.*, 22, 63, 1984.
54. Rossignol, J. F. and Maisonneuve, H., Nitazoxanide in the treatment of *Taenia saginata* and *Hymenolepsis nana*, *Am. J. Trop. Med. Hyg.*, 33, 511, 1984.
55. Murphy, J. and Friedman, J. C., Preclinical toxicology of nitazoxanide — a new antiparasitic comnpound, *J. Applied Toxicol.*, 5, 49, 1985
56. Doumbo, O., Rossignol, J. F., Pichard, E., Traore, H., Dembele, M., Diakite, M., Traore, F., and Dialio, A., D., Nitazoxanide in the treatment of cryptosporidiosis in 24 AIDS patients with chronic diarrhea in Mali, (submitted).
57. Rolston, K. V .I., Fainstein, V., and Bodey, G. P., Intestinal cryptosporidiosis treated with efluornithine: a prospective study among patients with AIDS, *J. AIDS,* 2, 426, 1989.
58. Soave, R., Sjoerdsma, A., and Cawein, M. J., Treatment of cryptosporidiosis in AIDS patients with DFMO, First International Congress on AIDS, Abstract 30, Atlanta, GA, 1985.
59. Gutteridge, W.E., 566C80, an antimalarial hydroxynaphthoquinone with broad spectrum: experimental activity against opportunistic parasitic infections of AIDS patients, *J. Protozool.*, 38, 141, 1991.
60. Greenberg, R. E., Mir, R., Bank, S., and Siegal, F. P., Resolution of intestinal cryptosporidiosis after treatment of AIDS with AZT, *Gastroenterol.*, 97, 1327, 1989.
61. Chandrasekar, P. H., "Cure" of chronic cryptosporidiosis during treatment with azidothymidine in a patient with acquired immunodeficiency syndrome, *Am. J. Med.*, 83, 187, 1987.
62. Kern, P., Toy, J., and Dietrich, M., Preliminary clinical observation with recombinant interleukin-2 in patients with AIDS or LAS, *Blut*, 50 1, 1985.

63. Romeu, J., Miro, J. M., Sirera, G., Mallolas, J., Arnal, J., Valls, M. E., Tortosa, F., Clotet, B., and Foz, M., Efficacy of octreotide in the management of chronic diarrhea in AIDS, *AIDS,* 5, 1495, 1991.
64. Fanning, M., Monte, M., Sutherland, L. R., Broadhead, M., Murphy, G. F., and Harris, A. G., Pilot study of Sandostatin® (octreotide) therapy of refractory HIV-associated diarrhea, *Dig. Dis. Sci.*, 36, 476, 1991.
65. Moroni, M., Esposito, R., Cernuschi, M., Franzetti, F., Carosi, G. P., and Fiori, G. P., Treatment of AIDS-related refractory diarrhea with octrotide, *Digestion*, 54, 30, 1993.
66. Clotet, B., Sirera, G., Cofran, F., Monterola, J. M., Tortosa, F., and Fox, M., Efficacy of the somatostatin analogue, Sandostatin®, for cryptosporidial diarrhea in patients with AIDS, *AIDS*, 3, 857, 1989
67. Cello, J. P., Grendell, J. H., Basuk, P., Simon, D., Weiss, L., Wittner, M., Rood, R. P., Wilcox, M., Forsmark C. E., Read, A. E., Satow, J. A., Weikel, C. S., and Beaumont, C., Effect of octreotide on refractory AIDS-associated diarrhea: a prospective, multicenter clinical trial, *Ann. Intern. Med.*, 115, 705, 1991.
68. Girard, P., Goldshmidt, E., Vittecoq, D., Massip, P., Gastiaburu, J., Meyohas, M. C., Couland, J. P., and Schally, A. V., Vapreotide, a somatostatin analogue, in cryptosporidiosis and other AIDS-related diarrheal diseases, *AIDS*, 6, 715, 1992.
69. St. Georgiev, V., Opportunistic infections: treatment and developmental therapeutics of cryptosporidiosis and isosporiasis, *Drug Devel. Res.*, 28, 445, 1993.
70. Ritchie D. J. and Becker, E. S., Update on the management of intestinal cryptosporidiosis in AIDS, *Ann. Pharmacother.*, 28, 767, 1994.
71. Laughon, B. E., Allaudeen, H. S., Becker, J. M., Current, W. L., Feinberg, J., Frenkel, J., Hafner, R., Hughes, W. T., Laughlin, C. A., Meyers, J. D., and Young, L. S., Summary of the workshop on furture directions in discovery and development of therapeutic agents for opportunistic infections associated with AIDS, *J. Infect. Dis.*, 164, 244, 1991.
72. Rasmussen, K. R., Healey, M. C., Cheng, L., and Yang, S., Effects of dehydroepiandrosterone in immunosuppressed adult mice infected with *Cryptosporidium parvum, J. Parasitol.,* 8, 429, 1995.
73. Mead, J. R., You, X., Pharr, J. E., Belenkaya, Y., Arrowood, M. J., Fallon, M. T., and Schinazi, R. F., Evaluation of maduramicin and alborixin in a SCID mouse model of chronic cryptosporidiosis, *Antimicrob. Agents Chemother.,* 39, 854, 1995.
74. Healey, M. C., Yang, S., Rasmussen, K. R., Jackson, M. K., and Du, C., Therapeutic efficacy of paromomycin in immunosuppressed adult mice infected with *Cryptosporidium parvum, J. Parasitol.,* 81, 114, 1995.
75. Kuhls, T. L., Mosier, D. A., Abrams, V. L., Crawford, D. L., and Greenfield, R. A., Inability of interferon-gamma and aminoguanidine to alter *Cryptosporidium parvum* infection in mice with severe combined immunodeficiency, *J. Parasitol.,* 80, 480, 1994.
76. Leitch, G. J. and He, Q., Arginine-derived nitric oxide reduces fecal oocyst shedding in nude mice infected with *Cryptosporidium parvum, Infect. Immun.,* 62, 5173, 1994.
77. Leitch, G. J. and He, Q., Putative anticryptosporidial agents tested in an immunodeficient mouse model, *Antimicrob. Agents Chemother.,* 38, 865, 1994.
78. Cama, V. A., Marshall, M. M., Shubitz, L. F., Ortega, Y. R., and Sterling, C. R., Treatment of acute and chronic *Cryptosporidium parvum* infections in mice using clarithromycin and 14-OH clarithromycin, *J. Euk. Microbiol.,* 41, 25S, 1994.
79. Fayer, R. and Ellis, W., Qinghaosu (Artemisinin) and derivatives fail to protect neonatal BALB/c mice against *Cryptosporidium parvum* (CP) infection, *J. Euk. Microbiol.,* 41, 41S, 1944.
80. Fayer, R. and Ellis, W., Glycoside antibiotics alone and combined with tetracyclines for prophylaxis of experimental cryptosporidiosis in neonatal BALB/c mice, *J. Parasitol.,* 79, 533, 1993.
81. Rohlman, V. C., Kuhls, T. L., Mosier, D. A., Crawford, D. L., Hawkins, D. R., Abrams, V. L., and Greenfield, R. A., Therapy with atovaquone for *Cryptosporidium parvum* infection in neonatal severe combined immunodeficiency mice, *J. Infect. Dis.,* 168, 258, 1993.
82. Kimata, I., Shigehiko, U., and Iseki, M., Chemotherapeutic effect of azithromycin and lasalocid on *Cryptosporidium* infection in mice, *J. Protozool.*, 38, 232S, 1991.
83. Blagburn, B. L., Sundermann, C. A., Lindsay, D. S., Hall, J. E., and Tidwell, R. R., Inhibition of *Cryptosporidium parvum* in neonatal Hsd:(ICR)BR Swiss mice by polyether ionophores and aromatic amidines, *Antimicrob. Agents Chemother.,* 35, 1520, 1991.
84. Gobel, E. and Bretschneider, M., Mikromorphologische untersuchungen zur wirksamkeit von lasalocid auf die entwicklungsstadien von *Cryptosporidium*, Kongresses der Deuteschen Veterinarmedizinischen Gesellschaft, Bad Nauheim, 17–20, April, 1985, 278–282.
85. Angus, K. W., Hutchison, G., Campbell, I., and Snodgrass, D. R., Prophylactic effects of anticoccidial drugs in experimental murine cryptosporidiosis, *Vet. Rec.,* 114, 166, 1984.
86. Tzipori, S. R., Campbell, I., and Angus, K., The therapeutic effect of 16 antimicrobial agents on *Cryptosporidium* infection in mice, *Aust. J. Exper. Biol. Med. Sci.*, 60, 187, 1982.
87. Fayer, R. and Fetterer, R., Activity of benzimidazoles against cryptosporidiosis in neonatal BALB/c mice, *J. Parasitol.,* 81, 794, 1995.
88. Rasmussen, K. R. and Healy, M. C., Dehydorepiandrosterone-induced reduction of *Cryptosporidium parvum* infection in aged Syrian golden hamsters, *J. Parasitol.,* 78, 554, 1992.
89. Kim, C. W., Chemotherapeutic effect of arprinocid in experimental cryptsporidiosis, *J. Parasitol.*, 73, 663, 1987.

90. Rehg, J. E., The activity of halofuginone in immunosuppressed rats infected with *Cryptosporidium parvum, J. Antimicrob. Chem.*, 35, 391, 1995.
91. Verdon, R., Polianski, J., Gaudebout, C., Marche, C., Garry, L., Carbon, C., and Pocidalo, J. J., Evaluation of high dose regimen of paromomycin against *Cryptosporidium* in the dexamethasone-treated rat model, *Antimicrob. Agents Chemother.*, 39, 2155, 1995.
92. Verdon, R., Polianski, J., Gaudebout, C., Marche, C., Garry, L., and Pocidalo, J. J., Evaluation of curative anticryptosporidial activity of paromomycin in a dexamethasone-treated rat model, *Antimicrob. Agents Chemother.*, 38, 1681, 1994.
93. Rehg, J. E., A comparison of anticryptosporidial activity of paromomycin with that of other aminoglycosides and azithromycin in immunosuppressed rats, *J. Infect. Dis.*, 170, 934, 1994.
94. Lemeteil, D., Roussel, R., Favennec, L., Ballet, J. J., and Brasseur, P., Assessment of candidate anticryptosporidial agents in an immunosuppressed rat model, *J. Infect. Dis.*, 167, 766, 1993.
95. Brasseur, P., Lemeteil, D., and Ballet, J. J., Curative and preventive anticryptosporidium activities of sinefungin in an immunosuppressed adult rat model, *Antimicrob. Agents Chemother.*, 37, 889, 1993.
96. Rasmussen, K. R., Arrowood, M. J., and Healey, M. C., Effectiveness of dehydroepiandrosterone in reduction of cryptosporidial activity in immunosuppressed rats, *Antimicrob. Agents Chemother.*, 36, 220, 1992.
97. Brasseur, P., Lemeteil, D., and Ballet, J. J., Anti-cryptosporidial drug activity screened with an immunosuppressed rat model, *J. Protozool.*, 38, 230S, 1991.
98. Rehg, J. E., Anticryptosporidial activity is associated with specific sulfonamides in immunosuppressed rats, *J. Parasitol.*, 77, 238, 1991.
99. Rehg, J. E., Anti-cryptosporidial activity of macrolides in immunosuppressed rats, *J. Protozool.*, 38, 228S, 1991.
100. Rehg, J. E., Activity of azithromycin against cryptosporidia in immunosuppressed rats, *J. Infect. Dis.*, 163, 1293, 1991.
101. Rehg, J. E. and Hancock, M. L., Effectiveness of arprinocid in the reduction of cryptosporidial activity in immunosuppressed rats, *Am. J. Vet. Res.*, 51, 1668, 1990.
102. Rehg, J. E., Anticryptosporidial activity of lasalocid and other ionophorous antibiotics in immunosuppressed rats, *J. Infect. Dis.*, 168, 1566, 1993.
103. Rehg, J. E., Hancock, M. L., and Woodmansee, D. B., Anticryptosporidial activity of sulfadimethoxine, *Antimicrob. Agents Chemother.*, 32, 1907, 1988.
104. Rehg, J. E., Effect of diethyldithiocarbamate on *Cryptosporidium parvum* infection in immunosuppressed rats, *J. Parasitol.*, 82, 158, 1996.
105. Fayer, R. and Ellis, W., Paromomycin is effective as prophylaxis for cryptosporidiosis in dairy calves, *J. Parasitol.*, 79, 771, 1993.
106. Peeters, J. E., Villacorta, I., Naciri, M, and Vanopdenbosch E., Specific serum and local antibody responses against *Cryptosporidium parvum* during medication of calves with halofuginone lactate, *Infect. Immun.*, 61, 4440, 1993.
107. Naciri, M., Mancassola, R., Yvore, P., and Peeters, J. E., The effect of halofuginone lactate on experimental *C. parvum* infections in calves, *Vet. Parasitol.*, 45, 199, 1993.
108. Fayer, R., Activity of sulfadimethoxine against cryptosporidiosis in dairy calves, *J. Parasitol.*, 78, 534, 1992.
109. Villacorta, I., Peeters, J. E., Vanopdenbosch, E., Ares-Mazas, E., and Theys, H., Efficacy of halofuginone lactate against *Cryptosporidium parvum* in calves, *Antimicrob. Agents Chemother.*, 35, 283, 1991.
110. Gobel, E., Diagnose und therapie der akuten Kryptosporidiose beim kalb, *Tierarztl. Umsch.*, 42, 863, 1987.
111. Fischer, O., Attempted therapy and prophylaxis of cryptosporidiosis in calves by administration of sulfadimidine, *Acta Vet. Brno.*, 52, 183, 1983.
112. Moon, H. W., Woode, G. N., and Ahrens, F. A., Attempted chemoprophylaxis of cryptosporidiosis in calves, *Vet. Rec.*, 11O, 181, 1982.
113. Gobel, E., Possibilities of therapy of cryptosporidiosis in calves in problematic farms, *Zbl. Bakt. Hyg.*, 265, 489, 1987.
114. Naciri, M. and Yvore, P., Efficite du lactate d'halofuginone dans le traitemtnt de le cryptosporidiose chez l'agneau, *Rec. Med. Vet.*, 165, 823, 1989.
115. Mancassola, R., Reperant, J. M., Naciri, M., and Chartier, C., Chemoprophylaxis of *Cryptosporidium parvum* infection with paromomycin in kids and immunological study, *Antimicrob. Agents Chemother.*, 39, 75, 1995.
116. Nagy, B., Bozso, M., Palfi, V., Nagy, G., and Sahiby, M.A., Studies on cryptosporidial infection of goat kids, Colloque International, Niort, France, 9–11 October, 1984, 443.
117. Naciri, M., Yvore, P., and Levieux, D. Cryptosporidiose experimentale du chevreau. Influence de la prise du colostrum. Essais de traitments, Colloque International, Niort, France, 9–11 October, 1984, 465.
118. Fayer, R., Phillips, L., Anderson, B. C., and Bush, M. Chronic cryptosporidiosis in a Bactrian camel, *J. Zoo Wildlife Dis.*, 22, 228, 1991.
119. Moon, H. W., Schwartz, A., Welch, M. J., McCann, P. P., and Runnels, P. L., Experimental fecal transmission of human cryptosporidia to pigs, and attempted treatment with an ornithine decarboxylase inhibitor, *Vet. Pathol.*, 19, 700, 1982.
120. Barr, S. C., Jamrosz, G. F., Hornbuckle, W. E., Bowman, D. D., and Fayer, R., Use of paromomycin for treatment of cryptosporidiosis in a cat, *J. Am. Vet. Med. Assoc.*, 205, 1742, 1994.
121. Varga, I., Sreter, T., and Bekesi, L., Potentiation of ionophorous antiCoccidials with duokin: battery trials against *Cryptosporidium baileyi* in chickens, *J. Parasitol.*, 81, 777, 1995.

122. Lindsay, D. S., Blagburn, B. L., Sundermann, C. A., and Ernest, J. A., Chemoprophylaxis of cryptosporidiosis in chickens using halofuginone, salinomycin, lasalocid, or monensin, *Am. J. Vet. Res.*, 48, 354, 1987.
123. Hoerr, F. J., Current, W. L., and Haynes, T. B., Fatal cryptosporidiosis in Quail, *Avian Dis.*, 30, 421, 1986.
124. Glisson, J. R., Brown, T. P., Brugh, Page, R. K., Kleven S. H., and Davis, R. B., Sinusitis in turkeys associated with respiratory cryptosporidiosis, *Avian Dis.*, 28, 783, 1984.
125. Mason, R. W. and Hartley, W. J., Respiratory cryptosporidiosis in a peacock chick, *Avian Dis.*, 24, 771, 1980.
126. Cranfield, M. R. and Graczyk, T. K., Experimental infection of elaphid snakes with *Cryptosporidium serpentis* (Apicomplexa: Cryptosporidiidae), *J. Parasitol.*, 80, 823, 1994.
127. Graczyk, T. K., Cranfield, M. R., and Hill, S. L., Therapeutic efficacy of halofuginone and spiromycin treatment against *Cryptosporidium serpentis* (Apicomplexa: Cryptosporidiidae) infections in captive snakes, *Parasitol. Res.*, 82, 143, 1996.
128. Arrowood, M. J., Long-Ti Xie, and Hurd, M. R., *In vitro* assays of maduramicin activity against *Cryptosporidium parvum, J. Euk. Microbiol.*, 41, 23S, 1994.
129. Favennec, L., Egraz-Bernard, M., Comby, E., Lemeteil, D., Ballet, J. J., and Brasseur, P., Immunofluorescence detection of *Cryptosporidium parvum* in Caco-2 cells: a new screening method for anticryptosporidial agents, *J. Protozool.*, 41, 39S, 1994.
130. Weist, P. M., Johnson, J. H., and Flanigan, T. P., Microtubule inhibitors block *Cryptosporidium parvum* infection of a human enterocyte cell line, *Infect. Immun.*, 61, 4888, 1993.
131. Marshall, R. J. and Flanigan, T. P., Paromomycin inhibits *Cryptosporidium* infection of a human enterocyte cell line, *J. Infect. Dis.*, 165, 772, 1992.
132. Arrowood, M. J., Jaynes, J. M., and Healey, M. C., *In vitro* activities of lytic peptides against sporozoites of *Cryptosporidium parvum, Antimicrob. Agents Chemother.*, 35, 224, 1991.
133. McDonald, V., Stables, R., Warhurst, D. C., Barer, M. R., Blewett, D. A., Chapman, H. D., Connolly, G. M., Chiodini, P. L., and McAdam, K. P. W. J., *In vitro* cultivation of *Cryptosporidium parvum* and screening for anticryptosporidial drugs, *Antimicrob. Agents Chemother.*, 34, 1498, 1990.
134. Angus, K.W., Cryptosporidiosis in ruminants, in *Cryptosporidiosis in Animals and Man*, Dubey, J. P., Speer, C. A., and Fayer, R., Eds., CRC Press, Boca Raton, FL, 1990, 83.
135. O'Donoghue, P. J., *Cryptosporidium* and cryptosporidiosis in man and animals, *Int. J. Parasitol.*, 25, 139, 1995.

Chapter 6

Immunology: Host Response and Development of Passive Immunotherapy and Vaccines

Michael W. Riggs

CONTENTS

0-8493-7695-5/97/$0.00+$.50

I. INTRODUCTION

Knowledge relating to the immunobiology of *Cryptosporidium* has been generated largely in the last 10 years. Research activity has paralleled the recognition of *Cryptosporidium* as a parasite of veterinary and medical importance, beginning with the first reported human, bovine, and avian cases.[1] The impact of *C. parvum* on decreased livestock production worldwide[2] and its role in the morbidity and mortality of patients with AIDS are now well defined[3] (see Chapter 2). Two additional species of *Cryptosporidium*, *C. muris* and *C. baileyi*, have received increasing attention as pathogens of cattle[4,5] and chickens,[6] respectively (see Chapter 1).

Immunologic control of cryptosporidiosis is indicated by the following observations. Immunocompetent humans,[7-12] nonhuman primates,[13,14] calves,[2] lambs,[2,15] goat kids,[2] and mice[16,17] usually develop self-limiting infections which remain localized to the gastrointestinal tract. Such hosts are resistant to reinfection and clinical disease after recovery.[12,13,17-22] In contrast, immunodeficient hosts, including those with congenital T-cell, B-cell, combined T- and B-cell, or other effector cell deficiencies;[7,12,23-31] immunodeficiency induced by drug therapy (see Chapter 9);[12,32] or AIDS,[3,7,12] usually develop persistent, progressive infections of greater severity. Infection in such hosts may disseminate to extraintestinal sites (see Chapters 1 and 2).[24,27,28,33] Persistent infection in reversibly immunosuppressed hosts is resolved upon restoration of immune function.[12,34] Further, upregulatory immunomodulators such as thymomodulin[35] or dehydroepiandrosterone[36,37] may increase resistance to infection and reduce patency in rodent models with compromised immune function related to aging or chemical suppression[36,37] or in neonatal mice (see Chapter 9).[35] Efforts to develop a precocious strain of *C. parvum* for immunization strategies were unsuccessful.[38] However, orally administered killed oocysts significantly reduced the duration of oocyst shedding and diarrhea in challenged calves.[39] Immune responses to this oral vaccination are under investigation.[39] Because cryptosporidiosis is resolved or prevented by normal immune responses,[34,40] active and passive immunization strategies have been investigated for its control. In this chapter, humoral and cellular immune responses to *Cryptosporidium*, including characterization of antigens recognized and mechanisms of passive and active immunity, will be reviewed.

II. HUMORAL IMMUNE RESPONSES

A. ROLE OF HUMORAL IMMUNITY

While cell-mediated immune responses are clearly important in resistance to coccidia,[41,42] including *Cryptosporidium*, several observations indicate that humoral immune responses against extracellular stages may contribute to protection against cryptosporidiosis. Serum and mucosal IgG, IgM, and IgA responses to asexual and sexual stages accompany resolution of diarrhea and oocyst shedding in all mammals examined, including humans,[43-49] calves,[20,47,50-55] lambs,[21,52,56] and foals.[47] In the absence of studies to determine if antibodies against neutralization-sensitive epitopes are produced during infection, these antibody responses may be only temporal markers of other protective immune responses. A role for neutralizing antibody in controlling infection is suggested by reports that congenitally hypogammaglobulinemic humans with no detectable antibody to *C. parvum* may develop persistent cryptosporidiosis.[7,43,57-60] Others with selective IgA and complement-dependent opsonin deficiencies may develop persistent infection with recurrent diarrhea.[61] Such patients lack IgA-positive lymphocytes in intestinal mucosa but have normal cell mediated immune responses.

The role of active humoral immune responses in controlling experimental *C. parvum* infection in mice has been investigated.[62] Normal, infected BALB/c neonates produced low levels of IgG and IgM antibody that reacted with sonicated oocysts in enzyme-linked immunosorbent assay (ELISA). Unlike in other mammals, the antibody response was not correlated with the duration or level of oocyst shedding. Oocyst shedding patterns did not differ between neonates treated with specific anti-IgM antibody, which depleted B cells in lymph nodes and spleen but left most T-cell functions intact, and controls. Neither B-cell-depleted adult mice nor normal adult controls shed oocysts after challenge, suggesting that age-associated resistance in adult BALB/c mice is not influenced by B cells. It was concluded that active antibody responses to infection in mice play only a minor role in resolution of cryptosporidiosis. Oocyst shedding was the index for infection in mice of this study and was determined for a 3-week period. In subsequent studies, differences in oocyst shedding and infection levels between immunodeficient and normal mice were not evident before 4 weeks postinfection (PI).[30] Infection levels in severe combined immune deficient (SCID) neonates, lacking functional T and B cells, were similar to controls until approximately 4 weeks PI. In adult infected SCID and control mice, differences in infection levels were not evident until approximately 2 months PI. It has been suggested[25,34] that humoral responses may represent a redundant, rather than irrelevant, mechanism of immunity against cryptosporidiosis, as demonstrated for IFN-γ and CD4 lymphocytes.[25]

Most AIDS patients with persistent cryptosporidiosis produce *C. parvum*-specific serum[34,43,44,63-67] and/or mucosal[34,64,65,68] IgG, IgM, and IgA detectable by ELISA, immunofluorescence assay (IFA), or western blot. Serum IgG, IgM, and IgA[64,65] and salivary secretory IgA[65] titers against disrupted oocyst preparations were higher in AIDS patients with persistent cryptosporidiosis than in AIDS patients or normal humans without cryptosporidiosis. Fecal IgA, but not IgM, titers against disrupted oocysts were higher in AIDS patients with persistent cryptosporidiosis than in AIDS patients without cryptosporidiosis.[64] There was no apparent correlation between serum or fecal IgG, IgM, or IgA (predominantly IgA_1) titers, or their oocyst antigen specificities, and oocyst shedding levels.[64] It was not clear what influence the semi-quantitative method used for evaluation of fecal oocyst levels or the variation in fecal volume among patients had on accurate quantitation of oocyst shedding. HIV-positive, pre-AIDS patients who cleared infection had higher titers of specific secretory IgA in saliva than AIDS patients with persistent cryptosporidiosis,[34] suggesting that *C. parvum*-specific secretory IgA in the intestine may be responsible for recovery from cryptosporidiosis in such patients or may accompany, or simply signal, other protective immune responses.

Qualitative and quantitative antibody responses to neutralization-sensitive *C. parvum* epitopes by AIDS patients who mount an antibody response to *C. parvum* have not been determined.[63-65,68] The epitope specificity of secretory IgA responses in such patients may be defective, and antibody against neutralizing epitopes of *C. parvum* may not be produced.[34] Attempts to derive heterohybrid monoclonal antibodies (mAbs) against *C. parvum* from infected humans failed with donor lymphocytes from AIDS patients but succeeded with donor lymphocytes from immunocompetent recovered patients.[69] Related to this, most Centers for Disease Control (CDC) Stage IV AIDS patients with cryptosporidiosis in one study produced serum and fecal antibodies reactive with 15- and 23-kDa antigens in western blots of freeze-disrupted oocysts.[64] Similar findings were obtained with serum antibody from infected AIDS and immunocompetent patients.[67] Sporozoite antigens of 15 and 23 kDa have neutralization-sensitive epitopes defined by mAbs,[70-72] but it was not determined if AIDS patients responded to these epitopes or neutralizing epitopes of other antigens. Specific antibodies reactive with *C. parvum* antigens known to express neutralization-sensitive epitopes may not recognize such epitopes.[73] These observations underscore the importance of evaluating neutralizing antibody when assessing the immunorelevance of antibody responses to specific antigens. Persistent infection in advanced AIDS patients suggested that neutralizing antibodies to 15- or 23-kDa or other antigens either were not generated or were insufficient to control infection in the presence of cellular and other immune dysfunctions.[64,67] Serum and fecal IgA responses to *C. parvum* in AIDS patients were predominantly IgA_1, the subclass typically represented in responses to peptide and glycoprotein antigens.[64,66] IgA_2, the predominant subclass responding to polysaccharide antigens, was considered poorly represented in AIDS patients with cryptosporidiosis.[66]

In that many neutralization-sensitive epitopes of *C. parvum* are carbohydrate/carbohydrate-dependent[71,73-76] and glycoconjugate antigens are common,[1,70,75,77] the IgA_2 response in AIDS patients warrants further investigation.

B. ACTIVE IMMUNIZATION — CHARACTERIZATION OF ANTIGENS

1. *Cryptosporidium parvum*

Identification and characterization of *C. parvum* antigens has been facilitated by optimized protocols for purifying oocysts,[71,78-80] sporozoites,[71,78,79] and merozoites[76,81] and development of neutralization assays for infectious stages. Difficulties in identification and isolation of the autoinfective stages have hampered study of their immunobiology and characterization of antigens. Autoinfective sporozoite and merozoite stages may allow persistent infection in immunocompromised hosts and complicate immune-based control strategies.[82] Difficulties isolating sexual stages have also precluded their antigenic characterization.

The protein,[77] carbohydrate,[74,77] and lipid[74,75,77,83] composition of *C. parvum* is diverse and complex (see Chapter 7). Many of these biochemical classes are antigenic and may be immune-response targets. Major and minor protein/glycoprotein molecules from disrupted oocysts,[77,80,84] purified oocyst shells,[80] purified sporozoites,[47,84,85] and purified merozoites,[85] ranging from <14 to >200 kDa, have been identified in protein-stained sodium dodecyl sulfate-polyacrylamide gel electrophoresis (SDS-PAGE) gels. Multiple comigrating proteins in sporozoite and merozoite preparations have been described.[85] About 20 to 25 radio-iodinated sporozoite surface proteins from <14 to >270 kDa in reducing SDS-PAGE have been identified.[75,84,86] Fewer iodinated proteins on the surface of intact oocysts ranged from 15.5 to 290 kDa in reducing SDS-PAGE.[80,87] Some radio-iodinated oocyst surface proteins comigrated with those of similar M_r from *C. muris* and *C. bayleyi*.[87] [^{35}S]methionine-labeled protein patterns in reducing SDS-PAGE suggested differential expression of sporozoite proteins in the <14- to >200-kDa range during and after excystation.[75]

Lectin binding to solubilized oocysts resolved in SDS-PAGE revealed approximately 15 glycoproteins in the 14- to >100-kDa range.[77] Multiple carbohydrate moieties were demonstrated on each glycoprotein, especially in the 72- to >100-kDa range. In another study, 19 oocyst derived proteins ranging from 15 to >300 kDa in reducing SDS-PAGE were glycosylated; many comigrated with iodinated sporozoite surface proteins.[84] Specific carbohydrate moieties on some glycoproteins, including those recognized by convalescent human serum or mAbs, have been identified (see Chapter 7).[70,77,84,88,89]

Using specific lectins or carbohydrates *in vitro* to characterize the role of putative parasite attachment molecules, it appeared that sporozoite glycoproteins with terminal *N*-acetyl-D-glucosamine residues may function in attachment or invasion.[90] The same carbohydrate was expressed on the oocyst surface,[89] and some glycoproteins recognized by immune serum.[77] Alternatively, *C. parvum*-derived proteins with lectin activity may attach to host cell carbohydrate receptors.[91,92] A sporozoite surface lectin with hemaglutinating activity and specificity for galactose and *N*-acetylgalactosamine residues has been identified.[91,92] Inhibition of lectin activity by specific glycoproteins reduced sporozoite attachment to fixed MDCK cells and invasion of live MDCK cells, suggesting a lectin role in the initial interaction of sporozoites with host cells.[92] The lectin, semi-purified from sporozoite membranes, was thought to be one or more proteins in the 15- to 60-kDa range.[91]

Kinetics and isotypes of serum[20,22,44,53,67,93-95] and fecal[20,53,96] antibody responses to infection have been studied in calves,[20,53,67] lambs,[67,93,94,96] goat kids,[22] rabbits,[67] pigs,[67] mice,[67] and immunocompetent[44,95] and immunodeficient humans.[44] The specificity of antibodies produced in response to infection or parenteral immunization has been studied in mice,[67,97,98] rats,[29] rabbits,[67,99] cattle,[20,55,67,73,75,81,86,100-104] sheep,[67] goats,[80] pigs,[67] and humans.[67,95] Many antigens with similar M_rs identified on specific life-cycle stages in these studies were likely the same. However, accurate comparisons between studies are difficult because of differences in electrophoretic or blotting conditions, radiolabeling methods and specificities, reducing agents in SDS-PAGE, isolates examined, hypochlorite in oocyst purification, purity of life-cycle stages, trypsin and taurocholate in excystation medium, protease and glycosidase inhibitors in parasite processing, detergents in antigen solubilization, method of solubilization, and amount of parasite material examined. The influence of these variables on M_r has been illustrated.[67,88,105-108] Additionally, differences in the use of isotype control mAbs or normal control sera in immunoassays designed to determine the specificity of anti-*Cryptosporidium* antibodies have made some studies difficult to interpret and further hampered comparative analysis.

A high degree of antigen conservation has been demonstrated between life-cycle stages of *Cryptosporidium*. Epitopes shared by sporozoites, merozoites, oocyst walls, and/or residual bodies have been defined with mAbs,[70,71,75,76,85] or polyclonal antibodies[81] raised against sporozoites or merozoites. mAbs raised against disrupted oocysts demonstrated broad conservation of epitopes among asexual and sexual stages.[105,109-112] Neutralizing hyperimmune bovine colostral antibodies prepared against oocysts recognized epitopes expressed by all life-cycle stages.[113] Sporozoite epitopes not expressed by merozoites or expressed by only a minor subpopulation of merozoites have been identified with mAbs raised against sporozoites.[76] Similarly, merozoite epitopes not expressed by sporozoites have been identified with mAbs.[85]

a. Laboratory Rodents

Serum IgG from Swiss-Webster mice inoculated orally or intraperitoneally with oocysts recognized peptide and carbohydrate epitopes in sonicated oocyst preparations, as determined by ELISA after protease, glycosidase, or neuraminidase treatment of antigen.[77] Four oocyst-derived glycoproteins, 72 to >100 kDa, were consistently recognized in western blots by serum IgG from infected mice, especially a 98-kDa glycoprotein. Some antigens had disulfide linkages based on comparison of M_rs in reducing and nonreducing SDS-PAGE. It was concluded that carbohydrates, alone or in association with lipids and proteins, are targets of the immune response to *C. parvum* and may influence vaccine development. Interestingly, immunodominant antigens recognized by many mammals, including BALB/c mice, were not reported for Swiss-Webster mice. These included one or more antigens in a broadly migrating zone of approximately 15 kDa,[47,56,67,70,80,101] a 23-kDa antigen,[47,67] and a 32-kDa antigen.[80] These may not be recognized by Swiss-Webster mice, but it is more likely the antigens were not present in the preparation due to the method of solubilization. Incomplete solubilization of oocysts may have accounted for recognition of only several 88- to 150-kDa antigens by convalescent rat serum antibody and lack of recognition of immunodominant 15- and 23-kDa antigens in western blots (nonreducing).[29]

Serum and fecal antibody responses in adult BALB/c mice orally inoculated with oocysts have been characterized.[67,97] Fecal IgG and IgA reactivity was limited to a 17- (nonreducing) or 18-kDa (reducing) broad antigen zone in western blots of freeze-disrupted oocysts.[97] Serum IgG and IgA reacted with a broad 15- to 17-kDa immunodominant antigen zone and others including 13-kDa and 23- to 24-kDa antigens in western blots (nonreducing).[67,97] The immunodominant 15- to 17-kDa antigen zone recognized by serum antibody might contain two distinct sporozoite-derived antigens.[67] One, approximately 15 kDa, comigrated with an antigen designated GP15, defined by mAb 5C3,[70] and expressed epitopes shared with 50- to 55-kDa antigens. The second was approximately 17 kDa.[67,97] Serum antibody from mice immunized with antigen electroeluted from SDS-PAGE gels in the 17-kDa zone reacted only with a 17-kDa antigen zone in western blots (nonreducing), suggesting this antigen did not share immunogenic epitopes with 50- to 55-kDa antigens.[97] This antibody reacted multifocally with the interior sporozoite mid-region in IFA, but not with oocysts. Other studies have characterized the western blot reactivity patterns of serum antibody produced by BALB/c mice in response to infection, multiple oral inoculations with viable oocysts, and IP/IV injection of soluble oocyst extracts[114] or subcutaneous immunization with excysted oocysts.[98]

Serum antibody from oocyst-immunized mice has been used to screen sporozoite gDNA libraries and identify genes encoding immunogenic apical or pellicle proteins.[115,116] Such proteins may be useful as recombinant antigens for vaccination. Genes encoding cognate sporozoite proteins of >500, 68/95, 45, 23, and 15 (doublet)/35 (doublet) kDa, as defined by recombinant-eluted antibodies in western blots of freeze-disrupted oocysts, have been cloned.[115] Other recombinant proteins recognized by anti-*C. parvum* antibodies[117-119,121] or possessing membrane-disrupting (hemolytic) activity[120] have been reported. Further characterization of recombinant proteins may show applicability to immunization strategies (see Chapter 10).

The murine humoral response to *C. parvum* has been dissected using antibody-secreting cells from immune mice to produce mAbs against specific antigens and to identify epitopes to which neutralizing antibodies are directed. mAbs have also identified other distinct native antigens and clones expressing recombinant proteins for evaluation as immunogens. Studies to define antigens as potentially useful targets for passive or active immunization are reviewed next.

i. Neutralization-Sensitive Antigens Defined by Monoclonal Antibodies

At least seven *C. parvum* antigens capable of inducing neutralizing antibody have been defined. BALB/c mice hyperimmunized with viable sporozoites produced IgG_3 mAb 18.44, and IgM mAb 17.41.[75] Each mAb recognized distinct molecules on sporozoites[75] and merozoites[76] and neutralized the infectivity of each stage. mAb 18.44 bound diffusely to live sporozoite and merozoite surface membranes and oocyst residual bodies in IFA and recognized an antigen designated CPS-500 which migrated with the dye front in SDS-PAGE.[75,76,122] CPS-500, not radiolabeled with ^{125}I or [^{35}S]-methionine and insensitive to proteinase K digestion,[75] was isolated and characterized as a polar glycolipid with a carbohydrate/carbohydrate-dependent neutralization-sensitive epitope.[74] Mouse immunization studies suggested the molecule was poorly immunogenic unless coupled to a carrier protein. Because the glycolipid composition of CPS-500 complicated recombinant approaches for subunit antigen production, anti-idiotypic antibody and synthetic strategies are being pursued with CPS-500 for evaluation as protective immunogens.[74] mAb 17.41 bound in a multifocal or focal polar pattern to live sporozoite and merozoite surface membranes, oocyst walls, and residual bodies in IFA.[75,76,122] In western blots (reducing), mAb 17.41 recognized sporozoite antigens of 75, 150, 175, 215, and >215 kDa (designated SA-1) and merozoite antigens of 75, 215, and >215 kDa.[123] mAb 17.41 also immunoprecipitated [^{35}S]-methionine-labeled sporozoite antigens of 28, 55, and 98 kDa.[75] The 28- and 55-kDa antigens were also iodinatable and exposed on the surface of viable sporozoites.[75] The neutralization-sensitive epitope recognized by mAb 17.41 was shown to be carbohydrate/carbohydrate-dependent.[73]

BALB/c mice immunized with excysted oocyst preparations produced mAbs with anticryptosporidial activity, including IgG_1 mAb C6B6, IgG_3 mAb C8C5, and IgM mAb C4A1.[47,71] C6B6 and C8C5 bound diffusely to the sporozoite and merozoite membrane in IFA and recognized a 23-kDa sporozoite antigen designated P23 in western blots (reducing).[47,71,124] This antigen was originally reported as 20 kDa, possibly from use of trypsin in excystation medium.[47,71] However, distinct 20- and 23-kDa antigens have been defined by mAbs.[47,67,71] Membrane protein biotinylation demonstrated P23 in the sporozoite surface membrane.[47] C6B6 and C8C5 recognized peptide epitopes on P23.[72] C4A1 bound to the anterior pole of sporozoites and merozoites in IFA, recognized sporozoite antigens of 25 to >200 kDa in western blots (reducing),[47,71] and recognized several merozoite antigens of M_r in western blots similar to those of sporozoites.[85] The epitope recognized by C4A1 was carbohydrate/carbohydrate-dependent.[74a]

BALB/c mice immunized with mAb C6B6-affinity chromatography-purified native P23 from sporozoites produced serum antibody reactive with a 23-kDa sporozoite antigen in western blots (reducing) and an expanded panel of mAbs reactive with P23.[72,74] From this panel, an IgG_1 mAb designated 7D10 had significant anti-cryptosporidial activity in mice, reacted with the surface membrane of sporozoites in IFA, and recognized a 23-kDa sporozoite antigen in western blots (reducing).[72] Screening sporozoite cDNA libraries with 7D10, C6B6, and hyperimmune bovine serum against sporozoites identified a gene encoding P23 epitopes.[72] Epitope mapping with nested sets of hexamer peptides deduced from the DNA sequence demonstrated a linear peptide epitope recognized by 7D10 distinct from the nonlinear (conformational) peptide epitope recognized by C6B6.[72] Serum antibody from mice immunized with a synthetic polypeptide containing the epitopes defined by 7D10 and C6B6 reacted with a 23-kDa antigen in western blots (reducing) of sporozoites, which comigrated with the 23-kDa antigen recognized by mAb 7D10.[72] Recombinant P23 and synthetic peptides representing neutralization-sensitive P23 epitopes are being evaluated for their ability to induce protective immune responses.[72]

Murine IgM mAb 3E2, produced against mAb C4A1-affinity chromatography-purified native sporozoite antigen, bound to excysted oocyst walls and the surface of live sporozoites and merozoites in IFA and neutralized sporozoite infectivity.[74a,74b] 3E2 reacted with sporozoite apical complex organelles, including electron dense granules, by immunoelectron microscopy (IEM). In western blots (reducing) of sporozoites and merozoites, 3E2 recognized multiple 46- to 230-kDa antigens and a high M_r antigen, designated CSL, which variably migrated between 1200 and 1400 kDa. The neutralization-sensitive CSL epitope recognized by 3E2 was repetitive and carbohydrate/carbohydrate-dependent.

A monomeric IgA mAb designated 5C3 had highly significant anticryptosporidial activity and bound diffusely to the sporozoite surface membrane, before and during excystation.[70,125] Binding to excysted sporozoites diminished, suggesting antigen was shed immediately postexcystation. By IFA and IEM, 5C3 also bound to meronts and free merozoites and recognized an approximately 15 (14 to 16) kDa, broadly migrating glycoprotein antigen (GP15) in western blots of sporozoites or merozoites.[70,125] GP15 from either sporozoites or merozoites had similar isoelectric points.[125] Epitopes on GP15 were also expressed on a 60-kDa antigen[126,127] in western blots (reducing) of freeze-disrupted oocysts or sporozoites. The apparent M_rs of GP15 and the 60 kDa antigen recognized by 5C3 in western blots of sporozoites were similar under reducing or nonreducing conditions, suggesting the 15-kDa species was not a disulfide-linked subunit of the 60-kDa species.[125,127]

Rats were immunized with antigen electroeluted from the region containing GP15 in SDS-PAGE-resolved oocyst proteins.[127] Rat antiserum recognized 15- and 60-kDa antigens in western blots of oocysts and *in vitro* translation products of sporozoite RNA and was used to identify recombinant *Escherichia coli* expressing epitopes shared by the two sporozoite antigens (rP15/60). Recombinant P15/60 was recognized by hyperimmune bovine colostral antibody prepared against oocysts and by convalescent serum antibody from calves. 5C3 also recognized 15- and 60-kDa *in vitro* translation products of sporozoite RNA and the recombinant *E. coli* clone expressing portions of 15- and 60-kDa proteins, indicating the cloned cDNA encoded the neutralization-sensitive epitope defined by 5C3.[127] 5C3 recognition of recombinant protein indicated that the target epitope was peptide, supporting reports of trypsin susceptibility[125] and clarifying earlier results suggesting carbohydrate-dependency.[70] Further, serum antibody from mice immunized with rP15/60 bound to surface and internal epitopes of sporozoites in IFA and recognized 15- and 60-kDa antigens in western blots of sporozoites.[127] Although 15- and 60-kDa antigens expressed common epitopes, the molecular relationship between the two native antigens is unknown.[127] The ability of rP15/60 to induce neutralizing antibody is under investigation.[126]

Another recombinant clone expressing epitopes of a 15-kDa protein corresponding to GP15 was identified in a sporozoite cDNA library.[128] The cDNA encoded epitopes of a 15-kDa protein not shared with a 60-kDa cross-reactive protein reported previously[127] but apparently shared with oocyst surface protein. Rabbit antiserum against the recombinant protein expressed in *E. coli* recognized an approximately 15-kDa antigen in western blots of disrupted oocysts. Rabbit antiserum against oocyst antigens recognized the recombinant 15-kDa protein, but hyperimmune bovine colostral antibody against oocyst antigens did not. After subcloning and expressing cDNA encoding the 15-kDa protein in yeast, both the rabbit and hyperimmune bovine antibody recognized the recombinant 15-kDa protein, suggesting that glycosylation of the recombinant protein in a eukaryotic expression system was required for recognition by hyperimmune bovine antibody. The ability of yeast-expressed recombinant P15 to induce neutralizing antibody is under investigation.[128]

BALB/c mice have been immunized with merozoites to identify and characterize antigens of this stage.[129] Merozoite-immunized mice produced IgG_3 mAb Cmg-3.[129] Cmg-3 reacted with both sporozoites and merozoites in IFA, possessed anti-cryptosporidial activity, and recognized a 3.5-kDa merozoite antigen in western blots.[129]

ii. Other Antigens Defined by Monoclonal Antibodies

Many studies of murine mAbs reactive with surface membrane antigens and/or apical complex organelles of sporozoites and merozoites did not report neutralization data for the mAbs. Because surface-exposed epitopes and apical complex components are involved in attachment, invasion, and development and may be targets of protective immune responses,[130-133] further evaluation of such mAbs and the antigens they recognize may identify new molecular targets for immunologic control.

mAbs produced from BALB/c mice orally immunized with oocysts reacted with sporozoites in IFA.[114] They recognized both peptide and carbohydrate/carbohydrate-dependent epitopes on antigen migrating at 14 to 17 kDa in western blots[114] and were considered reactive with GP15, the neutralization-sensitive surface antigen defined by mAb 5C3.[70] Additional mAbs produced in this study reacted with peptide epitopes on antigen migrating at 23 to 27 kDa, designated GP25, considered to be the same as P23 defined by mAb C6B6.[47,71,108] mAbs against GP15 and GP25 reacted with sporozoite and merozoite surface membranes by IEM. mAbs against GP15 also recognized internal zoite structures and oocyst residual bodies. Five mAbs produced in another study from mice immunized with excysted oocysts all reacted with sporozoite surface membrane in IFA and recognized a 47-kDa antigen in western blots (reducing) of oocysts, with variable recognition of several higher M_r antigens.[112,134] Some sporozoite surface-reactive mAbs produced against *C. parvum* variably cross-reacted with *C. baileyi* and *C. muris* sporozoites in IFA or western blots, while other mAbs produced against *C. parvum* were species specific.[99] mAbs 2C3 and 2B2 reacted with the entire sporozoite by IFA.[135] 2B2 recognized a carbohydrate/carbohydrate-dependent epitope on a high M_r antigen not resolved in 10% SDS-PAGE gels, a >190-kDa doublet, and a 40-kDa antigen in western blots (nonreducing) of freeze-disrupted oocysts. Under reducing conditions, a 43-kDa antigen was recognized and reactivity with the 40-kDa antigen did not change, but the high M_r antigen was only faintly recognized. mAb 2C3 recognized a carbohydrate/carbohydrate-dependent epitope but was unreactive in western blots of 8 to 12% SDS-PAGE gels. mAb C2A3 produced against disrupted oocysts bound to both sporozoites and merozoites in IFA and recognized a 16-kDa comigrating antigen in western blots of sporozoites and merozoites and a 50-kDa antigen of merozoites.[85] mAb 11A5 produced against disrupted oocysts bound to the surface membrane of sporozoites in IFA and recognized a carbohydrate/carbohydrate-dependent epitope on a 15-kDa glycoprotein in western blots of Triton X-100 extracted oocysts.[88]

mAbs reactive with the apical region of sporozoites[104,105,109,110,116] and merozoites[105,109,110] were produced from BALB/c mice immunized with disrupted oocyst preparations. mAbs 10C6, 7B3, and E6 reacted with apical antigen in acetone-fixed sporozoites but not unfixed sporozoites by IFA, indicating recognition of an internal cytoplasmic antigen.[104,116] 10C6 also bound to merozoites in MDCK cells but had a diffuse, not apical, immunofluorescent pattern. The three mAbs recognized at least two carbohydrate/carbohydrate-dependent epitopes on a >900 kDa sporozoite antigen, designated GP900, in western blots (reducing). One GP900-reactive mAb, 7B3, also recognized a 38-kDa antigen. The previously undescribed GP900 antigen was identified by use of a low-percentage polyacrylamide resolving gel (5%) and an extended western blot transfer period.[104] Five mAbs in another study — TOU, HAD, ABD, BAX, and SPO — bound the apical interior of sporozoites and merozoites by IFA.[109,110] TOU recognized a carbohydrate/carbohydrate-dependent epitope on 130- and 210-kDa antigens and several lower M_r antigens in western blots (nonreducing) of freeze disrupted oocysts. HAD recognized a peptide epitope on 63- to 210-kDa antigens, many of which co-migrated with those recognized by TOU. ABD, BAX, and SPO recognized carbohydrate/carbohydrate-dependent epitopes on a 100-kDa antigen and several <100-kDa antigens.[110] BAX also recognized a >900-kDa antigen in western blots (reducing) of freeze-disrupted oocysts, which comigrated with GP900 defined in an earlier study.[104] Other mAbs reactive with the sporozoite apical region in IFA were produced from BALB/c mice immunized by oral inoculation with oocysts.[114] Peptide-reactive mAbs recognized 50- to 200-kDa antigens in western blots of disrupted oocysts. mAbs recognizing carbohydrate/carbohydrate-dependent epitopes were unreactive in western blots or recognized an antigen of >205 kDa.

Immunoelectron microscopy has been used to determine the subcellular location of antigens recognized by mAbs having apical reactivity. Many recognized micronemes.[105,109,110] BAX, previously shown to recognize a >900-kDa antigen[104] reacted with micronemes of both sporozoites and merozoites, suggesting that GP900 was a microneme antigen. Microneme-reactive mAbs described in other studies recognized a >500-kDa sporozoite glycoprotein in western blots of disrupted oocysts.[105] The relationship between the high M_r apical complex antigens reported in several studies is not clear.[74,74a,104,105,110,] LOI and BKE mAbs that bound to the apical interior of sporozoites in IFA did not recognize micronemes in IEM[111] but recognized a subset of electron dense granules in the merozoite mid-region, which was distinct from dense granules typically associated with the apical complex. In western blots (reducing) of freeze-disrupted oocysts, each recognized carbohydrate/carbohydrate-dependent epitope(s) on a 110-kDa antigen.

Conservation of microneme[109,110] and other organelle epitopes[110,111] of life-cycle stages and subcellular structures of *C. parvum* has been studied comprehensively. TOU, ABD, BAX, and SPO, which recognized micronemes of sporozoites and merozoites by IEM, also labeled peripheral cytoplasm and parasitophorous vacuole of trophozoites and developing macrogametes in rat gut.[109,110] ABD, BAX, and SPO recognized distinct electron dense cytoplasmic granules in macrogametes thought to contain fibrillar material exocytosed into the parasitophorous vacuole. Functional relationships between these microneme and macrogamete epitopes were unclear.[110]

2C3 and 2B2, which reacted with sporozoites in IFA, recognized cytoplasm and parasitophorous vacuole membrane of sexual and asexual stages, surface membrane of zoites, and electron-dense granules in macrogametocytes thought to be oocyst wall-forming bodies.[135] The oocyst wall was not recognized by either 2C3 or 2B2, despite reactivity with macrogamete granules. LOI and BKE also recognized cytoplasmic inclusions in macrogamonts and parasitophorous vacuole membrane of intracellular sexual and asexual stages.[111] These observations suggested that merozoite granule and macrogamont inclusion contents were incorporated into parasitophorous vacuole membrane. Antigen in parasitophorous vacuole membrane was Triton X-100 insoluble.

Oocyst wall antigens recognized by mice immunized with oocyst preparations have been defined by mAbs.[112,135,136] 181B5 reacted with the oocyst wall of *C. parvum* by IFA and cross-reacted with the walls of *C. baileyi* and *C. muris* oocysts.[99,112] Antigens on which the conserved epitope was expressed varied in M_r for each *Cryptosporidium* species examined.[99] mAbs raised against *C. parvum* also cross-reacted with *C. wrairi* and vice versa.[137,138] 181B5 recognized 41- and 44-kDa and several higher M_r *C. parvum* antigens in western blots (reducing) of oocysts.[112,135] IEM confirmed mAb 181B5 reactivity with oocyst wall as well as macrogametes and the parasitophorous vacuole of microgametocytes in mice.[135] Specific reactivity with macrogametes included the oocyst wall-forming bodies, cytoplasm, pellicle, and parasitophorous vacuole space. OW-IGO bound specifically to the oocyst wall of *C. parvum* in IFA.[136] The antigen recognized by mAb OW-IGO was localized by IEM to electron-lucent vesicles in the peripheral cytoplasm of macrogametes, associated fibrillar material within the parasitophorous vacuole, and the wall of both thick- and thin-walled oocysts. Accompanying development of the macrogamete, labeled vesicular material decreased as free fibrillar material in the parasitophorous vacuole increased to eventually surround the macrogamete. These observations suggested that macrogamete vesicles, distinct from oocyst wall-forming bodies,[135] contain precursor material for thick- and thin-walled oocyst walls released during gametogenesis into the parasitophorous vacuole. OW-IGO recognized a carbohydrate-dependent epitope on 250- and 40-kDa antigens and several minor antigens in nonreducing western blots of freeze-disrupted oocysts. In reducing western blots, a similar but more complex pattern was obtained, suggesting the presence of disulfide-linked subunits in the recognized antigen.

Neutralizing bovine colostral antibody prepared against *C. parvum* oocysts demonstrated a high degree of epitope conservation among all asexual and sexual stages.[100,113] Some antigens recognized by neutralizing bovine colostral antibody may be the same as those defined by the mAbs above. These molecules, therefore, warrant further characterization as potentially protective immunogens.

b. Rabbits

Adult rabbits orally inoculated with oocysts produced serum antibody reactive with antigens of 15 to 17, 22 to 23, 30, and 40 kDa and many antigens of >40 kDa and others migrating in the dye front in western blots (nonreducing) of freeze-disrupted oocysts.[67] The 15-kDa antigen recognized by serum antibody[67] comigrated with GP15 defined by mAb 5C3,[70] whereas the 17-kDa antigen was considered distinct. Although infection was not documented in rabbits,[67] the antibody response suggested they were infected. Alternatively, the antibody response to *C. parvum* in rabbits may result from mucosal exposure without infection.

Antigen specificity of rabbit antibodies in response to parenteral immunization with disrupted oocysts,[54,99,134,139] sporozoites,[54] or semipurified 23-kDa *C. parvum* antigen[140] has been characterized by western blot. Rabbit anti-whole *C. parvum* polyclonal antibody recognized multiple *C. parvum* antigens in western blots and also reacted with antigens of similar or different M_r from disrupted *C. muris* and *C. baileyi* oocysts.[99] Rabbit polyclonal antibody produced against a 23-kDa antigen electro-eluted from *C. parvum* oocyst blots recognized a 23-kDa antigen and higher M_r antigens in western blots of oocysts, suggesting conserved epitopes among the antigens recognized.[140] This antibody recognized sporozoite and merozoite membranes and, to a lesser extent, cytoplasmic components of both stages by IEM, suggesting that one or more of the antigens was in these sites. Rabbit polyclonal antibody against whole *C. parvum* has been useful for identification and isolation of clones expressing recombinant antigens for evaluation as subunit immunogens.[121,128]

c. Humans

Many antigens recognized by serum[45,47,64,65,67,80,84,95,141-143] and fecal[64] antibodies from humans infected with *C. parvum* range from <14 to >200 kDa. Most oocyst-derived glycoproteins and sporozoite surface proteins were immunogenic in humans and specifically recognized by convalescent serum antibody.[84] Serum IgM or IgG from most infected humans recognized an immunodominant 23-kDa sporozoite antigen in western blots (reducing) of disrupted oocysts[45,84] or purified sporozoites.[47] Antigens of M_r higher or lower than 23 kDa were less consistently recognized, with some heterogeneity between responses of individuals.[45,47,80] Others reported a similar broad array of antigens recognized by infected humans. Convalescent serum antibody reacted in western blots with major oocyst derived antigens of 15.5, 20, 33, 42, 67, and 200 kDa (reducing);[142] 15.5, 23, and 32 kDa (reducing);[80] 15, 21, 22 to 25, 29, 33, 37, 44, 48, 54, 59, 65, 76, 102, 117, 126, 133, 141, 155, 185, and >200 kDa (reducing);[84] 15, 25, 75, 125 to 175, and >205 kDa (reducing);[143] and 15 to 17, 22 to 23, 27, 30, 40, 53, and 58 kDa (nonreducing).[67,95] In one study, antigen migrating in or near the dye front in SDS-PAGE as well as antigens of >66 kDa were recognized.[67]

Antigens of approximately 15, 20, 23, and 32 kDa reported in several studies above may be the same, although the influence of proteolysis in some studies[67,80,143] must be considered. The 15-kDa component of the 15- to 17-kDa antigen recognized by convalescent human serum antibody[67] comigrated with GP15, defined by mAb 5C3,[70] whereas the 17-kDa component was thought to be a distinct antigen.[67,97] Sporozoite surface proteins of 20 and 23 kDa, considered to be the same in many studies, also may be distinct.[67,84] While the possible effect of trypsin used to excyst sporozoites must be considered, a 20-kDa sporozoite protein was heavily glycosylated and iodinatable, whereas a highly immunogenic 23-kDa sporozoite surface protein was not detectably glycosylated and only weakly iodinatable.[84] Shared epitopes between a 23-kDa antigen and six higher M_r antigens have been demonstrated in western blots of disrupted oocyst preparations with polyclonal rabbit antibody raised against the 23-kDa antigen.[140] These antigens appear distinct from the 25- to >200-kDa antigen complex defined by mAb C4A1 and other apical-reactive mAbs.[71,74,81] Proteins of approximately 15, 20, and 32 kDa have been identified in purified oocyst walls[80] and sporozoites,[84] but those of oocyst wall origin may be poorly immunogenic in humans, while those of sporozoite origin were not.[80] Convalescent serum antibody recognized antigens of 47.5, 55, 66, and 130 kDa but not 15, 20, or 32 kDa in western blots of purified oocyst walls.[80] The investigators concluded that the 15-, 20-, and 32-kDa antigens recognized by human serum antibody in western blots of disrupted oocysts, containing both sporozoites and oocyst walls, were therefore sporozoite derived.[80]

Coproantigens of 18- and 20-kDa in reducing SDS-PAGE gels have been described in the soluble fraction of feces from infected calves and humans.[141] The 18-kDa coproantigen was likely a proteolytic degradation product of the 20-kDa coproantigen. The relationship between the two coproantigens and antigens of similar M_r in solubilized oocyst/sporozoite preparations described by others was unclear. Rabbit polyclonal antibody prepared against coproantigen electro-eluted from the 20-kDa region of reducing SDS-PAGE gels recognized both 18- and 20-kDa coproantigens in western blots (reducing). Rabbit antibody against the 18- to 20-kDa coproantigens reacted with the apical end of sporozoites but weakly with oocysts by IFA, suggesting that the coproantigens may be released from the apical end of infectious sporozoites or represent degraded products from nonviable sporozoites. It is unlikely that these 18- to 20-kDa coproantigens are the 20- and 23-kDa sporozoite antigens described by others,[71,72,84] because antibodies against the latter recognize diffusely distributed sporozoite surface membrane antigens, not apical antigens.

Heterohybrid monoclonal antibodies have been produced from *C. parvum*-infected humans having high titers of serum antibody reactive with sonicated oocysts in ELISA.[69] Peripheral blood lymphocytes from a recovered immunocompetent patient or an AIDS patient with persistent cryptosporidiosis were fused with a mouse-human heteromyeloma cell line. Fusions from the immunocompetent but not the AIDS patient yielded positive hybridomas. Two resulting mAbs, 17B-D4 and 17A-1D, reacted with sporozoites by IFA and reduced their infectivity *in vitro*. mAb 17B-D4 recognized 24- and 44-kDa antigens in western blots of sonicated oocysts, while mAb 17A-1D was unreactive.

Serum antibodies from AIDS patients with persistent cryptosporidiosis possessed considerable heterogeneity in western blot recognition of disrupted oocyst antigens.[64] When antigen recognition patterns by sera from AIDS patients were compared to those of convalescent sera from immunocompetent patients, differences and similarities were demonstrated.[45,67] Serum antibody from most AIDS patients in one study recognized antigens of 15 to 17, 22 to 23, 30, 40, 53, and 58 kDa, with variable recognition of antigens <14 and >66 kDa in western blots (nonreducing).[67] This recognition pattern was similar to that of convalescent serum antibody from immunocompetent patients.[67]

d. Cattle

Serum antibody from most *C. parvum*-infected calves recognized an immunodominant 23-kDa antigen in western blots (reducing) of purified sporozoites which comigrated with P23 defined by mAb C6B6 and other antigens of higher and lower M_r.[47] Serum antibody from colostrum-deprived, infected calves recognized antigens of 15 to 17 and 23 kDa and multiple other antigens of <14 and >66 kDa in western blots (nonreducing) of freeze-disrupted oocysts.[67] The 15-kDa antigen recognized[67] comigrated with GP15 defined by mAb 5C3,[70] while the 17-kDa antigen was considered distinct. In other studies, serum antibody from infected cattle recognized multiple oocyst-derived antigens in western blots (reducing), including those of 11 to 14, 17 to 20, 20 to 23, 38, 44, 47 to 49, 55 to 69, 90, 140, and 200 kDa.[20,54,55,143a] Fecal IgA from infected calves also recognized multiple antigens in western blots (reducing) of oocysts, the most prominent being 11, 15, 23, 44, and 66 to 200 kDa.[20] Proteolysis was likely in some of these studies[20,54,55,67] and the degree of antigen solubilization varied, but antigens of approximately 11, 15, and 23 kDa may be the same and appear immunodominant in *C. parvum*-infected calves.

The antigen specificity of bovine serum antibodies induced by parenteral immunization with *C. parvum* has been characterized.[73,75] Neutralizing serum from cows hyperimmunized with infectious sporozoite preparations has been used to identify native[75] or recombinant[72,144] antigens potentially useful for passive or active immunization strategies. Multiple (20) radio-iodinated sporozoite surface antigens from <14 to >200 kDa in reducing SDS-PAGE were specifically precipitated by hyperimmune serum antibody, including antigens of 16 to 18, 23, 28, and 55 kDa and a >200-kDa antigen not resolved in 7.5 to 17.5% gradient SDS-PAGE.[75] Fewer [^{35}S]-methionine-labeled sporozoite proteins were immunoprecipitated.[75] Among these were seven antigens in the 28- to 150-kDa range which comigrated in reducing SDS-PAGE with radio-iodinated sporozoite surface antigens of similar M_r.

The serum antibody response to immunization with a semipurified preparation containing native sporozoite and oocyst wall antigens defined by neutralizing mAb 17.41 (SA-1) has been examined in cattle[73] and mice.[98] SA-1 immune bovine and murine serum antibody reacted with the surface of viable sporozoites in IFA and recognized an antigen profile similar to that of mAb 17.41 in western blots (reducing) of purified sporozoites. Additional antigens in the 80- to >210-kDa range and a <46-kDa antigen were also recognized by the immune sera.

The antigen specificity of bovine colostral antibody prepared by parenteral hyperimmunization with *C. parvum* oocysts and sporozoites has been examined.[81,86,94,100,101,103,104,171] An antibody-enriched fraction of neutralizing hyperimmune colostral whey bound to surface and internal epitopes of sporozoites and oocyst walls in IFA.[81] Antigens from 14 to >270 kDa were recognized by the hyperimmune antibody in western blots (reducing) of excysted oocysts, purified sporozoites, or purified merozoites.[81] Antigens common to merozoites, sporozoites, and excysted oocysts, as well as stage-specific antigens, were recognized. Specifically recognized antigens migrating at approximately 23 kDa, 25 to >200 kDa (multiple), and in the dye front were common to the oocyst, sporozoite, and merozoite stages.[81] These antigens comigrated with those of similar M_r defined by neutralizing mAbs C6B6, C4A1, and 18.44, respectively.[47,71,75]

The isotype-specific colostral antibody responses of cows to intramammary hyperimmunization with freeze-disrupted or intact oocysts have been characterized.[100,101,171] While various immunization regimens influenced specific antibody responses, neutralizing colostral IgG_1, IgG_2, IgA and IgM each recognized multiple antigens from <14 to >180 kDa in western blots of freeze-disrupted oocysts or purified sporozoites. Colostral IgA and IgG_1 recognized the greatest diversity of antigens; IgG_2 and IgM recognized fewer antigens. IgA, IgG_1, IgG_2, or IgM in an antibody lot with *C. parvum* neutralizing activity prominently recognized antigens of 9 to 147 kDa. A sporozoite antigen migrating at 9 to 10 kDa in SDS-PAGE, possibly in the dye front, was recognized only by IgA[101] and IgM,[171] while 15- and 23-kDa sporozoite antigens were recognized by all colostral isotypes. Additional antigens were unique to oocyst walls or residual bodies.

Sporozoite surface antigens recognized by neutralizing colostral antibodies from cows hyperimmunized with excysted oocysts have been defined by radio-immunoprecipitation.[86] Reducing SDS-PAGE profiles of 17 sporozoite surface antigens in the 14- to 200-kDa range precipitated by neutralizing bovine colostral antibodies[86] were similar to those reported for neutralizing bovine serum antibodies.[75] However, M_rs of the >200-kDa sporozoite surface antigens were better defined as 250 and >900 kDa using a 5% SDS-PAGE system.[86,104] Similarly, in western blots (reducing) of solubilized oocysts resolved in a 5% SDS-PAGE system, multiple 46- to >900-kDa antigens were recognized by neutralizing colostral antibodies.[104] The 250- and >900-kDa antigens were the immunodominant species recognized by neutralizing colostral antibody in western blots (reducing) of excysted oocysts or in western blots of antigens immunoprecipitated from excysted oocysts by neutralizing colostral antibody.[86] The >900-kDa antigen was a heavily *N*-glycosylated, Triton X-100-soluble sporozoite glycoprotein.[104] Hyperimmune mouse serum and bovine colostral antibodies, eluted from a recombinant fusion protein corresponding to the >900-kDa antigen, recognized a <190-kDa protein in western blots (reducing) of *N*-deglycosylated oocyst preparations.[104] This observation suggested that the >900-kDa glycoprotein had a polypeptide backbone of <190 kDa. Further, the >900-kDa sporozoite apical antigen immunoprecipitated by mAb 10C6 (GP900) was specifically recognized by neutralizing colostral antibody in western blots.[104] Based on these collective observations, the >900-kDa glycoprotein recognized by neutralizing colostral antibody was considered the same as GP900 defined by mAb 10C6.[86,104]

e. Sheep

Serum and fecal antibody responses to *C. parvum* infection in lambs have been evaluated. Antigen profiles recognized by individual IgG, IgM, and IgA isotypes in convalescent lamb sera have been reported to be similar[56] or different.[93] Serum and fecal antibody responses to *C. parvum* infection in colostrum-deprived lambs similarly recognized antigens of 15, 23, and 180 kDa in western blots of excysted or freeze-disrupted oocysts.[56,94] In another report, serum antibody from colostrum-deprived, infected lambs recognized antigens of <14, 15, 18, 23, and >66 kDa (multiple) in western blots (nonreducing) of freeze-disrupted oocysts.[67] The 15-kDa antigen recognized by serum antibody[67] comigrated with GP15 defined by mAb 5C3.[70] Others have reported that serum antibody from colostrum-deprived, infected lambs recognized antigens of 15 to 17, 28 to 30, 30 to 94 (multiple), and >94 kDa in western blots (reducing) of sonicated oocysts.[93] Hyperimmune lamb serum prepared by parenteral immunization with oocysts recognized antigens of 20- and 23-kDa but serum antibody from infected lambs did not.[93] The lack of 20- and 23-kDa antigen recognition possibly reflected incomplete solubilization of oocysts used for immunoblot analysis.[93] Antigens of approximately 15, 17, 20, 23, 30, and 180 kDa described in these studies may be the same.

Antigens recognized by ovine serum or colostral antibodies after parenteral hyperimmunization with oocysts included the major antigens recognized in response to infection (15, 17, 20, 23, and 30 kDa) and others in western blots of oocysts.[56,93,94,139] Colostral and serum antibody responses were induced in sheep by jet injection of a novel recombinant plasmid containing a 569 base-pair insert encoding epitopes of 15- and 60-kDa *C. parvum* sporozoite surface antigens.[126] Recombinant-immunized sheep mounted a dose and injection route-dependent IgG response which recognized 15- and 60-kDa antigens in western blots (reducing) of freeze-disrupted oocysts. Serum and colostral antibodies from recombinant-immunized sheep eluted from the 15-kDa antigen recognized in western blots bound to the surface of sporozoites in IFA, further demonstrating specificity of the response.

f. Goats

Convalescent serum antibody from *C. parvum*-infected goat kids recognized antigens from <14 to >200 kDa (principally 15.5, 23, and 32 kDa) in western blots (reducing) of disrupted oocysts.[80] Convalescent serum IgG from infected goat kids prominently recognized a 15- to 17-kDa antigen and several higher M_r antigens in western blots (nonreducing) of freeze-disrupted oocysts.[22]

g. Pigs

Serum antibody from *C. parvum*-inoculated piglets recognized antigens of 15 to 17, 21, 23, and >66 kDa (multiple) in western blots (nonreducing) of freeze-disrupted oocysts.[67,145] The 15-kDa antigen recognized by serum antibody[67] comigrated with GP15 defined by mAb 5C3;[70] the 17-kDa antigen was considered distinct.[67]

h. Horses

Serum antibody from *C. parvum*-infected foals recognized antigens of <14, 14, and 23 kDa and several antigens of higher M_r in western blots (reducing) of sporozoites.[47] The 23-kDa antigen recognized by serum antibody was considered immunodominant and comigrated with P23 defined by mAb C6B6.[47]

i. Isolate Variation

Different *C. parvum* isolates have shown conservation and variation of antigens or epitopes they express. Neutralization-sensitive epitopes defined by mAbs 18.44 (CPS-500), C6B6 (P23), 7D10 (P23), 17.41 (SA-1), C4A1 (GP25-200), and 3E2 (CSL) were conserved among geographically diverse human or bovine isolates originating from North and South America as determined by IFA, immunoblot, and sporozoite neutralization assays.[72,74,74a,122] Expression of these epitopes by both infective life-cycle stages[47,71,75,76] and conservation of epitope structure and function among different isolates suggest an important biological role, which may be relevant to active and passive immunization strategies.

Epitopes recognized by multistage reactive mAbs TOU, HAD, ABD, BAX, SPO, LOI, and BKE were conserved among different human, ovine, or bovine isolates of sporozoites as determined by IFA.[109,111] Other comparisons of *C. parvum* isolates from humans and ruminants reported variation[112,134] or conservation[99,134] of molecules recognized by mAbs and polyclonal antibodies in IFA or western blots.

Little variation in M_r was observed between proteins of disrupted oocysts or purified oocyst shells from human and caprine isolates, as determined by SDS-PAGE and protein staining.[80] The major radio-iodinated oocyst surface proteins of different isolates from humans also varied little in M_r, although some proteins were considered isolate specific.[80] A 32-kDa oocyst surface protein was highly conserved among isolates from humans.[80]At the genomic level, both human and bovine isolates contained homologous genes for the sporozoite antigen GP900.[104]

2. *Cryptosporidium muris*

Humoral immune responses to infection, including persistent abomasal infection in cattle, have not been reported. Radio-iodinated oocyst surface proteins ranged from 14 to 290 kDa in reducing SDS-PAGE.[87] Some comigrated with those of similar M_r from *C. baileyi* and *C. parvum*. Rabbit serum prepared against *C. muris* oocysts which was used to define antigens by western blot also recognized *C. parvum* and *C. baileyi* oocyst antigens of similar or different M_r in western blots, suggesting antigenic relationships.[99]

3. *Cryptosporidium baileyi*

Relatively few reports have evaluated humoral immune responses to *C. baileyi* infection in chickens. Primary bursal and cloacal infections were self limiting in orally inoculated broilers,[146,146a] and induced serum[146,147] and bile[147] antibodies reactive with oocyst antigens of <14 to 198 kDa in western blots. Recovered chickens were resistant to reinfection.[146] Patent periods in surgically bursectomized broilers having significantly lowered serum antibody responses were similar to those of controls following oral inoculation with oocysts.[146a] Bursectomy did not increase susceptibility to reinfection.[146a] Clinical signs and gross lesions of respiratory cryptosporidiosis did not correlate with serum antibody responses and were similar in testosterone- or cyclophosphamide-bursectomized and untreated broilers, although respiratory lesions and infection levels were not evaluated histologically.[149] The effect of bursectomy on persistence of respiratory infection was not examined.

Radio-iodinated oocyst surface proteins ranged from 15 to 290 kDa in reducing SDS-PAGE.[87] Some co-migrated with those of similar M_r from *C. muris* and *C. parvum*. Antigenic relationships were suggested by reactivity of serum[146,147] or bile[147] antibodies from *C. baileyi*-recovered chickens with *C. parvum* oocysts,[146,147] and reactivity of serum or bile antibodies from *C. parvum*-inoculated chickens with *C. baileyi* antigens.[147] Rabbit serum prepared against *C. parvum* oocysts recognized *C. baileyi* oocyst antigens in western blots, some of which co-migrated with *C. parvum* oocyst antigens of similar M_r.[146]

C. PASSIVE IMMUNIZATION

Passive immunization against cryptosporidiosis by oral administration of neutralizing antibodies has been investigated in neonates or hosts with acquired or congenital immunodeficiencies. In these groups, suboptimal active immune responses heighten susceptibility to infection by *C. parvum* and delay or preclude termination of established infections. In this section, polyclonal and monoclonal antibody-based passive immunization studies in such hosts are reviewed.

1. Polyclonal Antibodies Against *Cryptosporidium parvum*

a. Laboratory Rodents

Numerous studies have demonstrated that high-titer polyclonal antibody preparations against oocysts and sporozoites have anticryptosporidial activity *in vitro* and *in vivo*. Hyperimmune bovine serum prepared against live sporozoites[79,172] or merozoites[28] contained antibody reactive with surface epitopes of viable sporozoites and merozoites and neutralized the infectivity of each stage *in vitro*. Oral administration of hyperimmune bovine serum also significantly reduced intestinal infection in oocyst-challenged neonatal BALB/c mice.[73] Serum from a lamb immunized by *C. parvum* infection and subsequent parenteral inoculation with excysted oocysts neutralized infectivity of sporozoite preparations for neonatal rats.[15]

Hyperimmune bovine colostrum (HBC) or whey fraction prepared against oocysts neutralized oocyst[103] or sporozoite[148,172] infectivity *in vitro* and provided highly significant protection against oocyst challenge in neonatal BALB/c mice.[148,172] In addition, HBC whey prepared against *C. parvum*[100] significantly neutralized infectivity of *C. wrairi* sporozoite preparations for guinea pigs when incubated *in vitro* prior to intraintestinal inoculation; however, orally administered colostral whey did not protect guinea pigs against heterologous challenge with *C. wrairi* oocysts or significantly reduce existing infection, based on the intensity of oocyst shedding.[137] The soluble, antibody-containing fraction of hyperimmune hen yolk prepared against oocysts significantly reduced existing infections in BALB/c neonates.[149a]

In these studies, anticryptosporidial activity was hypothesized to be due principally to specific neutralizing antibodies induced by immunization. In the case of HBC and lamb serum, this hypothesis was tested and accepted.[15,86,100] Specific immunoglobulin isotypes (IgG_1, IgG_2, IgM, IgA) were purified from HBC whey by monoclonal-antibody affinity chromatography and evaluated for therapeutic efficacy in neonatal BALB/c mice.[100] Each purified immunoglobulin isotype provided a highly significant treatment effect against existing infection that was greater than or equal to that provided by HBC whey. An enriched immunoglobulin fraction from HBC and purified *C. parvum*-specific colostral antibody eluted from oocyst immunoblots each significantly inhibited infectivity of oocysts for MDCK cells *in vitro*.[86] This observation indicated that *C. parvum*-specific polyclonal antibody alone can neutralize parasite infectivity. Similarly, protein G-purified IgG from immune lamb serum neutralized sporozoite infectivity for neonatal rats.[15]

BALB/c or CF1 mice infected with oocysts as neonates and re-exposed as adults by oral and/or parenteral immunization with oocysts, provided minimal or no passive lacteal immunity against infection for their nursing pups.[71,150] The reason for these observations is not clear in that BALB/c mice can produce neutralizing serum or mAb against *C. parvum*.[70-72,74-76,125,129] However, in mice and other rodents, a substantial amount of maternal to fetal passive immunoglobulin transfer occurs transplacentally, bypassing the intestine.[151,152] Whey or serum antibodies from immunized mice reacted with oocyst antigen in ELISA at 1:4 or 1:10 dilutions, respectively;[150] however, it was not known whether neutralizing antibodies were present. In the second study, milk from mouse dams was not examined for antibody to *C. parvum*.[71]

b. Immunodeficient Animal Models

An enriched immunoglobulin fraction from HBC prepared against sonicated oocysts and purified sporozoites significantly neutralized infectivity of isolated sporozoites for BALB/c neonates after *in vitro* incubation, compared to enriched immunoglobulin from sham-immunized controls.[81] To evaluate therapeutic efficacy *in vivo*, adult SCID mice infected with oocysts and allowed to develop persistent infection were treated orally for 10 days with hyperimmune or control antibody. Oocyst shedding and infection scores in the gastrointestinal and biliary tracts were significantly lower than in controls. Neutralization of zoites released from the pyloric-duodenal junction may have led to reduced common bile duct infection anatomically downstream.[30,81] In another report, an immunoglobulin-enriched fraction from HBC prepared against oocysts was tested for efficacy against persistent infection in SCID mice infected as neonates.[153] Oocyst shedding levels were no different between hyperimmune and control antibody-treated mice, but intestinal infection scores were significantly reduced in the hyperimmune antibody-treated mice 1 week after a 10-day treatment regimen.

Pooled colostral whey from nonimmunized cows was tested for therapeutic efficacy against persistent cryptosporidiosis in adult C57Bl/6 mice with cellular and humoral immune deficiencies resulting from murine leukemia retrovirus.[154] Oocyst shedding and intestinal infection scores were significantly reduced in mice treated for 10 days compared to untreated controls.

Monospecific bovine serum against semipurified native sporozoite and oocyst wall antigens defined by neutralizing mAb 17.41 (SA-1)[75,76] was tested for efficacy against infection in neonatal BALB/c and SCID mice.[73] Orally administered SA-1 immune serum did not protect BALB/c neonates or 4-week-old SCID mice from oocyst challenge and did not reduce intestinal infection in 9-week-old SCID mice with persistent cryptosporidiosis. In contrast, hyperimmune bovine serum prepared against whole sporozoites neutralized sporozoite infectivity and significantly protected neonatal mice against oocyst challenge.[73,79] Binding inhibition assays determined that SA-1 immune bovine serum lacked detectable antibody to the neutralization-sensitive epitope defined by mAb 17.41, and the concentration and specificity of other neutralizing antibodies in SA-1 immune serum may have been lower than those of hyperimmune bovine serum.

Foals with SCID have been used as an outbred animal model of severe clinical cryptosporidiosis.[26,28] They lack functional B and T cells and are incapable of antigen-specific immune responses.[157] Hyperimmune bovine plasma prepared against isolated merozoites, which neutralized merozoite infectivity *in vitro*,[28] and mAb 18.44, which neutralized sporozoite[75] and merozoite[76] infectivity *in vitro*, were orally administered to persistently infected SCID foals for 7 days but did not significantly reduce oocyst shedding or intestinal and biliary infections compared to control antibody-treated foals.

Pooled normal human serum IgG, containing low levels of antibody against oocyst antigens, significantly reduced intestinal infection scores in neonatal SCID mice when administered orally 6 hours before oocyst challenge and daily thereafter for 4 days.[143] When IgG was administered 3 weeks following initiation of infection, no reduction of intestinal infection was observed.

Bile from orally immunized adult rats contained IgA reactive with oocyst antigen in ELISA and was tested for efficacy in adult nude mice with chronic cryptosporidiosis.[155] After two oral treatments 10 days apart, intestinal infection was significantly lower and lesions less severe in mice treated with immune bile compared to control bile.

c. Domestic Livestock

Because calves, lambs, goat kids, piglets, and foals are usually infected during the early neonatal period, before their immune system is mature and capable of specific responses, passive immunization against cryptosporidiosis with maternal colostrum has been investigated in these species. Ruminants, pigs, and foals are normally protected from infectious diseases during the neonatal period by passively transferred colostral immunity; immunoglobulins are not transplacentally transferred in these species.[156,157] Passive colostral immunization strategies for cryptosporidiosis have been based largely on successful application of this approach against viral and bacterial enteropathogens.[51,157-160,168] In neonatal calves, passive protection against pathogens may be mediated locally by colostral antibody in transit through the intestinal tract during the first few days of life. Intestinal-level protection after this period may be mediated by colostrum-derived serum antibody secreted into the gastrointestinal tract during the first month of life.[159,160]

Calves fed HBC, prepared against oocysts, at 4 hours of age and challenged with oocysts 3 days later had significantly reduced patent periods, oocyst shedding, and days of diarrhea, compared to calves fed normal colostrum.[161] Similarly prepared HBC with specific antibody that neutralized infectivity of *C. parvum in vitro* was fed to calves at 4 hours of age and once daily thereafter for 7 days; calves were challenged with oocysts at 12 hours of age.[86,104] Calves fed HBC had significantly reduced diarrhea and dehydration during peak patency and significantly reduced oocyst shedding compared to calves fed control colostrum.

Colostrum from cows immunized with oocysts, but not hyperimmune, did not protect oocyst-challenged calves as defined by onset and duration of diarrhea and presence or absence of fecal oocysts, although diarrhea severity and oocyst numbers were not actually quantitated.[162] The ELISA titer of *C. parvum*-specific whey antibody from immunized and nonimmunized cows was not significantly different. When calves were protected against cryptosporidiosis by HBC, the ELISA titer of *C. parvum*-specific antibody was 10- to 200-fold higher than normal colostrum.[86,161] Different immunization protocols may account for different antibody responses and levels of protection.

Colostrum from ewes infected as neonates or hyperimmunized as adults was tested for efficacy against cryptosporidiosis in lambs.[94] Lambs acquired colostrum by normal suckling and were challenged with oocysts at 2 days of age. Lambs that suckled hyperimmune colostrum had no diarrhea, had longer prepatent periods, and shed fewer oocysts than those that suckled non-hyperimmune colostrum. HBC whey and whey from nonimmunized cows provided partial protection against cryptosporidiosis in lambs deprived of maternal colostrum.[94]

Piglets allowed to suckle colostrum from nonimmunized, nonexposed sows had infections similar to colostrum-deprived piglets after oocyst challenge.[163] An enriched immunoglobulin fraction from HBC,[86] fed daily for 5 to 8 days to gnotobiotic piglets infected at 2 days of age and commencing at the onset of diarrhea and oocyst shedding, significantly reduced oocyst excretion compared to control immunoglobulin, but did not affect diarrhea.[153] Significantly, >98% of *C. parvum*-specific antibody activity in the hyperimmune preparation was destroyed during gastrointestinal transit, highlighting the effects of acidity and proteolysis on orally administered antibody and the need to control these effects in passive immunotherapeutic approaches. Because the *C. parvum* isolate in this study induced severe diarrhea, accelerated gastrointestinal transit may have reduced intraluminal concentration of intact antibody to subtherapeutic levels.

d. Immunodeficient Humans

The first report of polyclonal antibodies to treat human cryptosporidiosis involved a hypogammaglobulinemic infant with diarrhea of approximately 3 week's duration.[59] The patient had normal T-cell function, but lacked IgA- and IgG-positive lymphocytes in duodenal lamina propria and normal serum immunoglobulin levels. Diarrhea resolved and oocyst shedding ceased after 12 days of continuous intraduodenal infusion with HBC prepared against oocysts and oral treatment for 4 additional days. Four months later the patient relapsed, and biliary tract infection was diagnosed. Shortly after this report, an AIDS patient and a chemotherapeutically immunosuppressed leukemic child with chronic cryptosporidial diarrhea were similarly treated[59] with HBC.[164] Diarrhea resolved in the AIDS patient, but intermittent oocyst shedding continued for 3 months before death from other complications. Diarrhea and oocyst shedding ceased in the leukemic child. Another AIDS patient with chronic fulminant cryptosporidial diarrhea was treated for 3.5 days by continuous intraduodenal infusion of HBC prepared against oocysts.[165] Diarrhea and oocyst shedding ceased for 3 months but relapsed. Positive treatment results in these early case reports led to a study involving five AIDS patients with persistent cryptosporidial diarrhea.[166] Three patients received HBC by continuous nasogastric tube infusion for 10 days, while two patients received normal bovine colostrum. Diarrhea improved and oocyst counts dropped in two of the three patients treated with HBC. Diarrhea also improved in patients treated with normal bovine colostrum but oocyst counts did not change. Definitive information obtained from this study was limited by heterogeneity in clinical and microbiologic parameters among the patients evaluated and by small treatment group size.

Normal bovine colostrum and serum often have low antibody titers against *C. parvum*, reflecting previous infection or continuing exposure,[52,143,143a] and may contain cytokines with anti-cryptosporidial activity.[148,154] Such preparations also contain nonimmune, nonimmunoglobulin microbicidal factors that nonspecifically reduce infectivity of *C. parvum*.[15,86,103,148,154] For example, IgG-depleted serum from gnotobiotic lambs possessed low-level, nonspecific, heat-stable, anticryptosporidial activity.[15] Further, normal bovine colostrum, whey,[103,161] or an immunoglobulin-enriched colostral fraction from adjuvant-immunized cows[81,86,153] had neutralizing activity against sporozoites[81,148] or oocysts[86,103] *in vitro*, and reduced infection levels in oocyst-challenged BALB/c neonates[148] or persistently infected SCID mice[153] vs. saline-treated controls. The principle nonspecific neutralizing factors in colostrum may be in the nonwhey fraction.[103]

Normal bovine colostrum (NBC) was used to treat two terminal adult AIDS patients and an adult hypogammaglobulinemic patient with persistent cryptosporidiosis.[60] Patients were treated orally for 5 to 7 days with two lots of colostrum. One lot had antibody titers against oocysts and sporozoites of 10 and 40, respectively; titers of the second lot were not reported. Diarrhea and oocyst shedding (not quantitated) persisted in all three patients. An immunodeficient child with persistent cryptosporidiosis treated orally with NBC had improved diarrhea and reduced oocyst shedding.[167] A child with AIDS and severe persistent cryptosporidial diarrhea was treated with an immunoglobulin-enriched fraction from NBC by continuous nasogastric tube infusion for 14 days.[168] Diarrhea eventually ceased, and feces and jejunal biopsies were free of *C. parvum* during a 6-month followup period. In an uncontrolled prospective study, seven AIDS patients with mean CD4 lymphocyte counts of 38/nl and persistent cryptosporidial diarrhea were treated orally with an enriched antibody preparation from NBC for 10 days.[158] In three patients, diarrhea resolved and did not recur during a 4-week followup period. In two patients, the frequency of diarrheal episodes was reduced by >50%. Two others had no initial response, but in one the frequency of diarrhea decreased after an additional 10 days of treatment at an increased dose.

A normal human serum immunoglobulin preparation was used to treat persistent cryptosporidial diarrhea in a chemotherapeutically immunosuppressed leukemic child.[169] Diarrhea improved after 2 days of oral therapy, and 2 weeks after treatment a duodenal biopsy was free of infection. The patient remained asymptomatic for a 6-month followup period.

Low levels of antibody against *C. parvum* and/or undefined anticryptosporidial factors in nonimmune preparations described in these studies may account for variable efficacy reported.

Patient and treatment variables that influenced the outcome and interpretation of many of the above studies included: immunization regimen for production of polyclonal antibody and neutralizing antibody concentration; use of whole colostrum containing concentrated IgG and factors which reduce gastrointestinal antibody degradation[59] vs. colostral whey, enriched colostral antibody, or serum; patient age and immune status; group sample size; definition of clinical and microbiologic response parameters, including quantitation of oocyst shedding and diarrhea indices; concurrent infections which cause diarrhea and degree of intestinal mucosal damage; concurrent noninfectious causes of diarrhea; oral vs. duodenal administration and intraintestinal neutralizing antibody concentration; timing of treatment relative to duration of infection; extraintestinal infection; anticryptosporidial treatment duration; and concurrent treatment with antimicrobial and antiacid drugs. Many of these variables also apply to mAb-based approaches. Additional controlled clinical trials designed to address these variables are required to clarify the utility of hyperimmune antibody treatment for cryptosporidiosis.[167,170] Microbial contamination and safety issues relating to polyclonal antibodies for treatment of cryptosporidiosis in immunodeficient hosts have been investigated. Gamma irradiation sterilized bovine colostrum without significant loss of specific antibody activity[94,165,166,171] or therapeutic efficacy.[94,165,166] Lyophilization reduced colostral volume and storage problems without appreciably altering antibody activity.[94,168,171]

2. Monoclonal Antibodies Against *Cryptosporidium parvum*

Monoclonal antibody preparations may be more desirable than polyclonal preparations, considering the potential for lot-to-lot heterogeneity with polyclonal antibodies.[158,168,171] Sporozoite and merozoite *in vitro* neutralization assays have been used to identify mAbs warranting evaluation in animal model systems for passive immunization against cryptosporidiosis.

a. Oral Treatment with Monoclonal Antibodies in Normal Mice

Monoclonal antibodies 17.41, or 17.41 and 18.44, which neutralized sporozoite and merozoite infectivity *in vitro*,[75,76] significantly reduced intestinal infection in oocyst-challenged BALB/c neonates compared to isotype-matched control mAbs.[172]

A single treatment with mAbs C6B6, C8C5 or C4A1 before oocyst challenge did not reduce intestinal infection in BALB/c neonates.[71] Daily treatment with mAb Cmg-3 significantly reduced infection by day 4, but not day 8 PI.[129] Daily treatment with pooled C6B6, C8C5, and C4A1 or C6B6, C4A1, and Cmg-3 significantly reduced intestinal infection, compared to untreated controls.[71,129]

Treatment with mAbs C6B6, 7D10, or C6B6 and 7D10, at oocyst challenge and twice daily thereafter significantly reduced intestinal infection in BALB/c neonates vs. control mAbs.[72,74] C6B6 was used at a dosage higher than previously evaluated.[71] Treatment with mAb 3E2 according to the same regimen[72] provided near complete protection against intestinal infection.[74a]

mAb 5C3 significantly reduced oocyst shedding in infected BALB/c neonates when administered twice daily for 5 days.[70] Neonates treated with 5C3 at 22 to 48 hours post-oocyst challenge had intestinal infections 75% lower than untreated controls.[70]

b. Oral Treatment with Monoclonal Antibodies in Immunodeficient Mice

Persistently infected adult, congenitally athymic nude mice treated with mAb 17.41 once daily for 10 days had significantly reduced intestinal, but not hepatobiliary or pancreatic duct, infections, suggesting that extra-intestinal sites may be inaccessible to orally administered mAb.[28,123] In adult SCID mice with persistent infection, daily treatment with 17.41 for 14 to 17 days significantly reduced oocyst shedding and intestinal, but not pyloric, infections.[173] Low pH may have reduced binding of 17.41 to zoites in the pyloric region. Heavy pyloric infection in SCID mice and more severe infection in duodenum than lower small intestine in nude mice[123] may be related to biliary and pancreatic infection, as suggested in other studies.[30,81]

D. MECHANISMS OF *CRYPTOSPORIDIUM PARVUM* NEUTRALIZATION

Neutralizing antibodies may reduce infectivity by agglutination or immobilization, direct blockade or alteration of apical invasion molecules or putative zoite ligands for host cells, lysis via C′-activation and cytotoxicity.[75,79,174-176a] Neutralizing antibodies may also enter host cells bound to invading zoites and arrest intracellular development.[176a-178]

Direct killing of sporozoites by nonspecific, antibody-independent cytotoxic factors was greater in HBC than NBC fractions.[81,148,172] Neutralization of sporozoites by hyperimmune bovine serum did not involve significant sporozoite killing and was C′-independent.[79,81] Sporozoites and merozoites invade host cells rapidly.[133] Neutralization of sporozoites by monoclonal or polyclonal antibodies is time and antibody concentration dependent *in vitro*.[172] Time-dependency of antibody-mediated neutralization may account in part for some reduction of antibody efficacy *in vivo*. Rapid gastrointestinal transit with severe cryptosporidial diarrhea reduces exposure time of zoites to neutralizing antibody and may contribute to some treatment failures or suboptimal responses.[28,158] Other studies have highlighted the importance of optimal antibody concentration in control of cryptosporidiosis *in vivo*[94,153,172] and neutralization *in vitro*.[86,172]

mAbs 18.44, 17.41, C6B6, 7D10, C4A1, and 3E2; hyperimmune bovine serum; and HBC antibody did not significantly agglutinate sporozoites as a mechanism of neutralization.[72,74a,75,81,172] However, HBC antibody[81] and mAb 3E2[74a] induced distinct morphologic changes in sporozoites and merozoites characterized by progressive formation and release of membranous antigen-antibody complexes from the posterior zoite, resembling the malaria circumsporozoite precipitate (CSP) reaction (Figure 1).[179] HBC antibody elicited the reaction within minutes of exposure, while mAb 3E2 did so within seconds. Specifically bound antibody demonstrated in membranous complexes by IEM indicated mechanistic involvement in the CSP-like reaction (Figure 2). After this reaction, sporozoites were noninfective *in vivo*[81] and did not attach to or invade Caco-2 cells *in vitro*.[74b] CSL, the 1200- to 1400-kDa antigen recognized by 3E2, was determined to be the principal antigen species mechanistically involved in the CSP-like reaction and was further characterized as a soluble exoantigen released from the apical sporozoite.[74a] Cytoskeletal inactivation with cytochalasin-D did not inhibit apical release of CSL but did inhibit its posterior translocation.[74a] Further, purified CSL bound specifically to live Caco-2 cells and significantly reduced permissiveness to sporozoite infection *in vitro*.[74b] Collectively, these observations suggested that CSL may function as a putative sporozoite ligand.

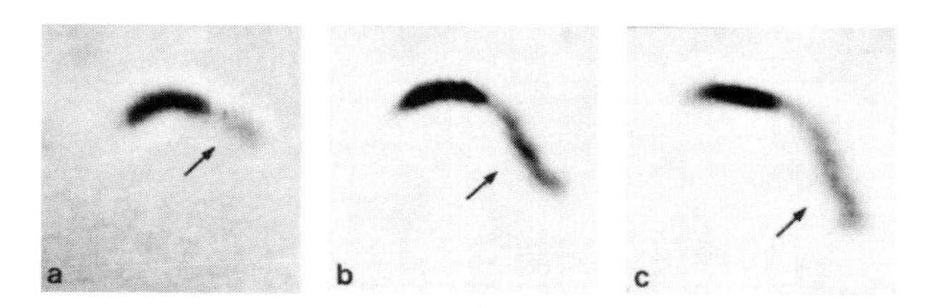

Figure 1 Videophotomicrographic demonstration of the progressive nature of the CSP-like reaction elicited by HBC antibody over time. Representative surface precipitate formation (arrows) at **(a)** 10 minutes, **(b)** 20 minutes, and **(c)** 30 minutes after incubation of sporozoites with HBC antibody in medium.

Sporozoites deposited surface antigens of 15 (GP15),[88,108] 23 (P23),[108,124] and 38 or >900 kDa[88] on glass substrate during gliding motility, forming "trails" visualized with mAbs specific for each antigen (Figures 3 and 4). GP15 and P23, expressed on the sporozoite surface or deposited in trails, were recognized by neutralizing mAbs.[70-72] Surface antigen trails deposited by gliding sporozoites were morphologically similar to those described for the circumsporozoite protein of *Plasmodium* sp.,[180] suggesting that specific *C. parvum* antigens may be involved in sporozoite motility and invasion. This type of motility involves attachment of zoite surface adhesin molecules to substrate, posterior capping via a cytoskeletal-dependent mechanism, and forward locomotion.[88] During this process, surface adhesins

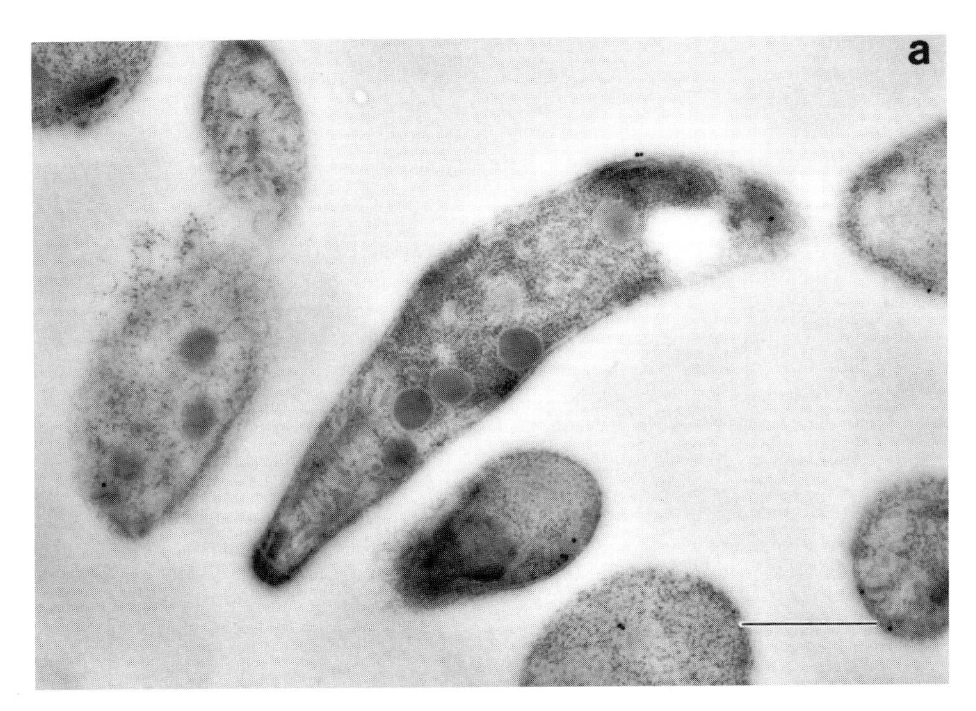

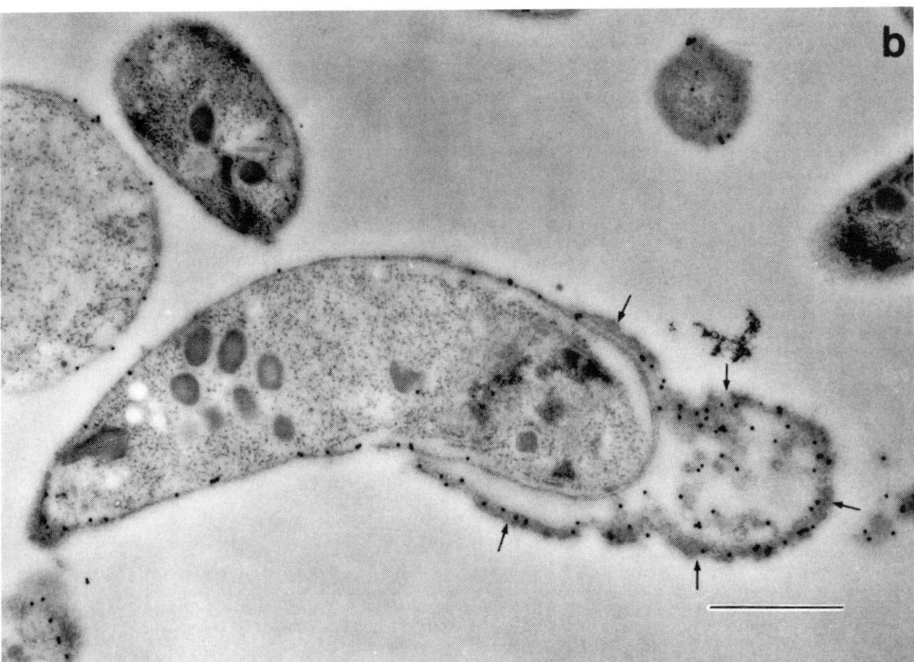

Figure 2 Transmission immunoelectron micrographs of *C. parvum* sporozoites after incubation with **(a)** NBC antibody or **(b)** HBC antibody. Note dense immunogold labeling of bovine antibody in membranous precipitate (arrows) at the posterior end of the sporozoite in the HBC antibody-treated preparation and the absence of this reaction in NBC antibody-treated sporozoites. Bars = 1 μm. (From Riggs, M. W. et al., *Infect. Immun.*, 62, 1927, 1994. With permission.)

are deposited as trails demarcating the path of travel. A similar phenomenon is associated with posterior translocation of the moving junction between zoite and host cell membranes during invasion.[88]

GP15 trails deposited on MDCK cells before sporozoite invasion were demonstrated by mAb 11A5 in IFA (Figure 4).[88] During invasion, GP15 was capped posteriorly and shed from sporozoites and not re-expressed until the meront stage.[88] Sporozoite motility and trail deposition were inhibited by cytochalasin

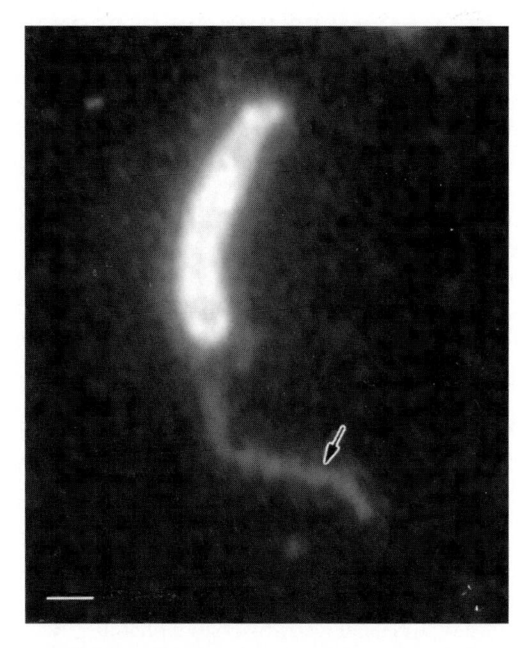

Figure 3 Immunofluorescence detection of P23 trail (arrow) deposited by a *C. parvum* sporozoite during gliding motility on glass slide substrate. Note trail extending from posterior end of sporozoite over course previously traveled. Trail and sporozoite visualized with mAb C6B6. Bar = 1 μm. (From Arrowood, M. J. et al., *J. Parasitol.*, 77, 315, 1991. With permission.)

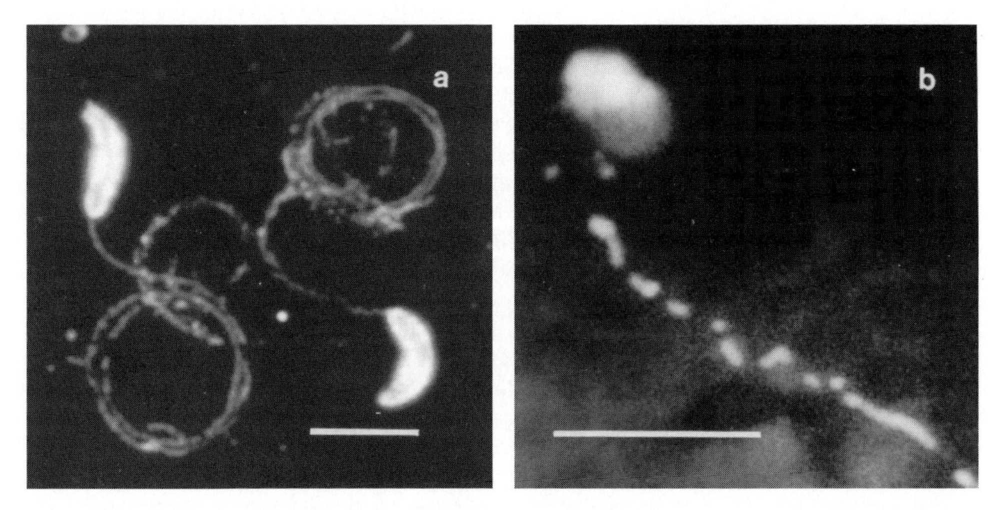

Figure 4 Immunofluorescence detection of GP15 trails deposited by *C. parvum* sporozoites during gliding motility on **(a)** glass slide substrate, or **(b)** living MDCK cells *in vitro.* Trails and sporozoites visualized with mAb 11A5. Bar = 5 μm. (From Gut, J. and Nelson, R. G., *J. Euk. Microbiol.*, 41, 42S, 1994. With permission.)

D or incubation at 0°C, suggesting that trail deposition by sporozoites is a biologically relevant, active metabolic process dependent on cytoskeletal microfilaments.[88]

The CSP-like reaction appears to be distinct from P23 trail deposition.[74,81] P23 reactive mAbs which detected sporozoite trails on substrate did not elicit the CSP-like reaction, and CSL-reactive mAbs which elicited the CSP-like reaction did not detect sporozoite trails. Microfilament (actin) and microtubule (α and β tubulin) genes have been cloned (see Chapter 10).[181] Both classes of cytoskeletal polymers have biomechanical roles in sporozoite gliding motility, invasion, and intracellular development.[181,182] Further study of the cytoskeletal apparatus and antigens involved in trail formation or the CSP-like reaction may reveal relationships.

III. CELL-MEDIATED IMMUNE RESPONSES

A. ROLE OF CELL-MEDIATED IMMUNITY

The importance of cell-mediated immune (CMI) responses in resistance to Coccidia closely related to *Cryptosporidium* has been well characterized.[41,42] Studies on the cellular immunology of cryptosporidiosis are more recent, derive largely from laboratory rodent models, and represent an area of growing emphasis. The central importance of CD4 T cells, IFN-γ, and IL-12 in protective immune responses against *Cryptosporidium* infection in mouse models has recently emerged.[24,25,183-187] The role of other cytokines and cell types is less clear. Differences in cell subset or cytokine depletion/enrichment protocols and their efficacy, cell number, administration route, and engraftment in reconstitution studies and duration of infection before immunologic manipulation make some results difficult to compare accurately. Consensus conclusions relating to cellular immunity in cryptosporidiosis are summarized in this section.

1. *Cryptosporidium parvum*

a. *Control of Infection in Laboratory Rodent Models*

Persistent, progressive intestinal and hepatobiliary infections in T- and B-cell-deficient neonatal[27,30] or adult[17,27,30,81,188] SCID mice and SCID foals[26,28] indicated that T cells or B cells, or both, are required for control of infection. Infection in adult SCID mice progressed more slowly than in SCID neonates, suggesting that T- and/or B-cell-independent mechanisms contribute to control of infection.

The first report of persistent cryptosporidiosis in neonatal or adult congenitally athymic nude mice indicated the central importance of T cells in recovery from infection.[23] Because nude mice lack mature CD4 and CD8 T cells, the role of these cells in protection against persistent intestinal and hepatobiliary infection has been studied further.[24,25] Infection in nude adults progressed more slowly than in nude neonates, suggesting that T-cell-independent mechanisms contribute to control of infection. NIH III athymic nude (bg/nu/xid) adults and neonates have additional cellular and humoral immune deficiencies and develop persistent intestinal and hepatobiliary cryptosporidiosis.[27]

Neonatally infected athymic nude rats developed persistent cryptosporidiosis and lacked specific, delayed-type hypersensitivity (DTH) skin reactions or serum IgG responses.[29] Infection was self limited in heterozygous littermates that developed specific DTH skin reactions and serum IgG.

Study of major histocompatibility complex (MHC) class I-deficient and MHC class II-deficient mice, lacking functional CD8 T cells and CD4 T cells, respectively, avoided variables associated with potentially incomplete depletion of CD4 or CD8 T cells by *in vivo* or *in vitro* treatment with mAbs in other model approaches.[189] Neonatal or adult C57BL/6J MHC class II-deficient, but not MHC class I-deficient or C57BL/6J wild-type mice, developed persistent intestinal infections and lesions after oocyst inoculation. From flow cytometric analysis of spleen cells and use of specific mAbs it was concluded that CD4 T cells, but not CD8 T cells, are required to prevent persistent infection in C57BL/6J neonates or adults.

αβ-T-cell-deficient and γδ-T-cell-deficient mice have been used to further define the role of T-cell subsets in control of infection.[189a] Neonatal and adult αβ-T-cell-deficient mice developed persistent intestinal infections following oocyst inoculation. Infection levels were higher than those in γδ-T-cell-deficient mice or C57BL/6 controls. γδ-T-cell-deficient neonates developed more severe infections of longer patency than controls but were able to resolve infection, whereas γδ-T-cell-deficient adults were completely resistant or developed very low-level, transient infections. It was concluded that αβ-T cells are critical for resistance to *C. parvum* and that γδ-T cells have an important but less critical role in mice.

i. Requirements for Protective Reconstitution by Naive Lymphocytes

Adult SCID mice injected i.p. with unfractionated spleen cells from naive (uninfected) BALB/c mice 24 hours before oocyst challenge shed no oocysts and had no evidence of intestinal or biliary infection 4 or 13 weeks later.[17] SCID mice injected with unfractionated human peripheral blood mononuclear cells (PBMC) before oocyst challenge shed as many oocysts as control mice, except for a transient reduction 4 to 6 weeks postinjection and had no reduction of intestinal infection at 8 or 12 weeks.[188] Persistently infected SCID mice ceased oocyst shedding within 2 weeks after receiving a mixture of unfractionated thymocytes, spleen, and bone marrow cells from naive BALB/c mice, had no intestinal infection 4 weeks later, and had splenic T- and B-cell numbers similar to controls.[188]

Persistent infection in adult SCID mice was significantly reduced by i.v. reconstitution with individual preparations of unfractionated spleen cells, bone marrow cells, or thymocytes from naive BALB/c adults.[185] Thymocytes alone reduced infection in reconstituted mice, confirming the critical role of T cells in immunity to *C. parvum*. Infected SCID mice also had significantly reduced oocyst shedding and intestinal infection 4 weeks postreconstitution with CD8-depleted or CD4-depleted preparations of naive spleen cells treated with specific mAb and C′ *in vitro*. No significant differences in infection were observed between mice reconstituted with CD8-depleted or CD4-depleted cells. However, when CD8 depletion and CD4 enrichment were greater, mice reconstituted with CD8-depleted spleen cells (0.27% CD8, 52% CD4) had significantly lower infections than mice reconstituted with CD4-depleted spleen cells (1.16% CD4, 28% CD8). The most consistent observation was a significant increase in CD4 cells in mice reconstituted with CD8-depleted spleen cells. It was concluded that CD4 cells are more critical than CD8 cells in mediating recovery from persistent infection in SCID mice.

Persistently infected adult SCID mice reconstituted with unfractionated spleen cells from naive donor mice recovered from infection.[183] Reconstituted mice depleted of either CD4 cells or IFN-γ with specific mAbs had reduced or no ability to control infection, but depletion of both did not increase infection levels above depletion of either alone. Depletion of CD8 or natural killer (NK) cells in reconstituted mice by specific mAbs did not affect the ability to resolve infection. It was concluded that complete recovery from an established infection in reconstituted SCID mice requires CD4 cells and IFN-γ. These findings[183,185] were extended when adult SCID mice reconstituted with CD8-depleted naive spleen cells recovered from persistent infection, but more slowly than mice reconstituted with unfractionated spleen cells.[186] It was concluded that CD4 cells are required for recovery from persistent infection but that CD8 cells also contribute to protection. Differences in cell depletion protocols and duration of infection prior to immunologic manipulations may account for different observations relating to CD8 cells in protection.[183,185,186]

ii. Effects of Cytokine Depletion

IFN-γ — Depletion of IFN-γ in BALB/c adults by mAb 1 day before oocyst inoculation and weekly thereafter greatly enhanced oocyst shedding compared to control mAb-treated mice or CD4- or CD8-depleted adults.[25] IFN-γ depletion also prolonged patency compared to control mAb-treated mice and resulted in more severe intestinal inflammation and lesions but did not prevent termination of infection.[25] In BALB/c neonates, IFN-γ depletion begun 2 days before oocyst challenge insignificantly increased infection but not intestinal inflammation.[187] In adult nude mice, IFN-γ depletion resulted in a nearly 75-fold increase in oocyst shedding for an extended time, suggesting IFN-γ produced by non-T cells contributed to resistance.[25] Adult SCID mice depleted of IFN-γ by mAb 4 hours before oocyst challenge and thrice weekly thereafter, developed severe intestinal infections by 3 weeks PI, whereas control mAb-treated mice did not.[184] A single injection of anti-IFN-γ mAb 2 hours before or 18 hours after oocyst challenge also increased susceptibility of SCID adults to infection. IFN-γ was not detected in serum, intestinal homogenates, or intestinal secretions from untreated SCID mice before oocyst challenge or up to 9 days after. However, nonspecific production of IFN-γ by spleen cells from uninfected SCID mice and SCID mice 4 weeks PI, was detected *in vitro* following culture with oocyst homogenate. It was concluded that endogenous IFN-γ from a non-T cell functions in the resistance of SCID mice to initial colonization with *C. parvum*.[184] Other oocyst-challenged adult SCID mice depleted of IFN-γ had reduced prepatent periods, increased intestinal infection and oocyst shedding, more severe intestinal lesions, and accelerated disease progression. [17,186,190] In SCID neonates, IFN-γ depletion with mAb begun 1 day before oocyst challenge increased intestinal infection 9 days later compared to controls.[187]

Interleukins — Depletion of IL-2 by mAb 1 day before oocyst challenge and weekly thereafter, did not allow infection in BALB/c adults or increase the severity of infection in IFN-γ-depleted mice.[25] IL-2 appeared not to be essential for T-cell-mediated protective responses or for induction of IFN-γ secretion by non-T cells. Depletion of IL-2 in BALB/c adults by mAb 2 days before and 5 days after oocyst challenge significantly increased oocyst shedding, but the increase was transient.[191] Sensitivity of oocyst detection techniques may account for the differences observed.[25,191] In BALB/c adults, depletion of IL-5, or IL-4 and IL-5, but not IL-4 alone, by mAbs 2 days before and 5 days after oocyst challenge resulted in significantly increased infection and oocyst shedding compared to control mAb-treated mice.[191] It was concluded that IL-5, complemented by IL-4, represented Th2-dependent responses involved in control of *C. parvum* infection. Depletion of IL-12 in BALB/c or SCID neonates beginning 1 day prior to oocyst challenge increased the severity of established infections and prolonged patency.[187]

Tumor necrosis factor (TNF) — Depletion of TNF-α in adult SCID mice by monospecific antibody 2 hours before and 5 days after oocyst challenge did not increase susceptibility to infection.[184] Another study reported similar findings,[18] suggesting that TNF does not contribute significantly to resistance against *C. parvum*.

iii. Effects of Lymphocyte Depletion

Depletion of CD8 and CD4 or only CD4 lymphocytes in neonatally infected BALB/c mice by weekly treatment with mAbs begun 2 days PI resulted in persistent and progressively symptomatic infections with marked oocyst shedding that resolved 2 to 4 weeks after treatments stopped.[24] In contrast, neonatally infected BALB/c mice depleted of CD8 lymphocytes cleared infection within 3 to 4 weeks PI, as did control antibody-treated or untreated mice, suggesting that CD4 lymphocytes were critical to control infection and that CD8 lymphocytes had a minor or no role. In BALB/c adults, depletion of CD8 and/or CD4 lymphocytes by mAbs 1 day before oocyst inoculation and weekly thereafter allowed persistent, low-grade, asymptomatic infection with sparse oocyst shedding for up to 100 days. CD4 and CD8 depletion resulted in slightly but significantly higher oocyst shedding than CD4 depletion alone. Oocyst shedding in CD4-depleted adults was much lower than in CD4-depleted neonates. When mAb treatment stopped, oocyst shedding eventually ceased. In contrast, CD8-depleted adults had only transient, low-grade infection and oocyst shedding, similar in duration to CD8-depleted neonates. Adults depleted of both CD4 lymphocytes and IFN-γ had persistent infections with oocyst shedding >100-fold higher than CD4-depleted alone and >7-fold higher than IFN-γ-depleted alone, but shedding declined to CD4-depleted levels after anti-IFN-γ treatment ceased.[25] IFN-γ depletion in persistently infected, CD4-depleted adults exacerbated infection, increasing oocyst shedding approximately 20-fold. It was concluded that (1) both CD4 lymphocytes and IFN-γ, of T-cell and non-T-cell origin, are required to prevent initiation of infection in BALB/c adults; (2) IFN-γ alone can limit the severity of infection but is not required to terminate infection; and (3) CD4 lymphocytes alone can limit the duration of infection and are required to terminate infection.[25]

iv. Role of Granulocytes, Macrophages, and Other Inflammatory Cells

Inflammatory cells that accompany intestinal lymphocyte and plasma cell responses to cryptosporidiosis in mice,[23,25,81,189,192] calves,[2] lambs,[2] pigs,[193,194] and humans[12] are predominantly neutrophils, eosinophils, and macrophages. Intracytoplasmic zoites seen in M cells overlying Peyer's patches and subjacent macrophages in guinea pigs suggested one route of antigen uptake and presentation to intestinal lymphoid tissue.[195]

Persistent cryptosporidiosis in SCID mice suggested that NK cells alone had a minor role but may have a more important T-cell-dependent role in competent hosts.[27] Although splenic NK cell depletion in adult nude,[25] SCID,[196] or BALB/c[196] mice by specific antibody did not alter infection and disease progression, suggesting no significant contribution to controlling infection, a role for intestinal NK cells was not excluded. Adult NK-cell-deficient C57BL/6J-bg^{j} [197] and C3H/HeJ/beige mice[36] were slightly more susceptible to infection than 18 other adult inbred strains but developed only mild, transient infections.

Infant W/W^{v} mice with normal T-cell function, but abnormal mast cells and other granulocytes, were similar in susceptibility and recovery to normal mice that had no intestinal mast cell accumulations during recovery, suggesting that these cells may not contribute significantly to control of infection in mice.[198]

b. Control of Infection in Humans

Cryptosporidiosis may persist in AIDS patients and other patients with cellular immune deficiencies.[12] AIDS patients with CD4 counts of ≥180 cells/mm^3 had self-limited infection, whereas most of those having <140 cells/mm^3 had persistent infection.[199] Similar CD4 count associations were reported in other AIDS patients[65,200] but were less consistent in children of developing countries.[201] Late-stage AIDS patients may have elevated serum IFN-γ,[202] but lower intestinal IFN-γ may exacerbate infection.[203]

2. *Cryptosporidium muris:* Control of Infection in Laboratory Mouse Models

a. Requirements for Protective Reconstitution by Naive Lymphocytes

Gastric *C. muris* infection persisted in adult SCID or nude mice but did not become extraintestinal.[17,18] In normal BALB/c or C57BL/6 adults, the infection was self limiting, after which mice were resistant to reinfection.[17,18] SCID mice injected with unfractionated spleen or mesenteric lymph node cells from naive BALB/c mice 24 hours before oocyst challenge became infected but shed significantly fewer oocysts than unreconstituted mice and ceased shedding by 42 days PI.[17] Depletion of Thy.1 cells, but not B cells, from spleen cell preparations with mAbs before injection significantly increased oocyst shedding levels and duration, compared to SCID mice given unfractionated spleen cells. Unreconstituted SCID mice shed significantly more oocysts than mice given Thy.1-depleted spleen cells. These observations suggested that T cells were required for recovery and that any role for B-cell responses was T-cell dependent.

b. Effects of Cytokine or Lymphocyte Depletion

BALB/c or C57BL/6 adults depleted of IFN-γ by specific mAb beginning 2 days before oocyst challenge shed significantly more oocysts than control mAb-treated mice but eventually recovered.[17] In contrast, TNF-depletion in BALB/c mice did not increase oocyst shedding and in C57BL/6 mice significantly decreased shedding, compared to control mAb-treated mice.[17]

BALB/c adults depleted of Thy.1 and Lyt.1 or CD4 T cells by treatment with antibodies had higher oocyst shedding and longer patency after primary infection than control antibody-treated mice.[18] CD8-depleted mice had higher oocyst shedding than control mice but did not have extended patency. It was concluded that CD4 cells, and to a lesser extent CD8 cells, are involved in resistance to primary *C. muris* infection in mice.[18]

3. *Cryptosporidium baileyi*

Surgically thymectomized broiler chickens having depressed DTH skin reactions and serum antibody responses shed more oocysts and had extended patent periods after oral oocyst inoculation compared to controls.[146a] Thymectomized broilers were slightly more susceptible to reinfection than controls but were able to resolve infection. Clinical signs and gross lesions of respiratory cryptosporidiosis were considered more severe in cyclosporin A-treated broilers with reduced DTH skin reactions than in untreated birds.[149] The effect of cyclosporin A treatment on persistence of respiratory infection was not reported.[149]

B. ACTIVE IMMUNIZATION

1. *Cryptosporidium parvum:* T-Lymphocyte Responses to Infection or Vaccination

T-cell clones against specific *Cryptosporidium* antigens have not been reported. Similarly, while CD4 T cells are critical in controlling infection, phenotypic characterization of protective responses has only recently begun.

a. Laboratory Mice

Peyer's patch lymphocyte responses in infected BALB/c neonates involved predominantly CD8 cells in ileum and CD4 cells in jejunum.[192] Mesenteric lymph node, but not splenic lymphocytes, from infected SWR/J H-2^q neonatal mice proliferated when cultured with oocyst extracts.[107] In contrast, splenic, but not mesenteric lymph node, lymphocytes from infected BALB/c mice proliferated when cultured with oocyst homogenate.[204] Spleen cells from BALB/c adults parenterally immunized with excysted oocysts or semi-purified SA-1 also proliferated when cultured with antigen.[98] Proliferative responses to oocyst homogenate in a T-cell-enriched spleen fraction from infected BALB/c mice were higher after CD8 depletion but not after CD4 depletion.[205] CD4-enriched spleen cells cultured with oocyst homogenate produced IFN-γ and IL-2, but not IL-4, suggesting a Th1 phenotype.[205] IFN-γ gene expression increased in mesenteric lymph node and ileum of BALB/c neonates 3 days PI, whereas IL-2, IL-4, and IL-10 gene expressions were similar to uninfected controls.[187] In another study, increased INF-γ production and CD4 T-cell percentages and proliferative responses of spleen cells to oocyst antigen during the first 4 weeks of life were similar in uninfected and infected BALB/c mice.[205a] Changes in these parameters were temporally associated with acquisition of resistance and were considered nonspecific and antigen-independent. Studies involving adoptive transfer of immune T-cell subsets or cytokines are reviewed in Section C.1.

b. Humans

In some recovered immunocompetent patients, blastogenic responses of PBMC cultures to native or recombinant oocyst antigen were largely antigen specific.[206] Stimulated PBMC cultures contained slightly elevated IL-10 and highly elevated IFN-γ.[206,206a] PBMC from late-stage AIDS patients with cryptosporidiosis did not proliferate when cultured with antigen.[206] It was suggested that lower IFN-γ production by PBMC of normal children compared to adults and reduced ability of some AIDS patients to produce IFN-γ may influence the severity of cryptosporidiosis. Further, persistent cryptosporidiosis in an HIV-negative infant was associated with deficient IFN-γ production by PBMC.[206a]

c. Cattle

Blastogenic responses of PBMC from infected calves to oocyst homogenate were highly variable but detected by 2 days PI and significantly increased by 16 days PI.[102] At 21 days PI, PBMC, spleen, and mesenteric and prescapular lymph nodes contained more non-T, non-B ($CD5^+/CD4^-/CD8^-$) null lymphocytes than uninfected calves, but only the prescapular lymph node increase was significant.[207] Spleen cells from infected calves contained a significantly higher percentage of CD8 cells and had a significantly lower proliferative response to oocyst homogenate than medium. The percentage of cells expressing CD2 (pan T-cell), CD4, B2 (B-cell), IL-2 receptor, and MHC class-II molecules did not differ between infected and uninfected calves. Null and CD8 lymphocytes may be involved in the bovine response to *C. parvum,* but relationships to protective immunity were not reported.

2. *Cryptosporidium muris:* T-Lymphocyte Responses to Infection in Laboratory Mice

Recovery from a primary infection in BALB/c adults was associated with increased oocyst antigen-dependent splenocyte proliferation from 21 days PI until cessation of oocyst shedding.[207a] Splenocyte production of IFN-γ and IL-2 was detected during peak shedding at 14 days PI and increased until shedding ceased. IL-4 was detected at 21 days PI. TNF production was not detected during the infection. Depletion of T cells or CD4 cells from splenocytes prior to incubation with antigen decreased production of IFN-γ, IL-2, and IL-4. It was concluded that recovery from a primary *C. muris* infection in mice correlated with antigen-dependent splenocyte proliferation and production of both Th1 and Th2 cytokines.

C. ADOPTIVE IMMUNIZATION

1. *Cryptosporidium parvum*

a. *Adoptive Transfer of Immune Lymphocytes in Mouse Models*

Unfractionated spleen and mesenteric lymph node cells from recovered (immune) BALB/c mice resolved persistent infection in adult nude mice after i.v. administration.[24] In contrast, unfractionated spleen or mesenteric lymph node cells from immune BALB/c adults, before or after culture with concanavalin A or supernatants from these cultures, did not prevent infection in BALB/c neonates when administered prior to challenge.[208] Adoptive transfer of (1) unfractionated spleen cells, (2) CD4- and B-enriched cells, (3) CD4-enriched cells, or (4) B-enriched cells from BALB/c adults immunized parenterally with excysted oocysts or semi-purified SA-1 led to splenic CD4 engraftment and *C. parvum*-specific serum antibody in SCID adults.[98] However, infection occurred in all groups challenged 7 days post-transfer. In contrast, persistently infected SCID adults adoptively transferred with either SA-1 immune or naive B- and/or CD4-enriched cells had near complete termination of infection by 18 days post-transfer. Splenic CD4 engraftment occurred in all mice, and specific serum antibody was produced by SA-1-immune cell-reconstituted mice, but all mice cleared infection at approximately the same rate. These studies suggest that mechanisms of resistance to initial infection and recovery from established infection may differ in neonatal and adult mice.[24,98,187,208]

b. *Treatment with Cytokines*

i. Transfer Factor

Transfer factor is a <30-kDa fraction from culture supernatant of unfractionated lymph node cells harvested from immune calves and thought to enhance CMI.[209] Eight AIDS patients with persistent cryptosporidiosis were treated by weekly oral administration of transfer factor for 6 weeks in an uncontrolled study.[209] During a 5-month followup period, five patients had decreased diarrhea, and four ceased oocyst shedding, but two relapsed. One patient's diarrhea and oocyst shedding resolved for a 2-year followup period. A prospective, placebo-controlled, double-blind trial involving 14 AIDS patients with persistent cryptosporidiosis was subsequently conducted. Seven patients were treated weekly for 8 weeks with a <10-kDa fraction of transfer factor from *C. parvum*-recovered and orally boosted calves, and seven were treated with a similar preparation from noninfected calves (placebo).[210] Six of seven patients given transfer factor had significantly decreased bowel movement frequency and increased fecal consistency. One of seven placebo-treated patients had significant improvement in bowel movement frequency. Of five placebo-treated patients crossed-over to transfer factor treatment, four had significantly decreased bowel movement frequency and increased fecal consistency. Overall, diarrhea significantly improved in 10 of 12 patients given transfer factor. Oocyst shedding ceased in some transfer factor-treated patients but was not significantly different from placebo-treated patients. Transfer factor given once weekly for 13 weeks had no effect on persistent cryptosporidiosis in an adult dysgammaglobulinemic patient.[211] Transfer factor from the same lot evaluated in AIDS patients[209] given to oocyst challenged-neonatal calves orally, or i.p. at twice the oral dose, had no effect on diarrhea or patent period compared to untreated control calves.[212]

ii. IFN-γ

Treatment of SCID neonates with IFN-γ 3 days prior to oocyst challenge did not significantly reduce susceptibility to initial infection.[190] SCID mice treated weekly with IFN-γ for up to 7 weeks, beginning 3 days prior to challenge as neonates, did not differ in clinical disease, median survival, or intestinal infection levels from control mice. These findings suggested that (1) deficient IFN-γ production by non-T cells did not account for the increased susceptibility of SCID neonates to infection, (2) IFN-γ may not be efficacious against established infection in SCID adults, and (3) endogenous IFN-γ production in SCID mice was insufficient to control infection. However, it was possible that prophylactic or therapeutic levels of IFN-γ were not obtained at the site of infection. Other studies have reported increased infection in IFN-γ-depleted adult SCID,[184,186] nude,[25] and BALB/c[25] mice and possibly BALB/c neonates.[187] Together, these observations suggest that redundant mechanisms contribute to control of infection.

iii. Interleukins

BALB/c or SCID neonates treated with murine recombinant IL-12 one day before oocyst challenge were completely protected from infection, while treatment on the day of challenge markedly reduced infection.[187] Treatment 1 day PI had a less pronounced effect on infection, and treatment 2 days PI had no effect. Prechallenge depletion of IFN-γ with mAb blocked the ability of IL-12 to prevent infection in BALB/c neonates. IFN-γ gene expression increased in BALB/c mesenteric lymph node and ileum during infection. IL-12 treatment further increased IFN-γ gene expression, but when administered 3 to 8 days PI did not reduce infection in BALB/c or SCID neonates. However, prechallenge depletion of endogenous IL-12 in these neonates increased infection severity and patency. Further, prechallenge depletion of IFN-γ in SCID neonates increased infection levels by 9 days PI. This was the first direct evidence that exogenous IL-12 can prevent initial *C. parvum* infection by an IFN-γ-dependent mechanism, independent of antigen-specific T- or B-cell responses, and that endogenous IL-12 is important in controlling infection in both immunocompetent and immunodeficient mice.

2. *Cryptosporidium muris*: Adoptive Transfer of Immune Lymphocytes in Mouse Models

Unfractionated immune spleen or mesenteric lymph node cells from *C. muris*-infected BALB/c mice at 29 to 97 days PI, when adoptively transferred prior to infection, significantly reduced mean oocyst production in SCID mice compared to controls given cells from naive BALB/c mice.[18] When Thy.1 cells were depleted by 94% from immune spleen-cell preparations before transfer to SCID mice, mean oocyst production and patent period significantly increased compared to SCID mice receiving unfractionated immune spleen cells. SCID mice reconstituted with Thy.1-depleted immune spleen cells recovered from infection but more slowly than mice reconstituted with unfractionated naive BALB/c spleen cells. When CD4 cells were depleted by >85% from immune spleen cell preparations before transfer to SCID mice, mean oocyst production was significantly higher than in mice receiving nondepleted immune spleen cells, and not significantly different than in mice receiving unfractionated naive spleen cells. When CD8 cells were depleted by >85% from immune spleen cell preparations before transfer to SCID mice, mean oocyst production was higher than in mice receiving nondepleted immune spleen cells, but the increase was not always significant; however, mean oocyst production in mice receiving CD8-depleted immune spleen cells was significantly lower than in mice receiving unfractionated naive spleen cells. It was concluded that unfractionated immune spleen or lymph node cells from *C. muris*-infected mice conferred greater resistance to infection than naive cells, and further that CD4 and, to a lesser extent, CD8 cells in immune spleen cell populations were specifically involved in this resistance.[18] The results also demonstrated that immune cells capable of transferring resistance are not limited to mucosal sites or draining lymph nodes but are also found in spleen.

D. MECHANISMS OF *CRYPTOSPORIDIUM PARVUM* NEUTRALIZATION

While IFN-γ is important in controlling cryptosporidial infections, the mechanisms involved are unclear. Mechanisms suggested include macrophage activation, direct toxicity to parasites, and inhibition of invasion and/or intracellular development.[17,25,187] IFN-γ can exert activity independent of antigen-specific immune responses, based on IFN-γ depletion[17,184,186,190] or augmentation[187] studies in SCID mice. It has been suggested that IL-12-mediated IFN-γ production may prevent zoite invasion or intracellular development directly or reduce host cell permissiveness.[187] Anticryptosporidial activity of IFN-γ was suggested not to involve enhanced generation of nitric oxide or other reactive nitrogen intermediates because infection was not detected in oocyst-challenged BALB/c adults after treatment with the nitric oxide synthase inhibitor aminoguanidine.[190] In similarly treated SCID adults, intestinal infection levels and lesions did not differ from controls. Nitric oxide synthase may not have been inhibited at the site of infection.[190] A role for nitric oxide in controlling infection has been suggested in other studies.[213]

Potential mechanisms by which T cells mediate protection against cryptosporidial infections include MHC class I-restricted CD8 cytotoxicity, MHC class II-restricted CD4 cytotoxicity, CD4 help for neutralizing antibody production, and protective cytokine production.[25,185] *Cryptosporidium* antigens incorporated into the host cell membrane may provide targets for immune cell recognition.[135] Host membrane immediately surrounding the parasitophorous vacuole was thought to contain *C. parvum* antigen recognized by mAbs in IEM studies.[135]

IV. AGE-ASSOCIATED SUSCEPTIBILITY TO INFECTION

Age-associated susceptibility to *C. parvum* infection in mice, calves, lambs, nonhuman primates, and humans and *C. baileyi* infection in chickens is reviewed in Chapters 1, 3, and 9. Neonatal immunocompetent[214] or immunodeficient[27,29,30] mice and rats are highly susceptible to *C. parvum* infection, but adults are comparatively resistant. Immaturity of antigen-specific and nonspecific immunologic effector systems in the young accounts in part for higher susceptibility. Increased susceptibility in the young may be related in part to underdevelopment of an accessory or effector cell and may involve an IL-12-dependent IFN-γ pathway.[187]

V. CONCLUSIONS

Research relating to the immunobiology of cryptosporidiosis has increased exponentially in recent years, and a fundamental knowledge base is beginning to emerge. Research priorities will likely continue to be driven by the need to develop effective strategies for passive immunization of immunodeficient or neonatal hosts and active mucosal immunization of immunocompetent hosts using novel approaches.[215-217]

REFERENCES

1. O'Donoghue, P. J., *Cryptosporidium* and cryptosporidiosis in man and animals, *Int. J. Parasitol.*, 25, 139, 1995.
2. Angus, K. W., Cryptosporidiosis in ruminants, in *Cryptosporidiosis of Man and Animals*, Dubey, J. P., Speer, C. A., and Fayer, R., Eds., CRC Press, Boca Raton, FL, 1990, chap. 5.
3. Petersen, C., Cryptosporidiosis in patients infected with the human immunodeficiency virus, *Clin. Infect. Dis.*, 15, 903, 1992.
4. Anderson, B. C., Abomasal cryptosporidiosis in cattle, *Vet. Pathol.*, 24, 235, 1987.
5. Esteban, E. and Anderson, B. C., *Cryptosporidium muris*: prevalence, persistency, and detrimental effect on milk production in a drylot dairy, *J. Dairy Sci.*, 78, 1068, 1995.
6. Blagburn, B. L., Lindsay, D. S., Hoerr, F. J., Davis, J. F., Giambrone, J. J., Pathobiology of cryptosporidiosis (*C. baileyi*) in broiler chickens, *J. Protozool.*, 38, 25S, 1991.
7. Current, W. L., Reese, N. C., Ernst, J. V., Bailey, W. S., Heyman, M. B., and Weinstein, W. M., Human cryptosporidiosis in immunocompetent and immunodeficient persons, *N. Engl. J. Med.*, 308, 1252, 1983.
8. Wolfson, J. S., Richter, J. M., Waldron, M. A., Weber, D. J., McCarthy, D. M., and Hopkins, C. C., Cryptosporidiosis in immunocompetent patients, *N. Engl. J. Med.*, 312, 1278, 1985.
9. Isaacs, D., Hunt, G. H., Phillips, A. D., Price, E. H., Raafat, F., and Walker-Smith, J. A., Cryptosporidiosis in immunocompetent children, *J. Clin. Pathol.*, 38, 76, 1985.
10. Carter, M. J. and Anzimlt, T., *Cryptosporidium*: an important cause of gastrointestinal disease in immunocompetent patients, *N. Z. Med. J.*, 99, 101, 1986.
11. Stehr-Green, J. K., McCaig, L., Remsen, H. M., Rains, C. S., Fox, M., and Juranek, D. D., Shedding of oocysts in immunocompetent individuals infected with *Cryptosporidium*, *Am. J. Trop. Med. Hyg.*, 36, 338, 1987.
12. Ungar, B. L. P., Cryptosporidiosis in humans (*Homo sapiens*), in *Cryptosporidiosis of Man and Animals*, Dubey, J. P., Speer, C. A., and Fayer, R., Eds., CRC Press, Boca Raton, FL, 1990, chap. 4.
13. Miller, R. A., Bronsdon, M. A., Kuller, L., and Morton, W. R., Clinical and parasitologic aspects of cryptosporidiosis in nonhuman primates, *Lab. Anim. Sci.*, 40, 42, 1990.
14. Riggs, M. W., Cryptosporidiosis in cats, dogs, ferrets, raccoons, opossums, rabbits, and nonhuman primates, in *Cryptosporidiosis of Man and Animals*, Dubey, J. P., Speer, C. A., and Fayer, R., Eds., CRC Press, Boca Raton, FL, 1990, chap. 7.
15. Hill, B. D., Dawson, A. M., and Blewett, D. A., Neutralisation of *Cryptosporidium parvum* sporozoites by immunoglobulin and nonimmunoglobulin components in serum, *Res. Vet. Sci.*, 54, 356, 1993.
16. Perryman, L. E., Cryptosporidiosis in rodents, in *Cryptosporidiosis of Man and Animals*, Dubey, J. P., Speer, C. A., and Fayer, R., Eds., CRC Press, Boca Raton, FL, 1990, chap. 8.
17. McDonald, V., Deer, R., Uni, S., Iseki, M., and Bancroft, G. J., Immune responses to *Cryptosporidium muris* and *Cryptosporidium parvum* in adult immunocompetent or immunocompromised (nude and SCID) mice, *Infect. Immun.*, 60, 3325, 1992.

18. McDonald, V., Robinson, H. A., Kelly, J. P., and Bancroft, G. J., *Cryptosporidium muris* in adult mice: adoptive transfer of immunity and protective roles of CD4 vs. CD8 cells, *Infect. Immun.*, 62, 2289, 1994.
19. Harp, J. A., Woodmansee, D. B., and Moon, H. W., Resistance of calves to *Cryptosporidium parvum*: effects of age and previous exposure, *Infect. Immun.*, 58, 2237, 1990.
20. Peeters, J. E., Villacorta, I., Vanopdenbosch, E., Vandergheynst, D., Naciri, M., Ares-Mazas, E., and Yvoré, P., *Cryptosporidium parvum* in calves: kinetics and immunoblot analysis of specific serum and local antibody responses (immunoglobulin A [IgA], IgG, and IgM) after natural and experimental infections, *Infect. Immun.*, 60, 2309, 1992.
21. Ortega-Mora, L. M., Troncoso, J. M., Rojo-Vazquez, F. A., and Gomez-Bautista, M., Serum antibody response in lambs naturally and experimentally infected with *Cryptosporidium parvum*, *Vet. Parasitol.*, 50, 45, 1993.
22. Mancassola, R., Répérant, J. M., Naciri, M., and Chartier, C., Chemoprophylaxis of *Cryptosporidium parvum* infection with paromomycin in kids and immunological study, *Antimicrob. Agents Chemother.*, 39, 75, 1995.
23. Heine, J., Moon, H. W., and Woodmansee, D. B., Persistent *Cryptosporidium* infection in congenitally athymic (nude) mice, *Infect. Immun.*, 43, 856, 1984.
24. Ungar, B. L. P., Burris, J. A., Quinn, C. A., and Finkelman, F. D., New mouse models for chronic *Cryptosporidium* infection in immunodeficient hosts, *Infect. Immun.*, 58, 961, 1990.
25. Ungar, B. L. P., Kao, T.-C., Burris, J. A., and Finkelman, F. D., *Cryptosporidium* infection in an adult mouse model. Independent roles for IFN-γ and CD4+ T lymphocytes in protective immunity, *J. Immunol.*, 147, 1014, 1991.
26. Bjorneby, J. M., Leach, D. R., and Perryman, L. E., Persistent cryptosporidiosis in horses with severe combined immunodeficiency, *Infect. Immun.*, 59, 3823, 1991.
27. Mead, J. R., Arrowood, M. J., Sidwell, R. W., and Healey, M. C., Chronic *Cryptosporidium parvum* infections in congenitally immunodeficient SCID and nude mice, *J. Infect. Dis.*, 163, 1297, 1991.
28. Perryman, L. E. and Bjorneby, J. M., Immunotherapy of cryptosporidiosis in immunodeficient animal models, *J. Protozool.*, 38, 98S, 1991.
29. Gardner, A. L., Roche, J. K., Weikel, C. S., and Guerrant, R. L., Intestinal cryptosporidiosis: pathophysiologic alterations and specific cellular and humoral immune responses in rnu/+ and rnu/rnu (athymic) rats, *Am. J. Trop. Med. Hyg.*, 44, 49, 1991.
30. Kuhls, T. L., Greenfield, R. A., Mosier, D. A., Crawford, D. L., and Joyce, W. A., Cryptosporidiosis in adult and neonatal mice with severe combined immunodeficiency, *J. Comp. Pathol.*, 106, 399, 1992.
31. Mead J. R., Ilksoy, N., You, X., Belenkaya, Y., Arrowood, M. J., Fallon, M. T., and Schinazi, R. F., Infection dynamics and clinical features of cryptosporidiosis in SCID mice, *Infect. Immun.*, 62, 1691, 1994.
32. Rasmussen, K. R., Martin, E. G., and Healey, M. C., Effects of dehydroepiandrosterone in immunosuppressed rats infected with *Cryptosporidium parvum*, *J. Parasitol.*, 79, 364, 1993.
33. Godwin, T. A., Cryptosporidiosis in the Acquired Immunodeficiency Syndrome: a study of 15 autopsy cases, *Hum. Pathol.*, 22, 1215, 1991.
34. Flanigan, T. P., Human immunodeficiency virus infection and cryptosporidiosis: protective immune responses, *Am. J. Trop. Med. Hyg.*, 50, 29, 1994.
35. Koudela, B. and Hermánek, J., Nonspecific immunomodulation influences the course and location of *Cryptosporidium parvum* infection in neonatal BALB/c mice, *Ann. Parasitol. Hum. Comp.*, 68, 3, 1993.
36. Rasmussen, K. R. and Healey, M. C., Experimental *Cryptosporidium parvum* infections in immunosuppressed adult mice, *Infect. Immun.*, 60, 1648, 1992.
37. Rasmussen, K. R. and Healey, M. C., Dehydroepiandrosterone-induced reduction of *Cryptosporidium parvum* infections in aged Syrian golden hamsters, *J. Parasitol.*, 78, 554, 1992.
38. Fayer, R., Development of a precocious strain of *Cryptosporidium parvum* in neonatal calves, *J. Euk. Microbiol.*, 41, 40S, 1994.
39. Harp, J. A. and Goff, J. P., Protection of calves with a vaccine against *Cryptosporidium parvum*, *J. Parasitol.*, 81, 54, 1995.
40. Zu, S-X., Fang, G-D., Fayer, R., and Guerrant, R. L., Cryptosporidiosis: pathogenesis and immunology, *Parasitol. Today*, 8, 24, 1992.
41. Lillehoj, H. S., Immune responses to coccidian parasites, *Proc. VIth Int. Coccidiosis Conf.*, 6, 11, 1993.
42. Lillehoj, H. S. and Trout, J. M., $CD8^+$ T cell-coccidia interactions, *Parasitol. Today*, 10, 10, 1994.
43. Campbell, P. N. and Current, W. L., Demonstration of serum antibodies to *Cryptosporidium* sp. in normal and immunodeficient humans with confirmed infections, *J. Clin. Microbiol.*, 18, 165, 1983.
44. Ungar, B. L. P., Soave, R., Fayer, R., and Nash, T. E., Enzyme immunoassay detection of immunoglobulin M and G antibodies to *Cryptosporidium* in immunocompetent and immunocompromised persons, *J. Infect. Dis.*, 153, 570, 1986.
45. Ungar, B. L. P. and Nash, T. E., Quantification of specific antibody response to *Cryptosporidium* antigens by laser densitometry, *Infect. Immun.*, 53, 124, 1986.
46. Casemore, D. P., The antibody response to *Cryptosporidium*: development of a serological test and its use in a study of immunologically normal persons, *J. Infect.*, 14, 125, 1987.
47. Mead, J. R., Arrowood, M. J., and Sterling, C. R., Antigens of *Cryptosporidium* sporozoites recognized by immune sera of infected animals and humans, *J. Parasitol.*, 74, 135, 1988.
48. Laxer, M. A., Alcantara, A. K., Javato-Laxer, M., Menorca, D. M., Fernando, M. T., and Ranoa, C. P., Immune response to cryptosporidiosis in Philippine children, *Am. J. Trop. Med. Hyg.*, 42, 131, 1990.

49. Kuhls, T. L., Mosier, D. A., Crawford, D. L., and Griffis, J., Seroprevalence of cryptosporidial antibodies during infancy, childhood, and adolescence, *Clin. Infect. Dis.*, 18, 731, 1994.
50. Lazo, A., Barriga, O. O., Redman, D. R., and Bech-Neilsen, S., Identification by transfer blot of antigens reactive in the enzyme-linked immunosorbent assay (ELISA) in rabbits immunized and a calf infected with *Cryptosporidium* sp., *Vet. Parasitol.*, 21, 151, 1986.
51. Hill, B. D., Enteric protozoa in ruminants: diagnosis and control of *Cryptosporidium*, the role of the immune response, *Rev. Sci. Tech.*, 9, 423, 1990.
52. Tzipori, S. and Campbell, I., Prevalence of *Cryptosporidium* antibodies in 10 animal species, *J. Clin. Microbiol.*, 14, 455, 1981.
53. Williams, R. O., Measurement of class specific antibody against *Cryptosporidium* in serum and faeces from experimentally infected calves, *Res. Vet. Sci.*, 43, 264, 1987.
54. Whitmire, W. M. and Harp, J. A., Characterization of bovine cellular and serum antibody responses during infection by *Cryptosporidium parvum*, *Infect. Immun.*, 59, 990, 1991.
55. Mosier, D. A., Kuhls, T. L., Simons, K. R., and Oberst, R. D., Bovine humoral immune response to *Cryptosporidium parvum*, *J. Clin. Microbiol.*, 30, 3277, 1992.
56. Hill, B. D., Blewett, D. A., Dawson, A. M., and Wright, S., Analysis of the kinetics, isotype and specificity of serum and coproantibody in lambs infected with *Cryptosporidium parvum*, *Res. Vet. Sci.*, 48, 76, 1990.
57. Lasser, K. H., Lewin, K. J., and Ryning, F. W., Cryptosporidial enteritis in a patient with congenital hypogammaglobulinemia, *Hum. Pathol.*, 10, 234, 1979.
58. Sloper, K. S., Dourmashkin, R. R., Bird, R. B., Slavin, G., and Webster, A. D. B., Chronic malabsorption due to cryptosporidiosis in a child with immunoglobulin deficiency, *Gut*, 23, 80, 1982.
59. Tzipori, S., Roberton, D., and Chapman, C., Remission of diarrheoa due to cryptosporidiosis in an immunodeficient child treated with hyperimmune bovine colostrum, *Br. Med. J.*, 293, 1277, 1986.
60. Saxon, A. and Weinstein, W., Oral administration of bovine colostrum-anti-cryptosporidia antibody fails to alter the course of human cryptosporidiosis, *J. Parasitol.*, 73, 413, 1987.
61. Jacyna, M. R., Parkin, J., Goldin, R., and Baron, J. H., Protracted enteric cryptosporidial infection in selective immunoglobulin A and saccharomyces opsonin deficiencies, *Gut*, 31, 714, 1990.
62. Taghi-Kilani, R., Sekla, L., and Hayglass, K. T., The role of humoral immunity in *Cryptosporidium* spp. infection. Studies with B cell-depleted mice, *J. Immunol.*, 145, 1571, 1990.
63. Kassa, M., Comby, E., Lemeteil, D., Brasseur, P., and Ballet, J.-J., Characterization of anti-*Cryptosporidium* IgA antibodies in sera from immunocompetent individuals and HIV-infected patients, *J. Protozool.*, 38, 179S, 1991.
64. Banhamou, Y., Kapel, N., Hoang, C., Matta, H., Meillet, D., Magne, D., Raphael, M., Gentilini, M., Opolon, P., and Gobert, J. G., Inefficacy of intestinal secretory immune response to *Cryptosporidium* in Acquired Immunodeficiency Syndrome, *Gastroenterology*, 108, 627, 1995.
65. Cozon, G., Biron, F., Jeannin, M., Cannella, D., and Revillard, J. P., Secretory IgA antibodies to *Cryptosporidium parvum* in AIDS patients with chronic cryptosporidiosis, *J. Infect. Dis.*, 169, 696, 1994.
66. Favennec, L., Coomby, E., Ballet, J. J., and Brasseur, P., Serum IgA antibody response to *Cryptosporidium parvum* is mainly represented by IgA1, *J. Infect. Dis.*, 171, 256, 1995.
67. Répérant, J. M., Naciri, M., Iochmann, S., Tilley, M., and Bout, D. T., Major antigens of *Cryptosporidium parvum* recognized by serum antibodies from different infected animal species and man, *Vet. Parasitol.*, 55, 1, 1994.
68. Kapel, N., Meillet, D., Buraud, M., Favennec, L., Magne, D., and Gobert, J. G., Determination of anti-*Cryptosporidium* coproantibodies by time-resolved immunofluorometric assay, *Trans. Roy. Soc. Trop. Med. Hyg.*, 87, 330, 1993.
69. Flanigan, T. P., Wisnewski, A., Wiest, P. M., Johnson, J., Tzipori, S., Hamer, D., Lam, N., and Kresina, T. F., Human monoclonal antibodies against *Cryptosporidium parvum* generated by hypo-osmolar electrofusion, *Trans. Assoc. Am. Phys.*, 106, 86, 1993.
70. Tilley, M., Upton, S. J., Fayer, R., Barta, J. R., Chrisp, C. E., Freed, P. S., Blagburn, B. L., Anderson, B. C., and Barnard, S. M., Identification of a 15-kilodalton surface glycoprotein on sporozoites of *Cryptosporidium parvum*, *Infect. Immun.*, 59, 1002, 1991.
71. Arrowood, M. J., Mead, J. R., Mahrt, J. L., and Sterling, C. R., Effects of immune colostrum and orally administered antisporozoite monoclonal antibodies on the outcome of *Cryptosporidium parvum* infections in neonatal mice, *Infect. Immun.*, 57, 2283, 1989.
72. Perryman, L. E., Jasmer, D. P., Riggs, M. W., Bohnet, S. G., McGuire, T. C., and Arrowood, M. J., A cloned gene of *Cryptosporidium parvum* encodes neutralization-sensitive epitopes, *Mol. Biochem. Parasitol.*, 80, 137, 1996.
73. Tatalick, L. M. and Perryman, L.E., Attempts to protect severe combined immunodeficient (SCID) mice with antibody enriched for reactivity to *Cryptosporidium parvum* surface antigen-1, *Vet. Parasitol.*, 58, 281, 1995.
74. Riggs, M. W., Stone, A. L., Arrowood, M. J., Langer, R. C., Bahle, J., Yount, P., Molecular targets for passive immunotherapy of cryptosporidiosis, in *Proc. 47th Annual Meeting of the Society of Protozoologists and 3rd Int. Workshop on* Pneumocystis, Toxoplasma, Cryptosporidium, *and* Microsporidia, Abstract C46, 76, 1994.
74a. Riggs, M. W., Yount, P. A., Stone, A. L., and Langer, R. C., Protective monoclonal antibodies define a distinct, conserved epitope on an apical complex exoantigen of *Cryptosporidium parvum* sporozoites, *J. Euk. Microbiol.*, 43, 74S, 1996.
74b. Langer, R. C. and Riggs, M. W., Neutralizing monoclonal antibody protects against *Cryptosporidium parvum* infection by inhibiting sporozoite attachment and invasion, *J. Euk. Microbiol.*, 43, 76S, 1996.

75. Riggs, M. W., McGuire, T. C., Mason, P. H., and Perryman, L. E., Neutralization-sensitive epitopes are exposed on the surface of infectious *Cryptosporidium parvum* sporozoites, *J. Immunol.*, 143, 1340, 1989.
76. Bjorneby, J. M., Riggs, M. W., and Perryman, L. E., *Cryptosporidium parvum* merozoites share neutralization-sensitive epitopes with sporozoites, *J. Immunol.*, 145, 298, 1990.
77. Luft, B. J., Payne, D., Woodmansee, D., and Kim, C. W., Characterization of the *Cryptosporidium* antigens from sporulated oocysts of *Cryptosporidium parvum*, *Infect. Immun.*, 55, 2436, 1987.
78. Current, W. L., Techniques and laboratory maintenance of *Cryptosporidium*, in *Cryptosporidiosis of Man and Animals*, Dubey, J. P., Speer, C. A., and Fayer, R., Eds., CRC Press, Boca Raton, FL, 1990, chap. 2.
79. Riggs, M. W. and Perryman, L. E., Infectivity and neutralization of *Cryptosporidium parvum* sporozoites, *Infect. Immun.*, 55, 2081, 1987.
80. Lumb, R., Lanser, J. A., and O'Donoghue, P. J., Electrophoretic and immunoblot analysis of *Cryptosporidium* oocysts, *Immunol. Cell Biol.*, 66, 369, 1988.
81. Riggs, M. W., Cama, V. A., Leary, H. L., Jr., and Sterling, C. R., Bovine antibody against *Cryptosporidium parvum* elicits a circumsporozoite precipitate-like reaction and has immunotherapeutic effect against persistent cryptosporidiosis in SCID mice, *Infect. Immun.*, 62, 1927, 1994.
82. Current, W. L. and Bick, P. H., Immunobiology of *Cryptosporidium* spp., *Pathol. Immunopathol. Res.*, 8, 141, 1989.
83. Mitschler, R. R., Welti, R., and Upton, S. J., A comparative study of lipid compositions of *Cryptosporidium parvum* (Apicomplexa) and Madin-Darby bovine kidney cells, *J. Euk. Microbiol.*, 41, 8, 1994.
84. Tilley, M. and Upton, S. J., Electrophoretic characterization of *Cryptosporidium parvum* (KSU-1 isolate) (Apicomplexa: Cryptosporidiidae), *Can. J. Zool.*, 68, 1513, 1990.
85. Regan, S., Cama, V., and Sterling, C. R., *Cryptosporidium* merozoite isolation and purification using differential centrifugation techniques, *J. Protozool.*, 38, 202S, 1991.
86. Doyle, P. S., Crabb, J., and Petersen, C., Anti-*Cryptosporidium parvum* antibodies inhibit infectivity *in vitro* and *in vivo*, *Infect. Immun.*, 61, 4079, 1993.
87. Tilley, M., Upton, S. J., Blagburn, B. L., and Anderson, B. C., Identification of outer oocyst wall proteins of three species of *Cryptosporidium* (Apicomplexa: Cryptosporidiidae) by ^{125}I surface labeling, *Infect. Immun.*, 58, 252, 1990.
88. Gut, J. and Nelson, R. G., *Cryptosporidium parvum* sporozoites deposit trails of 11A5 antigen during gliding locomotion and shed 11A5 antigen during invasion of MDCK cells *in vitro*, *J. Euk. Microbiol.*, 41, 42S, 1994.
89. Llovo, J., Lopez, A., Fabregas, J., and Munoz, A., Interaction of lectins with *Cryptosporidium parvum*, *J. Infect. Dis.*, 167, 1477, 1993.
90. Kuhls, T. L., Mosier, D. A., and Crawford, D. L., Effects of carbohydrates and lectins on cryptosporidial sporozoite penetration of cultured cell monolayers, *J. Protozool.*, 38, 74S, 1991.
91. Thea, D. M., Miercio, E. A., Pereira, M. E., Kotler, D., Sterling, C. R., and Keusch, G. T., Identification and partial purification of a lectin on the surface of the sporozoite of *Cryptosporidium parvum*, *J. Parasitol.*, 78, 886, 1992.
92. Joe, A., Hamer, D. H., Kelley, M. A., Pereira, M. E. A., Keusch, G. T., Tzipori, S., and Ward, H. D., Role of a Gal/GalNAc-specific sporozoite surface lectin in *Cryptosporidium parvum*-host cell interaction, *J. Euk. Microbiol.*, 41, 44S, 1994.
93. Ortega-Mora, L. M., Troncoso, J. M., Rojo-Vazquez, F. A., and Gomez-Bautista, M., Identification of *Cryptosporidium parvum* oocyst/sporozoite antigens recognized by infected and hyperimmune lambs, *Vet. Parasitol.*, 53, 159, 1994.
94. Naciri, M., Mancassola, R., Répérant, J. M., Canivez, O., Quinique, B., and Yvoré, P., Treatment of experimental ovine cryptosporidiosis with ovine or bovine hyperimmune colostrum, *Vet. Parasitol.*, 53, 173, 1994.
95. Moss, D. M., Bennett, S. N., Arrowood, M. J., Hurd, M. R., Lammie, P. J., Wahlquist, S. P., and Addiss, D. G., Kinetic and isotypic analysis of specific immunoglobulins from crew members with cryptosporidiosis on a U.S. Coast Guard cutter, *J. Euk. Microbiol.*, 41, 52S, 1994.
96. Ortega-Mora, L. M. and Wright, S. E., Age-related resistance in ovine cryptosporidiosis: patterns of infection and humoral immune response, *Infect. Immun.*, 62, 5003, 1994.
97. Répérant, J. M., Naciri, M., Chardes, T., and Bout, D. T., Immunological characterization of a 17-kDa antigen from *Cryptosporidium parvum* recognized early by mucosal IgA antibodies, *FEMS Microbiol. Lett.*, 99, 7, 1992.
98. Tatalick, L. M. and Perryman, L. E., Effect of surface antigen-1 (SA-1) immune lymphocyte subsets and naive cell subsets in protecting SCID mice from initial and persistent infection with *Cryptosporidium parvum*, *Vet. Immunol. Immunopathol.*, 47, 43, 1995.
99. Nina, J. M. S., McDonald, V., Dyson, D. A., Catchpole, J., Uni, S., Iseki, M., Chiodini, P. L., McAdam, K. P. W. J., Analysis of oocyst wall and sporozoite antigens from three *Cryptosporidium* species, *Infect. Immun.*, 60, 1509, 1992.
100. Fayer, R., Guidry, A., and Blagburn, B. L., Immunotherapeutic efficacy of bovine colostral immunoglobulins from a hyperimmunized cow against cryptosporidiosis in neonatal mice, *Infect. Immun.*, 58, 2962, 1990.
101. Tilley, M., Fayer, R., Guidry, A., Upton, S. J., and Blagburn, B. L., *Cryptosporidium parvum* (Apicomplexa: Cryptosporidiidae) oocyst and sporozoite antigens recognized by bovine colostral antibodies, *Infect. Immun.*, 58, 2966, 1990.
102. Whitmire, W. M. and Harp, J. A., Characterization of bovine cellular and serum antibody responses during infection by *Cryptosporidium parvum*, *Infect. Immun.*, 59, 990, 1991.
103. Flanigan, T., Marshall, R., Redman, D., Kaetzel, C., and Ungar, B., *In vitro* screening of therapeutic agents against *Cryptosporidium*: hyperimmune cow colostrum is highly inhibitory, *J. Protozool.*, 38, 225S, 1991.

104. Petersen, C., Gut, J., Doyle, P. S., Crabb, J. H., Nelson, R. G., and Leech, J. H., Characterization of a >900,000-M(r) *Cryptosporidium parvum* sporozoite glycoprotein recognized by protective hyperimmune bovine colostral immunoglobulin, *Infect. Immun.*, 60, 5132, 1992.
105. Robert, B., Antoine, H., Dreze, F., Coppe, P., and Collard, A., Characterization of a high molecular weight antigen of *Cryptosporidium parvum* micronemes possessing epitopes that are cross-reactive with all parasitic life-cycle stages, *Vet. Res.*, 25, 384, 1994.
106. Wisher, M. H. and Rose, M. E., *Eimeria tenella* sporozoites: the method of excystation affects the surface membrane proteins, *Parasitol.*, 95, 479, 1987.
107. Moss, D. M. and Lammie, P. J., Proliferative responsiveness of lymphocytes from *Cryptosporidium parvum*-exposed mice to two separate antigen fractions from oocysts, *Am. J. Trop. Med. Hyg.*, 49, 393, 1993.
108. Tilley, M. and Upton, S. J., Both CP15 and CP25 are left as trails behind gliding sporozoites of *Cryptosporidium parvum* (Apicomplexa), *FEMS Microbiol. Lett.*, 120, 275, 1994.
109. Bonnin, A., Dubremetz, J. F., and Camerlynck, P., Characterization of microneme antigens of *Cryptosporidium parvum*, *Infect. Immun.*, 59, 1703, 1991.
110. Bonnin, A., Dubremetz, J. F., and Camerlynck, P., A new antigen of *Cryptosporidium parvum* micronemes possessing epitopes cross-reactive with macrogamete granules, *Parasitol. Res.*, 79, 8, 1993.
111. Bonnin, A., Gut, J., Dubremetz, J. F., Nelson, R., and Camerlynck, P., Monoclonal antibodies identify a subset of dense granules in *Cryptosporidium parvum* zoites and gamonts, *J. Euk. Microbiol.*, 42, 395, 1995.
112. McDonald, V., Deer, R. M. A., Nina, J. M. S., Wright, S., Chiodini, P.L., and McAdam, K. P. W. J., Characteristics and specificity of hybridoma antibodies against oocyst antigens of *Cryptosporidium parvum* from man, *Parasite Immunol.*, 13, 251, 1991.
113. Fayer, R., Barta, J. R., Guidry, A. J., and Blagburn, B. L., Immunogold labeling of stages of *Cryptosporidium parvum* recognized by immunoglobulins in hyperimmune bovine colostrum, *J. Parasitol.*, 77, 487, 1991.
114. Tilley, M., Eggleston, M. T., and Upton, S. J., Multiple oral inoculations with *Cryptosporidium parvum* as a means of immunization for production of monoclonal antibodies, *FEMS Microbiol. Lett.*, 113, 235, 1993.
115. Petersen, C., Gut, J., Leech, J. H., and Nelson, R. G., Identification and initial characterization of five *Cryptosporidium parvum* sporozoite antigen genes, *Infect. Immun.*, 60, 2343, 1992.
116. Petersen, C., Gut, J., Nelson, R. G., and Leech, J. H., Characteristics of a *Cryptosporidium parvum* sporozoite glycoprotein, *J. Protozool.*, 38, 20S, 1991.
117. Khramtsov, N. V., Tilley, M., Blunt, D. S., Montelone, B. A., Upton, S. J., Cloning and analysis of a *Cryptosporidium parvum* gene encoding a protein with homology to cytoplasmic form Hsp 70, *J. Euk. Microbiol.*, 42, 416, 1995.
118. Lally, N. C., Baird, G. D., McQuay, S. J., Wright, F., and Oliver, J. J., A 2359-base pair DNA fragment from *Cryptosporidium parvum* encoding a repetitive oocyst protein, *Mol. Biochem. Parasitol.*, 56, 69, 1992.
119. Ranucci, L., Muller, H. M., La Rosa, G., Reckmann, L., Morales, M. A., Spano, F., Pozio, E., and Crisanti, A., Characterization and immunolocalization of a *Cryptosporidium* protein containing repeated amino acid motifs, *Infect. Immun.*, 61, 2347, 1993.
120. Steele, M. I., Kuhls, T. L., Nida, K., Meka, C. S. R., Halabi, I. M., Mosier, D. A., Elliott, W., Crawford, D. L., and Greenfield, R. A., A *Cryptosporidium parvum* genomic region encoding hemolytic activity, *Infect. Immun.*, 63, 3840, 1995.
121. Mead, J. R., Lloyd, R. M. Jr., You, X., Tucker-Burden, C., Arrowood, M. J., and Schinazi, R. F., Isolation and partial characterization of *Cryptosporidium* sporozoite and oocyst wall recombinant proteins, *J. Euk. Microbiol.*, 41, 51S, 1994.
122. Uhl, E. W., O'Connor, R. M., Perryman, L. E., and Riggs, M. W., Neutralization-sensitive epitopes are conserved among geographically diverse isolates of *Cryptosporidium parvum*, *Infect. Immun.*, 60, 1703, 1992.
123. Bjorneby, J. M., Hunsaker, B. D., Riggs, M. W., and Perryman, L. E., Monoclonal antibody immunotherapy in nude mice persistently infected with *Cryptosporidium parvum*, *Infect. Immun.*, 59, 1172, 1991.
124. Arrowood, M. J., Sterling, C. R., and Healey, M. C., Immunofluorescent microscopical visualization of trails left by gliding *Cryptosporidium parvum* sporozoites, *J. Parasitol.*, 77, 315, 1991.
125. Tilley, M. and Upton, S. J., Sporozoites and merozoites of *Cryptosporidium parvum* share a common epitope recognized by a monoclonal antibody and two-dimensional electrophoresis, *J. Protozool.*, 38, 48S, 1991.
126. Jenkins, M., Kerr, D., Fayer, R., and Wall, R., Serum and colostrum antibody responses induced by jet-injection of sheep with DNA encoding a *Cryptosporidium parvum* antigen, *Vaccine*, 13, 1658, 1996.
127. Jenkins, M. C., Fayer, R., Tilley, M., and Upton, S. J., Cloning and expression of a cDNA encoding epitopes shared by 15- and 60-kilodalton proteins of *Cryptosporidium parvum* sporozoites, *Infect. Immun.*, 61, 2377, 1993.
128. Jenkins, M. C. and Fayer, R., Cloning and expression of cDNA encoding an antigenic *Cryptosporidium parvum* protein, *Mol. Biochem. Parasitol.*, 71, 149, 1995.
129. Cho, M.- H., Passive transfer of immunity against *Cryptosporidium* infection in neonatal mice using monoclonal antibodies, *Kor. J. Parasitol.*, 31, 223, 1993.
130. Dubremetz, J. F., Apical organelles (rhoptries, micronemes, dense granules) and host cell invasion by coccidia: what do we know now?, *Proc. VIth. Int. Coccidiosis Conf.*, 6, 3, 1993.
131. Perkins, M. E., Rhoptry organelles of apicomplexan parasites, *Parasitol. Today*, 8, 28, 1992.
132. Petersen, C., Cellular biology of *Cryptosporidium parvum*, *Parasitol. Today,* 9, 87, 1993.
133. Lumb, R., Smith, K., O'Donoghue, P. J., and Lanser, J. A., Ultrastructure of the attachment of *Cryptosporidium* sporozoites to tissue cultures, *Parasitol. Res.,* 74, 531, 1988.

134. Nina, J. M. S., McDonald, V., Deer, R. M. A., Wright, S. E., Dyson, D. A., Chiodini, P. L., and McAdam, K. P. W. J., Comparative study of the antigenic composition of oocyst isolates of *Cryptosporidium parvum* from different hosts, *Parasite Immunol.*, 14, 227, 1992.
135. McDonald, V., McCrossan, M. V., and Petry, F., Localization of parasite antigens in *Cryptosporidium parvum*-infected epithelial cells using monoclonal antibodies, *Parasitol.*, 110, 259, 1995.
136. Bonnin, A., Dubremetz, J. F., and Camerlynck, P., Characterization and immunolocalization of an oocyst wall antigen on *Cryptosporidium parvum* (Protozoa: Apicomplexa), *Parasitol.*, 103, 171, 1991.
137. Hoskins, D., Chrisp, C. E., Suckow, M. A., and Fayer, R., Effect of hyperimmune bovine colostrum raised against *Cryptosporidium parvum* on infection of guinea pigs by *Cryptosporidium wrairi*, *J. Protozool.*, 38, 185S, 1991.
138. Chrisp, C. E., Mason, P., and Perryman, L. E., Comparison of *Cryptosporidium parvum* and *Cryptosporidium wrairi* by reactivity with monoclonal antibodies and ability to infect severe combined immunodeficient mice, *Infect. Immun.*, 63, 360, 1995.
139. Ortega-Mora, L. M., Troncoso, J. M., Rojo-Vazquez, F. A., and Gomez-Bautista, M., Cross-reactivity of polyclonal serum antibodies generated against *Cryptosporidium parvum* oocysts, *Infect. Immun.*, 60, 3442, 1992.
140. Lumb, R., Smith, P. S., Davies, R., O'Donoghue, P. J., Atkinson H. M., and Lanser, J. A., Localization of a 23,000-MW antigen of *Cryptosporidium* by immunoelectron microscopy, *Immunol. Cell Biol.*, 67, 267, 1989.
141. El-Shewy, K., Kilani, R. T., Hegazi, M. M., Makhlouf, L. M. and Wenman, W. M., Identification of low-molecular-mass coproantigens of *Cryptosporidium parvum*, *J. Infect. Dis.*, 169, 460, 1994.
142. Ditrich, O., Kopáček, P., and Kučerova, Z., Antigenic characterization of human isolates of cryptosporidia, *Fol. Parasitol.*, 40, 301, 1993.
143. Kuhls, T. L., Orlicek, S. L., Mosier, D. A., Crawford, D. L., Abrams, V. L., and Greenfield, R. A., Enteral human serum immunoglobulin treatment of cryptosporidiosis in mice with severe combined immunodeficiency, *Infect. Immun.*, 63, 3582, 1995.
143a. Lorenzo, M. J., Ben, B., Mendez, F., Villacorta, I., and Ares-Mazas, M. E., *Cryptosporidium pavum* oocyst antigens recognized by sera from infected asymptomatic adult cattle, *Vet. Parasitol.*, 60, 17, 1995.
144. Dykstra, C. C., Blagburn, B. L., and Tidwell, R. R., Construction of genomic libraries of *Cryptosporidium parvum* and identification of antigen-encoding genes, *J. Protozool.*, 38, 76S, 1991.
145. Arnault, I., Répérant, J. M., and Naciri, M., Humoral antibody response and oocyst shedding after experimental infection of histocompatible newborn and weaned piglets with *Cryptosporidium parvum*, *Vet. Res.*, 25, 371, 1994.
146. Current, W. L. and Snyder, D. B., Development of and serologic evaluation of acquired immunity to *Cryptosporidium baileyi* by broiler chickens, *Poult. Sci.*, 67, 720, 1988.
146a. Sréter, T., Varga, I., and Békési, L., Effects of bursectomy and thymectomy on the development of resistance to *Cryptosporidium baileyi* in chickens, *Parasitol. Res.*, 82, 174, 1996.
147. Naciri, M., Mancassola, R., Répérant, J. M., and Yvoré, P., Analysis of humoral immune response in chickens after inoculation with *Cryptosporidium baileyi* or *Cryptosporidium parvum*, *Avian Dis.*, 38, 832, 1994.
148. Fayer, R., Perryman, L. E., and Riggs, M. W., Hyperimmune bovine colostrum neutralizes *Cryptosporidium* sporozoites and protects mice against oocyst challenge, *J. Parasitol.*, 75, 151, 1989.
149. Hatkin, J., Giambrone, J. J., and Blagburn, B. L., Correlation of circulating antibody and cellular immunity with resistance against *Cryptosporidium baileyi* in broiler chickens, *Avian Dis.*, 37, 800, 1993.
149a. Cama, V. A. and Sterling, C. R., Hyperimmune hens as a novel source of anti-*Cryptosporidium* antibodies suitable for passive immune transfer, *J. Protozool.*, 38, 42S, 1991.
150. Moon, H. W., Woodmansee, D. B., Harp, J. A., Abel, S., and Ungar, B. L. P., Lacteal immunity to enteric cryptosporidiosis in mice: immune dams do not protect their suckling pups, *Infect. Immun.*, 56, 649, 1988.
151. Brambell, F. W. R., Transmission of immunity in the rat and the mouse before birth, in *The Transmission of Passive Immunity from Mother to Young*, Neuberger, A. and Tatum, E. L., Eds., American Elsevier, New York, 1970, chap. 4.
152. Brambell, F. W. R., Transmission of immunity in the rat and the mouse after birth, in *The Transmision of Passive Immunity from Mother to Young*, Neuberger, A. and Tatum, E. L., Eds., American Elsevier, New York, 1970, chap. 5.
153. Tzipori, S., Rand, W., Griffiths, J., Widmer, G., and Crabb, J., Evaluation of an animal model system for cryptosporidiosis: therapeutic efficacy of paromomycin and hyperimmune bovine colostrum-immunoglobulin, *Clin. Diag. Lab. Immunol.*, 1, 450, 1994.
154. Watzl, B., Huang, D. S., Alak, J., Darban, H., Jenkins, E. M., and Watson, R. R., Enhancement of resistance to *Cryptosporidium parvum* by pooled bovine colostrum during murine retroviral infection, *Am. J. Trop. Med. Hyg.*, 48, 519, 1993.
155. Albert, M. M., Rusnak, J., Luther, M. F., and Graybill, J. R., Treatment of murine cryptosporidiosis with anti-cryptosporidial immune rat bile, *Am. J. Trop. Med. Hyg.*, 50, 112, 1994.
156. Brambell, F. W. R., Transmission of immunity in the ruminants, in *The Transmision of Passive Immunity from Mother to Young*, Neuberger, A. and Tatum, E. L., Eds., American Elsevier, New York, 1970, chap. 8.
157. Riggs, M. W., Evaluation of foals for immune deficiency disorders, in *The Veterinary Clinics of North America: Equine Practice, Clinical Pathology*, Vol. 3, Brobst, D. F., Ed., W. B. Saunders, Philadelphia, PA, 1987, 515.
158. Plettenberg, A., Stoehr, A., Stellbrink, H. J., Albrecht, H., and Meigel, W., A preparation from bovine colostrum in the treatment of HIV-positive patients with chronic diarrhea, *Clin. Invest.*, 71, 42, 1993.
159. Besser, T. E., McGuire, T. C., Gay, C. C., Evermann, J. F., and Pritchett, L. C., Transfer of functional IgG antibody into the gastrointestinal tract accounts for IgG clearance in calves, *J. Virol.*, 62, 2234, 1988.

160. Besser, T. E., Gay, C. C., McGuire, T. C., and Evermann, J. F., Passive immunity to bovine rotavirus infection associated with transfer of serum antibody into the intestinal lumen, *J. Virol.*, 62, 2238, 1988.
161. Fayer, R., Andrews, C., Ungar, B. L. P., and Blagburn, B., Efficacy of hyperimmune bovine colostrum for prophylaxis of cryptosporidiosis in neonatal calves, *J. Parasitol.*, 75, 393, 1989.
162. Harp, J. A., Woodmansee, D. B., and Moon, H. W., Effects of colostral antibody on susceptibility of calves to *Cryptosporidium parvum* infection, *Am. J. Vet. Res.*, 50, 2117, 1989.
163. Vitovec, J. and Koudela, B., Pathogenesis of intestinal cryptosporidiosis in conventional and gnotobiotic piglets, *Vet. Parasitol.*, 43, 25, 1992.
164. Tzipori, S., Roberton, D., Cooper, D. A., and White, L., Chronic cryptosporidial diarrhoea and hyperimmune cow colostrum, *Lancet*, 2, 344, 1987.
165. Ungar, B. L., Ward, D. J., Fayer, R., and Quinn, C. A., Cessation of *Cryptosporidium*-associated diarrhea in an Acquired Immunodeficiency Syndrome patient after treatment with hyperimmune bovine colostrum, *Gastroenterology*, 98, 486, 1990.
166. Nord, J., Ma, P., DiJohn, D., Tzipori, S., and Tacket, C. O., Treatment with bovine hyperimmune colostrum of cryptosporidial diarrhea in AIDS patients, *AIDS*, 4, 581, 1990.
167. Heaton, P., Cryptosporidiosis and acute leukaemia, *Arch. Dis. Child.*, 65, 813, 1990.
168. Shield, J., Melville, C., Novelli, V., Anderson, G., Scheimberg, I., Gibb, D., and Milla, P., Bovine colostrum immunoglobulin concentrate for cryptosporidiosis in AIDS, *Arch. Dis. Child.*, 69, 451, 1993.
169. Borowitz, S. M. and Saulsbury, F. T., Treatment of chronic cryptosporidial infection with orally administered human serum immune globulin, *J. Pediatrics*, 119, 593, 1991.
170. Cooperstock, M., DuPont, H. L., Corrado, M. L., Fekety, R., and Murray, D. M., Evaluation of new anti-infective drugs for the treatment of diarrhea caused by *Cryptosporidium*, *Clin. Infect. Dis.*, 15, S249, 1992.
171. Fayer, R., Tilley, M., Upton, S. J., Guidry, A. J., Thayer, D. W., Hildreth, M., and Thomson, S., Production and preparation of hyperimmune bovine colostrum for passive immunotherapy of cryptosporidiosis, *J. Protozool.*, 38, 38S, 1991.
172. Perryman, L. E., Riggs, M. W., Mason, P. H., and Fayer, R., Kinetics of *Cryptosporidium parvum* sporozoite neutralization by monoclonal antibodies, immune bovine serum, and immune bovine colostrum, *Infect. Immun.*, 58, 257, 1990.
173. Perryman, L. E., Kegerreis, K. A., and Mason, P. H., Effect of orally administered monoclonal antibody on persistent *Cryptosporidium parvum* infection in SCID mice, *Infect. Immun.*, 61, 4906, 1993.
174. Hamer, D. H., Ward, H., Tzipori, S., Pereira, M. E., Alroy, J. P., and Keusch, G. T., Attachment of *Cryptosporidium parvum* sporozoites to MDCK cells *in vitro*, *Infect. Immun.*, 62, 2208, 1994.
175. Schwartzman, J. D., Inhibition of penetration-enhancing factor of *Toxoplasma gondii* by monoclonal antibodies specific for rhoptries, *Infect. Immun.*, 55, 760, 1986.
176. Cerami, C., Frevert, U., Sinnis, P., Takacs, B., Clavijo, P., Santos, M. J., and Nussenzweig, V., The basolateral domain of the hepatocyte plasma membrane bears receptors for the circumsporozoite protein of *Plasmodium falciparum* sporozoites, *Cell*, 70, 1021, 1992.
176a. Nudelman, S., Renia, L., Charoenvit, Y., Yuan, L., Miltgen, F., Beaudoin, R. L., and Mazier, D., Dual action of anti-sporozoite antibodies *in vitro*, *J. Immunol.*, 143, 996, 1989.
177. Kara, U. A. K., Stenzel, D. J., Ingram, L. T., Bushell, G. R., Lopez, J. A., and Kidson, C., Inhibitory monoclonal antibody against a myristylated small-molecular-weight antigen from *Plasmodium falciparum* associated with the parasitophorous vacuole membrane, *Infect. Immun.*, 56, 903, 1988.
178. Mazier, D., Mellouk, S., Beaudoin, R. L., Texier, B., Druilhe, P., Hockmeyer, W., Trosper, J., Paul, C., Charoenvit, Y., Young, J., Miltgen, F., Chedid, L., Chigot, J. P., Galley, B., Brandicourt, O., and Gentilini, M., Effect of antibodies to recombinant and synthetic peptides on *P. falciparum* sporozoites *in vitro*, *Science*, 231, 156, 1986.
179. Cochrane, A. H., Aikawa, M., Jeng, M., and Nussenzweig, R. S., Antibody-induced ultrastructural changes of malarial sporozoites, *J. Immunol.*, 116, 859, 1976.
180. Stewart, M. J. and Vanderberg, J. P., Malaria sporozoites leave behind trails of circumsporozoite protein during gliding motility, *J. Protozool.*, 35, 389, 1988.
181. Nelson, R. G., Kim, K., Gooze, L., Petersen, C., and Gut, J., Identification and isolation of *Cryptosporidium parvum* genes encoding microtubule and microfilament proteins, *J. Protozool.*, 38, 52S, 1991.
182. Wiest, P. M., Johnson, J. H., and Flanigan, T. P., Microtubule inhibitors block *Cryptosporidium parvum* infection of a human enterocyte cell line, *Infect. Immun.*, 61, 4888, 1993.
183. Chen, W., Harp, J. A., and Harmsen, A. G., Requirements for CD4+ cells and gamma interferon in resolution of established *Cryptosporidium parvum* infection in mice, *Infect. Immun.*, 61, 3928, 1993.
184. Chen, W., Harp, J. A., Harmsen, A. G., and Havell, E. A., Gamma interferon functions in resistance to *Cryptosporidium parvum* infection in severe combined immunodeficient mice, *Infect. Immun.*, 61, 3548, 1993.
185. Perryman, L. E., Mason, P. H., and Chrisp, C. E., Effect of spleen cell populations on resolution of *Cryptosporidium parvum* infection in SCID mice, *Infect. Immun.*, 62, 1474, 1994.
186. McDonald, V. and Bancroft, G. J., Mechanisms of innate and acquired resistance to *Cryptosporidium parvum* infection in SCID mice, *Parasite Immunol.*, 16, 315, 1994.
187. Urban, J. F. Jr., Fayer, R., Chen, S.-J., Gause, W. C., Gately, M. K., and Finkelman, F. D., IL-12 protects immunocompetent and immunodeficient neonatal mice against infection with *Cryptosporidium parvum*, *J. Immunol.*, 156, 263, 1996.

188. Mead, J. R., Arrowood, M. J., Healey, M. C., and Sidwell, R. W., Cryptosporidial infections in SCID mice reconstituted with human or murine lymphocytes, *J. Protozool.*, 38, 59S, 1991.
189. Aguirre, S. A., Mason, P. H., and Perryman, L. E., Susceptibility of major histocompatibility complex (MHC) class I- and MHC class II-deficient mice to *Cryptosporidium parvum* infection, *Infect. Immun.*, 62, 697, 1994.
189a. Waters, W. R. and Harp, J. A., *Cryptosporidium parvum* infection in T-cell receptor (TCR)-α- and TCR-δ-deficient mice, *Infect. Immun.,* 64, 185, 1996.
190. Kuhls, T. L., Mosier, D. A., Abrams, V. L., Crawford, D. L., and Greenfield, R. A., Inability of interferon-gamma and aminoguanidine to alter *Cryptosporidium parvum* infection in mice with severe combined immunodeficiency, *J. Parasitol.*, 80, 480, 1994.
191. Enriquez, F. J. and Sterling, C. R., Role of CD4+ TH1- and TH2-cell-secreted cytokines in cryptosporidiosis, *Fol. Parasitol.*, 40, 307, 1993.
192. Boher, Y., Perez-Schael, I., Caceres-Dittmar, G., Urbina, G., Gonzalez, R., Kraal, G., and Tapia, F. J., Enumeration of selected leukocytes in the small intestine of BALB/c mice infected with *Cryptosporidium parvum*, *Am. J. Trop. Med. Hyg.*, 50, 145, 1994.
193. Argenzio, R. A., Lecce, J., and Powell, D. W., Prostanoids inhibit intestinal NaCl absorption in experimental porcine cryptosporidiosis, *Gastroenterology*, 104, 440, 1993.
194. Kandil, H. M., Berschneider, H. M., and Argenzio, R. A., Tumour necrosis factor α changes porcine intestinal ion transport through a paracrine mechanism involving prostaglandins, *Gut*, 35, 934, 1994.
195. Marcial, M. A. and Madara, J. L., *Cryptosporidium*: cellular localization, structural analysis of absorptive cell-parasite membrane-membrane interactions in guinea pigs, and suggestion of protozoan transport by M cells, *Gastroenterology*, 90, 583, 1986.
196. Rohlman, V. C., Kuhls, T. L., Mosier, D. A., Crawford, D. L., and Greenfield, R. A., *Cryptosporidium parvum* infection after abrogation of natural killer cell activity in normal and severe combined immunodeficiency mice, *J. Parasitol.*, 79, 295, 1993.
197. Enriquez, F. J. and Sterling, C. R., *Cryptosporidium* infections in inbred strains of mice, *J. Protozool.*, 38, 100S, 1991.
198. Harp, J. A. and Moon, H. W., Susceptibility of mast cell-deficient W/W^v mice to *Cryptosporidium parvum*, *Infect. Immun.*, 59, 18, 1991.
199. Flanigan, T., Whalen, C., Turner, J., Soave, R., Toerner, J., Havlir, D., and Kotler, D., *Cryptosporidium* infection and CD4 counts, *Ann. Intern. Med.*, 116, 840, 1992.
200. Crowe, S. M., Carlin, J. B., Stewart, K. I., Lucas, C. R., and Hoy, J. F., Predictive value of CD4 lymphocyte numbers for the development of opportunistic infections and malignancies in HIV-infected persons, *J. AIDS*, 4, 770, 1991.
201. Mølbak, K., Lisse, I. M., Højlyng, N., and Aaby, P., Severe cryptosporidiosis in children with normal T-cell subsets, *Parasite Immunol.*, 16, 275, 1994.
202. Fuchs, D., Hansen, A., Reibnegger, G., Werner, E. R., Werner-Felmayer, G., Dierich, M. P., and Wachter, H., Interferon-gamma concentrations are increased in sera from individuals with human immunodeficiency virus type 1, *J. AIDS*, 2, 158, 1989.
203. Reinecker, H. C., Steffen, M., Doehn, C., Petersen, J., Pfluger, I., Voss, A., and Raedler, A., Proinflammatory cytokines in intestinal mucosa, *Immunol. Res.*, 10, 247, 1991.
204. Whitmire, W. M. and Harp, J. A., *In vitro* murine lymphocyte blastogenic responses to *Cryptosporidium parvum*, *J. Parasitol.*, 76, 450, 1990.
205. Harp, J. A., Whitmire, W. M., and Sacco, R., *In vitro* proliferation and production of gamma interferon by murine CD4+ cells in response to *Cryptosporidium parvum* antigen, *J. Parasitol.*, 80, 67, 1994.
205a. Harp, J. A. and Sacco, R. E., Development of cellular immune functions in neonatal to weanling mice: relationship to *Cryptosporidium parvum* infection, *J. Parasitol.,* 82, 245, 1996.
206. Gomez-Morales, M. A., Ausiello, C. M., Urbani, A. F., and Pozio, E., Crude extract and recombinant protein of *Cryptosporidium parvum* oocysts induce proliferation of human peripheral blood mononuclear cells *in vitro*, *J. Infect. Dis.*, 172, 211, 1995.
206a. Morales, M. A. G., Ausiello, C. M., Guarino, A., Urbani, F., Spagnuolo, M. I., Pignata, C., and Pozio, E., Severe, protracted intestinal cryptosporidiosis associated with interferon γ deficiency: pediatric case report, *Clin. Infect. Dis.*, 22, 848, 1996.
207. Harp, J. A., Franklin, S. T., Goff, J. P., and Nonnecke, B. J., Effects of *Cryptosporidium parvum* infection on lymphocyte phenotype and reactivity in calves, *Vet. Immunol. Immunopathol.*, 44, 197, 1995.
207a. Tilley, M., McDonald, V., and Bancroft, G. J., Resolution of cryptosporidial infection in mice correlates with parasite-specific lymphocyte proliferation associated with both T_h1 and T_h2 cytokine secretion, *Parasite Immunol.*, 17, 459, 1995.
208. Harp, J. A. and Whitmire, W. M., *Cryptosporidium parvum* infection in mice: inability of lymphoid cells or culture supernatants to transfer protection from resistant adults to susceptible infants, *J. Parasitol.*, 77, 170, 1991.
209. Louie, E., Borkowsky, W., Klesius, P. H., Haynes, T. B., Gordon, S., Bonk, S., and Lawrence, H. S., Treatment of cryptosporidiosis with oral bovine transfer factor, *Clin. Immunol. Immunopathol.*, 44, 329, 1987.
210. McMeeking, A., Borkowsky, W., Klesius, P. H., Bonk, S., Holzmann, R. S., and Lawrence, H. S., A controlled trial of bovine dialyzable leukocyte extract for cryptosporidiosis in patients with AIDS, *J. Infect. Dis.*, 161, 108, 1990.
211. Chng, H. H., Shaw, D., Klesius, P., and Saxon, A., Inability of oral bovine transfer factor to eradicate cryptosporidial infection in a patient with congenital dysgammaglobulinemia, *Clin. Immunol. Immunopathol.*, 50, 402, 1989.

212. Fayer, R., Klesius, P. H., and Andrews, C., Efficacy of bovine transfer factor to protect neonatal calves against experimentally induced clinical cryptosporidiosis, *J. Parasitol.*, 73, 1061, 1987.
213. Leitch, G. J. and He, Q., Arginine-derived nitric oxide reduces fecal oocyst shedding in nude mice infected with *Cryptosporidium parvum*, *Infect. Immun.*, 62, 5173, 1994.
214. Novak, S. M. and Sterling, C. R., Susceptibility dynamics in neonatal BALB/c mice infected with *Cryptosporidium parvum*, *J. Protozool.*, 38, 102S, 1991.
215. Rabinovich, N. R., McInnes, P., Klein, D. L., and Hall, B. F., Vaccine technologies: view to the future, *Science*, 265, 1401, 1994.
216. Taylor, C. E., Cytokines as adjuvants for vaccines: antigen-specific responses differ from polyclonal responses, *Infect. Immun.*, 63, 3241, 1995.
217. Shalaby, W. S. W., Development of oral vaccines to stimulate mucosal and systemic immunity: barriers and novel strategies, *Clin. Immunol. Immunopathol.*, 74, 127, 1995.

Chapter 7

Biochemistry of *Cryptosporidium*

Michael Tilley and Steve J. Upton

CONTENTS

I. INTRODUCTION

Although the life cycles of *Cryptosporidium* spp. are similar to that of other coccidia, what little is known of the basic biochemistry suggests there are major differences between *Cryptosporidium* and other coccidia. Based on small-subunit ribosomal RNA similarities, *Cryptosporidium* spp. appear more closely related to malaria than to Eimeriidae or Sarcocystidae.[1,2] Many pharmaceuticals active against Eimeriidae or Sarcocystidae have little or no effect on *C. parvum*.[3] Mitochondria, present in *C. muris* and other coccidia,[4,5] have not been seen in *C. parvum*.[6-8] *Cryptosporidium* spp., with few or no introns, have a genome approximately 12 to 15% that of other coccidia. Collectively, these findings suggest *Cryptosporidium* spp. possess a different, and perhaps unusual, biochemistry. Because elucidation of surface antigen structure and function, as well as metabolic pathways, are essential to discovery of novel mechanisms of chemotherapeutic control, these areas of biochemistry are a priority of the National Cooperative Drug Discovery Group for Treatment of Opportunistic Infections (NCDDG-OI) at the National Institutes of Health.[9]

II. METABOLISM

Overall, coccidia are not amenable to biochemical studies due, predominately, to their intracellular existence, which makes purification of endogenous stages difficult. Most studies have focused on the exogenous stages of the life cycle, the oocysts and sporozoites, which cannot represent metabolism throughout the life cycle. Studies as diverse as zymography, gene sequencing, pharmacology, and *in vitro* supplementation, have provided clues about biochemical pathways. However, all reports (through July 1995) should be considered preliminary and fragmentary, and the majority of information summarized below should be considered hypothetical.

0-8493-7695-5/97/$0.00+$.50

A. FOLATE METABOLISM

Based on sequenced parasite DNA, it was suggested that *C. parvum* had a functional thymidylate synthetase.[10] Later, it was shown that both folic acid and calcium pantothenate enhanced *in vitro* development.[11] These results and data showing that several sulfonamides inhibit parasite development[12-17] suggest that *C. parvum* synthesizes its own folic acid and also incorporates this molecule. If true, it is likely that sulfonamides and diaminopyrimidines may act synergistically.

B. ISOZYMES

Isoenzymes from extracts of 10^7 to 10^8 oocysts each of *C. baileyi*, *C. muris*, and *C. parvum* were detected by Starch-gel zymography.[18] All species had high levels of activity associated with glucose-6-phosphate isomerase (GPI; EC 5.3.1.9) and phosphoglucomutase (PGM; EC 5.4.2.7), but were weak in malate dehydrogenase (MDH; EC 1.1.1.37), carboxylesterase (EC 3.1.1.1), and lactate dehydrogenase (LDH; EC 1.1.1.27). No bands were detected for alanine aminotransferase (equal to glutamic-pyruvic transaminase; EC 2.6.1.2), aspartate aminotransferase (equal to glutamic-oxaloacetic transaminase; EC 2.6.1.1), glucose-6-phosphate dehydrogenase (EC 1.1.1.49), mannose-6-phosphate isomerase (EC 5.3.1.8), or proline iminopeptidase (EC 3.4.11.5), but lack of zymogram activity does not always equate with lack of an enzyme. In addition, different stages will possess varying enzyme activities, and the aforementioned enzymes were studied only from sporozoites. Significant differences in electrophoretic mobility of both GPI and PGM were observed among species, and *C. muris* appeared to have two forms of GPI. One of five isolates of *C. parvum* had PGM of different electrophoretic mobility than the others.

The discovery of MDH in *C. parvum* and *C. baileyi* is intriguing. MDH catalyzes the interconversion of L-malate and oxaloacetate and is usually found within the mitochondrial matrix. Since mitochondria appear to be absent in *C. parvum*, one might speculate that they possess remnant respiratory membranes. However, MDH is also used in the glyoxylate cycle, which utilizes acetate as the sole carbon source and converts lipids into carbohydrates. This cycle, common in plants, fungi, and some nematodes, protozoa, and prokaryotes, appears to be absent in vertebrates.[19,20] Both isocitrate lyase and malate synthase, key enzymes in this pathway, also appear to be absent in animal cells[19,20] and other coccidia.[21] Acetyl-CoA synthetase (EC 6.2.1.1), which catalyzes the formation of acetyl-CoA from acetate and coenzyme A, is necessary in the glyoxylate pathway and has recently been cloned and sequenced from *C. parvum* (GenBank™, U24082).

C. MANNITOL CYCLE

The mannitol cycle, common in both plants and fungi, appears to be absent in animals and most animal-like protozoa. However, mannitol comprises nearly 25% of the dry weight of the unsporulated oocyst of eimerians.[22-24] *Cryptosporidium parvum* might also utilize the mannitol cycle as both mannitol-1 phosphate dehydrogenase (M1PDH, EC 1.1.1.17) and mannitol-1-phosphatase (M1Pase, EC 3.1.3.22) have been reported.[22] Mannitol reserves are significantly depleted during sporulation of *Eimeria tenella*, suggesting that mannitol is used as an energy source.[22,24] With the corresponding reduction of NAD, mannitol would be oxidized to fructose by MDH followed by the phosphorylation of fructose into fructose-6-phosphate (F6P), which could enter glycolysis or the pentose phosphate pathway. Mannitol may also serve as an electron reservoir during macrogametogenesis.[22,24] NADH would be oxidized during conversion of F6P to M1P, and M1P should be converted into mannitol by M1Pase. Since M1P and MDH are absent in vertebrates, both represent potential therapeutic targets.

D. NUCLEIC ACID METABOLISM

Nucleic acid metabolism has not yet been studied in *Cryptosporidium* spp. However, studies to date suggest that coccidia are incapable of synthesizing purines and must rely on salvage enzymes.[21,25] Glycine, formate, and glucose are poorly incorporated, whereas exogenous adenosine, adenoside nucleotides, hypoxanthine, and inosine are utilized.[21,25-32] *Cryptosporidium* spp. might also use such salvaging systems as arprinocid-1-N-oxide, the metabolized form of arprinocid, which inhibits uptake of hypoxanthine and guanine by coccidia.[32,33] Arprinocid is known to inhibit *C. parvum in vivo*.[34,35]

Pyrimidines can be salvaged or synthesized *de novo* by coccidia.[21,36-39] Coccidia possess the enzyme uracil phosphoribosyl transferase, which converts uracil into uridylic acid.[40-43] Various *in vitro* assays rely on ^{3}H-uracil incorporation as a measure of coccidian development,[25,44,45] and preliminary studies have suggested that *C. parvum* may be capable of incorporating free uracil as well.[46-48]

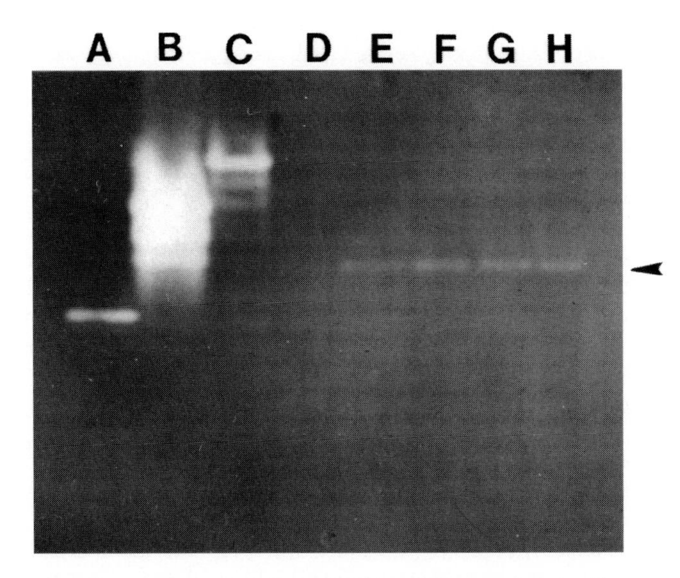

Figure 1 Native PAGE of superoxide dismutase (SOD) activity associated with oocysts of *C. parvum* and compared to SODs from other organisms. Lane A, Fe^{++}-containing SOD from *E. coli*; Lane B, Cu^{++}/Zn^{++} SOD from *Bos taurus*; Lane C, Mn^{++}-containing SOD from *E. coli*; Lane D, spacer lane without protein; Lanes E to H, Fe^{++}-containing SOD from *C. parvum* with 50, 100, 150, and 200 $\times$ 10^6 oocysts per lane, respectively. The Fe^{++}-containing SOD of *C. parvum* migrates with the lower isotype of Cu^{++}/Zn^{++}, but its activty is inhibited by H_2O_2. (Zymograms courtesy B. Oppert, Kansas State University.)

E. SUPEROXIDE DISMUTASE

Sporozoites of many coccidia possess scant amounts of oxygen-scavenging enzymes; low levels of superoxide dismutase (SOD) are found in dormant sporozoites.[49-51] Sporozoites of *C. parvum* apparently possess very low levels of a single cytosolic Fe^{++}-like SOD (Figure 1).[18,51] On western blots, antiserum against *Escherichia coli* Fe^{++} SOD recognizes *C. parvum* bands at 15 and 30 kDa.[51] Although this enzyme is inhibited by H_2O_2 (as are most Fe^{++}-containing SODs), it comigrates on native gels with the lower isotype of bovine Cu^{++}/Zn^{++} SOD.[51,52] SOD activity is so low that it generally requires >5 $\times$ 10^7 fresh oocysts per lane of a minigel to detect activity.[52]

F. MISCELLANEOUS ENZYMES AND PROTEINS

Miscellaneous proteins have been reported from *C. parvum*. The gene encoding a protein homologous to cytoplasmic hsp70 with a predicted size of 674 amino acids and 73.4 kDa was sequenced.[53] Antiserum against the recombinant protein recognized a single band on western blots of oocyst/sporozoite proteins. An oocyst wall-associated antigen >200 kDa was found encoded by a gene with partial homology to hsp70.[54] Some additional gene sequences reported for *C. parvum* include the complete sequence for elongation factor-2,[55] a partial sequence of the type II topoisomerase gene,[56] and a partial sequence that has an area of homology with the gene encoding mitochondrial ATPase.[57] Whether this latter finding actually represents a functional mitochondrial ATPase or remnant of such a gene is uncertain. A phospho-intermediate ATPase has also been reported.[58] Analogs of arginine and arginine decarboxylase inhibited *C. parvum in vitro* whereas ornithine decarboxylase inhibitors had no effect,[59] suggesting a plant-/fungal-like polyamine biosynthetic pathway, another unexplored avenue of therapeutic intervention.

III. LIPIDS AND GLYCOLIPIDS

Only the basic lipid composition of sporulated *C. parvum* oocysts has been documented.[7,8] Of approximately 1.2 $\times$ 10^{-9} μmol phospholipid per oocyst, phosphatidylcholine (PC) comprised about 66%. This high level of PC appears to be typical of coccidia,[60-64] and especially high levels are thought to be localized in the rhoptries.[61] Since PC tends to remain in a stable bilayer phase and not form fusogenic structures,[65] extrusion of high amounts of rhoptry-associated PC into the parasitophorous vacuole may inhibit lysosomal fusion.[7,8,66] Sphingomyelin (24.5%), phosphatidylethanolamine (7.3%), and small amounts of phosphatidylinositol/phosphatidylserine (1.8%) and cardiolipin (0.9%) were also detected. This very low level of cardiolipin supports the hypothesis that *C. parvum* may have no functional mitochondria.

The major fatty acids detected in *C. parvum* were palmitic (16:0), stearic (18:0), oleic (18:1), and linoleic (18:2) with smaller amounts of myristate (14:0) and palmitoleate (16:1) but only trace amounts of other fatty acids.[7,8] Although amounts vary slightly, these same fatty acids are dominant in tachyzoites of *Toxoplasma gondii*.[60,61] Arachidonic acid (20:4) was detected in tachyzoites of *T. gondii*,[61] whereas little 20:4 was found in dormant sporozoites of *C. parvum*.[7,8] Although *E. tenella* has 16:0 and 18:1 as major acyl chains, like *C. parvum*, very little 18:2 was noted.[64]

The only sterol detected in *C. parvum* was cholesterol, at about 1.7×10^{-10} mol per oocyst.[7,8] All coccidia have a low cholesterol-to-phospholipid ratio. This, along with the dominance of short chain fatty acids, significant amounts of unsaturated fatty acids, and high amounts of PC, should contribute significantly to a high membrane fluidity.[62]

Sporozoites and merozoites of *C. parvum* appear to possess a nonprotein surface antigen which may be glycolipid or phospholipid. First recognized by monoclonal antibody 18.44 (mAb 18.44) which bound diffusely over the sporozoite surface in immunofluorescent assays (IFA),[67] the antigen did not label with ^{125}I or with ^{35}S-methionine incorporation and co-migrated with the dye front in SDS-PAGE.[67] It eluted with the void volume in Bio Gel A columns with an exclusion limit of 500 kDa and was insensitive to proteinase K digestion.[67] Sporozoites incubated with mAb 18.44 had diminished ability to infect suckling mice.[67-71]

IV. CYTOSKELETAL PROTEINS

The first complete gene sequence published for the genus *Cryptosporidium* was for actin.[72] It was found as a single-copy on a 1200-kb chromosome and had no introns. The predicted 376 amino acid protein, with a size of 42.1 kDa, was 85% homologous to human γ-actin and *Plasmodium falciparum* actin I.[72,73]

Portions of the α and β tubulin genes of *C. parvum* have been sequenced,[73,74] and a partial sequence of the β-tubulin gene has been entered in GenBank™ (L31806). Both appear to represent single-copy genes, which hybridize with chromosomes of 1200 and 1400 kb, respectively. In immunoblots, a mAb to β-tubulin recognized 54- and 58-kDa bands in *C. parvum* extracts which correspond in size to α- and β-tubulin.[75] Because organisms with glutamic acid and phenylalanine at residues 198 and 200 in the β-tubulin sequence, respectively, tend to be benzimidazole sensitive,[74] it was speculated that *C. parvum*, with Ala198 and Gln200, would be benzimidazole insensitive.[74] Indeed, benzimidazoles tested prophylactically in neonatal mice failed to inhibit parasite development.[76]

Tubulin from *C. parvum* sporozoites bound ^{3}H-colchicine,[75] and microtubule inhibitors failed to prevent excystation. However, both colchicine and vinblastine significantly reduced invasion of sporozoites in HT29.74 cells *in vitro*.[75,77]

V. THE OOCYST WALL

The oocyst wall of *Cryptosporidium* is bilayered and serves as a barrier to the sporozoites within. Due to the relatively inert nature of the oocyst wall, little is known about its structure. The oocyst wall of *Eimeria tenella* is composed of 67% protein, 19% carbohydrate, and 14% lipid.[63] The protein component is easily dissociated with reducing agents under denaturing conditions following hypochlorite treatment,[63] and it has been suggested that some of the lipid may mask disulfide bridges, which become susceptible to reduction only after pretreatment.[78]

The oocyst walls of *Cryptosporidium* spp. appear to be unique. Although rich in disulfide bonds (below), no fatty alcohols have been detected,[7,8] unlike the eimerians.[63,64] In addition, the number and sizes of proteins differ from those of *E. tenella*.[63] *C. parvum* oocysts stored in $K_2Cr_2O_7$ solution and then exposed to trypsin/bile salts had 21 antigen bands in 11- to 20%-gradient SDS-PAGE and Coomassie staining.[79] Bands ranged in size from 14 to 200 kD, with seven major bands at 190, 91, 65, 40, 19, 18, and 14 kD.[79] Acid/detergent-treated oocysts had only four bands in 7.5% SDS-PAGE at 190, 85, 55, and 30 kDa.[80] Oocysts stored in $K_2Cr_2O_7$ were bleach pretreated and excysted at 37°C. Oocyst walls purifed from sporozoites using Percoll® had bands at 190, 95, 60 to 65, 45, 27 to 28, and 23 kDa (Figure 2).[81] Some oocyst wall components appeared too large to enter the acrylamide, suggesting that one or more additional molecules were present or that undenatured forms of the other molecules remained intact.

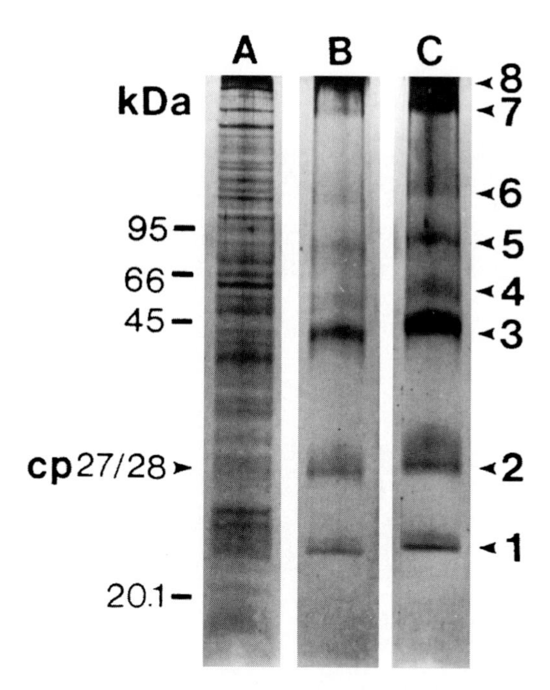

Figure 2 Silver staining of SDS-PAGE of proteins associated with purified oocysts and oocyst walls of *C. parvum*. Lane A, bands associated with 2 μg oocyst per sporozoite homogenate; Lanes B and C, bands associated with 0.2 μg (B) and 0.4 μg (C) Percoll®-purified walls. Positions of molecular size markers shown on left; numbers on right emphasize positions of distinct bands. cp27/28 represents position of sporozoite surface molecule that reacts with Mab 2D7. (SDS-PAGE courtesy of M. V. Nesterenko, Kansas State University.)

The mAb 2D7, previously reported to react with a 50-kDa inner oocyst wall antigen,[82] recognized a 45-kDa wall protein[81] (Figure 2, band 3). The mAb 3E3,[82] which recognized a 27- to 28-kDa sporozoite surface band,[82] reacted with a 27/28-kDa band from oocyst walls,[81] suggesting that this molecule is a sporozoite surface molecule shed by the parasites during excystation (Figure 2, band 2). Other studies have shown mAb 1B5, which reacts both with the outer wall and gametes, binds a doublet on SDS-PAGE at 41 and 44 kDa.[83-85]

Radiolabeling of intact, $K_2Cr_2O_7$-treated oocysts of *C. parvum* with ^{125}I revealed multiple surface proteins (Figure 3). In one study, 6 to 10 bands were noted in western blots.[79] In another, 17 to 18 bands were seen but only about 10 were prominent.[86] When mAbs and radiolabeling were used to study the outer wall components of five *Cryptosporidium* spp., significant differences were detected among species and isolates.[79,85-89] A distinct pattern was seen in *C. wrairi* with highly labeled proteins clustered between 40 and 65 kDa (Figure 3).[89] The high number of ^{125}I-labeled antigens from intact oocysts vs. lower numbers associated with unlabeled walls is bothersome. $K_2Cr_2O_7$, ether, and detergents can affect the outer wall so that mAbs to the outer wall react only to oocysts stored in $K_2Cr_2O_7$ solutions but are not reactive against newly passed or formalin fixed oocysts.[90] Another study found that both periodate and $K_2Cr_2O_7$ treatment affected recognition by some mAbs.[87] Other possible explanations include differences in the strain of parasite,[79,87] adherent host-cell antigens or bacteria, oocyst residuum components, suture proteins, lysed sporozoite antigens, and the distinct possibility that some ^{125}I penetrates the oocyst wall and labels inner components.

The inner and outer walls of *C. parvum* are immunologically distinct and mAbs can be specific for either surface. The inner wall is strongly immunogenic, and at least three antigens are associated with this surface. One protein is the 27- to 28-kDa sporozoite surface protein shed during excystation that adheres to the inner wall surface.[81] Another, about 45 to 50 kDa on western blots,[82] may be the same molecule recognized by mAb 1B5.[83-85] The third protein was termed *Cryptosporidium* oocyst wall protein 190 (COWP-190); antiserum and mAbs against the recombinant peptide recognized a *C. parvum* protein with a molecular size of about 190 kDa.[91,92] The inserts encoded a portion of a *C. parvum* oocyst wall protein with repetitive motifs,[80,91,92] and the complete sequence was entered into GenBank™ (Z22537). The deduced 1622 amino acid sequence suggests a molecule rich in cysteine, proline, and histidine and

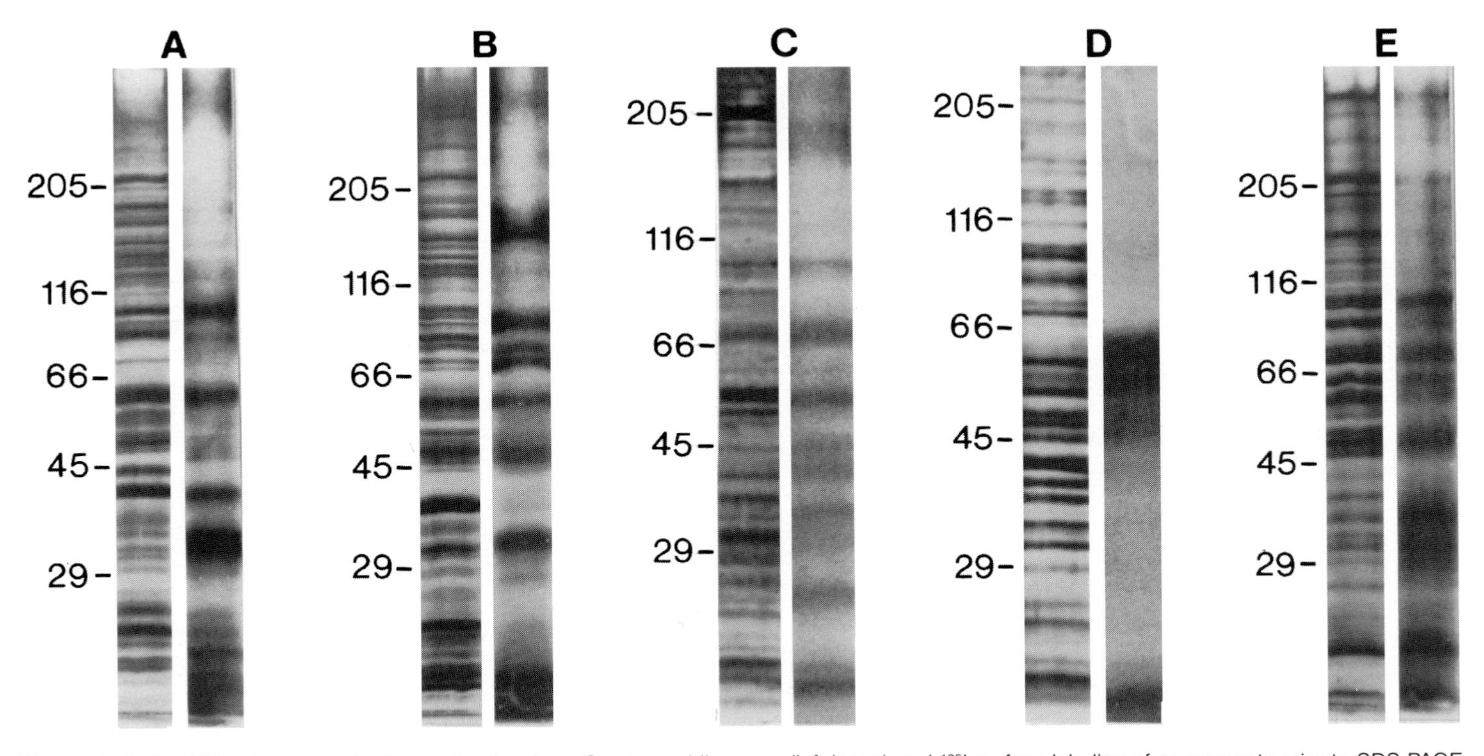

Figure 3 SDS-PAGE of homogenized oocysts of various *Cryptosporidium* spp. (left lanes) and ^{125}I surface labeling of components prior to SDS-PAGE (right lanes). Lane A, *C. baileyi* from chickens; Lane B, *C. parvum* from goat; Lane C, *C. serpentis* from a rat snake; Lane D, *Cryptosporidium* sp. from a guinea pig; Lane E, *Cryptosporidium* sp. from gastric glands of a cow. (Adapted from References 88, 89, and 106. With permission.)

capable of forming numerous disulfide bonds. No reactivity was observed on western blots under nonreducing conditions.[92] Incomplete denaturation may explain why only small amounts of the oocyst wall protein enter the gel even on SDS-PAGE. Analysis of the predicted amino acid sequence revealed numerous *N*-glycosylation consensus sequences (NXS/T) and potential *O*-glycosylation motifs. The C-terminus encodes a cysteine followed by NAA, suggesting that the cysteine is methylated prior to removal of NAA, at which time a lipid tail might be added to the modified cysteine.[93] Confocal microscopy has suggested that the protein occurs as granules on the inner surface of the oocyst wall and oocyst residuum; it is likely that these antibodies detect either the 190-kDa oocyst wall protein or a precursor.[92]

Recombinant protein from a construct encoding amino acids 371 to 1156 of COWP-190, termed cpRL3/COWP, as well as intact oocysts adhered to Caco-2 cells *in vitro*, and oocysts formed rosettes on the cell surface.[94] Heparin inhibited oocyst binding to host cells and had a high affinity for cpRL3/COWP. It is unclear whether this molecule, identified only with the inside surface of the oocyst, truly mediates binding with the host cell under natural conditions or is interacting nonspecifically.

The outer oocyst walls of *Cryptosporidium* spp. appear highly glycosylated. Periodate treatment influences binding of some mAbs to the outer wall. Several lectins agglutinated *C. parvum* oocysts, including CFL (*Codium fragile*; galNAc, glcNAc), tomentine (*C. tomentosum*; glcNAc), and UEA II (*Ulex europaeus* II; glcNAc$_2$).[87,95] In another study, jacalin (*Artocarpus integrifolia*; a-gal→OMe) and UEA II reacted strongly, whereas ABA (*Agaricus bisporus*; β-gal(1→3)galNAc), ACL (*Amaranthus caudatus*; gal, β-gal(1→3)galNAc), PNA (*Arachis hypogaea*; β-gal(1→3)galNAc), and VVA (*Vicia villosa*; galNAc) reacted weakly.[96] MPA (*Maclura pomifera*; gal, galNAc) was a potent agglutinator, whereas ConA (*Canavalia ensiformis*; glc, man) agglutinated weakly.[90] Collectively, these studies show that the outer wall of *C. parvum* possesses galactose/galactosamine and glucose/glucosamine residues, possibly with both *N*- and *O*-linked glycosylation.

The mAb OW3 identified *C. parvum* sporozoite cDNA that encoded a putative 670 amino acid sequence with significant homology to both hsp70 and a 78 kDa glucose regulation protein.[54] This mAb recognized a >200-kD *C. parvum* oocyst wall antigen on western blots.[54] Although this molecule may be a component of the oocyst wall, the Gly-Gly-Met-Pro repeat of *C. parvum* hsp70 should be highly immunogenic and may simply have an antigenic feature in common with COWP-190. The mAb OW-IGO, developed to the outer wall of *C. parvum* oocysts, also labeled the fibrillar material in the parasitophorous vacuole of developing macrogametes, microgametocytes, and sporulating oocysts.[97] On western blots, the antibody recognized major bands at 250 and 40 kDa and additional smaller bands. Most western blot activity was abolished by periodate treatment, again suggesting carbohydrate on the outer wall.

VI. SPOROZOITES AND MEROZOITES

A variety of proteins and glycoproteins appear to be shared by *Cryptosporidium* sporozoites and merozoites, including those associated with the surface and apical organelles.[68,70,98-106] Most studies have focused on sporozoite surface, rhoptry, and microneme proteins, but some progress has also been made toward understanding the types and kinetics of proteases from zymograms.

A. SURFACE PROTEINS AND GLYCOPROTEINS

SDS-PAGE profiles of ^{125}I- or biotin-surface-labeled sporozoites, as well as mAb IFA assays of intact sporozoites, have revealed from 5 to over 20 surface molecules ranging from 11 to 900 kD.[67,101,102,107-110] Limited consensus has been reached between laboratories on the numbers and sizes of these molecules, particularly those <30 kDa. Factors that can influence results on western blots include excystation conditions, acrylamide concentrations, pH, purity of the final sporozoite preparation, the labeling procedure, denaturation conditions, variability in the quality of molecular size markers, and even the addition of protease during excystation.[111]

Percoll®-purified sporozoites, labeled by ^{125}I and repurified and washed extensively, were run on SDS-PAGE. An autoradiogram revealed 10 to 12 surface antigens (Figure 4).[108] Additional studies using mAbs or polyclonal antiserum on immunoblots and in IFA assays revealed four surface antigens <30 kDa (Figure 4). The mAbs 2B3 and 3E3, which react with the highly immunogenic 15- and 25-kDa-sloughed surface antigens, respectively,[82,112] detected two molecules on SDS-PAGE that labeled only weakly with ^{125}I. Two additional surface molecules, at 20- and 24-kDa, labeled strongly with ^{125}I. Further regression analysis of multiple immunoblots has revealed these four bands at 17 to 18, 20 to 21, 24 to 25, and 27 to 30 kDa (Figure 4).[108]

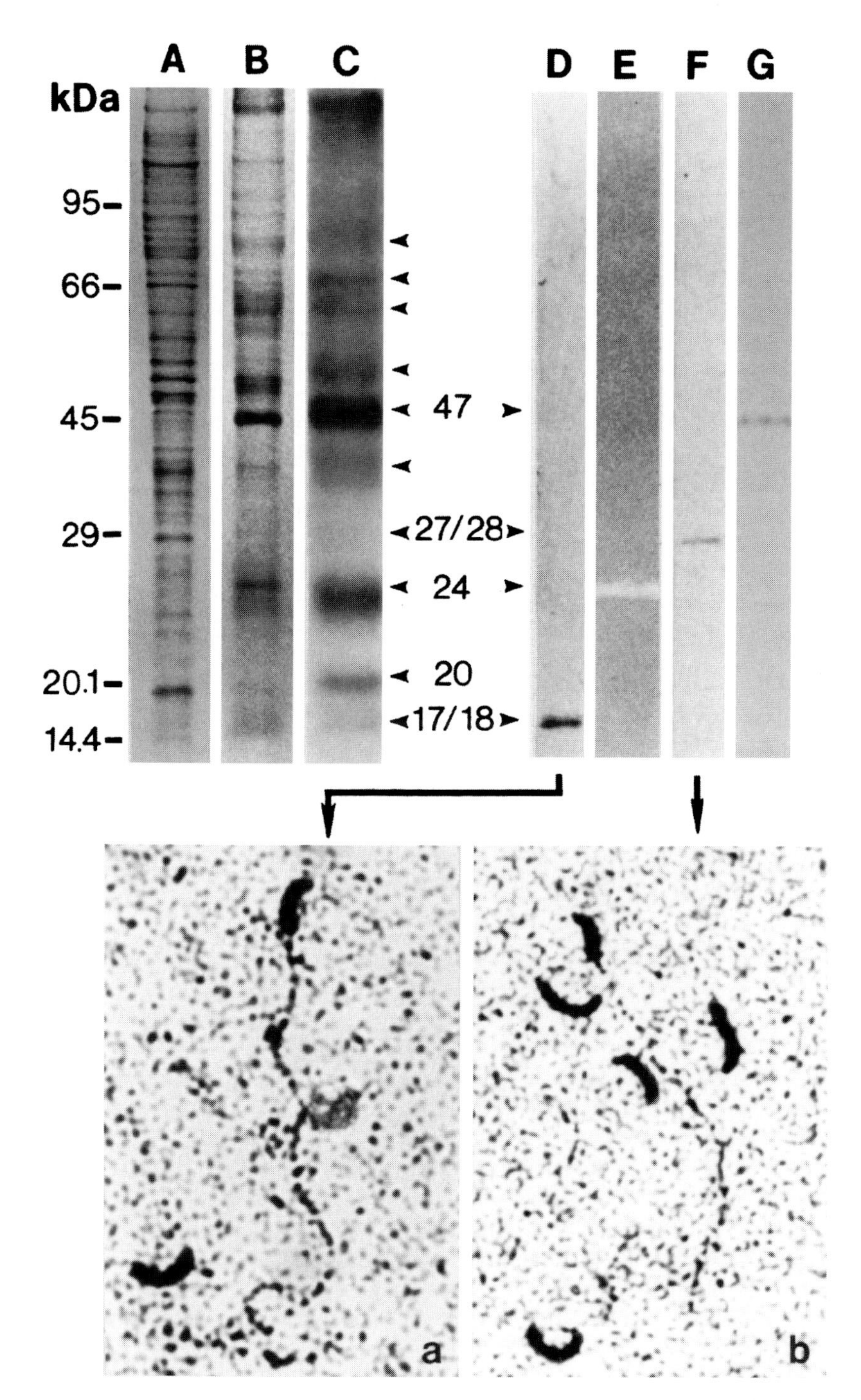

Figure 4 SDS-PAGE of *C. parvum* proteins. Lane A, silver stain of total sporozoite proteins; Lane B, silver stain of detergent extracted membrane fraction from sporozoites; Lane C, ^{125}I-labeled sporozoite surface proteins; Lane D, immunoblot of sporozoite membrane fraction probed with Mab 2B3 to the 17- to 18-kDa surface molecule; Lane E, PAGE-zymogram of protease activity associated with the 24-kDa surface protein; Lane F, immunoblot of sporozoite membrane fraction probed with Mab 3E3 to the 27- to 28-kDa surface molecule; Lane G, immunoblot of sporozoite membrane fraction probed with Mab 5B3.47 to the 47-kDa surface molecule (portions adapted from Reference 108, with permission). Insets (bottom), immunogold silver enhancement of sloughed sporozoites surface antigens: (a) Mab 2Bs against the 17- to 18-kDa surface antigen; (b) Mab 3E3 against the 27- to 28-kDa surface antigen (portions adapted from Reference 112, with permission).

To resolve differences in reported sizes of sporozoite surface antigens <30 kDa, several mAbs developed at Kansas State University were forwarded to the Centers for Disease Control in Atlanta. The mAb 4B10,[82] which recognized a 15-kDa sporozoite surface molecule, bound the same band on western blots as mAb C6C1.[113,114] The mAb C6C1 also recognized a second band at 47 kDa,[54] unlike mAb 4B10. The mAb 3E3,[82] which recognized a 25-kDa sporozoite surface antigen, recognized the same region of the gel as mAbs C6B6 and C3B4.[98,102,113-115] Extensive immunoblot analysis by both laboratories has provided more precise molecular sizes of these two sporozoite surface antigens.[81,116] We now concur that mAbs 4B10 and C6C1 bind a diffuse band with an upper molecular size at 17 to 18 kDa; 3E3, C6B6, and C3B4 identify a band at 27 to 28 kDa.

Based on the above data, and the available literature, the following can be deduced. A major surface molecule on both sporozoites and merozoites,[106] with a very diffuse appearance on western blots, has reported ranges from 11 to 23 kDa.[67,79,82,101,102,105,106,110,112,113,117-130] One study reported the size to be 1 to 2 kDa larger under nonreducing conditions.[127] This is one of the most immunogenic *C. parvum* antigens and, in most cases, the lowest band identified on immunoblots (occasionally, a 9- to 10-kDa antigen may also appear following 15 to 18% SDS-PAGE[128,131]). This 17- to 18-kDa molecule labels weakly with ^{125}I,[108] possesses both glucose/glucosamine and galactose/galactosamine residues, has a pI of 4.25,[105] and requires detergent extraction for proper removal from the sporozoite membrane.[81] Individual mAbs sometimes recognize two or three bands, and it is likely that this difference is due to the number of the sugars retained following processing for SDS-PAGE. This doublet/triplet effect has been clearly demonstrated.[105,114,117,119,126-128] The lower band of the doublet has a pI of about 4.15,[105] and total chemical deglycosylation yields a protein 10 to 12 kDa in size.[81] This 17- to 18-KDa sporozoite surface molecule may be a substrate adhesion molecule shed during movement (Figure 4),[112,132] similar to other apicomplexan sporozoites.[133-135] Immunoelectron microscopy and colloidal gold labeling have localized this antigen on the sporozoite surface and also within the sporozoite and in electron-lucent vesicles thought to be residual amylopectin granules in the oocyst residuum.[90]

A second major surface antigen on *C. parvum* sporozoites and merozoites is approximately 27 to 28 kDa. It has also been reported to be 20,[102,115] 23,[117,119,128,130,136,137] 25,[82,112] 27,[114] and 28 to 30[108,126] kDa. This molecule is also highly immunogenic and may also be a substrate adhesion molecule (Figure 4).[112,135] Although the protein usually appears as a single band following SDS-PAGE, it has also been reported as a doublet.[114] It labels poorly with ^{125}I[108] and weakly with biotin, and has a pI of 4.5 to 4.7.[81] During excystation, considerable amounts attach to the inner surface of the oocyst wall or are shed into the medium.[81] Although the function of this molecule appears similar to the 17- to 18-kDa surface molecule, it differs in size, has considerably less glycosylation, and is water soluble.[81]

Far less immunogenic than the former two molecules are surface molecules at 20 and 24 to 25 kDa. They label readily with ^{125}I.[108] This has caused investigators to confuse the ^{125}I- or biotin-labeled bands with those recognized by immune serum. Although the function of the 20-kDa molecule is unknown, the 24- to 25-kDa band is a surface-active, metallo-dependent cysteine proteinase (Figure 4).[108] It optimally hydrolyses azocasein, bovine serum albumin, casein, and gelatin at pH 6.5 to 7.0. Hydrolysis can be inhibited by ethylenediaminotetraacetic acid (EDTA), iodoacetic acid (IAA), trans-epoxysuccinyl-L-leucylamido (4-guanido)butane (E-64), and phosphoramidon. Both serine and aspartate protease inhibitors fail to inhibit enzyme activity. An IFA assay using polyclonal antiserum to the proteinase revealed the molecule on the surface of sporozoites. Putative functions include dissolving the oocyst wall suture proteins during excystation, cleaving antibody or complement along the mucosa, or enhancing the ability of sporozoites to pass through mucins or the host cell glycocalyx.

Because some sporozoite surface molecules are heavily glycosylated, sporozoites can be affected by various sugars and lectins. In a rabbit erythrocyte hemagglutination assay in which sporozoites formed rosettes around erythrocytes,[138,139] agglutination was inhibited by several glycoproteins, especially bovine submaxillary mucin (BSM) and the blood group antigen-related P_1 glycoprotein. Agglutination was also inhibited by galactose and *N*-acetylgalactosamine, but dissacharides Gal(β1-3)GalNAc and Gal(α1-4)Gal were the most potent inhibitors.[138] Both BSM and fetuin significantly reduced sporozoite attachment to MDCK cells *in vitro*.[138] *N*-acetyl-D-glucosamine, chitobiose, chitotriose, and wheat germ agglutinin (WGA) acted similarly. WGA significantly reduced penetration of host cells by sporozoites, whereas ConA enhanced penetration and increased the numbers of developmental stages *in vitro*.[140]

B. MICRONEME AND RHOPTRY PROTEINS AND GLYCOPROTEINS

Many microneme and rhoptry components are highly glycosylated and extremely immunogenic, and some even aggregate. The mAbs to these molecules often recognize multiple, periodate-sensitive epitopes associated with the secretory molecules, as well as unrelated glycoproteins that possess similar sugar moieties. On western blots, mAbs binding these common epitopes detect multiple bands, possibly because sugars may be lost during preparation, resulting in subpopulations of a molecule with different migration rates. With immunoelectron microscopy, these cross-reacting epitopes result in labeling patterns associated with the secretory organelles, additional granules, and membranes.

Microneme/rhoptry antigens have sometimes been found associated with the outer surface of the sporozoite or merozoite. Some microneme/rhoptry glycoproteins may become associated with host-cell membranes during extrusion and may possess adhesive domains or sugars similar to those found associated with the parasite surface molecules. Thus, some antibodies may simply bind epitopes that are shared between microneme/rhoptry and parasite surface molecules. Also, manipulation of Coccidian sporozoites and merozoites has often resulted in extrusion of microneme/rhoptry contents. Because these organelles contain high amounts of membrane material, including phosphatidylcholine,[7,8,60,61] lipid-associated glycoproteins probably have become associated with the outer membrane nonspecifically. Finally, some microneme/rhoptry glycoproteins may be secreted, translocated along the parasite surface and sloughed during motility, as hypothesized for the malarial circumsporozoite proteins.[29,134,141-143] GP900 (below) may represent such a molecule.

The mAbs, IFA, immunoelectron microscopy, and western blotting were used to characterize the first reported microneme antigens of *C. parvum*.[99] One antibody, termed TOU, bound periodate sensitive epitopes 40 to 210 kDa on western blots. Another mAb, HAD, bound a ladder of periodate insensitive epitopes ranging from 63 to 210 kDa. Because many bands on western blots were recognized by both antibodies, they appeared to be different residues of the same glycoprotein. This same ladder effect of microneme antigens was seen with mAbs 3E6 and 6G10[82] and AT7G10,[144] which recognized periodate sensitive epitopes.

A second type of microneme glycoprotein was reported by Bonnin et al.[100] The mAbs ABD, BAX, and SPO all recognized a 100-kDa, periodate-sensitive antigen on western blot. Immunoelectron microscopy revealed the epitope to be associated with the sporozoite and merozoite micronemes, a heterologous population of granules in macrogametes, and the parasitophorous vacuole membrane. The mAb AT7G10 also appeared to recognize carbohydrates associated with these same structures, although this mAb also recognized a ladder of bands that extended higher than 500 kDa.[144]

A recurring pattern of some microneme proteins appears to be areas of similarity or homology with adhesive domains. The WSPCSVTCG cell adhesion domain common to thrombospondin and several other adhesive proteins were used to design degenerate PCR primers.[145] This sequence is also a component of region II of the microneme/rhoptry-derived circumsporozoite protein of *Plasmodium* spp.[135,141-143,146,147] involved in binding to hepatocytes[148] and microneme protein Et100 of *E. tenella*.[149,150] A microneme protein analogue of Et100 found in *Eimeria maxima* also has an area of similarity to this adhesive motif.[151] Fragments of two *C. parvum* gene sequences identified from a genomic library contained five tandem repeats of approximately 60 amino acids each. Each repeat carried one copy of the thrombospondin-like motifs.

GP900, a heavily glycosylated, detergent-soluble rhoptry glycoprotein associated with both sporozoites and merozoites of *C. parvum*,[103,107,152-154] may be identical with P25-200.[70,155] Upon extrusion, this glycoprotein appeared to incorporate into the plasma membrane, especially in the anterior ½ of sporozoites, and was shed like the malarial circumsporozoite protein. Many mAbs have been made to this glycoprotein. Most tend to bind multiple bands on immunoblots ranging from 25 to greater than 900 kDa. Treatment of GP900 with *N*-glycosidase F revealed a core protein <190 kDa.[103,152] GP900 appeared to be encoded by a single copy gene, and sequence analysis suggested 10 potential *N*-glycosylation sites as well as putative *O*-glycosylation motifs.[103,152] The major consituent was threonine, comprising 39% of the sequence.[152] The mAbs against this molecule inhibited invasion of host cells by sporozoites *in vitro*.[152]

C. ENDOGENOUS PROTEINASES

Membrane-associated aminopeptidase (AP) activity has been detected in sporozoites of *C. parvum*.[156] The topographic localization of AP activity was examined using arginyl-6-amino-2-styrylquinoline, an insoluble fluorescent substrate that precipitates at the site of reactivity. Reactivity, best discerned in the apical complex of sporozoites during excystation, may be associated with rhoptries. When AP activity was examined using detergent-extracted proteins and amino acids bound to the fluorescent substrate 7-amino-4-trifluoromethyl coumarin, highest activity was found associated with arginine, followed by alanine, phenylalanine, glycine, and leucine. No activity was found associated with intact oocysts.

Several proteases found associated with the sporulated oocysts have not been fully characterized. Six neutral proteinases, found in gelatin substrate gels, included serine and cysteine proteinases. Portions of the genes encoding one or more of the cysteine proteinase(s) were sequenced.[157] At least one of the serine proteinases is water soluble.[81]

VII. ENTEROTOXINS

For several decades, investigators have attempted to correlate the diarrhea associated with enteric Coccidial infections with an enterotoxin.[158-160] Although the fluids within the oocyst are toxic to host cells,[46,47] the term enterotoxin implies production of a molecule whose role is to induce intestinal fluid secretion by specifically increasing intracellular concentrations of either Ca^{++}, cAMP, or cGMP. Cell-free medium containing extracts of "cultured" *C. parvum* oocysts caused an increase in short-circuit current when added to the serosal surface of rabbit ileal mucosa.[161] Injection of ligated rabbit ileal loops with the extracts resulted in marked distention after 24 hours when compared to controls. A more recent study used centrifuged and filtered extracts of diarrheic stool samples from *C. parvum*-infected and noninfected calves.[162] When applied to the serosal surface of human jejunum in Ussing chambers, the addition of 2.5 mg protein of fecal supernatant from infected animals resulted in an increase in short-circuit current with no effects on tissue conductance. This effect was both chloride and calcium dependent and was heat sensitive.

Although both studies suggest *C. parvum* affected electrical activity in the gut, they do not conclusively establish that *C. parvum* possesses enterotoxin. Sporozoites and merozoites slough a variety of surface antigens during motility, and others are secreted by rhoptries, micronemes, and dense granules during penetration of host cells. It has not been determined if any of these could cause the results seen in traditional enterotoxin assays. Many of the coccidia and host cells are also lysed during an infection, resulting in liberation of proteases, phospholipases, and other molecules that may account for an increase in short circuit effect.

VIII. ISOLATE VARIATION

Although many antigens associated with *Cryptosporidium* spp. appear to be conserved, several studies have reported differences among species and among isolates within a species. One study clearly delineated differences in isolates of *C. parvum* sporozoites by visually comparing two-dimensional electrophoretic patterns of five isolates.[163] A narrow subset of spots ranging from 95 to 120 kDa, with pIs of 5.4 to 6.4, varied among some isolates. In particular, differences in a 105-kDa protein served to distinguish three of the isolates from one another. A 40-kDa spot was unique to an isolate from Mexico. Other techniques used to differentiate isolates or species include isozyme typing,[18,164] restriction length fragment polymorphism,[165,166] DNA sequencing,[167-169] immunoblotting with mAbs or polyclonal antiserum,[84,85,87,125,170-172] and ^{125}I surface labeling of oocysts.[79,86,88,89] Variations in excystation rates of oocysts from different isolates, as well as incorporation of vital dyes, has also been reported.[173,174] One study suggested that isolates missing 15.5- and 33-kDa bands on western blots were those incapable of infecting suckling mice.[169] Such studies search for appropriate methods of typing isolates and species, while suggesting that detectible differences in isolates may eventually correlate with infectivity *in vivo*.

ACKNOWLEDGMENTS

The authors are indebted to the following individuals for allowing the use of their previously unpublished data: M. J. Arrowood, N. V. Khramtsov, D. M. Moss, M. V. Nesterenko, B. Oppert, D. S. Roos, and K. M. Woods. M. Tilley is supported by a Hitchings-Elion Fellowship from the Burroughs-Wellcome Fund. Much of the research on surface molecules was sponsored by NIH grant AI30881 to SJU.

FURTHER READING

Since this chapter was written, several additional papers concerning *Cryptosporidium* biochemistry have been published, including a decription of the glycolytic enzymes phosphofructokinase and pyruvate kinase activities in oocyst extracts (Denton et al., *Mol. Biochem. Parasitol.,* 76, 23, 1996); cloning of the gene encoding acetyl-CoA synthetase (Khramtsov et al., *J. Parasitol.,* 82, 423, 1996); immunolocalization of actin and myosin in developmental stages (Yu and Chai, *Kor. J. Parasitol.*, 33, 155, 1995); development of monoclonal antibodies which recognize dense granule-like structures in zoites and gamonts (Bonnin et al., *J. Euk. Microbiol.*, 42, 395, 1995); further evidence for a sporozoite surface proteinase (Foney et al., *J. Parasitol.*, 82, 496, 1996); and additional data concerning genetic heterogeneity among isolates of *C. parvum* (Bonnin et al., *FEMS Microbiol. Lett.*, 137, 207, 1996 and Carraway et al., *Appl. Environ. Micro.*, 62, 712, 1996).

REFERENCES

1. Barta, J. R., Jenkins, M. C., and Danforth, H. D., Evolutionary relationships between avian *Eimeria* species among other apicomplexan protozoa: monophyly of the apicomplexa is supported, *Mol. Biol. Evol.*, 8, 345, 1991.
2. Johnson, A. M., Fielke, R., Lumb, R., and Baverstock, P. R., Phylogenetic relationships of *Cryptosporidium* determined by ribosomal RNA sequence comparison, *Int. J. Parasitol.*, 20, 141, 1990.
3. Fayer, R., Speer, C. A., and Dubey, J. P., General biology of *Cryptosporidium*, in *Cryptosporidiosis of Man and Animals*, Dubey, J. P., Speer, C. A., and Fayer, R., Eds., CRC Press, Boca Raton, FL, 1990, 1.
4. Moriya, K., Ultrastructural observations on oocysts, sporozoites and oocyst residuum of *Cryptosporidium muris* (strain RN 66), *J. Osaka City Med. Ctr.*, 38, 177, 1989.
5. Uni, S., Iseki, M., Maekawa, T., Moriya, K., and Takada, S., Ultrastructure of *Cryptosporidium muris* (strain RM 66) parasitizing the murine stomach, *Parasitol. Res.*, 74, 123, 1987
6. Current, W. L. and Reese, N. C., A comparison of endogenous development of three isolates of *Cryptosporidium* in suckling mice, *J. Protozool.*, 33, 98, 1986.
7. Mitschler, R. R., Lipid Biology of *Cryptosporidium parvum* and *Eimeria nieschulzi* (Apicomplexa), Ph.D. dissertation, Kansas State University, 1994, 1.
8. Mitschler, R. R., Welti, R., and Upton, S. J., A comparative study of lipid compositions of *Cryptosporidium parvum* (Apicomplexa) and Madin-Darby bovine kidney cells, *J. Euk. Microbiol.*, 41, 8, 1994.
9. Fairfield, A. and Laughon, B., Research priorities and resources for new therapies against cryptosporidiosis, in *Proc. Int. Workshop: Microsporidiosis and Cryptosporidiosis in Immunodeficient Patients*, Czech Academy of Sciences and Centers for Disease Control, 28 Sept.–1 Oct., 1993, Ceské Budejovice, Czech Republic, *Folia Parasitol.*, 40 (Abstr.), 250, 1993.
10. Goozé, L., Kim, K., Peterson, C., Gut, J., and Nelson, R. G., Amplification of a *Cryptosporidium parvum* gene fragment encoding thymidylate synthetase, *J. Protozool.*, 38, 56s, 1991.
11. Upton, S. J., Tilley, M., and Brillhart, D. B., Effects of select medium supplements on *in vitro* development of *Cryptosporidium parvum* in HCT-8 cells, *J. Clin. Microbiol.*, 33, 371, 1995.
12. Brasseur, P., Leméteil, D., and Ballet, J. J., Anti-cryptosporidial drug activity screened with an immunosuppressed rat model, *J. Protozool.*, 38, 230s, 1991.
13. Leméteil, D., Roussel, F., Favennec, L., Ballet, J. J., and Brasseur, P., Assessment of candidate anticryptosporidial agents in an immunosuppressed rat model, *J. Infect. Dis.*, 167, 766, 1993.
14. Rehg, J. E., Anticryptosporidial activity associated with specific sulfonamides in immunosuppressed rats, *J. Parasitol.*, 77, 238, 1991.
15. Rehg, J. E., Hancock, M. L., and Woodmansee, D. B., Anticryptosporidial activity of sulfadimethoxine, *Antimicrob. Agents Chemother.*, 32, 1907, 1988.
16. Upton, S. J., *In vitro* cultivation, in Cryptosporidium *and Cryptosporidiosis*, Fayer, R., Ed., CRC Press, Boca Raton, FL, 1997.
17. Woods, K. M., Nesterenko, M. V., and Upton, S. J., Development of a microtitre ELISA to quantify development of *Cryptosporidium parvum in vitro*, *FEMS Microbiol. Lett.*, 128, 89, 1995.
18. Ogunkolade, B. W., Robinson, H. A., McDonald, V., Webster, K. and Evans, D. A., Isozyme variation within the genus *Cryptosporidium*, *Parasitol. Res.*, 79, 385, 1993.
19. Cioni, M., Pinzauti, G., and Vanni, P., Comparative biochemistry of the glyoxylate cycle, *Comp. Biochem. Physiol.*, 70B, 1, 1981.
20. Vanni, P., Giachetti, E., Pinzauti, G., and McFadden, B. A., Comparative structure, function and regulation of isocitrate lyase, an important assimilatory enzyme, *Comp. Biochem. Physiol.*, 95B, 431, 1990.
21. Wang, C. C., Biochemistry and physiology of Coccidia, in *The Biology of the Coccidia*, Long, P. L., Ed., University Park Press, Baltimore, 1982, 167.
22. Schmatz, D. M., The mannitol cycle — a new metabolic pathway in the coccidia, *Parasitol. Today*, 5, 205, 1989.
23. Schmatz, D. M., Arison, B. H., Dashkevicz, M. P., Liesch, J. M., and Turner, M. J., Identification and possible role of D-mannitol and 2-*O*-methyl-chiro-inositol (quebrachitol) in *Eimeria tenella*, *Mol. Biochem. Parasitol.*, 29, 29, 1988.
24. Schmatz, D. M., Baginsky, W. F., and Turner, M. J., Evidence for and characterization of a mannitol cycle in *Eimeria tenella*, *Mol. Biochem. Parasitol.*, 32, 263, 1989.
25. Schwartzman, J. D. and Pfefferkorn, E. R., *Toxoplasma gondii*: purine synthesis and salvage in mutant host cells and parasites, *Exp. Parasitol.*, 53, 77, 1982.

26. Morgan, K. and Canning, E. U., Incorporation of ^{3}H-thymidine and ^{3}H-adenosine by *Eimeria tenella* grown in chick embryos *J. Parasitol.*, 60, 364, 1974.
27. Perrotto, J., Keister, D. B., and Gelderman, A. H., Incorporation of precursors into *Toxoplasma* DNA, *J. Protozool.*, 18, 470, 1971.
28. Pfefferkorn, E. R. and Pfefferkorn, L. C., *Toxoplasma gondii*: specific labeling of nucleic acids of intracellular parasites in Lesch-Nyhan cells, *Exp. Parasitol.*, 41, 95, 1977.
29. Pfefferkorn, E. R. and Pfefferkorn, L. C., The biochemical basis for resistance to adenine arabinoside in a mutant of *Toxoplasma gondii, J. Parasitol.*, 64, 486, 1978.
30. Wang, C. C. and Simashkevich, P. M., A comparative study of the biological activities of arprinocid and arprinocid-1-N-oxide, *Mol. Biochem. Parasitol.*, 1, 335, 1980.
31. Wang, C. C. and Simashkevich, P. M., Purine metabolism in a protozoan parasite *Eimeria tenella*, *Proc. Natl. Acad. Sci. U.S.A.,* 78, 6618, 1981.
32. Wang, C. C., Simashkevich, P. M., and Stotish, R. L., Mode of antiCoccidial action of arprinocid, *Biochem. Pharmacol.*, 28, 2241, 1979.
33. Wang, C. C., Tolman, R. L., Simashkevich, P. M., and Stotish, R. L., Arprinocid, an inhibitor of hypoxanthine-guanine transport, *Biochem. Pharmacol.*, 28, 2249, 1979.
34. Kim, C. W., Chemotherapeutic effect of arprinocid in experimental cryptosporidiosis, *J. Parasitol.*, 73, 663, 1987.
35. Rehg, J. E. and Hancock, M. L., Effectiveness of arprinocid in the reduction of cryptosporidial activity in immunosuppressed rats, *Am. J. Vet. Res.*, 51, 1668, 1990.
36. Iltzsch, M. H., Pyrimidine salvage pathways of *Toxoplasma gondii*, *J. Euk. Microbiol.*, 40, 24, 1993.
37. Ouellette, C. A., Strout, R. G., and McDougald, L. R., Incorporation of radioactive pyrimidine nucleosides into DNA and RNA of *Eimeria tenella* (coccidia) cultures *in vitro*, *J. Protozool.*, 20, 150, 1973.
38. Ouellette, C. A., Stout, R. G., and McDougald, L. R., Thymidylic acid synthesis in *Eimeria tenella* (coccidia) cultures *in vitro*, *J. Protozool.*, 21, 398, 1974.
39. Schwartzman, J. D. and Pfefferkorn, E. R., Pyrimidine synthesis by intracellular *Toxoplasma gondii*, *J. Parasitol.*, 67, 150, 1981.
40. Donald, R. G. K. and Roos, D. S., Insertional mutagenesis and marker rescue in a protozoan parasite: cloning of the uracil phosphoribosyl transferase locus from *Toxoplasma gondii*, *Proc. Natl. Acad. Sci. U.S.A.,* 5749, 1995.
41. Pfefferkorn, E. R., *Toxoplasma gondii*: the enzymatic defect of a mutant resistant to 5-fluorodeoxyuridine, *Exp. Parasitol.*, 44, 26, 1978.
42. Pfefferkorn, E. R. and Pfefferkorn, L. C., Arabinosyl nucleosides inhibit *Toxoplasma gondii* and allow the selection of resistant mutants, *J. Parasitol.*, 62, 993, 1976.
43. Pfefferkorn, E. R. and Pfefferkorn, L. C., *Toxoplasma gondii*: characterization of a mutant resistant to 5-fluorodeoxyuridine, *Exp. Parasitol.*, 42, 44, 1977.
44. Pfefferkorn, E. R. and Pfefferkorn, L. C., Specific labeling of intracellular *Toxoplasma gondii* with uracil, *J. Protozool.*, 24, 449, 1977.
45. Schmatz, D. M., Crane, M. S. J., and Murray, P. K., *Eimeria tenella*: parasite-specific incorporation of ^{3}H-uracil as a quantitative measure of intracellular development, *J. Protozool.*, 33, 109, 1986.
46. Eggleston, M. T., Development of a ^{3}H-Uracil Incorporation Assay to Monitor *In Vitro* Development of *Cryptosporidium parvum*, M.S. thesis, Kansas State University, 1993, 1.
47. Eggleston, M. T., Tilley, M., and Upton, S. J., Enhanced development of *Cryptosporidium parvum in vitro* by removal of oocyst toxins from infected cell monolayers, *J. Helminthol. Soc. Wash.*, 61, 122, 1994.
48. Upton, S. J., Tilley, M., Mitschler, R. R., and Oppert, B. S., Incorporation of exogenous uracil by *Cryptosporidium parvum in vitro*, *J. Clin. Microbiol.*, 29, 1062, 1991.
49. Hughes, H. P. A., Bolk, R. J., Gerhardt, S. A., and Speer, C. A., Susceptibility of *Eimeria bovis* and *Toxoplasma gondii* to oxygen intermediates and a new mathematical model for parasite killing, *J. Parasitol.*, 75, 489, 1989.
50. Michalski, W. P., and Prowse, S. J., Superoxide dismutases in *Eimeria tenella*, *Mol. Biochem. Parasitol.*, 47, 189, 1991.
51. Wack, M. F., Greenfield, R. A., Mosier, D. A., and Kuhls, T. L., Superoxide dismutase activity in *Cryptosporidium parvum*, in *Proc. 47th Ann. Meet., Soc. Protozool.*, Cleveland State University, 24–29 June, 1994, Society of Protozoologists, Allen Press, Lawrence, KS, Abstract C37, 75, 1994.
52. Oppert, B., unpublished data.
53. Khramtsov, N. V., Tilley, M., Blunt, D. S., Montelone, B. A., and Upton, S. J., Cloning and analysis of a *Cryptosporidium parvum* gene encoding a protein with homology to cytoplasmic form hsp70, *J. Euk. Microbiol.*, 42, 416, 1995.
54. Mead, J. R., Lloyd, R. M., You, X., Tucker-Burden, C., Arrowood, M. J., and Schinazi, R. F., Isolation and partial characterization of *Cryptosporidium* and oocyst wall recombinant proteins, *J. Euk. Microbiol.*, 41, 51s, 1994.
55. Jones, D. E., Tu, T. D., Mathur, S., Sweeney, R. W., and Clark, D. P., Molecular cloning and characterization of a *Cryptosporidium parvum* elongation factor-2 gene, *Mol. Biochem. Parasitol.*, 71, 143, 1995.
56. Christopher, L. J. and Dykstra, C. C., Identification of a type II topoisomerase gene from *Cryptosporidium parvum*, *J. Euk. Microbiol.*, 41, 28s, 1994.
57. Dykstra, C. C., Blagburn, B. L., and Tidwell, R. R., Construction of genomic libraries of *Cryptosporidium parvum* and identification of antigen-encoding genes, *J. Protozool.*, 38, 76s, 1991.

58. Zhu, G., and Keithly, J., Isolation of a P-ATPase from *Cryptosporidium parvum*, in *Proc. 6th Ann. Meet., East Coast Soc. Protozool.*, 21–23 June, 1995, Schenectady, NY.
59. Woods, K. M., Zhu, G., Yarlett, N., Upton, S. J., and Keithly, J., Polyamine metabolism: a rational drug target for *Cryptosporidium parvum*, in *Proc. 6th Ann. Meet., East Coast Soc. Protozool.*, 21–23 June, 1995, Schenectady, NY.
60. Foussard, F., Gallois, Y., Girault, A., and Menez, J. F., Lipids and fatty acids of tachyzoites and purified pellicles of *Toxoplasma gondii*, *Parasitol. Res.*, 77, 475, 1991.
61. Foussard, F., Leriche, M. A., and Dubremetz, J. F., Characterization of the lipid content of *Toxoplasma gondii* rhoptries, *Parasitology*, 102, 367, 1991.
62. Gallois, Y., Foussard, F., Girault, A., Hodbert, J., Tricaud, A., Mauras, G., and Motta, C., Membrane fluidity of *Toxoplasma gondii*: a fluorescence polarization study, *Biol. Cell*, 62, 11, 1988.
63. Stotish, R. L., Wang, C. C., and Meyenhofer, M., Structure and composition of the oocyst wall of *Eimeria tenella*, *J. Parasitol.*, 64, 1074, 1978.
64. Weppelman, R. M., Vandenheuvel, W. J. A., and Wang, C. C., Mass spectrometric analysis of the fatty acids and nonsaponifiable lipids of *Eimeria tenella* oocysts, *Lipids*, 11, 209, 1976.
65. Cullis, P. R. and Hope, M. J., Physical properties and functional roles of lipids in membranes, in *Biochemistry of Lipids and Membranes*, Vance, D. E. and Vance, J. E., Eds., Benjamin/Cummings, Menlo Park, CA, 1985, 25.
66. Joiner, K. A., Rhoptry lipids and parasitophorous vacuole formation: a slippery issue, *Parasitol. Today*, 7, 226, 1991.
67. Riggs, M. W., McGuire, T. C., Mason, P. H., and Perryman, L. E., Neutralization-sensitive epitopes are exposed on the surface of infectious *Cryptosporidium parvum* sporozoites, *J. Immunol.*, 143, 1340, 1989.
68. Bjorneby, J. M., Riggs, M. W., and Perryman, L. E., *Cryptosporidium parvum* merozoites share neutralization-sensitive epitopes with sporozoites, *Immunology*, 145, 298, 1990.
69. Perryman, L. E., Riggs, M. W., Mason, P. H., and Fayer, R., Kinetics of *Cryptosporidium parvum* sporozoite neutralization by monoclonal antibodies, immune bovine serum, and immune bovine colostrum, *Infect. Immun.*, 58, 257, 1990.
70. Riggs, M. W., Cama, V. A., Leary, H. L., Jr., and Sterling, C. R., Bovine antibody against *Cryptosporidium parvum* elicits a circumsporozoite precipitate-like reaction and has immunotherapeutic effect against persistent cryptosporidiosis in SCID mice, *Infect. Immun.*, 62, 1927, 1994.
71. Uhl, E. W., O'Connor, R. M., Perryman, L. E., and Riggs, M. W., Neutralization-sensitive epitopes are conserved among geographically diverse isolates of *Cryptosporidium parvum*, *Infect. Immun.*, 60, 1703, 1992.
72. Kim, K., Goozé, L., Petersen, C., Gut, J., and Nelsen, R. G., Isolation, sequence and molecular karyotype analysis of the actin gene of *Cryptosporidium parvum*, *Mol. Biochem. Parasitol.*, 50, 105, 1992.
73. Nelsen, R. G., Kim, K., Goozé, L., Petersen, C., and Gut, J. Identication and isolation of *Cryptosporidium parvum* genes encoding microtubule and microfilament proteins, *J. Protozool.*, 38, 52s, 1991.
74. Edlind, T., Visvesvara, G., Li, G., and Katiyari, S., *Cryptosporidium* and microsporidial β-tubulin sequences: predictions of benzimidazole sensitivity and phylogeny, *J. Euk. Microbiol.*, 41, 38s, 1993.
75. Wiest, P. M., Dong, K. L., Johnson, J. H., Tzipori, S., Boeklheide, K., and Flanigan, T. P., Effect of colchicine on microtubules in *Cryptosporidium parvum*, *J. Euk. Microbiol.*, 41, 66s, 1994.
76. Fayer, R. and Fetterner, R., Activity of benzimidazoles against cryptosporidiosis in neonatal BALB/c mice, *J. Parasitol.*, 81, 794, 1995.
77. Wiest, P. M., Johnson, J. H., and Flanigan, T. P., Microtubule inhibitors block *Cryptosporidium parvum* infection of a human enterocyte cell line, *Infect. Immun.*, 61, 4888, 1993.
78. Jolley, W. R., Burton, S. D., Nyberg, P. A., and Jenson, J. B., Formation of sulfhydryl groups in the walls of *Eimeria stiedae* and *E. tenella* oocysts subjected to *in vitro* excystation, *J. Parasitol.*, 62, 199, 1976.
79. Lumb, R., Lanser, J. A., and O'Donoghue, P. J., Electrophoretic and immunoblot analysis of *Cryptosporidium*, *Immunol. Cell Biol.*, 66, 369, 1988.
80. Lally, N. C., Baird, G. D., McQuay, S. J., Wright, F., and Oliver, J. J., A 2359 base pair fragment from *Cryptosporidium parvum* encoding a repetitive oocyst protein, *Mol. Biochem. Parasitol.*, 56, 69, 1992.
81. Nesterenko, M. V., unpublished data.
82. Tilley, M., Eggleston, M. T., and Upton, S. J., Multiple oral infection with *Cryptosporidium parvum* as a means of immunization for production of monoclonal antibodies, *FEMS Microbiol. Lett.*, 113, 235, 1993.
83. McDonald, V., McCrossan, M. V., and Petry, F., Localization of parasite antigens in *Cryptosporidium parvum*-infected epithelial cells using monoclonal antibodies, *Parasitology*, 110, 259, 1995.
84. McDonald, V., Deer, R. M. A., Nina, J. M. S., Wright, S., Chiodini, P. L., and McAdam, K. P. W. J., Characteristics and specificity of hybridoma antibodies against oocyst antigens of *Cryptosporidium parvum* from man, *Parasite Immunol.*, 13, 251, 1991.
85. Nina, J. M. S., McDonald, V., Dyson, D. A., Catchpole, J., Uni, S., Iseki, M., Chiodini, P. L., and McAdam, K. P. W. J., Analysis of oocyst wall and sporozoite antigens from three *Cryptosporidium* species, *Infect. Immun.*, 60, 1509, 1992.
86. Tilley, M., Upton, S. J., Blagburn, B. L., and Anderson, B. C., Identification of outer oocyst wall proteins of three *Cryptosporidium* (Apicomplexa: Cryptosporidiidae) species by ^{125}I surface labeling, *Infect. Immun.*, 58, 252, 1990.

87. Nichols, G. L., McLauchlin, J., and Samuel, D., A technique for typing *Cryptosporidium* isolates, *J. Protozool.*, 38, 237s, 1991.
88. Tilley, M., Upton, S. J., and Freed, P. S., A comparative study on the biology of *Cryptosporidium serpentis* and *Cryptosporidium parvum* (Apicomplexa: Cryptosporidiidae), *J. Zoo Wildlf. Med.*, 21, 463, 1990.
89. Tilley, M., Upton, S. J., and Chrisp, C. E., A comparative study on the biology of *Cryptosporidium parvum* and *Cryptosporidium* sp. from guinea pigs, *Can. J. Microbiol.*, 37, 949, 1991.
90. Tilley, M., unpublished data.
91. Ranucci, L., Muller, H. M., La Rosa, G., Gomez Morales, M. A., Groppo, G. P., Pozio, E., and Crisanti, A., Cloning of *Cryptosporidium* oocyst surface antigens, *Parassitologia*, 34, 104, 1992.
92. Ranucci, L., Muller, H. M., La Rosa, G., Reckmann, I., Gomez Morales, M. A., Spano, F., Pozio, E., and Crisanti, A., Characterization and immunolocalization of a *Cryptosporidium* protein containing repeated amino acid motifs, *Infect. Immun.*, 61, 2347, 1993.
93. Khramtsov, N. V., personnal communication.
94. Ranucci, L., Spano, F., Muller, H. M., Catteruccia, F., Saccheo, S., and Crisanti, A., *Cryptosporidium* oocyst wall protein (COWP): a role in the pathogenesis of parasite infection, *Parassitologia*, 36, 122s, 1994.
95. Llovo, J., Lopez, A., Fabreges, J., and Muñoz, A., Interaction of lectins with *Cryptosporidium parvum*, *J. Infect. Dis.*, 167, 1477, 1993.
96. Gut, J. and Nelson, R. G., Lectin reactivity of *Cryptosporidium parvum* oocysts and sporozoites, in *Proc. 47th Ann. Meet., Soc. Protozool.*, Cleveland State University, 24–29 June, 1994, Society of Protozoologists, Allen Press, Lawrence, KS, Abstract C31, 74, 1994.
97. Bonnin, A., Dubremetz, J. F., and Camerlynck, P., Characterization and immunolocalization of an oocyst wall antigen of *Cryptosporidium parvum* (protozoa: Apicomplexa), *Parasitology*, 103, 171, 1991.
98. Arrowood, M. J., Mead, J. R., Mahrt, J. L., and Sterling, C. R., Effects of immune colostrum and orally administered antisporozoite monoclonal antibodies on the outcome of *Cryptosporidium parvum* infections in neonatal mice, *Infect. Immun.*, 57, 2283, 1989.
99. Bonnin, A., Dubremetz, J. F., and Camerlynck, P., Characterization of microneme antigens of *Cryptosporidium parvum* (protozoa: Apicomplexa), *Infect. Immun.*, 59, 1703, 1991.
100. Bonnin, A., Dubremetz, J. F., and Camerlynck, P., A new antigen of *Cryptosporidium parvum* microneme possessing epitopes cross-reactive with macrogamete granules, *Parasitol. Res.*, 79, 8, 1993.
101. Lumb, R., Smith, P. S., Davies, R., O'Donoghue, P. J., and Lanser, J. A., Localization of a 23,000 MW antigen of *Cryptosporidium* by immunoelectron microscopy, *Immunol. Cell Biol.*, 67, 267, 1989.
102. Mead, J. R., Arrowood, M. J., and Sterling, C. R., Antigens of *Cryptosporidium* sporozoites recognized by immune sera of infected animals and humans, *J. Parasitol.*, 74, 135, 1988.
103. Petersen, C., Gut, J., Doyle, P. S., Crabb, J. H., Nelsen, R.G., and Leech, J. H., Characterization of a >900,000-M_r *Cryptosporidium parvum* sporozoite glycoprotein recognized by protective hyperimmune bovine colostral immunoglobulin, *Infect. Immun.*, 60, 5132, 1992.
104. Regan, S., Cama, V., and Sterling, C. R., *Cryptosporidium* merozoite isolation and purification using differential centrifugation techniques, *J. Protozool.*, 38, 202s, 1991.
105. Tilley, M. and Upton, S. J., Sporozoites and merozoites of *Cryptosporidium parvum* share a common epitope recognized by a monoclonal antibody and two-dimensional electrophoresis, *J. Protozool.*, 38, 48s, 1991.
106. Tilley, M., Upton, S. J., Fayer, R., Barta, J. R., Chrisp, C. E., Freed, P. S., Blagburn, B. L., Anderson, B. C., and Barnard, S. M., Identification of a 15-kilodalton surface glycoprotein on sporozoites of *Cryptosporidium parvum*, *Infect. Immun.*, 59, 1002, 1991.
107. Doyle, P. S., Crabb, J., and Petersen, C., Anti-*Cryptosporidium parvum* antibodies inhibit infectivity *in vitro* and *in vivo*, *Infect. Immun.*, 61, 4079, 1993.
108. Nesterenko, M. V., Tilley, M., and Upton, S. J., A metallo-dependent cysteine proteinase of *Cryptosporidium parvum* associated with the surface of sporozoites, *Microbios*, 83, 77, 1995.
109. Petersen, C., Cellular biology of *Cryptosporidium parvum*, *Parasitol. Today*, 9, 87, 1993.
110. Tilley, M. and Upton, S. J., Electrophoretic characterization of *Cryptosporidium parvum* (KSU-1 isolate) (Apicomplexa: Cryptosporidiidae), *Can. J. Zool.*, 68, 1513, 1990.
111. Wisher, M. H. and Rose, M. E., *Eimeria tenella* sporozoites: the method of excystation affects the surface membrane proteins, *Parasitology*, 95, 479, 1987.
112. Tilley, M. and Upton, S. J., Both Cp15 and Cp25 are left as trails behind gliding sporozoites of *Cryptosporidium parvum*, *FEMS Microbiol. Lett.*, 120, 275, 1994.
113. Arrowood, M. J., *Cryptosporidium*: oocyst production and hybridoma generation for examining colostrum and monoclonal antibody roles in cryptosporidial infections, Ph.D. dissertation, University of Arizona, 1988, 1.
114. Moss, D. M., Bennett, S. N., Arrowood, M. J., Hurd, M. R., Lammie, P. J., Wahlquist, S. P., and Addiss, D. G., Kinetic and isotypic analysis of specific immunoglobulins from crew members with cryptosporidiosis on a U.S. Coast Guard cutter, *J. Euk. Microbiol.*, 41, 52s, 1994.
115. Mead, J. R., *Cryptosporidium*: isolate variation and humoral responses to sporozoite antigens, Ph.D. dissertation, University of Arizona, 1988, 1.

116. Arrowood, M. J. and Moss, D. M., unpublished data.
117. Arnault, I., Répérant, J. M., and Naciri, M., Humoral antibody response and oocyst shedding after experimental infection of histocompatible newborn and weaned piglets with *Cryptosporidium parvum*, *Vet. Res.*, 25, 371, 1994.
118. El-Shewy, K., Kilani, R. T., Hegazi, M. M., Makhlouf, L. M., and Wenman, W. M., Identification of low-molecular-mass coproantigens of *Cryptosporidium parvum*, *J. Infect. Dis.*, 169, 460, 1994.
119. Hill, B. D., Blewett, D. A., Dawson, A. M., and Wright, S., Analysis of the kinetics, isotype, and specificity of serum and coproantibody in lambs infected with *Cryptosporidium parvum*, *Res. Vet. Sci.*, 48, 76, 1990.
120. Luft, B. J., Payne, D., Woodmansee, D., and Kim, C. W., Characterization of the *Cryptosporidium* antigens from sporulated oocysts of *Cryptosporidium parvum*, *Infect. Immun.*, 55, 2436, 1987.
121. Mancassola, R., Reperant, J. M., Naciri, M., and Chartier, C., Chemoprophylaxis of *Cryptosporidium parvum* infection with paromomycin in kids and immunological study, *Antimicrob. Agents Chemother.*, 39, 75, 1995.
122. Mosier, D. A., Kuhls, T. L., Simons, K. R., and Oberst, R. D., Bovine humoral immune response to *Cryptosporidium parvum*, *J. Clin. Microbiol.*, 30, 3277, 1992.
123. Moss, D. M. and Lammie, P. J., Proliferative responsiveness of lymphocytes from *Cryptosporidium parvum*-exposed mice to two separate antigen fractions from oocysts, *Am. J. Trop. Med. Hyg.*, 49, 393, 1993.
124. Naciri, M., Mancassola, R., Répérant, J. M., and Yvoré, P., Analysis of humoral immune response in chickens after inoculation with *Cryptosporidium baileyi*, or *Cryptosporidium parvum*, *Avian. Dis.*, 38, 832, 1994.
125. Ortega-Mora, L. M., Troncoso, J. M., Rojo-Vázquez, F. A., and Gómez-Bautista, M., Cross-reactivity of polyclonal serum antibodies generated against *Cryptosporidium parvum* oocysts, *Infect. Immun.*, 60, 3442, 1992.
126. Ortega-Mora, L. M., Troncoso, J. M., Rojo-Vázquez, F. A., and Gómez-Bautista, M., Identification of *Cryptosporidium parvum* oocyst/sporozoite antigens recognized by infected and hyperimmune lambs, *Vet. Parasitol.*, 53, 159, 1994.
127. Répérant, J. M., Naciri, M., Chardes, T., and Bout, D. T., Immunological characterization of a 17-kDa antigen from *Cryptosporidium parvum* recognized early by mucosal antibodies, *FEMS Microbiol. Lett.*, 99, 7, 1992.
128. Répérant, J. M., Naciri, M., Iochmann, S., Tilley, M., and Bout, D. T., Major antigens of *Cryptosporidium parvum* recognised by serum antibodies from different infected animals species and man, *Vet. Parasitol.*, 55, 1, 1994.
129. Tilley, M., An analysis of *Cryptosporidium parvum* sporoozites using lectins, ^{125}I surface labeling and monoclonal antibodies, Ph.D. dissertation, Kansas State University, 1991, 1.
130. Whitmire, W. M. and Harp, J. A., Characterization of bovine cellular and serum antibody responses during infection by *Cryptosporidium parvum*, *Infect. Immun.*, 59, 990, 1991.
131. Tilley, M., Fayer, R., Guidrey, A., Upton, S. J., and Blagburn, B. L., *Cryptosporidium parvum* (Apicomplexa: Cryptosporidiidae) oocyst and sporozoite antigens recognized by bovine colostral antibodies, *Infect. Immun.*, 58, 2966, 1990.
132. Gut, J. and Nelson, R. G., *Cryptosporidium parvum* sporozoites deposit trails of 11A5 antigen during gliding locomotion and shed 11A5 antigen during invasion of MDCK cells *in vitro*, *J. Euk. Microbiol.*, 41, 42s, 1994.
133. Entzeroth, R., Zgrzebski, G., and Dubremetz, J. F., Secretion of trails during gliding motility of *Eimeria nieschulzi* (Apicomplexa: Coccidia) sporozoites visualized by a monoclonal antibody and immuno-gold-silver enhancement, *Parasitol. Res.*, 76, 174, 1989.
134. Stewart, M. J. and Vanderberg, J. P., Malaria sporozoites leave behind trails of circumsporozoite protein during gliding motility, *J. Protozool.*, 35, 389, 1988.
135. Stewart, M. J. and Vanderberg, J. P., Malaria sporozoites release circumsporozoite protein from their apical end and translocate it along their surface, *J. Protozool.*, 38, 411, 1991.
136. Arrowood, M. J., Sterling, C. R., and Healy, M. C., Immunofluorescent microscopical visualization of trails left by gliding *Cryptosporidium parvum* sporozoites, *J. Parasitol.*, 77, 315, 1991.
137. Ungar, B. L. P. and Nash, T. E., Quantification of specific antibody response to *Cryptosporidium* antigens by laser densitometry, *Infect. Immun.*, 53, 124, 1986.
138. Joe, A., Hamer, D. H., Kelley, M. A., Pereira, M. E. A., Keusch, G. T., Tzipori, S., and Ward, H. D., Role of gal/galNAc-specific sporozoite surface lectin in *Cryptosporidium parvum*-host cell interaction, *J. Euk. Microbiol.*, 41, 44s, 1994.
139. Thea, D. M., Pereira, M. E. A., Kotler D., Sterling, C. R., and Keusch, G. T., Identification and partial purification of a lectin on the surface of the sporozoite of *Cryptosporidium parvum*, *J. Parasitol.*, 78, 886, 1992.
140. Kuhls, T. L., Mosier, D. A., and Crawford, D. L., Effects of carbohydrates and lectins on cryptosporidial sporozoite penetration of cultured cell monolayers, *J. Protozool.*, 38, 74s, 1991
141. Cochrane, A. H., Uni, S., Maracic, M., DiGiovanni, L., Aikawa, M., and Nussenzweig, R. S., A circumsporozoite-like protein is present in micronemes of mature blood stages of malaria parasites, *Exp. Parasitol.*, 69, 351, 1989.
142. Fine, E., Aikawa, M., Cochrane, A. H., and Nussenzweig, R. S., Immunoelectron microscopic observations on *Plasmodium knowlesi* sporozoites: localization of protective antigen and its precursors, *Am. J. Trop. Med. Hyg.*, 33, 220, 1984.
143. Nagasawa, H., Aikawa, M., Procell, P. M., Campbell, G. H., Collins, W. E., and Campbell, C. C., *Plasmodium malariae*: distribution of circumsporozoite protein in midgut oocysts and salivary gland sporozoites, *Exp. Parasitol.*, 66, 27, 1988.

144. Robert, B., Antoine, H., Dreze, F., Coppe, P., and Collard, A., Characterization of a high molecular weight antigen of *Cryptosporidium* micronemes possessing epitopes that are cross-reactive with all parasitic life-cycle stages, *Vet. Res.*, 25, 384, 1994.
145. Spano, F., Rannucci, L., Naitza, S., and Crisanti, A., Cloning and characterisation of a putative micronemal adhesion protein of *Cryptosporidium parvum*, *Parassitologia*, 36, 139s, 1994.
146. Posthuma, G., Meis, J., Verhave, J., Gigengack, S., Hollingdale, M., Ponnudurai, T., and Geuze, H., Immunogold determination of *Plasmodium falciparum* circumsporozoite protein in *Anopholes stephensi* salivary gland cells, *Eur. J. Cell Biol.*, 49, 66, 1989.
147. Robson, K. J. H., Hall, J. R. S., Jennings, M. W., Harris, T. J. R., Marsh, K., Newbold, C. I., Tate, V. E., and Weatherall, D. J., A highly conserved amino-acid sequence in thrombospondin, properdin and in proteins from sporozoites and blood stages of a human malaria parasite, *Nature*, 335, 79, 1988.
148. Cerami, C., Frevert, U., Sinnis, P., Takacs, B., Clavijo, P., Santos, M. J., and Nussenzweig, V., The basolateral domain of the hepatocyte plasma membrane bears receptors for the circumsporozoite protein of *Plasmodium falciparum* sporozoites, *Cell*, 70, 1021, 1992.
149. Clarke, L. E., Tomley, F. M., Wisher, M. H., Foulds, I. J., and Boursnell, M. E. G., Regions of an *Eimeria tenella* antigen contain sequences which are conserved in circumsporozoite proteins from *Plasmodium* spp. and which are related to the thrombospondin gene family, *Mol. Biochem. Parasitol.*, 41, 269, 1990.
150. Tomley, F. M., Clarke, L. E., Kawazoe, U., Dijkema, R., and Kok, J. J., Sequence of the gene encoding an immunodominant microneme protein of *Eimeria tenella*, *Mol. Biochem. Parasitol.*, 49, 277, 1991.
151. Pasamontes, L., Hug, D., Hümbelin, M., and Weber, G., Sequence of a major *Eimeria maxima* antigen homologous to the *Eimeria tenella* microneme protein Et100, *Mol. Biochem. Parasitol.*, 57, 171, 1993.
152. Barnes, D. A., Doyle, P., Lewis, S., and Petersen, C., Surface antigens as targets for protective antibodies in cryptosporidiosis, in *Proc. 47th Ann. Meet., Soc. Protozool.*, Cleveland State University, 24–29 June, 1994, Society of Protozoologists, Abstract C45, Allen Press, Lawrence, KS, 1994, 76.
153. Petersen, C., Gut, J., Nelsen, R. G., and Leech, J. H., Characterization of a *Cryptosporidium parvum* sporozoite glycoprotein, *J. Protozool.*, 38, 20s, 1991.
154. Petersen, C., Gut, J., Leech, J. H., and Nelsen, R. G., Identification and initial characterization of five *Cryptosporidium parvum* sporozoite antigen genes, *Infect. Immun.*, 60, 2343, 1992.
155. Riggs, M. W., Stone, A. L., Arrowood, M. J., Langer, R. C., Bahle, J., and Yount, P., Molecular targets for passive immunotherapy of cryptosporidiosis, in *Proc. 47th Ann. Meet., Soc. Protozool.*, Cleveland State University, 24–29 June, 1994, Society of Protozoologists, Abstract C46, Allen Press, Lawrence, KS, 1994, 76.
156. Okhuysen, P. C., DuPont, H. L., Sterling, C. R., and Chappell, C. L., Arginine aminopeptidase, an integral membrane protein of the *Cryptosporidium parvum* sporozoite, *Infect. Immun.*, 62, 4667, 1994.
157. Peterson, C., Doyle, P., and Azouaou, N., Surface active neutral proteinases of *Cryptosporidium parvum* sporozoites, *Proc. Mol. Parasitol. Meet.*, 18–22 Sept., 1994, Abstract 157A, Woods Hole, MA, 1994.
158. Burns, W. C., The lethal effect of *Eimeria tenella* extracts on rabbits, *J. Parasitol.*, 45, 38, 1959.
159. Rickimaru, M. T., Galysh, F. T., and Shumard, R. F., Some pharmacological aspects of a toxic substance from oocysts of the coccidium *Eimeria tenella*, *J. Parasitol.*, 47, 407, 1961.
160. Sharma, N. N., and Foster, J. W., Toxic substance in various consituents of *Eimeria tenella* oocysts, *Am. J. Vet. Res.*, 25, 211, 1964.
161. Garza, D. H., Fedorak, R. N., and Soave, R., Enterotoxin-like activiy in cultured cryptosporidia: role in diarrhea, *Gastroenterology*, 90 (Abstr.), 1424, 1986.
162. Guarino, A., Canani, R. B., Pozio, E., Terracciano, L., Albano, F., and Mazzeo, M., Enterotoxic effect of stool supernatant of *Cryptosporidium*-infected calves on human jejunum, *Gastroenterology*, 106, 28, 1994.
163. Mead, J. R., Humphries R. C., Sammons, D. W., and Sterling, C. R., Identification of isolate-specific sporozoite proteins of *Cryptosporidium parvum* by two-dimensional gel electrophoresis, *Infect. Immun.*, 58, 2071, 1990.
164. Awad-El-Kariem, F. M., Robinson, H. A., Dyson, D. A., Evans, D., Wright, S., Fox, M. T., and McDonald, V., Differentiation between human and animal strains of *Cryptosporidium parvum* using isoenzyme typing, *Parasitology*, 110, 129, 1995.
165. Awad-el-Kariem, F. M., Warhurst, D. C., and McDonald, V., Detection and species identification of *Cryptosporidium* oocysts using a system based on PCR and endonuclease restriction, *Parasitology*, 109, 19, 1994.
166. Ortega, Y. R., Sheehy, R. R., Cama, V. A., Oishi, K. K., and Sterling, C. R., Restriction fragement length polymorphism analysis of *Cryptosporidium parvum* isolates of bovine and human origin, *J. Protozool.*, 38, 40s, 1991.
167. Cai, J., Collins, M. D., McDonald, V., and Thompson, D. E., PCR cloning and nucleotide sequence determination of the 18S rRNA genes and internal transcribed spacer 1 of the protozoan parasites *Cryptosporidium parvum* and *Cryptosporidium muris*, *Biochim. Biophys. Acta*, 1131, 317, 1992.
168. Carraway, M., Widmer, G., and Tzipori, S., Genetic markers differentiate *C. parvum* isolates, *J. Euk. Microbiol.*, 41, 26s, 1994.
169. Taghi-Kilani, R. T. and Wenman, W. M., Geographical variation in 18S rRNA gene sequence of *Cryptosporidium parvum*, *Int. J. Parasitol.*, 24, 303, 1994.

170. Ditrich, O., Kopácek, P., and Zucerová, Z., Antigenic characterization of human isolates of cryptosporidia, *Folia Parasitol.*, 40, 301, 1993.
171. Nina, J., McDonald, V., Deer, R., Dyson, D. A., Chiodini, L., and McAdam, K., Antigenic differences between oocyst isolates of *Cryptosporidium parvum*, *Trans. Roy. Soc. Trop. Med. Hyg.*, 85, 315, 1991.
172. Nina, J. M. S., McDonald, V., Deer, R. M. A., Wright, S. E., Dyson, D. A., Chiodini, P. L., and McAdam, K. P. W. J., Comparative study of the antigenic composition of oocyst isolates of *Cryptosporidium parvum* from different hosts, *Parasite Immunol.*, 14, 227, 1992.
173. Campbell, A. T., Robertson, L. J., and Smith, H. V., Viability of *Cryptosporidium parvum* oocysts: correlation of *in vitro* excystation with inclusion or exclusion of fluorogenic vital dyes, *Appl. Environ. Microbiol.*, 58, 3488, 1992.
174. Robertson, L. J., Campbell, A. T., and Smith, H. V., *In vitro* excystation of *Cryptosporidium parvum*, *Parasitology*, 106, 13, 1993.

Chapter 8

In Vitro Cultivation

Steve J. Upton

CONTENTS

I. INTRODUCTION

Prior to 1948, suitable methods of isolating and maintaining host cells were lacking, which prevented successful cultivation of most species of coccidia in cell culture.[1-3] The only coccidian cultivated up to that time was *Toxoplasma gondii*.[4-9] By the late 1940s, however, the National Institutes of Health had defined some of the conditions necessary to successfully propagate many host cells *in vitro*.[3] Eventually, these techniques were applied to other coccidia, which led to the successful cultivation of asexual stages of parasites such as *Besnoitia besnoiti*, *Eimeria tenella*, and *Sarcocystis* spp.[10-12]

0-8493-7695-5/97/$0.00+$.50

In 1983, sporozoites of *Cryptosporidium parvum* were reported to develop into mature meronts after inoculation atop human rectal tumor cells.[13,14] Shortly thereafter, development from sporozoite to oocyst in human embryonic, porcine kidney, and primary chicken kidney cells was achieved.[15] Both studies utilized a relatively crude sporozoite inoculum derived from sucrose gradient purified oocysts and relied on antibiotics in cultures to prevent overgrowth of microbial contaminants. For over a decade thereafter, major advances in *C. parvum* cultivation systems were sparse and consisted of new oocyst and sporozoite concentration techniques. Although these techniques helped reduce microbial contamination in cultures, they did little to enhance parasite development. The majority of papers were descriptive and often reported the degree of parasite development in various cell lines only (Table 1). Most recently, studies have defined and optimized some of the conditions necessary for development of *C. parvum in vitro*. These studies, and new techniques for better manipulating oocysts and sporozoites, have allowed for more rapid evolution of *C. parvum in vitro* systems. Utilizing these systems, antibiotics are now optional, a high volume of pharmaceuticals can be screened, and receptor/ligand interactions can be studied.

Table 1 Chronology of Some *In Vitro* and *In Ovo* Research on *Cryptosporidium* spp.[a]

Date	Host Cell(s)[b]	Brief Comments[c]	Ref.
1983	Avian embryos	First report of cultivation in avian embryos	198
	HRT	First reports of *in vitro* cultivation; asexual development only	13,14
1984	HFL, PCK, PK	First report of complete development *in vitro*	15
1986	—	Description of Percoll® gradients for oocysts	48
	Avian embryos, BHK	Complete development *in vitro*; 45 successful serial passages *in ovo*	174
	HF	Complete development *in vitro*.	186
	Avian embryos	Description of *in ovo* cultivation of *C. baileyi*.	197
	IE	Enterotoxin activity postulated	200
1987	—	Description of Percoll® gradient for oocysts and sporozoites	26
	—	Description of anion exchange chromatography for purification of sporozoites; first use of vital staining for sporozoites	35
	—	Description of CsCl gradients for oocyst purification; description of Percoll® gradients for sporozoite purification	47
1988	MDCK	Attachment of sporozoites to host cells studied ultrastructurally	184
	Avian embryos	Successful cultivation of *C. bailey* in eggs of various host species; attempts to cultivate *C. baileyi in vitro*	199
1989	Caco-2	Complete development *in vitro*	201
1990	—	Summary of technqiues	16
	L929	First *in vitro* screen with pharmaceuticals	34
	LGA	Asexual development *in vitro*	202
1991	HT-29,HT29.74	Comparative study between differentiated and undifferentiated HT-29 cells	48
	—	Merozoite isolation and concentration techniques reported	63
	MDBK, MDCK	Multiple cell lines examined; supplemented medium used; importance of insulin and rabbit bile emphasized	90
	407	Effects of carbohydrates and lectins on penetration of host cells examined	141
	MDCK	Efficacy of MDCK cells as model	171
	MDBK	Immunofluorescent localization of a 15-kDa sporozoite surface molecule to stages *in vitro*	180
	HT29.74	Penetration and asexual development examined ultrastructurally	181
	Caco-2	Ultrastructural observations of sexual stages *in vitro*	182
	HT29.74	TEM of development	183
	MDBK	Use of reduce oxygen atmosphere; incorporation of exogenous ^{3}H-uracil	185
	Avian embryos, HeLa, McCoy, MDCK, Vero	Comparative development in various host cells and *in ovo*	203
	HT-29,HT29.74	Model for pharmaceutical screening	204
	407	Effects of carbohydrates and lectins on penetration of host cells examined	205
	Caco-2	Description of Caco-2 cells as model	206

Table 1 (continued) Chronology of Some *In Vitro* and In Ovo Research on *Cryptosporidium* spp.[a]

Date	Host Cell(s)[b]	Brief Comments[c]	Ref.
1992	HT29.74	Use of paromomycin as an inhibitor *in vitro*	172
	Mϕ	Use of mouse macrophages for *in vitro* assays	173
1993	MDBK	Development of ^{3}H-uracil assay	42
	RL95-2	Description of development *in vitro*; use of low parasite to host cell ratios for best results	124
	HT29.74	Effects of microtubule inhibitors on invasion	150
	MDCK	Inhibition of *in vitro* development by colostrum	170
	HeLa, McCoy, MDCK, Vero	Comparative development between four cell lines	175
1994	MDBK	Rapid methods to process oocysts and infect monolayers	49
	T84	Use of polarized cells to demonstrate injury of cell monolayer by oocysts	71
	MDCK	Effects of pH and various divalent cations on attachment of sporozoites to host cells	75
	BALB/3T3, BT-549, Caco-2, HCT-8, Hs-700T, HT-29, HT-1080, LS-174T, MDBK, MDCK, RL95-2	Comparative development in 11 cell lines	89
	MDBK, HCT-8	Comparative development in seven defined atmospheres	108
	MDBK	Use of ^{3}H-uracil incorporation; demonstration that washing monolayers after infections increases parasite development	115
	MDCK	Alcian blue-giemsa staining for infected monolayers	176
	Caco-2	Use of immunofluorescent screening for *in vitro* efficacy of pharmaceuticals	177
	Caco-2, MDBK	Induction of apical defect and increase in transmonolayer permeability in response to infection	178
	MDCK	Deposition of 15-kDa surface antigen from sporozoites during gliding atop and penetration of cells	179
	MDCK	Effects of lectins and glycoproteins on penetration of host cells by sporozoites	207
	Caco-2	Use of sporozoites purified by anion exchange chromatography for infections	208
	—	Use of fetal rabbit intestinal xenografts as a model	209
1995	HCT-8	Effects of 25 medium supplements on development *in vitro*; formulation of special parasite growth medium	136
	HCT-8	Development of 96-well ELISA and application for pharmaceutical screening	123,195
	Caco-2	Pharmaceutical screening assay	169
	Caco-2	Effects of lectins on sporozoite infectivity; microtiter ELISA binding assay	210
	BFTE, MDCK	Complete development *in vitro*; use of *in vitro*-derived oocysts for murine infections	211

[a] Through July 1995 only.

[b] Abbreviations: (407) human embryonic intestine; (BALB/3T3) BALB/c mouse embryo; (BFTE) bovine fallopian tube epithelial primary cells; (BHK) baby hamster kidney; (BT-549) human breast infiltrating ductal carcinoma; (Caco-2) human colonic adenocarcinoma; (HCT-8) human ileocaecal adenocarcinoma; (HeLa) human cervical carcinoma; (HF) human foreskin; (HFL) human fetal lung; (HRT) human rectal tumor; (Hs-700T) human pelvic adenocarcinoma; (HT-29) human colonic adenocarcinoma; (HT29.74) human colonic adenocarcinoma, galactose adapted; (HT-1080) human fibrosarcoma; (IE) intestinal epithelium; (L929) murine fibroblasts; (LGA) rat intestinal carcinoma; (LS-174T) human colonic adenocarcinoma; (McCoy) human synovial knee joint (some lines are contaminated with murine fibroblasts); (MDBK) Madin-Darby bovine kidney; (MDCK) Madin-Darby canine kidney; (Mϕ) mouse peritoneal macrophages; (T84) human colonic carcinoma; (PCK) primary chicken kidney; (PK) porcine kidney; (RL-95-2) human endometrial carcinoma; (Vero) African green monkey kidney.

[c] Unless stated otherwise, *C. parvum* was utilized in the assays.

II. OOCYST AND SPOROZOITE PURIFICATION TECHNIQUES

A reliable source of highly viable, clean parasites is essential to cultivate *Cryptosporidium* spp. successfully *in vitro*. A variety of techniques has been used to clean *C. parvum* oocysts and sporozoites for biochemical and *in vitro* use, including diethyl ether, sucrose, Percoll®, and cesium chloride gradients.[16] The overall quality of an oocyst and sporozoite preparation is also strongly influenced by the type of *in vivo* system employed, proper animal husbandry, fecal collection and storage procedures, and quality of reagents. Therefore, an overview of some *in vivo* procedures is provided to improve the quality of oocysts and sporozoites destined for *in vitro* use.

A. SOURCE OF PARASITES

Since the factors necessary to induce significant oocyst production *in vitro* are still unknown, it remains necessary to generate *C. parvum* oocysts *in vivo*. A variety of animal models can be used, including suckling mice, immunosuppressed adult rodents, goats, lambs, and calves. Some investigators surface sterilize oocysts with sodium hypochlorite or peracetic acid prior to animal inoculations;[17-24] however, these procedures are not necessary except for specific viability or immunological assays or if human infections are desired. Although many *in vivo* systems are described in Chapter 9, suggestions are presented below that may help increase the numbers and quality of *in vivo*-derived oocysts.

1. Rodents

Adequate numbers of oocysts for most *in vitro* studies can be generated in rodents. However, adult rodents also generate large amounts of particulate matter in the feces. Generally, immunosuppressed adult rodents experimentally infected with *C. parvum* are allowed to defecate through the wire mesh bottoms of cages. Collecting pans beneath the cages collect feces, which are kept moist either with 2 to 3% w/v aqueous potassium dichromate ($K_2Cr_2O_7$) solution or by wet paper toweling. As long as feces are kept moist, oocysts remain viable even when collected and processed after several days. Oocysts are easily separated from the feces by the following procedure.

a. Removal of Fecal Debris by Sieving

1. Homogenize feces in tap water or 1 to 3% $K_2Cr_2O_7$ solution with a blender or other homogenizer. Homogenization will not damage oocysts or the sporozoites within.
2. Strain the homogenate through a series of sieves, e.g., 60, 100, 150, and 200 mesh, to a final pore exclusion size <100 µm. Residual particulates in the screens should be rehomogenized and resieved several times to liberate as many oocysts as possible.
3. Pellet oocysts by centrifugation, 800 to 1500× *g*, 15 to 20 minutes. Decant or aspirate the supernate. Add fresh 2 to 3% aqueous $K_2Cr_2O_7$ or other storage medium; store at 4 to 8°C until needed.

It is particularly important to employ centrifugation speeds that will adequately pellet the oocysts, and investigators are urged to properly correlate rotor type with rpm and *g* force. Centrifugation speeds of <800× *g* for 5 minutes have been used to pellet oocysts by some investigators, but many oocysts remain in the supernatant at these speeds.

Adult immunosuppressed mice or rats are the most common rodent models used to generate *C. parvum* oocysts; however, suckling mice or rats are more useful because of less fecal debris. The strain of rodent will determine the optimal age for oocyst production, but 6 to 9 days of age is best for the inexpensive ICR outbred mouse. To obtain oocysts from ICR suckling mice, the following technique can be employed.

b. Obtaining Oocysts from Suckling Mice

1. Orally inoculate each 6- to 9-day old mouse with 1.0×10^4 oocysts in 5 to 10 µl of water or phosphate buffered saline (PBS) by placing the oocyst suspension in the back of the mouth with a micropipette or by gastric intubation using a 24-gauge gavage needle fitted onto a syringe. Inoculations with more oocysts will not result in proportionately higher numbers of oocysts. Return mice to the lactating dam. Discard very small mice (runts); they produce few oocysts.
2. Remove the intestine at 7 to 8 days post-inoculation (DPI), place it in 1 to 2 ml PBS and homogenize using a 15- to 30-ml, motor-driven, Teflon-coated tissue grinder or a hand-held, ground-glass tissue grinder. Even though regions in and near the stomach contain foci of developing parasites, the presence of undigested whey makes harvesting from these sites undesirable.

3. Dilute the homogenate in 2 to 3% $K_2Cr_2O_7$ solution, and quantitate the oocysts directly using a hemacytometer. Approximately 2 to 8 × 10^6 oocysts are generated per mouse using this technique. Mice with final body weights <4.5 g usually produce few oocysts.

2. Ruminants

If large numbers of oocysts are required, ruminants are needed. Calves 2 to 9 days of age inoculated with 1 × 10^6 to 2.5 × 10^7 oocysts can generate tens of billions of oocysts 5 to 8 DPI. Goats of the same age should be given only 1 to 2 × 10^6 oocysts each. Goats or lambs can produce several billion oocysts each. Feed all ruminants a commercially available milk replacer until 4 DPI, then switch to an oral electrolyte solution 4 times per day.[16] About 8 to 10 hours after feeding electrolyte solution, feces becomes free of lipid, allowing for greatly enhanced oocyst recovery. Beginning 5 DPI, collect all feces daily for 4 to 5 days and immediately process as described above for rodents. Generally, 10 to 12 l of processed fecal suspension per animal yield a final concentration of 2 to 12 × 10^6 oocysts per ml.

It is particularly important that oocysts be sieved, centrifuged, and placed in storage medium at 4°C as soon as possible as sludge. Long-term exposure to nitrogenous wastes may reduce parasite viability;[25] however, further processing of the fecal suspension beyond that described above should be avoided until oocysts are needed. Oocysts purified by sucrose or CsCl gradients have a somewhat shorter shelf-life than those stored as above.

B. REMOVAL OF FECAL LIPID USING DIETHYL ETHER

Prevention of lipid in the feces improves oocyst yield. Some investigators add detergents such as 0.5% (v/v) Nonidet P-40, 1% (v/v) sodium dodecyl sulfate, or 1% Tween-20 to aid in oocyst recovery.[26-31] However, if significant amounts of lipid are present, it may be necessary to use the chloroform/diethyl ether technique.[16] Ether works well, but chloroform is not particularly useful as oocysts become distributed in both the aqueous and chloroform phases. The following diethyl ether technique has been modified from Current[16] by Nesterenko:[32]

1. Vigorously mix 35 ml oocyst/fecal suspension in PBS with 15 ml diethyl ether. If oocysts are in $K_2Cr_2O_7$ solution, wash at least once in PBS before adding ether.
2. Centrifuge at 1000× *g*, 10 minutes.
3. Discard the supernatant. Although three bands should be evident, only the pellet contains oocysts.
4. Wash oocysts 2 times by centrifugation in PBS to remove residual ether.

It is likely that the ether affects at least some outer-wall antigens, and it should be avoided anytime outer oocyst wall antigens are needed. Sporozoites excysted from ether-treated oocysts are motile, appear viable, and establish successful infections both *in vivo* and *in vitro*.[33-36]

C. FLOTATION TECHNIQUES

Solutions of sodium chloride, sucrose, and zinc sulfate with specific gravities of about 1.20 have traditionally been used to concentrate Coccidian oocysts. Ryley et al.[37] provided an overview of many of these procedures, and although they concluded that salt flotation had slight advantages over sucrose, variations of the latter have become standard in most Coccidian biology. Although none of the flotation techniques purify oocysts completely from fecal and bacterial debris, all help enrich oocyst numbers.

The original Sheather's sugar solution technique consisted of sugar mixed with water to a specific gravity of 1.20.[38] Modifications consist of different specific gravities (Figure 1) and the addition of preservatives such as formaldehyde or phenol to prevent growth of bacteria and fungi. Whether laboratory-grade sucrose and distilled water or commercial sugar and tap water are used, satisfactory solutions will result, provided the specific gravity of the solution is above 1.15 and excess lipid is not present. Saturated salt solutions are also suitable to purify *Cryptosporidium* oocysts.[39-41]

Current[16] described a modified sucrose flotation technique (see below) for obtaining enriched populations of oocysts (below). This technique has an advantage over the commonly used coverslip flotation method, because oocysts are cleaner and larger numbers can be recovered.

1. Sucrose Flotation Technique[16]

1. Mix 30-ml sucrose solution (specific gravity 1.15 to 1.30) in a 50-ml centrifuge tube with 5 ml fecal suspension containing oocysts. Some investigators also add 0.2% Tween-20 to the fecal suspension.
2. Centrifuge at 800 to 1500× *g*, 15 to 20 minutes.

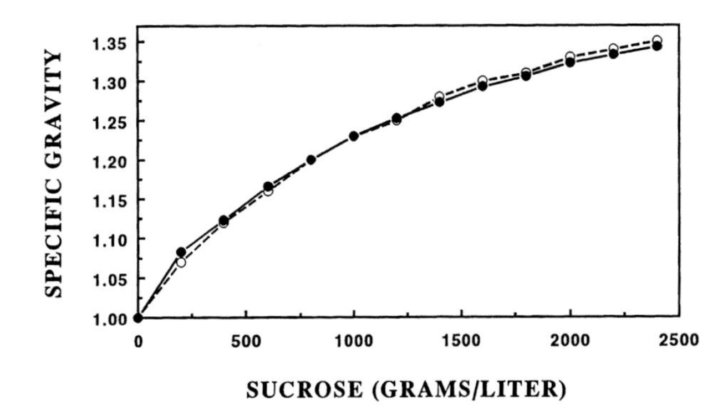

Figure 1 Specific gravities of various aqueous sucrose concentrations. Open circles (○) represent predicted specific gravities based on textbook calculations. Solid circles (●) represent specific gravities obtained using various amounts of laboratory grade sucrose heated in water and cooled. (Data courtesy of R. Najarian, Kansas State University.)

3. Carefully overlay 5 ml water.
4. Gently swirl the water layer with the tip of a pipet. A wisp of oocysts will become suspended in the water layer above the sucrose solution.
5. Aspirate the water layer to remove oocysts; save aspirate in a clean centrifuge tube.
6. Remix the sucrose solution and repeat steps 2 to 5 two times. Add oocysts to those obtained from the first centrifugation.
7. Pellet newly recovered oocysts 800 to 2000× *g*, 10 to 15 minutes.

Several sucrose density gradient procedures have also been published that yield similar or even greater numbers of enriched oocysts.[26,42,43] In most cases, oocyst purity is slightly inferior to the flotation technique described above. Because billions of oocysts are sometimes needed for some studies, however, an even more efficient and large scale sucrose purification procedure may be needed. The following sucrose solution has an unusually high specific gravity. At room temperature (approximately 22°C), the specific gravity is 1.33 (5-lb bag of commercial sugar; 1450 l tap water; 10 ml liquid phenol). It is designed to recover a high percentage of oocysts with only a moderate amount of debris prior to CsCl or Percoll® purification.

2. Sucrose Suspension Technique[32]

1. Place 40 to 45 ml of sieved fecal suspension in a 50-ml centrifuge tube.
2. Centrifuge at 1000× *g*, 10 minutes.
3. Resuspend the 15 to 20 ml of pellet in water, and centrifuge at 1000× g, 10 minutes.
4. Resuspend the pellet in 50 ml water, pour ½ into a second tube, and to each add an equal volume of sucrose solution (specific gravity 1.30 to 1.35); mix gently.
5. Centrifuge the tubes at 1000× *g*, 5 minutes.
6. The supernatant contains 80 to 90% of the oocysts. Dilute it with >2 parts PBS, and centrifuge at 2000× *g*, 15 to 20 minutes, to pellet oocysts.

The above one-step technique is fast and provides a high percent yield of oocysts. This procedure may eventually prove useful for increasing the percentage of oocysts collected from sediments following filtration of surface and ground waters.

D. PURIFICATION OF OOCYSTS

Percoll® gradients have been useful for purifying oocysts, sporozoites, and merozoites of *T. gondii* and *Eimeria* spp.,[44-46] as well as oocysts of *Cryptosporidium* spp.[26,47,48] The Percoll® procedure is most effective when fecal suspensions contain little debris. If overloaded by debris, purification with a second gradient may be necessary. The following procedure has been useful. As per the manufacturer's recommendation, use 1.5 *M* NaCl rather than Alsever's solution,[26] the latter of which has a short shelf-life; centrifuge at 400× *g*.

1. Isopycnic Percoll® Gradient Centrifugation[26,32]

1. Prepare two fresh Percoll® solutions consisting of the following:
   A. *Solution A*: 5.70 ml Percoll®, 1 ml 1.5 *M* aqueous NaCl solution, and deionized H_2O to 10 ml. The density must be 1.08 g/ml.
   B. *Solution B*: 2.63 ml Percoll®, 1 ml 1.5 *M* aqueous NaCl solution, and deionized H_2O to 10 ml. The density must be 1.04 g/ml.
2. In a 15-ml centrifuge tube, pipette 5 ml solution A. Carefully overlay with 3 ml solution B. Then, carefully overlay with 1 ml oocyst suspension in PBS (previously concentrated over sucrose). Oocysts resuspended in water are not suitable, nor are suspensions containing considerable debris.
3. Centrifuge at 400× *g*, 30 minutes.
4. Oocysts are in a light-colored band about ⅓ the distance from the bottom, slightly beneath the interface of the two Percoll® concentrations. The pellet contains debris, the PBS overlay is relatively clear, and bacteria are banded beneath the PBS overlay.
5. Collect oocysts by aspiration, then wash with water or PBS by centrifugation at 1000× *g* for 10 to 15 minutes.

Cesium chloride gradients can be used with higher concentrations of debris than Percoll®, and final oocyst suspensions are of higher purity. Up to 500 million highly purified oocysts can be collected from a single band in a 50-ml tube, although 100 to 300 million oocysts/tube are more routine (Figure 2). It is particularly important that the CsCl is of biotechnology grade with ≥99.5% purity.

2. Cesium Chloride Gradient Centrifugation[47]

1. Prepare two stock solutions.
   A. *Solution A*: Dissolve CsCl in deionized water at a concentration of 1.8 g/ml. Stir gently until the salt is completely dissolved, up to 1 hour. This solution is stable for about a month at 4°C.
   B. *Solution B*: Before use, prepare TRIS stock buffer solution consisting of 50 m*M* Tris and 10 m*M* EDTA in deionized water (6.055 g/l Tris, 2.923 g/l EDTA, pH 7.2).
2. From the two stock solutions above, prepare the following three solutions:
   A. *Solution 1* (1.40 g/ml): 1 part CsCl stock with 1 part TRIS stock.
   B. *Solution 2* (1.10 g/ml): 1 part CsCl stock with 7 parts TRIS stock.
   C. *Solution 3* (1.05 g/ml): 1 part CsCl stock with 15 parts TRIS stock.
3. In a 50-ml tube, pipette 8 ml solution 1. Overlay solution 1 with 10 ml solution 2. Overlay solution 2 with 10 ml solution 3. Overlay solution 3 with 5 to 8 ml oocyst suspension that has been prewashed and resuspended 1× in TRIS buffer stock solution. Best results are achieved when this oocyst suspension derived from fecal debris is enriched by one of the sucrose purification techniques, although sieved feces containing high numbers of oocysts and low amounts of lipid can also be applied directly to the gradient.
4. Centrifuge at 16,000× *g*, 15°C, 60 minutes.
5. Only the band between solutions 2 and 3 contains viable oocysts.
6. Aspirate oocysts, dilute with ≥2 parts water, and pellet by centrifugation at 2000× *g*, 20 minutes, to remove residual CsCl.

E. EXCYSTATION PROCEDURES

Sporozoites of most species of coccidia are liberated from oocysts or sporocysts by incubation in an excystation solution, usually consisting of a protease combined with a bile salt. This step is often preceeded by compromising the oocyst wall with a mechanical grinder, glass beads, sodium hypochlorite, or a reducing agent. Because oocysts of *Cryptosporidium* spp. lack sporocysts and because sporozoites exit through a suture at one pole of the oocyst, mechanical methods for oocyst wall disruption are not necessary.

Once oocysts are pipetted from sucrose, Percoll®, or CsCl gradients, they should be washed immediately in water or PBS to prevent damage to sporozoites. The following represents a suggested rapid protocol for processing oocysts following gradient purification.

1. Rapid Processing of Oocysts Prior to Excystation[49]

1. In a 15-ml tube, add 4 to 5 ml oocyst suspension derived from CsCl, Percoll®, or sucrose gradient to 9 to 10 ml water. Pellet oocysts by centrifugation at 2000× *g* for 20 minutes.

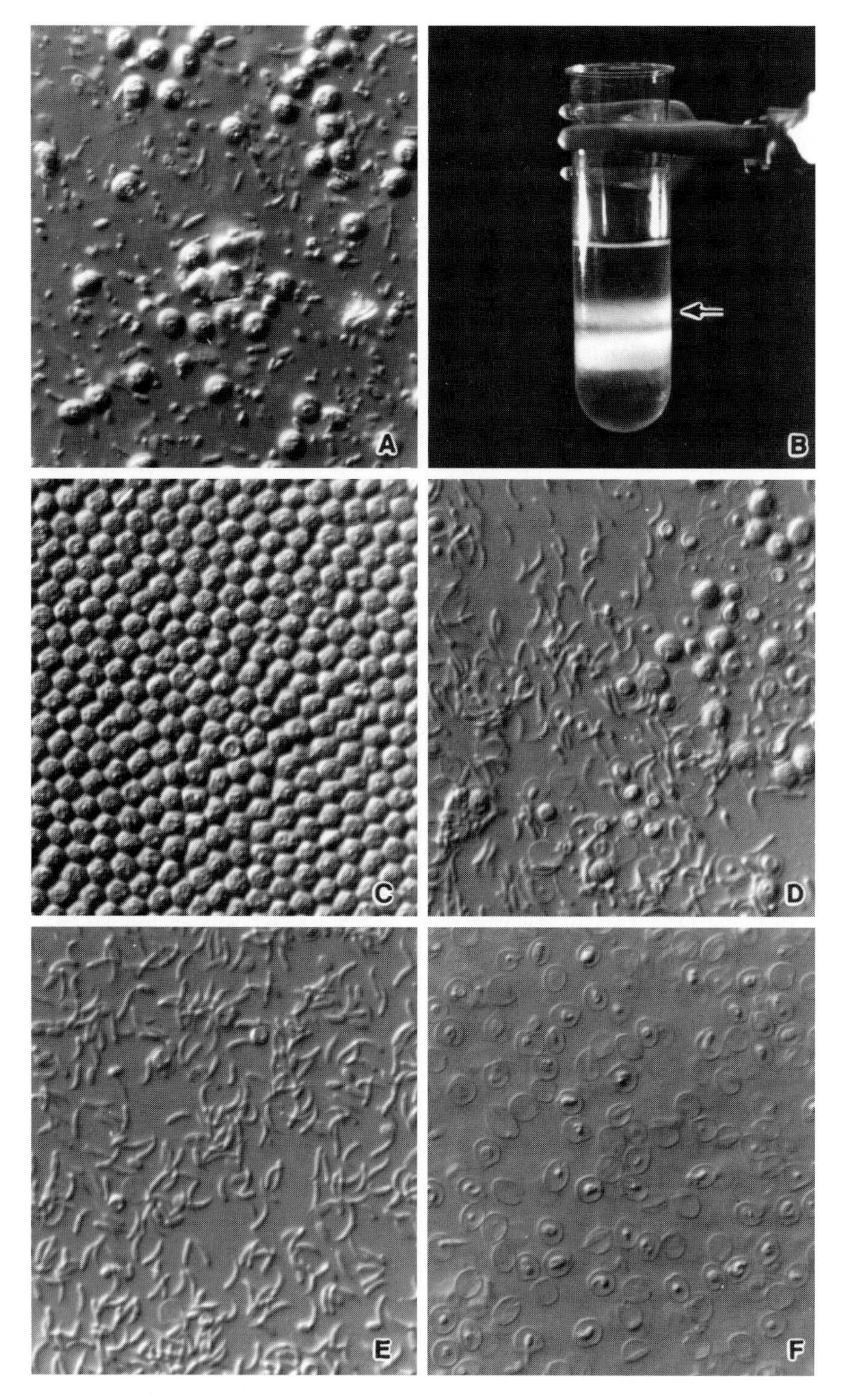

Figure 2 Photographs demonstrating different steps in the purification of *C. parvum* oocysts and sporozoites by CsCl and Percoll® procedures. **(A)** Oocysts concentrated from bovine feces by sucrose flotation; **(B)** centrifuge tube containing band of oocysts (arrow) following CsCl gradient of sucrose purified oocysts; **(C)** oocysts recovered from CsCl gradient; **(D)** excystation mixture of oocysts and sporozoites following a 60-minute incubation of bleach-pretreated oocysts in PBS at 37°C; **(E)** enriched population of sporozoites following rapid Percoll® procedure; **(F)** oocyst walls following rapid Percoll® procedure. All figures are interference contrast photomicrographs at an original magnification of 940×, except (B), which was taken using a macrolens at an original magnification of 0.55×.

2. Transfer pellet to 1.5-ml microfuge tube.
3. Wash one time with water by centrifugation in microfuge, 5000× *g*, 3 minutes.
4. Aspirate and discard supernatant. Oocysts to be stored or shipped should remain as a pellet stored at 4°C after addition of fresh PBS, cell culture medium, or $K_2Cr_2O_7$ solution. If oocysts are to be excysted or if surface sterilized oocysts are needed, oocysts should be resuspended in 10% fresh, aqueous commercial bleach solution at room temperature for 10 minutes.
5. Centrifuge at 5000× *g*, 3 minutes.
6. To maintain sterility, pipette under a hood using sterile solutions and pipettes. Aspirate supernatant, wash pellet two times in PBS by centrifugation, 5000× *g*, 3 minutes each.
7. Resuspend oocysts in PBS or cell culture medium. A small aliquot can be removed, diluted 1:100 with water, and oocysts quantitated using a platelet hemacytometer.

Pretreatment of *C. parvum* oocysts with sodium hypochlorite results in separation of the inner and outer oocyst walls,[50] enabling oocysts to excyst spontaneously when warmed to 37°C[35,49,51,52] with maximal excystation near 90 minutes (Figures 2 and 3).[49,51] However, even oocysts not pretreated with bleach excyst somewhat when warmed to 37°C.[30,50,51,53,54] The use of trypsin and bile salts, or bile salts alone,[51,55] increases excystation of unbleached oocysts. Although it is unknown why *C. parvum* sporozoites are capable of excysting when the temperature is elevated, sporozoites of other coccidia become active when the temperature is raised following mechanical manipulation.[56,57] It is likely that a protease, such as the 24-kDa cysteine proteinase on the sporozoite surface,[58] becomes active and affects the suture from within.

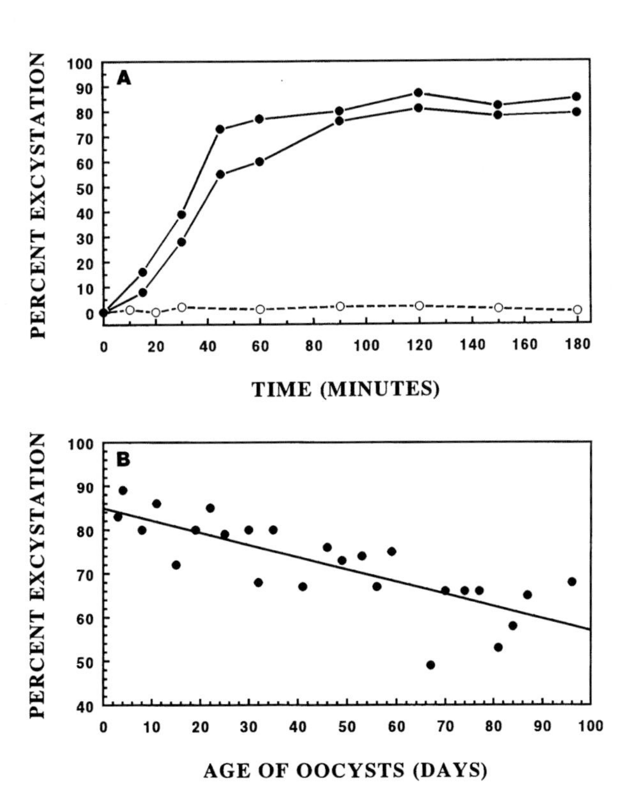

Figure 3 Excystation of CsCl purified, bleach-pretreated oocysts of *C. parvum* in RPMI 1640 supplemented with 10% FBS, 15 m*M* HEPES, and antibiotics. Percent excystation is expressed as the number of empty oocysts per number of empty + intact oocysts. **(A)** Comparative excystation of 30-day-old oocysts at 37°C (●) and 24°C (○); **(B)** scatter plot of excystation with time. Oocysts were stored in 2.5% (w/v) $K_2Cr_2O_7$ solution at 4°C and were purified and excysted at various intervals. Although numbers harvested from the gradient drop significantly around 8 weeks, excystation of those collected still lies along the linear path. (Portions adapted from Upton, S. J. et al., *FEMS Microbiol. Lett.,* 118, 45, 1994. With permission.)

Pretreating oocysts with a reducing agent is not necessary when excysting *C. parvum*, although a slight increase in the rapidity of excystation occurs when bleach treated oocysts are incubated in PBS containing 0.01 *M* cysteine HCl during the 37°C excystation process.[32] However, protease treatment of coccidian sporozoites significantly affects the migration of sporozoite surface antigens on SDS-PAGE.[59] These proteases should be avoided in most cases involving studies on sporozoite surface antigens. The effect of bile salts on sporozoite surface antigens is not yet known.

2. Excystation Procedure[49]

1. Purify, bleach, and wash oocysts with PBS as outlined above.
2. Resuspend pellet in fresh PBS or cell culture medium. Oocysts not pretreated with bleach should be resuspended in filter sterilized PBS containing 0.25% (w/v) trypsin and 0.75% (w/v) crude sodium taurocholate or the bile salt alone.
3. Incubate the oocyst suspension at 37°C.
4. Examine the oocyst suspension microscopically; at 90 minutes, 80 to 95% excystation should occur. If most oocysts remain intact after 90 minutes, they should be discarded. If oocysts are empty but sporozoites are lacking, suspect osmolarity problems.
5. If excystation solution was used, the parasite suspension should be washed in PBS $\geq 2\times$ by centrifugation at $5000\times$ *g*, 3 to 4 minutes each.
6. Sporozites can be further purified (see below).

F. PURIFICATION OF SPOROZOITES

Occasionally, it is necessary to separate excysted *Cryptosporidium* sporozoites from empty oocysts and other debris. Current[16] described a one-step filtration technique in which excysted suspension is pushed through a 2.0-µm Nucleopore filter. This technique has application only when few sporozoites are needed, because the empty and intact oocysts tend to rapidly clog the pores. Anion-exchange chromatography has been used successfully to purify sporozoites of *C. parvum*[31,35] and other coccidia.[60-62] Although yields can be low, up to 90% of the sporozoites initially applied to the column have been recovered by some investigators.[35]

1. DE-52 Anion Exchange Chromatography[35]

1. Prepare column buffer (10.78 g anhydrous Na_2HPO_4, 0.62 g $NaH_2PO_4 \cdot H_2O$, 3.4 g NaCl, and 10 g glucose to 1 l of deionized water [ionic strength 0.290, pH 8.0]).
2. Add 5 g of Whatman® DE-52 preswollen matrix to 50 ml column buffer. Wash the matrix gently by manual swirling 3 to 4 times, allow it to settle each time, and decant the supernatant.
3. The column consists of a 3-cc syringe with a small piece of nylon wool to plug the barrel. Clamp the syringe to a stand. Attach flexible tubing equipped with a stop clamp to the barrel. The tube runs into a 50-ml centrifuge tube.
4. Add approximately 1 ml of matrix solution in buffer to the syringe and allow to settle onto the nylon wool. Enough column buffer is added to submerge the matrix.
5. An excystation mixture consisting of sporozoites/empty oocysts in a 1.5-ml Eppendorf tube, derived from bleach-treated oocysts, should be washed two times in column buffer by centrifugation, $5000\times$ *g*, 4 minutes each. Only oocysts pretreated with bleach possess appropriate surface charge to be retained on the column.
6. Resuspend the sporozoite/oocyst wall suspension in 10-ml column buffer.
7. Add 1 ml matrix suspension to parasite suspension. Swirl gently for 1 to 2 minutes, then add to a column.
8. Elute with 30- to 50-ml column buffer at a flow rate of 2 to 3 ml per minute. Sporozoites should pass through column, whereas empty oocysts should become trapped in matrix and nylon wool.
9. Pellet sporozoites by centifugation at $2000\times$ *g*, 20 minutes.

Percoll® procedures are probably superior to the anion exchange method for purification of both sporozoites and merozoites.[26,47,63] Modifications allow sporozoites to be separated from oocyst walls in only 3 minutes using a microcentrifuge (Figure 2).[64]

2. Rapid Percoll® Separation[64]

1. Excyst sporozoites for 60 minutes at 37°C in PBS in a 1.5-ml Eppendorf tube.
2. Pellet suspension by centrifugation at $5000\times$ *g* for 3 minutes in microcentrifuge.

3. Discard supernatant and resuspend pellet in 1.2 to 1.5 ml isotonic Percoll® solution. This solution consists simply of 26.3% (v/v) commercial Percoll® solution in an aqueous 0.15 *M* NaCl solution. The specific gravity should be 1.04 (see solution B under Percoll® procedure for oocysts).
4. Centrifuge at 5000× *g*, 3 minutes.
5. The pellet contains some sporozoites and intact oocysts. Immediately above this pellet will be a buffy coat consisting of highly enriched sporozoites. Oocysts walls lie atop the Percoll®.
6. After aspirating sporozoites or empty oocysts by pipette, dilute in ≥4 volumes of PBS. Centrifuge at 5000× *g*, 10 minutes to pellet.

III. FACTORS INFLUENCING DEVELOPMENT *IN VITRO*

Cultivation of *Cryptosporidium parvum* has been challenging but significant progress has been made in the last few years so that today very large numbers of asexual and sexual stages can be generated in microbial free systems. Neither continuous propagation nor large-scale production of oocysts *in vitro* has been reported.

Good reviews by Doran,[1,2] McDougald,[65] Schmatz,[66] Speer,[3] and Stout and Schmatz[67] offer valuable suggestions concerning technique and provide insight into some important growth requirements for coccidia in general. Truly inspiring series of biochemistry studies in the 1970s and 1980s from the laboratories of C.C. Wang and E.R. and L.C. Pfefferkorn have revealed some of the metabolic pathways of coccidia. However, *in vitro* cultivation systems routinely used for propagation of *T. gondii* have been of limited value for studying most other coccidia.

A. AGE AND STRAIN OF OOCYSTS

Prolonged storage of coccidian oocysts results in reduced viability and ability to establish infections *in vitro* or *in vivo*.[68,69] The duration of infectivity is highly variable and depends upon such factors as species, temperature, and storage medium.[69] *Cryptosporidium* spp. undergo a significant decrease in parasite infectivity, sometime after 2 months.[50,70] However, with proper storage at 4 to 8°C, enough oocysts may remain viable to establish infections *in vivo* for up to 12 months.[16]

Members of the genus *Cryptosporidium* have a tremendous potential for bioamplification. Thus, it is little wonder that precise measurements of oocyst viability under crude storage conditions are lacking. If experiments are not designed properly, the bioamplification results in an all or nothing phenomenon. Excystation studies are also misleading, as some of the concentration techniques such as CsCl select for viable oocysts (Figure 3).[177]

Oocysts destined for *in vitro* use are best used at ≤6 weeks of age. Using excystation to determine sporozoite viability at the time of inoculation, a small aliquot of oocysts in cell culture medium can be set aside at 37°C at the same time cultures are being infected. If excystation levels are >70% at 90 minutes, cultures can be washed and the experiment allowed to continue.

Fluorogenic dyes or other reagents have been used to quantitate oocyst or sporozoite viability, which generally correlates well with percent excystation.[33] Some investigators find these techniques useful,[26,29,30,33,71-77] whereas others have found them unsuitable for viability studies.[78] Both motility and host cell penetration can occur after sporozoites presumably have accumulated lethal levels of oxygen intermediates.[79] Thus, many sporozoites destined to die within minutes or hours may still be motile and capable of picking up vital dyes. Despite the method used, some viability standard is helpful whenever performing *in vitro* assays, as they allow some measurement of protocol success.

Different isolates of *C. parvum* have different levels of excystation.[33] Sometimes an entire batch of oocysts of an isolate from calves excysts poorly. Such poorly excysting oocysts perform poorly *in vitro* but are suitable for *in vivo* immunization, biochemical analysis, and SDS-PAGE. Although it is unknown why some isolates excyst better than others, molecular and biochemical studies have shown differences among isolates.[30,80-84]

B. STORAGE CONDITIONS

Oocysts of *Cryptosporidium* spp. are best stored as a sludge in fecal debris in 2 to 4% (w/v) $K_2Cr_2O_7$ solution. The oxidative action of this solution inhibits bacterial and fungal overgrowth and prevents total anaerobia. Once oocysts are purified from the fecal debris on sucrose, Percoll®, or CsCl gradients, however, their longevity appears to be somewhat shortened.

Other storage media have been useful. Hanks balanced salt solution (HBSS) supplemented with antibiotics is the medium of choice for sporocysts of *Sarcocystis* spp.[69,85] Several laboratories use HBSS or dilute acid solutions for *Cryptosporidium* spp. where environmental concerns or disposal policies limit chromium use.[17,86-88]

C. HOST CELL TYPE

The most successful host cells used to study *C. parvum* are epithelial-like (Table 1). These include human colonic adenocarcinoma (Caco-2), human endometrial carcinoma (RL95-2), galactose-adapted human colonic carcinoma (HT29.74), human ileocaecal adenocarcinoma (HCT-8), and Madin-Darby canine kidney (MDCK). When development of *C. parvum* was compared in 11 different host cells, it was concluded that the yield of parasites in HCT-8 cells was superior to all other cells based on counts of live parasites grown in monolayers on coverslips (Figure 4).[89] Both Caco-2 and MDCK cells are more delicate than HCT-8 cells, however, and host cells as well as parasites are easily disrupted when coverslips are inverted onto microscope slides, compromising observation and counting. MDCK cells were found superior to other cells in a similar study.[90] Nevertheless, these studies demonstrate that one of the most important factors determining successful *in vitro* cultivation of *C. parvum* is the choice of host cell.

Selection of a particular host cell line depends upon factors such as length of the assay and type of detection system employed. Cell lines with a rapid doubling time can overgrow a culture prior to the completion of a study. Delicate cell lines, suitable for detection assays involving fixation and staining, become disrupted in a coverslip assay involving live cells or dislodged in an *in situ* ELISA. Some cell lines, such as Caco-2, grow at an exceptionally slow rate and can be useful in assays where minimal host cell growth is desired. Growth of HCT-8 cells is contact dependent and the normally rapid rate of growth slows once the monolayer has become established.

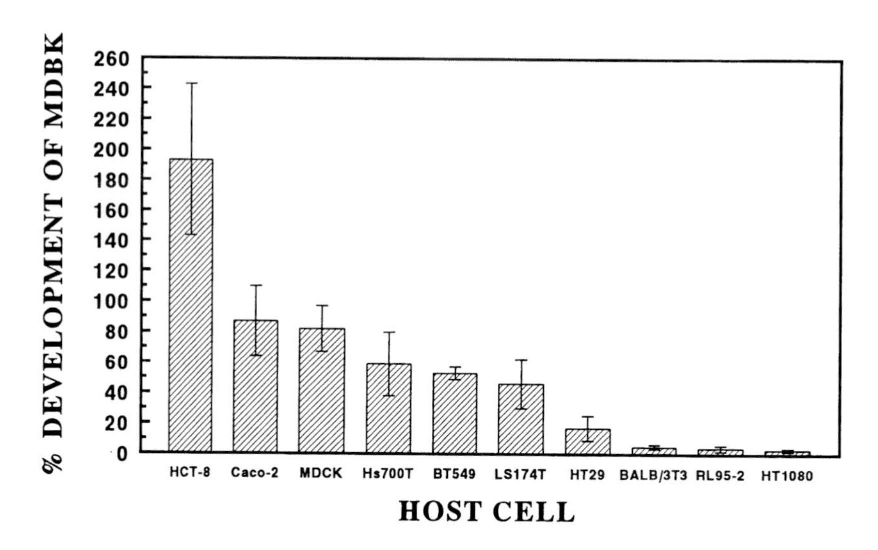

Figure 4 Comparative development of *C. parvum* in 10 host cell lines to that in Madin-Darby bovine kidney (MDBK) cells. Each data point represents 4 to 6 replicates and is expressed as a percentage of development of MDBK cells, with ±SD of the mean also expressed as a percentage. (Data adapted from Upton, S. J. et al., *FEMS Microbiol. Lett.*, 118, 233, 1994. With permission.)

Cell lines such as HCT-8 and MDCK are highly adherent and must be trypsinized to remove them from the substrate for proper dispersal into new cultures. Other cell lines may require only the addition of a calcium chelator or agitation for removal. Whichever procedure is employed, the method of host cell dispersal prior to plating is critical and has a pronounced effect on parasite distribution. If cells are plated as clumps, *Cryptosporidium* develops in a ring around an overgrown central core. This phenomenon is due, in part, to the sloughing of parasites and cells in the overgrown central core and the fact that younger cells at the leading edge of the cell cluster are lightly infected because of insufficient time for infections to occur.

Because liberated merozoites travel finite distances, subsequent development becomes focal. Older cells are thought to be poorly susceptible to infection with coccidia,[91-94] so areas with aged cells may have fewer parasites. Cells in G_2 phase are thought to be most susceptible to infections by *T. gondii*.[95] Because irradiation blocks mitosis and cells remain in G_2, some investigators have irradiated monolayers prior to addition of coccidia.[96] S-phase HeLa cells are more heavily parasitized by *Besnoitia jellisoni* and it was concluded that this parasite may also lengthen the first cell cycle.[142] The relationship between the patchy distribution of *C. parvum* in monolayer cultures and the phase of cells remains unknown.

Effects of trypsin treatment on host cells and the optimal time after protease treatment to infect host cells have not yet been reported. Cells inoculated soon after protease treatment may be more susceptible to *T. gondii* infections because the surface glycocalyx has been disrupted.[92] The binding of *T. gondii* to laminin receptors has been well studied,[97,98] and it is possible that additional binding sites would be exposed following trypsinization of the host cell. Although a similar hypothesis has been proposed for *C. parvum* sporozoites,[30] the presence of a host cell receptor for *Cryptosporidium* sporozoites has not yet been demonstrated.

D. ATMOSPHERE

The development of successful cultivation systems for *Plasmodium falciparum* provides a classic example of a few elementary experiments that have influenced subsequent research.[99,100] The mere mention that a reduced oxygen atmosphere enhanced *in vitro* development of the rabbit coccidian, *Eimeria magna*[100,101] led to followup studies demonstrating that a reduced oxygen atmosphere often has a pronounced effect on Coccidial development *in vitro*, either by affecting penetration of host cells or by enhancing subsequent development.[79,102-106] However, one study found no correlation between O_2 concentration and parasite growth.[107]

A study of *C. parvum* in both MDBK and HCT-8 cells under seven select atmospheres found that the effects of atmosphere were mediated through the type of host cell rather than by atmosphere alone.[108] Thus, a 16% oxygen (1.5% CO_2, 81.8% N_2, 0.7% other) atmosphere, similar to that in candle jars,[108] was optimal when MDBK cells were used, but a 5% CO_2/95% air incubator was superior when HCT-8 cells were employed. This phenomenon may involve levels of superoxide dismutases or oxygen intermediates in either the parasite or host cell.

E. SIZE AND TYPE OF INOCULUM

Host cells are inoculated with *C. parvum* sporozoites, oocysts, or a mixture of the two. If sporozoites are used, they must first be purified from debris (see above). Use of free sporozoites eliminates cell monolayer contact with oocyst walls and exposure to toxic fluids from within oocysts. Thus, high numbers of sporozoites can be used in assays, which is highly desirable for receptor/ligand assays.[75] However, there are two major disadvantages associated with using free sporozoites as inoculum for long-term cultures. First, sporozoites must be kept relatively free of other microbial contaminants throughout the purification procedure. Second, free sporozoites are delicate and adversely affected by the various concentration techniques, time of exposure to atmospheric O_2, and perhaps other environmental conditions.[30,33,51,109] Some authors have used fluorogenic dyes to correlate sporozoite viability,[30,33,35,75] but caution must be used in the interpretation of these results. Sporozoites, but not merozoites, of most coccidia possess very low levels of oxygen scavenging enzymes.[110-112] Once exposed to atmospheric O_2, these sporozoites may accumulate lethal levels of oxygen intermediates. Even though these sporozoites are highly motile and readily penetrate host cells, some may fail to survive long enough to establish long-term infections.[79]

There is a simplified method of establishing infections *in vitro* using only oocysts.[49] Oocysts pretreated for 10 minutes with 10% bleach are washed 2 times in PBS, resuspended in cell culture medium containing 10% FBS, counted by hemacytometer, then added to cell monolayers. When cultures are placed in a 37°C incubator, sporozoites excyst from the oocysts and have direct access to the host cells. In addition to simplicity, oocysts are far easier to keep free of microbial contamination than are sporozoites, there is minimal opportunity for oxidative damage to sporozoites, and oocysts settle onto monolayers faster than sporozoites, so fewer parasites remain suspended in the culture medium. The main disadvantage of this procedure is that cultures must be washed briefly with PBS after 60 to 90 minutes to remove residual oocysts and the fluids released from oocysts. These fluids and residual debris are highly toxic to host cells and can have a pronounced negative impact on the course of development.[1,12,42,113-118]

Whatever method is used to infect a cell monolayer, it is important to distribute the inoculum evenly over the cells. Even if host cells are distributed uniformly in culture (see above), distribution of coccidia will be patchy. Merozoites released from meronts infect nearby cells so failure to uniformly infect the monolayer will be accentuated even more.

Numbers of parasites used to infect cultures will greatly influence the final number of developmental stages. Generally lower parasite-to-host cell ratios result in proportionally higher levels of infection.[119-123] An oocyst-to-host cell ratio of about 1:1 to 1:2 is generally optimal for *C. parvum*.[42,115,123,124] A higher ratio should be employed for short-term binding assays.[75]

F. pH AND CATIONS

Both pH and extracellular ions significantly effect motility and penetration of host cells by Coccidian sporozoites. Extracellular cations affect the internal pH of *Toxoplasma* tachyzoites, and addition of acidic buffers significantly increases parasite motility.[125,126] $CaCl_2$, $MgCl_2$, and ATP all increase parasite motility.[127] Some cations, such as 1 m*M* calcium with a pH optimum of 7.0 to 7.6, enhanced attachment of *C. parvum* sporozoites to MDCK cells.[75]

N-[2-hydroxyethyl]piperazine-N′[2-ethanesulfonic acid] (HEPES), the most commonly used buffer for *in vitro* cultivation of coccidia, has a pKa of 7.5 at 25°C and useful buffering in a pH range of 6.8 to 8.2. Whenever cultures of host cells are placed in a CO_2 incubator, the pH falls well into the acidic range. Because *C. parvum* sporozoites are most infective in a pH range of 7.2 to 7.6,[75] other types of biological buffers may prove more useful. Considering that *Plasmodium* spp. gametocyte emergence and exflagellation can be induced by increasing the pH to 7.7 to 8.0[128,129] or by ion exchange mechanisms,[130-132] the reason that so few *C. parvum* oocysts are generated *in vitro* may, in part, reflect microgamete inactivity at suboptimal pH.

G. MEDIUM SUPPLEMENTS

Few *in vitro* studies have examined the role of supplements on coccidian growth. However, supplementing the cell culture medium with additional serum, vitamins, and sugars often leads to greatly enhanced development of coccidia *in vitro*.

When placed in close proximity to host cells, sporozoites of *Eimeria* spp. had increased motility, presumably due to substances released by the host cells.[133,134] In addition, supplementation of coccidial cultures with fetal bovine serum (FBS) or other serum proteins affected coccidian motility and development (Figure 5). Sporozoites incubated in 10% FBS exhibited significantly higher motility than those in medium or PBS alone.[135] Whereas 10% FBS resulted in nearly double the numbers of *C. parvum* parasites after 68 hours compared to cultures with only 5% FBS, concentrations beyond 10% FBS failed to increase parasite numbers.[136] Likewise, the serum proteins, albumin and fetuin, enhanced motility of eimerian sporozoites, and supplementation of medium containing 5% FBS with albumin and fetuin resulted in increased numbers of parasites in host cells.[135] However, supplementation of eimerian-infected cultures with either of these proteins did not enhance parasite development if the medium contained 10% (rather than 5%) FBS.[137] Even though some lots of FBS will probably be shown to be less supportive of development than others, there are no data as yet to indicate that FBS quality affects *C. parvum in vitro*. However, it is important to heat-inactivate FBS to prevent sporozoite lysis.[138]

The addition of glucose, galactose, maltose, and mannose at concentrations of 20 m*M* and 50 m*M* had a positive effect on development of *C. parvum in vitro* (Figure 5),[136] similar to that observed for other coccidia.[139,140] The same concentrations of mannitol, sorbitol, sorbose, *N*-acetyl-glucosamine, and *N*-acetyl-galactosamine had no effect,[136] although 10 m*M* *N*-acetyl-glucosamine slighlty inhibited parasite growth.[141] Insulin, too, had a slight positive effect on coccidial numbers *in vitro*,[90,136,142] possibly by facilitating sugar uptake by host cells.

Vitamins have positive effects on growth of coccidia *in vitro*, including *C. parvum*.[136,143-148] Supplementation of cell culture medium with ascorbic acid, calcium pantothenate, folic acid, and para-aminobenzoic acid all had a positive effect on growth of *C. parvum in vitro* (Figure 5), whereas biotin, niacin, nicotinamide, and riboflavin had little or no effect.[136] Because some basal level of each vitamin was present either within the cell culture medium and/or within the FBS, it would be erroneous to assume in these studies that a lack of effect equates with the vitamin being unnecessary for parasite growth.

Although pyruvate, choline chloride, β-mercaptoethanol, and hypoxanthine-thymidine are sometimes used as supplements for coccidia *in vitro*,[90,149,150,] none appear to enhance development of *C. parvum*.[136] Because *C. parvum* is not known to possess mitochondria,[137,151,152] pyruvate may not be used as an energy source. Coccidia are also thought to incorporate hypoxanthine.[153-157] There is a lack of *de novo* purine nucleotide synthesis in the coccidia, but all seem to have high levels of enzymes involved in purine

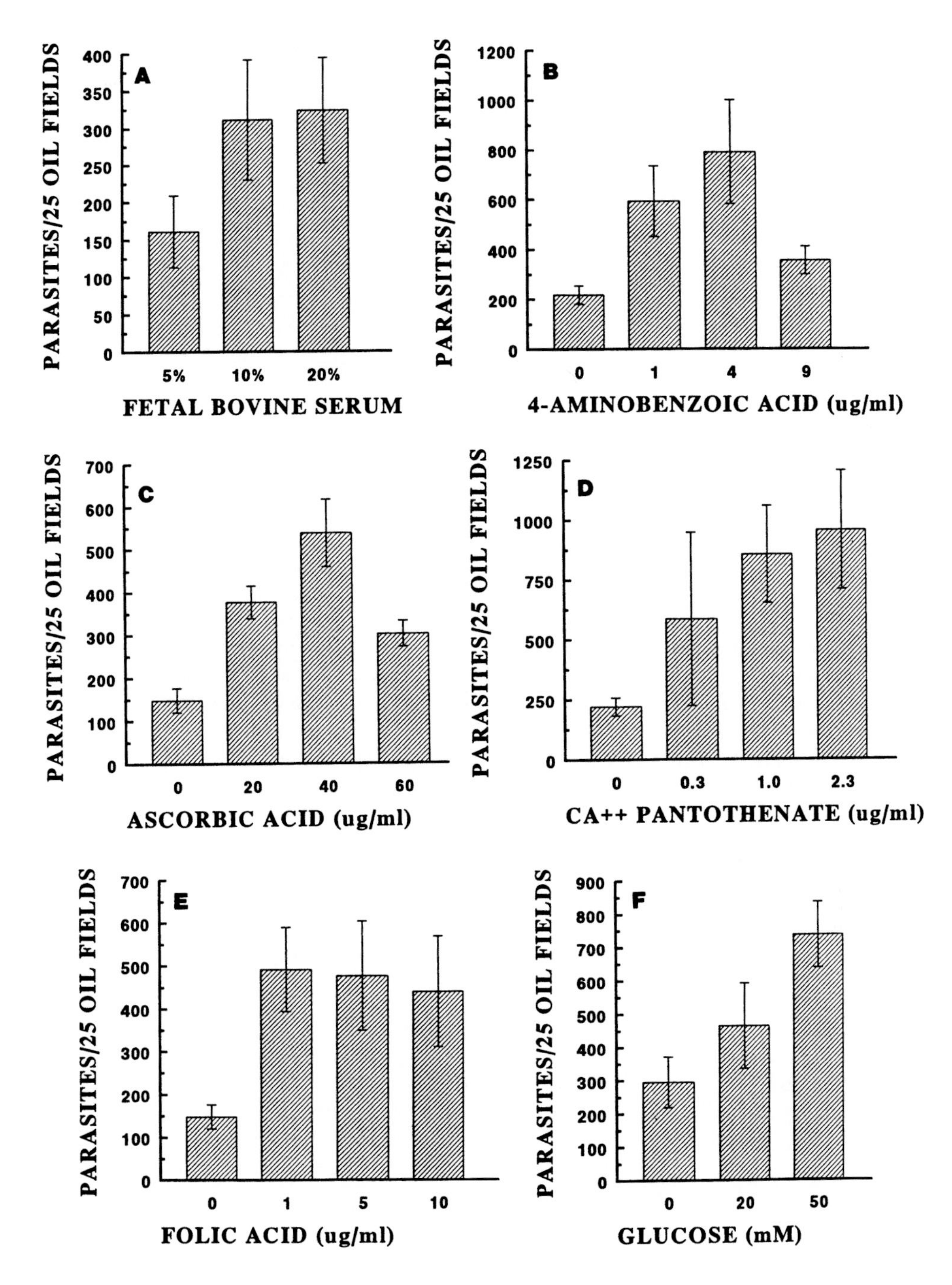

Figure 5 Effects of select medium supplements on development of *C. parvum* in HCT-8 cells grown on coverslips *in vitro*. Parasite infected cultures were incubated for 68 hours in RPMI 1640 supplemented with 10% FBS, 15 m*M* HEPES, and antibiotics at 37°C in a 5% CO_2/95% air incubator. Various medium supplements were added and numbers of parasites per 25 random oil fields were quantitated using Nomarski interference contrast optics. (Adapted from Upton, S. J. et al., *J. Clin. Microbiol.*, 33, 371, 1995. With permission.)

salvage.[154,155,158,159] However, it is likely that the intracellular pool is adequate so that *C. parvum* is unaffected by supplementation even if it is eventually found to incorporate this macromolecule.

Whereas coccidia have weak thymidine kinase activity and incorporate exogenous thymidine poorly,[155,160-165] they have the ability to synthesize and salvage pyrimidines in general.[153,155,162,166-168] Thus,

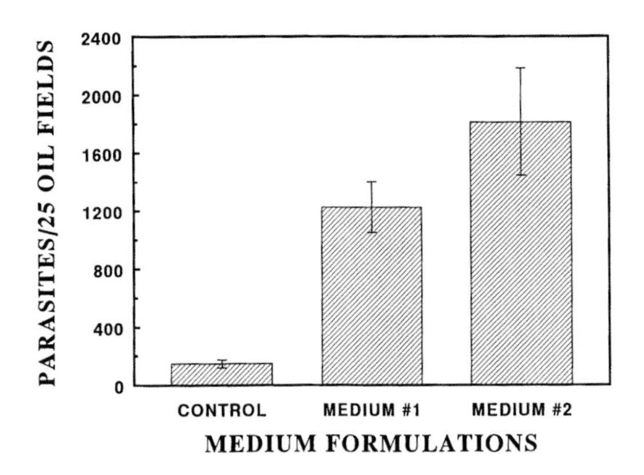

Figure 6 Effects of two medium formulations on development of *C. parvum* in HCT-8 cells grown on coverslips *in vitro*. Control cultures were incubated for 68 hours in RPMI 1640 supplemented with 10% FBS, 15 m*M* HEPES, and antibiotics at 37°C in a 5% CO_2/95% air incubator. Medium formulation #1 consisted of the control medium supplemented with 50 m*M* glucose, 35 µg/ml ascorbic acid, 1.0 µg/ml folic acid, 0.1 U/ml insulin, and 4% hypoxanthine-thymidine supplement. Medium formulation #2 consisted of the control medium supplemented with 50 m*M* glucose, 35 µg/ml ascorbic acid, 1.0 µg/ml folic acid, 4.0 µg/ml 4-aminobenzoic acid, 2.0 µg/ml calcium pantothenate, and 0.1 U/ml insulin. Parasites per 25 random oil fields were quantitated using Nomarski interference contrast optics. (Portions adapted from Upton, S. J. et al., *J. Clin. Microbiol.,* 33, 371, 1995. With permission.)

supplementation with a pyrimidine should have minimal effect on development of *C. parvum in vitro*. No effect was observed on growth when thymidine or uracil were used as supplements.[42,136]

Based on the above information, a medium was formulated that enhanced numbers of *C. parvum in vitro* tenfold (Figure 6).[136] This formula consisted of RPMI 1640 containing L-glutamine, supplemented with an additional 2 m*M* L-glutamine, 15 m*M* HEPES buffer, 50 m*M* glucose, 35 µg/ml ascorbic acid, 4.0 µg/ml para-aminobenzoic acid, 2.0 µg/ml calcium pantothenate, and 1.0 µg/ml folic acid. After adjusting the pH to 7.4 and filter sterilizing the mixture, FBS was added to a concentration of 10%. Insulin and antibiotics were eliminated; however, 100 U/ml penicillin, 100 µg/ml streptomycin, and 0.25 µg/ml amphotericin B can be used, if desired. Macrolides at high concentrations should be avoided as they may inhibit *C. parvum* development *in vitro*.

IV. QUANTITATION OF PARASITE DEVELOPMENT

Methods employed to quantitate development of *C. parvum in vitro* include visual counting using microscopy, ^{3}H-uracil incorporation, and ELISA. Several assays are discussed below.

A. VISUAL QUANTITATION USING COVERSLIPS

The most common method of quantifying development of *C. parvum in vitro* has been the coverslip assay. Infected host cells grown on coverslips have been fixed using formalin, acetone, methanol, Bouin's, paraformaldehyde, or glutaraldehyde and stained with dyes such as Giemsa, acid fast, hematoxylin and eosin, Hoescht 33258, Papanicolaou, Ziehl-Neelsen, or Alcian blue.[34,90,124,141,149,150,169-176] Parasites are then quantitated, usually as the ratio of parasites to host cells or number per field. The advantage of this technique is that quantitation can be performed as time permits, and a permanent record is available for future reference. Disadvantages include loss of some host cells and parasites during processing, difficulty distinguishing parasite stages from one another, and the length of time necessary to perform a large number of assays.

Modifications of the fix and stain technique include the use of indirect immunofluorescence assay (IFA)[177-180] and transmission electron microscopy (TEM).[71,149,174,181-184] IFA provides a rapid means of counting intracellular parasites, although it is difficult to distinguish developmental stages. TEM allows identification of stages and structure but is tedious and impractical for parasite quantitation.

Parasite development on coverslips has been quantitated using living cells and interference contrast microscopy.[2,10,15,49,89,108,136,185,186] At the time of inspection, coverslips are washed with PBS, then gently placed cell-side down onto a glass microscope slide. Parasites are generally quantitated per 100× field,

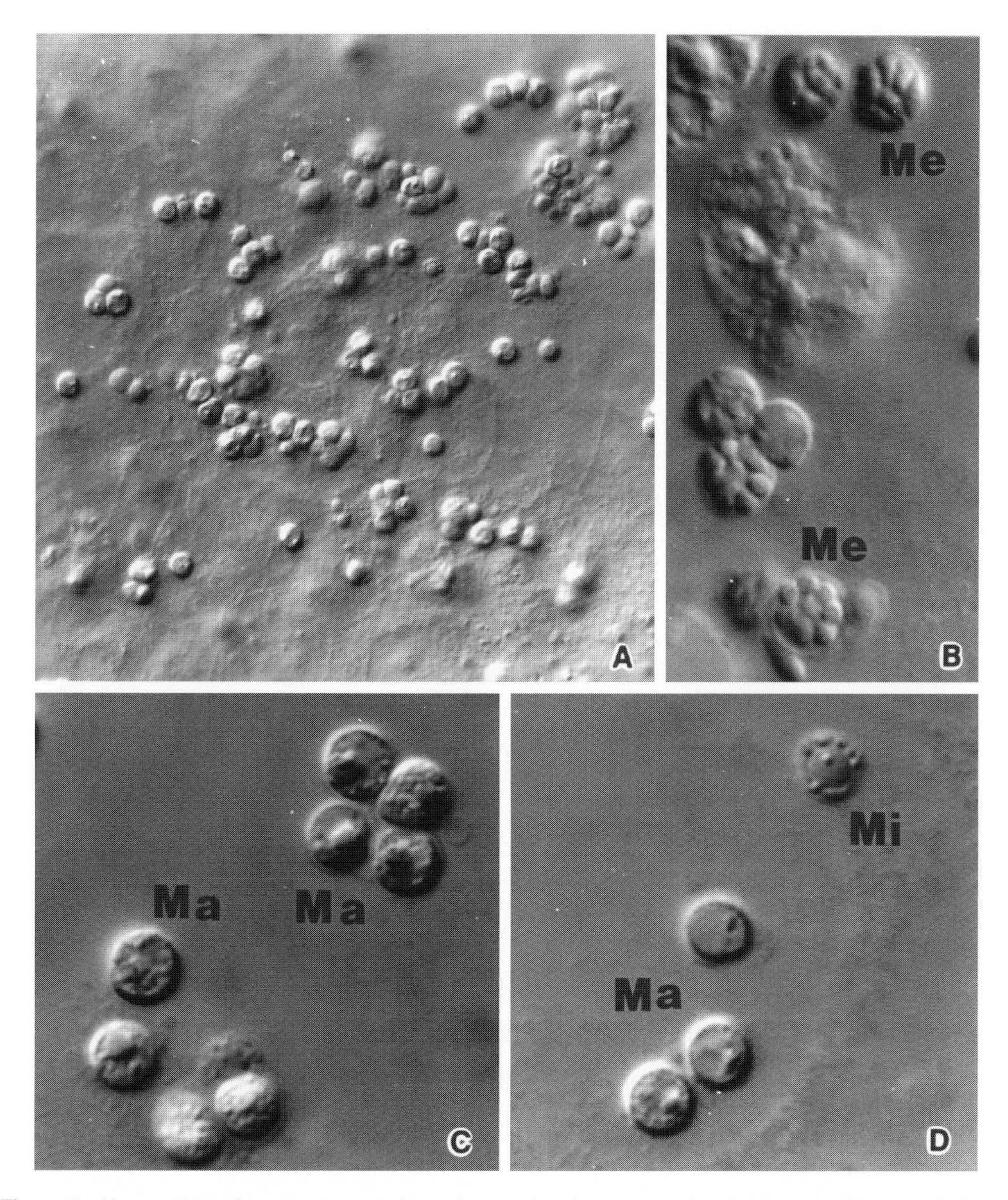

Figure 7 Nomarski interference contrast photomicrographs of representative developmental stages of *C. parvum* in HCT-8 cells *in vitro.* **(A)** Various developmental stages, predominately developing gametes, 48 hours post-infection, original magnification 900×; **(B)** Type I meronts (Me), original magnification 2500×; **(C)** macrogametes (Ma), original magnification 2500×; **(D)** microgametocyte (Mi) and developing macrogametes (Ma), original magnification 2500×.

and the different stages are easily recognized (Figure 7). However, both the live parasites and host cells are easily distorted by coverslip pressure, and slides must be examined immediately. In addition, no permanent record remains. Storage of coverslips in 10% neutral buffered formalin (NBF) prior to examination reduces fragility, leaving higher numbers of parasites per unit area and allows one to examine coverslips over many days.

All of the above techniques suffer from the inherent problem of focal parasite development. One method of overcoming this flaw is to use a high number of random microscope fields, but this is a very time consuming. Simpler, faster, and more reliable nonvisual methods of parasite quantitation for assays involving pharmaceutical screening and receptor/ligand interactions are described below.

B. ^{3}H-URACIL INCORPORATION

Coccidia must phosphorylyze uridine and deoxyuridine to uracil before conversion into uridylic acid by uracil phosphoribosyl transferase,[153,158,187,188] an enzyme not present in eukaryotes. Several coccidial assays based on ^{3}H-uracil incorporation have both research and some commercial application.[154,189,190] However, ^{3}H-uracil assays would not be useful for screening compounds that inhibit parasite replication but not total RNA synthesis.[191] In these assays, one must also utilize *Mycoplasma*-free cultures, as coccidia are not the only organisms with uracil phosphoribosyl transferase.[42,192]

Initial crude assays suggested *C. parvum* could incorporate free uracil.[185] Improved harvesting procedures and ^{3}H-uracil concentrations helped optimize the system;[42,115] however, ^{3}H-uracil incorporation in a 96-well format met with limited success. Possibly the large intracellular pool of free, unlabeled uracil dilutes the radionucleotide so that only low amounts are incorporated into the parasiste, or possibly *C. parvum* fails to incorporate uracil as effectively as other coccidia.

C. 96-WELL ELISA

An *in situ* ELISA may have the greatest application for rapid, large-scale assays in coccidial research. Such an immunoassay has been developed for *E. tenella*[193] and *T. gondii*[194] and more recently for *C. parvum*.[123,195] Keys to the *C. parvum* assay were increased parasite development with an enriched medium[136] and use of a primary antiserum with low cross-reactivity to the host cells.[58,123]

Microtiter wells must be seeded with a highly adherent host cell line, such as 3 to 4 × 10^4 HCT-8 cells per well (Figure 8). The next day, seed each well with 2 to 3 × 10^4 previously bleached oocysts (see above). After 90 minutes at 37°C, wash wells with PBS and incubate with new growth medium in a 5% CO_2/95% air humidified incubator. Terminate the experiment by adding 100 µl of 4% formalin in PBS to each well.[123] Wash wells 3 times with PBS, block in 1% bovine serum albumin (BSA)/0.002% Tween-20 in PBS, and add anti-*C. parvum* polyclonal antiserum in blocking buffer. Use antiserum against purified sporozoite membrane fractions, which is rich in surface antigens, microneme, and rhoptry proteins.[123] After 30 to 60 minutes, wash plates and add goat anti-rat polyvalent antiserum conjugated to horseradish peroxidase. After 20 to 60 minutes, wash plates and develop using 3,3″,5,5″-tetramethylbenzidine solution as substrate.[123]

High numbers of meronts and gametes were found at 48 (Figure 7)[123] and 68 to 72 hours in culture.[136] This procedure is particularly useful for screening pharmaceutical compounds (Figure 9), studying receptor/ligand interactions, and for examining the effects of medium supplements on parasite growth. The main disadvantages are the lack of a permanent record and inability to distinguish between the types of developmental stages.

V. CULTIVATION IN AVIAN EMBRYOS

Avian embryos support development of several species of coccidia,[1,65,196] and both *C. parvum* and *C. baileyi* complete development in chicken embryos.[174,197-199] Because culturing of *C. parvum* in monolayers has been successful, the use of avian embryos for this species appears to offer no advantage. In contrast, attempts to successfully cultivate *C. baileyi* in cells have met with failure and avian embryos remain useful. *Cryptosporidium baileyi* readily infects a wide range of eggs, including those of chicken, turkey, duck, chukar, guinea fowl, pheasant, and quail.[199]

Methods of inoculation and handling of avian embryos vary, but generally 10- to 12-day-old avian embryos are used. It is helpful to candle the eggs (view the contents by passing a strong light through the egg) to find an area relatively devoid of major blood vessels. Mark the shell above this area, swab with 70% ethanol, and punch a small opening through the shell with a 18- or 20-gauge needle. Inject parasites through this opening into the allantoic cavity. Punctures are sealed with hot wax or tape. Most investigators injected penicillin (1000 IU) and streptomycin (1 to 10 mg) into the allantoic cavity 24 hours prior to inoculations. Antifungals often kill the embryo. Either 1.5 × 10^5 sporozoites of *C. parvum*[198] or 4 × 10^5 bleach-pretreated oocysts of *C. baileyi*[199] were used as inoculum. Eggs were incubated at 37 to 38°C for *C. parvum* or 40 to 41°C for *C. baileyi*. After 6 to 7 days, portions of the chorioallantoic membrane (CAM) were removed and examined immediately as a squash preparation using interference contrast optics. Alternatively, the CAM was fixed and processed for routine sectioning and staining. Newly formed, viable oocysts were harvested from the allantoic fluid by pipette or collected from urates or CAM tissues.[198,199]

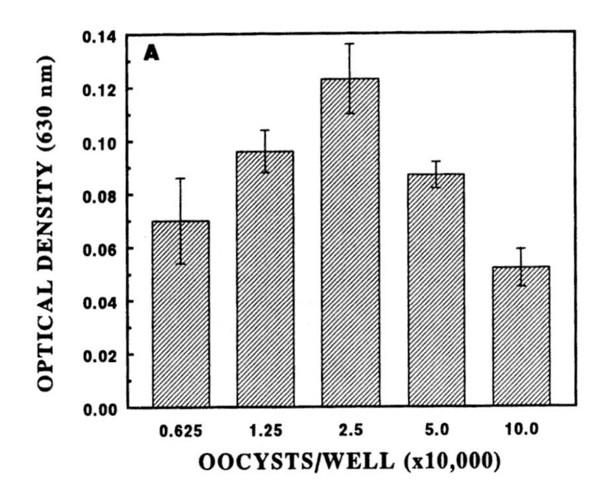

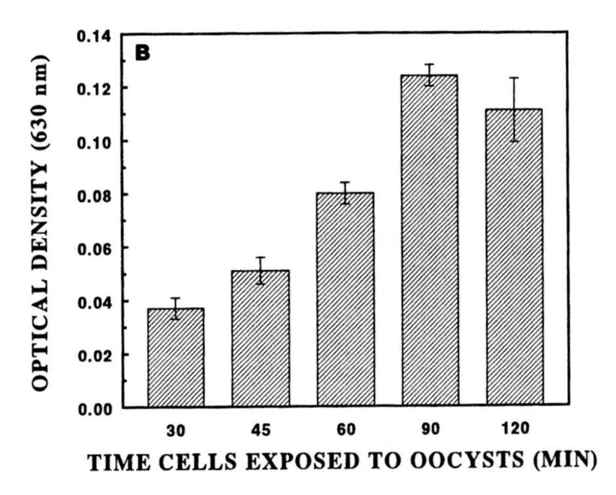

Figure 8 Effects of inoculating dose and time host cells exposed to oocysts on development of *C. parvum in vitro.* HCT-8 cells were seeded into microtiter wells at 4.0×10^4 cells per well, then exposed to CsCl purified, bleach-pretreated oocysts 24 hours later. After excystation, plates were washed, reincubated with fresh parasite growth medium at 37°C for 48 hours, then processed for ELISA. Results are expressed as means ±SD of 4 to 6 wells. **(A)** Dose response of numbers of oocysts required to obtain maximal infection. Cells were exposed to varying numbers of oocysts for 2 hour prior to washing and reincubation. **(B)** Time course to determine minimum time required for maximum infection of host cells. Cells were exposed to 3×10^4 oocysts for various intervals prior to being washed free of debris, then reincubated. (Adapted from Woods, K. M. et al., *FEMS Microbiol. Lett.*, 128, 89, 1995. With permission.)

ACKNOWLEDGMENTS

I thank Drs. K. M. Woods and M. V. Nesterenko for helpful suggestions during preparation of the manuscript. Much of our early research on *in vitro* cultivation was sponsored by NIH grant AI31774, and I am particularly indebted to Pfizer Animal Health for funding development of the microtiter ELISA.

FURTHER READING

Since this chapter was written, several additional papers dealing with *in vitro* cultivation of *C. parvum* have appeared. These include inhibitory effects of cycloheximide and anti-fibronectin serum on parasite infections in mock cells (Rosales et al., *Acta Trop.,* 60, 211, 1995); complete development in bovine fallopian tube epithelial cells (Yang et al., *Infec. Immun.,* 64, 349, 1996); and efficacy of 101 antimicrobials and other agents on parasite development in HCT-8 cells (Woods et al., *Ann. Trop. Med., Parasitol.,* 90, in press).

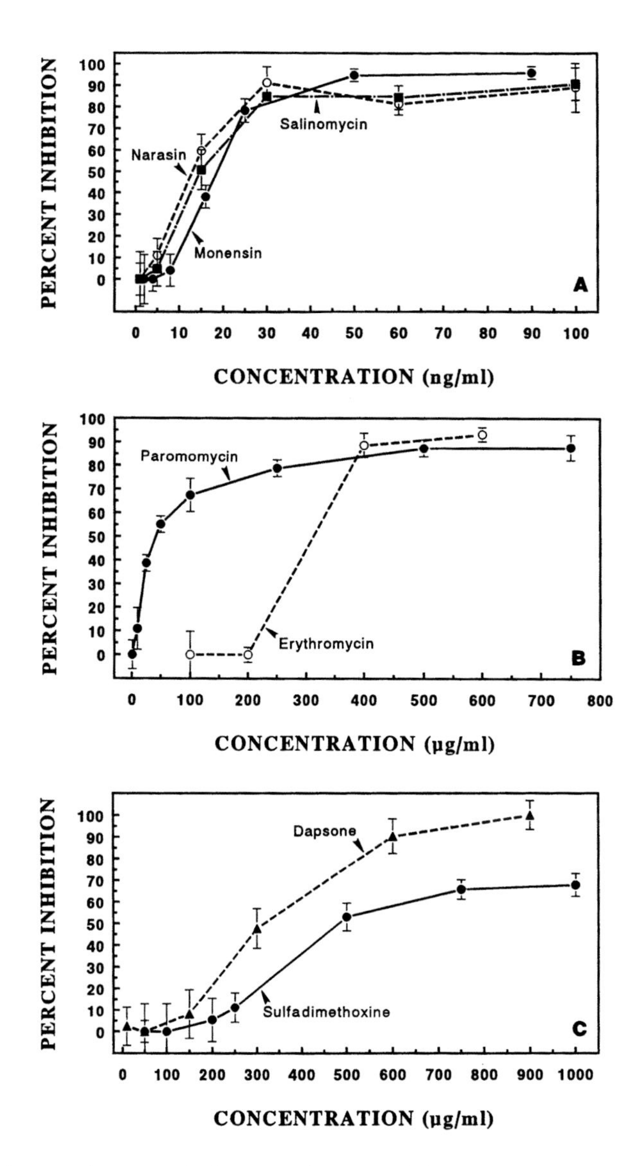

Figure 9 Efficacy of seven pharmaceuticals on development of *C. parvum in vitro.* Monolayers of HCT-8 cells in microtiter wells were each inoculated with 3.0×10^4 oocysts per well for 90 minutes at 37°C in the absence of any compound. Infected cells were washed, incubated at 37°C for 48 hours with various concentrations of pharmaceutical, then harvested for ELISA. Results are expressed as percent inhibition of the positive control. Error bars represent standard deviations that have been converted into percentages. (Portions adapted from Woods, K. M. et al., *FEMS Microbiol. Lett.,* 128, 89, 1995. With permission.)

REFERENCES

1. Doran, D. J., Cultivation of coccidia in avian embryos and cell culture, in *The Coccidia, Eimeria, Isospora, Toxoplasma, and Related Coccidia*, Hammond, D. M. and Long, P. L., Eds., University Park Press, Baltimore, 1973, 183.
2. Doran, D. J., Behavior of Coccidia *in vitro*, in *The Biology of the Coccidia*, Long, P. L., Ed., University Park Press, Baltimore, 1982, 229.
3. Speer, C. A., The coccidia, in In Vitro *Cultivation of Protozoan Parasites*, Jensen, J.B., Ed., CRC Press, Boca Raton, FL, 1983, 1.
4. Bland, J. O. W., Cultivation of a protozoan (*Toxoplasma*) in tissue cultures, *Arch. Exp. Zellforsch*, 14, 345, 1934.
5. Guimarães, F. N. and Meyer, H., Cultivo de "*Toxoplasma*" Nicolle e Manceaux, 1909, em culturas de tecidos, *Rev. Brasil. Biol.*, 2, 123, 1942.

6. Levaditi, C., Sanchis-Bayarri, V., Lepine, P., and Schoen, R., Étude sur l″encephalocyélite provoquee par le *Toxoplasma cuniculi*, *Ann. Inst. Pasteur.*, Paris, 43, 673, 1929.
7. Meyer, H. and de Oliveira, M. X., Observacões sobre divisoes mitoticas em células parasitadas, *Ann. Acad. Brasil. Cienc.*, 14, 289, 1942.
8. Meyer, H. and de Oliveira, M. X., Concervacoes de protozoáires em culteras de tecido mantidas a temperátura ambiente, *Rev. Brasil. Biol.*, 3, 341, 1943.
9. Sabin, A.B. and Olitsky, P.K., *Toxoplasma* and obligate intracellular parasitism, *Science*, 85, 336, 1937.
10. Bigalke, R.D., Preliminary communication on the cultivation of *Besnoitia besnoiti* (Marotel, 1912) in tissue culture and embryonated eggs, *J. S. Afr. Med. Assoc.*, 23, 523, 1962.
11. Fayer, R., *Sarcocystis*: development in cultured avian and mammalian cells, *Science*, 168, 1104, 1970.
12. Patton, W. H., *Eimeria tenella*: cultivation of the asexual stages in cultured animal cells, *Science*, 150, 767, 1965.
13. Woodmansee, D. B., Isolation, *in vitro* excystation, and *in vitro* development of *Cryptosporidium* sp. from calves, Ph.D. dissertation, Iowa State University, 1986, 1.
14. Woodmansee, D. B. and Pohlenz, J. F. L., Development of *Cryptosporidium* sp. in a human rectal tumor cell line, *Proc. Fourth Int. Symp. Neonatal Diarrhea,* 3–5 Oct. 1983, University of Saskatchewan, Canada, Veterinary Infectious Diseases Organization (VIDO), Saskatoon, Canada, 1983, 306–319.
15. Current, W. L. and Haynes, T. B., Complete development of *Cryptosporidium* in cell culture, *Science*, 224, 603, 1984.
16. Current, W. L., Techniques and laboratory maintenance of *Cryptosporidium*, in *Cryptosporidiosis of Man and Animals*, Dubey, J. P., Speer, C. A., and Fayer, R., Eds., CRC Press, Boca Raton, FL, 1990, 31.
17. Bjorneby, J. M., Leach, D. R., and Perryman, L. E., Persistent cryptosporidiosis in horses with severe combined immunodeficiency, *Infect. Immun.*, 59, 3823, 1991.
18. Fayer, R., Effect of sodium hypochlorite exposure on infectivity of *Cryptosporidium parvum* oocysts for neonatal BALB/c mice, *Appl. Environ. Microbiol.*, 61, 844, 1995.
19. Harp, J. A., Wannemuehler, M. W., Woodmansee, D. B., and Moon, H. W., Susceptibility of germfree or antibiotic-treated adult mice to *Cryptosporidium parvum*, *Infect. Immun.*, 56, 2006, 1988.
20. Heine, J., Pohlenz, J. F. L., Moon, M. W., and Woode, G. N., Enteric lesions and diarrhea in gnotobiotic calves monoinfected with *Cryptosporidium* species, *J. Infect. Dis.*, 150, 768, 1984.
21. Liebler, E. M., Pohlenz, J. F., and Woodmansee, D. B., Experimental intrauterine infection of adult BALB/c mice with *Cryptosporidium* sp., *Infect. Immun.*, 54, 255, 1986.
22. Meulbroek, J. A., Novilla, M. N., and Current, W. L., An immunosuppressed rat model of respiratory cryptosporidiosis, *J. Protozool.*, 38, 113s, 1991.
23. Perryman, L. E., Mason, P. H., and Chrisp, C. E., Effect of spleen cell populations on resolution of *Cryptosporidium parvum* infection in SCID mice, *Infect. Immun.*, 62, 1474, 1994.
24. Perryman, L. E., Riggs, M. W., Mason, P. H., and Fayer, R., Kinetics of *Cryptosporidium parvum* sporozoite neutralization by monoclonal antibodies, immune bovine serum, and immune bovine colostrum, *Infect. Immun.*, 58, 257, 1990.
25. Ruxton, G. D., Mathematical modeling of ammonia volatilization from slurry stores and its effect on *Cryptosporidium* oocyst viability, *J. Agr. Sci.*, 124, 55, 1995.
26. Arrowood, M. J. and Sterling, C. R., Isolation of *Cryptosporidium* oocysts and sporozoites using discontinuous sucrose and isopycnic percoll gradients, *J. Parasitol.*, 73, 314, 1987.
27. Mead, J. R., Arrowood, M. J., and Sterling, C. R., Antigens of *Cryptosporidium* sporozoites recognized by immune sera of infected animals and humans, *J. Parasitol.*, 74, 135, 1988.
28. Mead, J. R., Arrowood, M. J., Current, W. L., and Sterling, C. R., Field inversion gel electrophoresis separation of *Cryptosporidium* spp. chromosome-sized DNA, *J. Parasitol.*, 74, 366, 1988.
29. Robertson, L. J., Campbell, A. T., and Smith, H. V., Survival of *Cryptosporidium parvum* oocysts under various environmental pressures, *Appl. Environ. Microbiol.*, 58, 3494, 1992.
30. Robertson, L. J., Campbell, A. T., and Smith, H. V., *In vitro* excystation of *Cryptosporidium parvum*, *Parasitology*, 106, 13, 1993.
31. Tilley, M. and Upton, S. J., Electrophoretic characterization of *Cryptosporidium parvum* (KSU-1 isolate) (Apicomplexa: Cryptosporidiidae), *Can. J. Zool.*, 68, 1513, 1990.
32. Nesterenko, M. V., Kansas State University, Manhattan, KS, unpublished data.
33. Campbell, A. T., Robertson, L. J., and Smith, H. V., Viability of *Cryptosporidium parvum* oocysts: correlation of *in vitro* excystation with inclusion or exclusion of fluorogenic vital dyes, *Appl. Environ. Microbiol.*, 58, 3488, 1992.
34. McDonald, V., Stables, R., Warhurst, D. C., Barer, M. R., Blewett, D. A., Chapman, H. D., Connolly, G. M., Chiodini, P. L., and McAdam, K. P. W. J., *In vitro* cultivation of *Cryptosporidium parvum* and screening for anticryptosporidial drugs, *Antimicrob. Agents Chemother.*, 34, 1498, 1990.
35. Riggs, M. W. and Perryman, L. E., Infectivity and neutralization of *Cryptosporidium parvum* sporozoites, *Infect. Immun.*, 55, 2081, 1987
36. Riggs, M. W., McGuire, T. C., Mason, P. H., and Perryman, L. E., Neutralization-sensitive epitopes are exposed on the surface of infectious *Cryptosporidium parvum* sporozoites, *J. Immunol.*, 143, 1340, 1989.
37. Ryley, J. F., Meade, R., Hazelhurst, J., and Robinson, T. E., Methods in coccidiosis research: separation of oocysts from faeces, *Parasitology*, 73, 311, 1976.

38. Sheather, L., The detection of intestinal protozoa and mange parasites by a flotation technique, *J. Comp. Pathol. Ther.*, 36, 266, 1923.
39. Gardner, A. L., Roche, J. K., Weikel, C. S., and Guerrant, R. L., Intestinal cryptosporidiosis: pathophysiologic alterations and specific cellular and humoral immune responses in rnu/+ and rnu/rnu (athymic) rats, *Am. J. Trop. Med. Hyg.*, 44, 49, 1991.
40. Ungar, B. L. P., Burris, J. A., Quinn, C. A., and Finkelman, F. D., New mouse models for chronic *Cryptosporidium* infection in immunodeficient hosts, *Infect. Immun.*, 58, 961, 1990.
41. Ungar, B. L. P., Kao, T. C., Burris, J. A., and Finkelman, F. D., *Cryptosporidium* infection in an adult mouse model. Independent roles for IFN-γ and $CD4^+$ T lymphocytes in protective immunity, *J. Immunol.*, 147, 1014, 1991.
42. Eggleston, M. T., Development of a ^{3}H-uracil incorporation assay to monitor *in vitro* development of *Cryptosporidium parvum*, M.S. thesis, Kansas State University, 1993, 1.
43. Heyman, M. B., Shigekuni, L. K., and Ammann, A. J., Separation of *Cryptosporidium* oocysts from fecal debris by density gradient centrifugation and glass bead columns, *J. Clin. Microbiol.*, 23, 789, 1986.
44. Dempster, R. P., *Toxoplasma gondii*: purification of zoites from peritoneal exudates by eight methods, *Exp. Parasitol.*, 57, 195, 1984.
45. Dulska, P. and Turner, M., The purification of sporocysts and sporozoites from *Eimeria tenella* oocysts using Percoll® density gradients, *Avian. Dis.*, 32, 235, 1988.
46. Masihi, K. N., A simple method for separating *Toxoplasma gondii* from infected mouse peritoneal exudate cells, *Zbl. Bakt. Hyg. I. Abt. Orig. A.*, 233, 556, 1975.
47. Taghi-Kilani, R. and Sekla, L., Purification of *Cryptosporidium* oocysts and sporozoites by cesium chloride and Percoll® gradients, *Am. J. Trop. Med. Hyg.*, 36, 505, 1987.
48. Waldman, E., Tzipori, S., and Forsyth, J. R. L., Separation of *Cryptosporidium* species oocysts from feces by using a Percoll® discontinuous density gradient, *J. Clin. Microbiol.*, 23, 199, 1986.
49. Upton, S. J., Tilley, M., Nesterenko, M. V., and Brillhart, D. B., A simple and reliable method of producing *in vitro* infections of *Cryptosporidium parvum* (Apicomplexa), *FEMS Microbiol. Lett.*, 118, 45, 1994.
50. Reduker, D. W., Speer, C. A., and Blixt, J. A., Ultrastructural changes in the oocyst wall during excystation of *Cryptosporidium parvum* (Apicomplexa: Eucoccidiorida), *Can. J. Zool.*, 63, 1892, 1985.
51. Woodmansee, D.B., Studies of *in vitro* excystation of *Cryptosporidium parvum* from calves, *J. Protozool.*, 34, 398, 1987.
52. Woodmansee, D. B., Powell, E. C., Pohlenz, J. F., and Moon, H. W., Factors affecting motility and morphology of *Cryptosporidium* sporozoites *in vitro*, *J. Protozool.*, 34, 295, 1987.
53. Fayer, R. and Leek, R. G., The effects of reducing conditions, medium, pH, temperature, and time on *in vitro* excystation of *Cryptosporidium*, *J. Protozool.*, 31, 567, 1984.
54. Reduker, D. W. and Speer, C. A., Factors influencing excystation in *Cryptosporidium* oocysts from cattle, *J. Parasitol.*, 71, 112, 1985.
55. Sundermann, C. A., Lindsay, D. S., and Blagburn, B. L., *In vitro* excystation of *Cryptosporidium baileyi* from chickens, *J. Protozool.*, 34, 28, 1987.
56. Marquardt, W.C., Observations on living *Eimeria nieschulzi* of the rat, *J. Parasitol.*, 49 (Abstr. 45), 28, 1963.
57. Upton, S. J., Current, W. L., Barnard, S. M., and Ernst, J. V., *In vitro* excystation of *Caryospora simplex* (Apicomplexa: Eimeriorina), *J. Protozool.*, 31, 293, 1984.
58. Nesterenko, M. V., Tilley, M., and Upton, S. J., A metallo-dependent cysteine proteinase of *Cryptosporidium parvum* associated with the surface of sporozoites, *Microbios.*, 83, 77, 1995.
59. Wisher, M. H. and Rose, M. E., *Eimeria tenella* sporozoites: the method of excystation affects the surface membrane proteins, *Parasitology*, 95, 479, 1987.
60. Hosek, J. E., Todd, K. S., and Kuhlenschmidt, M. S., Improved method for high-yield excystation and purification of infective sporozoites of *Eimeria* spp., *J. Protozool.*, 35, 583, 1988.
61. Ono, K., Horn, K., and Heydorn, A. O., Purification of *in vitro* excysted *Sarcocystis* sporozoites by passage through a modified DE 52 anion-exchange column, *Parasitol. Res.*, 77, 717, 1991.
62. Schmatz, D. M., Crane, M. S. J., and Murray, P. K., Purification of *Eimeria tenella* sporozoites by DE-52 anion-exchange chromatography, *J. Protozool.*, 31, 181, 1984.
63. Regan, S., Cama, V., and Sterling, C. R., *Cryptosporidium* merozoite isolation and purification using differential centrifugation techniques, *J. Protozool.*, 38, 202s, 1991.
64. Nesterenko, M. V. and Upton, S. J., A rapid microcentrifuge procedure for purification of *Cryptosporidium* sporozoites, *J. Microbiol. Meth.*, 25, 87, 1996.
65. McDougald, L. D., The growth of avian *Eimeria in vitro*, in *Avian Coccidiosis*, Long, P. L., Boorman, K. N., and Freeman, B. M., Eds., *Brit. Poult. Sci. Ltd.*, Edinburgh, 1978, 185.
66. Schmatz, D. M., *In vitro* cultivation of the avian coccidia, *Adv. Cell Cult.*, 5, 241, 1987.
67. Strout, R. G. and Schmatz, D. M., Recent advances in the *in vitro* cultivation of the coccidia, in *Coccidiosis of Man and Domestic Animals*, Long, P. L., Ed., CRC Press, Boca Raton, FL, 1990, 221.
68. Doran, D. J. and Vetterling, J. M., Influence of storage period on excystation and development in cell culture of sporozoites of *Eimeria meleagrimitis* Tyzzer, 1929, *Proc. Helminthol. Soc. Wash.*, 36, 33, 1969.
69. Leek, R. G. and Fayer, R., Survival of sporocysts of *Sarcocystis* in various media, *Proc. Helminthol. Soc. Wash.*, 46, 151, 1979.

70. Sherwood, D., Angus, K. W., Snodgrass, D. R., and Tzipori, S., Experimental cryptosporidiosis in laboratory mice, *Infect. Immun.*, 38, 471, 1982.
71. Adams, R. B., Guerrant, R. L., Zu, S., Fang, G., and Roche, J. K., *Cryptosporidium parvum* infection of intestinal epithelium: morphologic and functional studies in an *in vitro* model, *J. Infect. Dis.*, 169, 170, 1994.
72. Arrowood, M. J., Jaynes, J. M., and Healey, M. C., *In vitro* activities of lytic peptides against the sporozoites of *Cryptosporidium parvum*, *Antimicrob. Agents Chemother.*, 35, 224, 1991.
73. Davis, T., Patten, K., Moyad, M., and Rose, J. B., Viability of *Cryptosporidium parvum* in marine waters, in *Proc. 95th Ann. Meet., Am. Soc. Microbiol.*, Washington, D.C., 21–25 May, 1995, American Society for Microbiology, Abstract Q-162, 428, 1995.
74. Fricker, C. R., Viability of *Cryptosporidium parvum* oocysts concentrated by calcium carbonate flocculation, *J. Appl. Bacteriol.*, 77, 120, 1994.
75. Hamer, D. H., Ward, H., Tzipori, S., Pereira, M. E. A., Alroy, J. P., and Keusch, G. T., Attachment of *Cryptosporidium parvum* sporozoites to MDCK cells *in vitro*, *Infect. Immun.*, 62, 2208, 1994.
76. Patton, K. L., Viabililty of *Cryptosporidium parvum* oocysts in beverages, in *Proc. 95th Ann. Meet., Am. Soc. Microbiol.*, Washington, D.C., 21–25 May, 1995, American Society for Microbiology, Abstract P-20, 385, 1995.
77. Smith, H. V., Grimason, A. M., Benton, C., and Parker, J. F. W., The occurrence of *Cryptosporidium* spp. oocysts in Scottish waters and the development of a fluorogenic viability assay for individual *Cryptosporidium* spp. oocysts, in *Health Related Water Microbiology 1990*, Grabow, W. O. K., Morris, R., and Botzenhart, E., Eds., Pergamon Press, Oxford, 1991, 169.
78. Korich, D. G., Mead, J. R., Madore, M. S., Sinclair, N. A., and Sterling, C. R., Effects of ozone, chlorine dioxide, chlorine, and monochloramine on *Cryptosporidium parvum* oocyst viability, *Appl. Environ. Microbiol.*, 56, 1423, 1990.
79. Upton, S. J. and Tilley, M., Effects of reduced oxygen atmosphere on motility, penetration of host cells, and intracellular survival of *Eimeria nieschulzi* sporozoites *in vitro*, *J. Helminthol. Soc. Wash.*, 62, 223, 1995.
80. Ditrich, O., Kopácek, P., and Zucerová, Z., Antigenic characterization of human isolates of cryptosporidia, *Folia Parasitol.*, 40, 301, 1993.
81. Fayer, R., Development of a precocious strain of *Cryptosporidium parvum* in neonatal calves, *J. Euk. Microbiol.*, 41, 40s, 1994.
82. Lumb, R., Lanser, J. A., and O'Donoghue, P. J. O., Electrophoretic and immunoblot analysis of *Cryptosporidium* oocysts, *Immunol. Cell Biol.*, 66, 369, 1988.
83. Mead, J. R., Humphreys, R. C., Sammons, D. W., and Sterling, C. R., Identification of isolate-specific sporozoite proteins of *Cryptosporidium parvum* by two-dimensional gel electrophoresis, *Infect. Immun.*, 58, 2071, 1990.
84. Nina, J. M. S., McDonald, V., Deer, R. M. A., Wright, S. E., Dyson, D. A., Chiodini, P. L., and McAdam, K. P. W. J., Comparative study of the antigenic composition of oocyst isolates of *Cryptosporidium parvum* from different hosts, *Parasite Immunol.*, 14, 227, 1992.
85. Dubey, J. P., Speer, C. A., and Fayer, R., Eds., *Sarcocystosis of Animals and Man*, CRC Press, Boca Raton, FL, 1989, 93.
86. Lindsay, D. S., Blagburn, B. L., and Ernest, J. A., Experimental *Cryptosporidium parvum* infections in chickens, *J. Parasitol.*, 73, 242, 1987.
87. Lindsay, D. S., Hendrix, C. M., and Blagburn, B. L., Experimental *Cryptosporidium parvum* infections in opossums (*Didelphis virginiana*). *J. Wildlf. Dis.*, 24, 157, 1988.
88. Uhl, E. W., O'Connor, R. M., Perryman, L. E., and Riggs, M. W., Neutralization-sensitive epitopes are conserved among geographically diverse isolates of *Cryptosporidium parvum*, *Infect. Immun.*, 60, 1703, 1992.
89. Upton, S. J., Tilley, M., and Brillhart, D. B., Comparative development of *Cryptosporidium parvum* (Apicomplexa) in 11 continuous host cell lines, *FEMS Microbiol. Lett.*, 118, 233, 1994
90. Gut, J., Petersen, C., Nelson, R. G., and Leech, J., *Cryptosporidium parvum*: *in vitro* cultivation in Madin-Darby canine kidney cells, *J. Protozool.*, 38, 72s, 1991.
91. Dvorak, J. and Howe, C., *Toxoplasma gondii*-vertebrate cell interactions. I. The influence of bicarbonate ion, CO_2, pH and host cell culture age on the invasion of vertebrate cells *in vitro*, *J. Protozool.*, 24, 416, 1977.
92. Hermentin, K. and Aspöck, H., Higher yields and increased purity of *in vitro* grown *Toxoplasma gondii*, *Zbl. Bakt. Parasit. Infek. Hyg. I, Abt. Orig. A*, 267, 272, 1987.
93. Porchet-Henneré, E., D'Hooghe, M. C., Sadak, A., and Frontier, S., Relations entre une coccidia *Besnoitia jellisoni*, et sa cellule-hote, *in vitro*, étudiées par microcinématographie, *Protistologica*, 21, 39, 1985.
94. Shkap, V., Bin, H., Lebovich, B., and Pipano, E., *Besnoitia besnoiti*: quantitative *in vitro* studies, *Vet. Parasitol.*, 39, 207, 1991.
95. Dvorak, J. and Crane, M. S. J., Vertebrate cell cycle modulates infection by protozoan parasites, *Science*, 214, 1034, 1981.
96. Crane, M. S. J., Schmatz, D. M., Stevens, S., Habbersett, M. C., and Murray, P. K., *Eimeria tenella*: *in vitro* development in irradiated bovine kidney cells, *Parasitology*, 88, 521, 1984.
97. Furado, G. C., Cao, Y., and Joiner, K. A., Laminin on *Toxoplasma gondii* mediates parasite binding to the $\beta1$ integrin receptor $\alpha6\beta1$ on human foreskin fibroblasts and Chinese hamster ovary cells, *Infect. Immun.*, 60, 4925, 1992.
98. Furado, G. C., Slowik, M., Kleinman, H. K., and Joiner, K. A., Laminin enhances binding of *Toxoplasma gondii* tachyzoites to J774 murine macrophage cells, *Infect. Immun.*, 60, 2337, 1992.

99. Jensen, J. B. and Trager, W., *Plasmodium falciparum* in culture: use of outdated erythrocytes and description of the candle jar method, *J. Parasitol.*, 63, 883, 1977.
100. Trager, W. and Jensen, J. B., Human malaria parasites in continuous culture, *Science*, 193, 673, 1976.
101. Jensen, J. B., *Plasmodium*, in In Vitro *Cultivation of Protozoan Parasites*, Jensen, J. B., Ed., CRC Press, Boca Raton, FL, 1983, 155.
102. Chandler-Conrey, N. E., *Eimeria tenella*: effect of oxygen concentration on parasite development in cell culture, M.S. thesis, University of New Hampshire, 1987.
103. Fukata, T., Sasai, K., Arakawa, A., and McDougald, L. R., Penetration of *Eimeria tenella* sporozoites under different oxygen concentrations *in vitro*, *J. Parasitol.*, 78, 537, 1992.
104. Kogut, M. H. and Patton, W. H., Effects of a low oxygen atmosphere on the asexual development of *Eimeria tenella* in a mammalian cell line, in *Proc. 54th Ann. Meet, Am. Soc. Parasitol.*, 29 July–3 Aug., 1979, Minneapolis, MN, Abstract 113, Allen Press, Lawrence, KS, 1979, 52.
105. Tilley, M. and Upton, S.J., A comparative study of the development of *Eimeria nieschulzi in vitro* under aerobic and reducing conditions, *J. Parasitol.*, 74. 1042, 1988.
106. Wrede, D., Salisch, H., and Siegmann, O., Oxygen concentration and asexual development of *Eimeria tenella* in cell cultures, *J. Vet. Med. B.*, 40, 391, 1993.
107. Ricketts, A. P., *Eimeria tenella*: growth and drug sensitivity in tissue culture under reduced oxygen, *Exp. Parasitol.*, 74, 463, 1992.
108. Upton, S. J., Tilley, M., and Brillhart, D. B., Comparative development of *Cryptosporidium parvum* in MDBK and HCT-8 cells under select atmospheres, *Biomed. Lett.*, 49, 265, 1994.
109. Cook, M. K. and Jacobs, L., Cultivation of *Toxoplasma gondii* in tissue cultures of various derivitives, *J. Parasitol.*, 44, 172, 1958.
110. Hughes, H. P. A., Bolk, R. J., Gerhardt, S. A., and Speer, C. A., Susceptibility of *Eimeria bovis* and *Toxoplasma gondii* to oxygen intermediates and a new mathematical model for parasite killing, *J. Parasitol.*, 75, 489, 1989.
111. Michalski, W. P. and Prowse, S. J., Superoxide dismutases in *Eimeria tenella*, *Mol. Biochem. Parasitol.*, 47, 189, 1991.
112. Wack, M. F., Greenfield, R. A., Mosier, D. A., and Kuhls, T. L., Superoxide dismutase activity in *Cryptosporidium parvum*, in *Proc. 47th Ann. Meet. Soc. Protozool.*, Cleveland State University, 24–29 June, 1994, Society of Protozoologists, Abstract C37, Allen Press, Lawrence, KS, 1994, 75.
113. Burns, W. C., The lethal effects of *Eimeria tenella* extracts on rabbits, *J. Parasitol.*, 45, 38, 1959.
114. Doran, D. J., Survival and development of *Eimeria adenoides* in cell cultures inoculated with sporozoites from cleaned and uncleaned syspensions, *Proc. Helminthol. Soc. Wash.*, 37, 45, 1970.
115. Eggleston, M. T., Tilley, M., and Upton, S. J., Enhanced development of *Cryptosporidium parvum in vitro* by removal of oocyst toxins from infected cell monolayers, *J. Helminthol. Soc. Wash.*, 61, 122, 1994.
116. Fayer, R. and Hammond, D. M., Development of first generation schizonts of *Eimeria bovis* in cultured bovine cells, *J. Protozool.*, 14, 764, 1967.
117. Rickimaru, M. T., Galysh, F. T., and Shumard, R. F., Some pharmacological aspects of a toxic substance from oocysts of the coccidium *Eimeria tenella*, *J. Parasitol.*, 47, 407, 1961.
118. Sharma, N. N. and Foster, J. W., Toxic substances in various constituents of *Eimeria tenella* oocysts, *Am. J. Vet. Res.*, 25, 211, 1964.
119. Doran, D. J., Increasing the yield of *Eimeria tenella* oocysts in cell culture, *J. Parasitol.*, 57, 891, 1971.
120. Hofmann, J. and Raether, W., Improved techniques for the *in vitro* cultivation of *Eimeria tenella* in primary chick kidney cells, *Parasitol. Res.*, 76, 479, 1990.
121. Reduker, D. W. and Speer, C. A., Effect of sporozoite inoculum size on *in vitro* production of merozoites of *Eimeria bovis* (Apicomplexa), *J. Parasitol.*, 73, 427, 1987.
122. Strout, R. G., Ouellette, C. A., and Gangi, D. P., Effect of inoculum size on development of *Eimeria tenella* in cell cultures, *J. Parasitol.*, 55, 406, 1969.
123. Woods, K. M., Nesterenko, M. V., and Upton, S. J., Development of a microtitre ELISA to quantify development of *Cryptosporidium parvum in vitro*, *FEMS Microbiol. Lett.*, 128, 89, 1995.
124. Rasmussen, K. R., Larsen, N. C., and Healey, M. C., Complete development of *Cryptosporidium parvum* in a human endometrial carcinoma cell line, *Infect. Immun.*, 61, 1482, 1993.
125. Endo, T. and Yagita, K., Effect of extracellular ions on motility and cell entry in *Toxoplasma gondii*, *J. Protozool.*, 37, 133, 1990.
126. Endo, T., Tokuda, K., Yagita, K., and Koyama, T., Effects of extracellular potassium on acid release and motility initiation in *Toxoplasma gondii*, *J. Protozool.*, 34, 291, 1987.
127. Mondragon, R., Meza, I., and Frixione, E., Divalent cation and ATP dependent motility of *Toxoplasma gondii* tachyzoites after mild treatment with trypsin, *J. Euk. Microbiol.*, 41, 330, 1994.
128. Martin, S. K., Miller, L. H., Nijhout, M. M., and Carter, R., *Plasmodium gallinaceum*: induction of male gametocyte exflagellation by phosphodiesterase inhibitors, *Exp. Parasitol.*, 44, 239, 1978.
129. Nijhout, M. M. and Carter, R., Gamete development in malaria organisms: bicarbonate-dependent stimulation by pH *in vitro*, *Parasitology*, 76, 39, 1978.
130. Kawamoto, F., Alejo-Blanco, R., Fleck, S. L., Kawamoto, Y., and Sinden, R. E., Possible roles of Ca^{2+} and cGMP as mediators of the exflagellation of *Plasmodium berghei* and *Plasmodium falciparum*, *Mol. Biochem. Parasitol.*, 42, 101, 1990.

131. Kawamoto, F., Alejo-Blanco, R., Fleck, S. L., and Sinden, R. E., *Plasmodium berghei*: ionic regulation and the induction of gametogenesis, *Exp. Parasitol.*, 72, 33, 1991.
132. Kawamoto, F., Kido, N., Hanaichi, T., Djamgoz, M. B. A., and Sinden, R. E., Gamete development in *Plasmodium berghei* regulated by ionic exchange mechanisms, *Parasitol. Res.*, 78, 277, 1992.
133. Kelly, G. L. and Hammond, D. M., Development of *Eimeria ninakohlyakimovi* from sheep in cultured cells, *J. Protozool.*, 17, 340, 1970.
134. Long, P. L. and Speer, C. A., Invasion of host cells by coccidia, in *Parasite Invasion, Proc. 15th Symp. Br. Soc. Parasitol.*, Taylor, A. E. R. and Muller, R., Eds., Blackwell Publications, Oxford, 1977, 1
135. Upton, S. J. and Tilley, M., Effects of select medium supplements on motility and development of *Eimeria nieschulzi in vitro*, *J. Parasitol.*, 78, 329, 1992.
136. Upton, S. J., Tilley, M., and Brillhart, D. B., Effects of select medium supplements on *in vitro* development of *Cryptosporidium parvum* in HCT-8 cells, *J. Clin. Microbiol.*, 33, 371, 1995.
137. Mitschler, R. R., Lipid biology of *Cryptosporidium parvum* and *Eimeria nieschulzi* (Apicomplexa), Ph.D. dissertation, Kansas State University, 1994, 1.
138. Bekhti, K. and Pery, P., *In vitro* interactions between murine macrophages and *Eimeria falciformis* sporozoites, *Res. Immunol.*, 140, 697, 1989.
139. Crane, M. S. J. and Dvorak, J. A., Influence of monosaccharides on the infection of vertebrate cells by *Trypanosoma cruzi* and *Toxoplasma gondii*, *Mol. Biochem. Parasitol.*, 5, 333, 1982.
140. Tanabe, K., Kimata, I., and Takada, S., Penetration of chick embryo erythrocytes by *Toxoplasma gondii* tachyzoites in simplified incubation media, *J. Parasitol.*, 66, 240, 1980.
141. Kuhls, T. L., Mosier, D. A., and Crawford, D. L., Effects of carbohydrates and lectins on cryptosporidial sporozoite penetration of cultured cell monolayers, *J. Protozool.*, 38, 74s, 1991.
142. Kyle, D. E. and McDougald, L. R., Effect of insulin and pancreatic polypeptide on *in vitro* development of *Eimeria tenella*, in *Proc. 60th Ann. Meet., Am. Soc. Parasitol.*, 4–8 August, 1985, Athens, GA, Abstract 113, Allen Press, Lawrence, KS, 1985, 48.
143. Doran, D. J. and Augustine, P. C., *Eimeria tenella*: vitamin requirements for development in primary cultures of chicken kidney cells, *J. Protozool.*, 25, 544, 1978.
144. James, S., Thiamine uptake in isolated schizonts of *Eimeria tenella* and the inhibitory effects of amprolium, *Parasitology*, 80, 313, 1980.
145. Latter, V. S. and Holmes, L. S., Identification of some nutrient requirements for the *in vitro* cultivation of *Eimeria tenella*, in *Proc. 19th Ann. Meet., Brit. Soc. Parasitol.*, Blackwell Publications, Oxford, 1985, 19.
146. Ryley, J. F. and Wilson, R. G., Growth factor antagonism studies with coccidia in tissue culture, *Z. Parasitenkd.*, 40, 31, 1972.
147. Sanford, J. C., Nutritional requirements of *Eimeria tenella* (coccidia) propagated in cell culture using a chemically defined medium, M.S. thesis, University of New Hampshire, 1983, 1.
148. Warren, E. W., Vitamin requirements of the Coccidia of the chicken, *Parasitology*, 58, 137, 1968.
149. Flanigan, T. P., Aji, T., Marshall, R., Soave, R., Aikawa, M., and Kaetzel, C., Asexual development of *Cryptosporidium parvum* within a differentiated human enterocyte cell line, *Infect. Immun.*, 59, 234, 1991.
150. Wiest, P. M., Johnson, J. H., and Flanigan, T. P., Microtubule inhibitors block *Cryptosporidium parvum* infection of a human enterocyte cell line, *Infect. Immun.*, 61, 4888, 1993.
151. Current, W. L. and Reese, N. C., A comparison of endogenous development of three isolates of *Cryptosporidium* in suckling mice, *J. Protozool.*, 33, 98, 1986.
152. Mitschler, R. R., Welti, R., and Upton, S. J., A comparative study of lipid compositions of *Cryptosporidium parvum* (Apicomplexa) and Madin-Darby bovine kidney cells, *J. Euk. Microbiol.*, 41, 8, 1994.
153. Pfefferkorn, E. R. and Pfefferkorn, L. C., Arabinosyl nucleosides inhibit *Toxoplasma gondii* and allow the selection of resistant mutants, *J. Parasitol.*, 62, 993, 1976.
154. Schwartzman, J. D. and Pfefferkorn, E. R., *Toxoplasma gondii*: purine synthesis and salvage in mutant host cells and parasites, *Exp. Parasitol.*, 53, 77, 1982.
155. Wang, C. C., Biochemistry and physiology of coccidia, in *The Biology of the Coccidia*, Long, P. L., Ed., University Park Press, Baltimore, 1982, 167.
156. Wang, C. C. and Simashkevich, P. M., A comparative study of the biological activities of arprinocid and arprinocid-1-N-oxide, *Mol. Biochem. Parasitol.*, 1, 335, 1980.
157. Wang, C. C., Simashkevich, P. M., and Stotish, R. L., Mode of anticoccidial action of arprinocid, *Biochem. Pharmacol.*, 28, 2241, 1979.
158. Donald, R. G. K. and Roos, D. S., Insertional mutagenesis and marker rescue in a protozoan parasite: cloning of the uracil phosphoribosyl transferase locus from *Toxoplasma gondii*, *Proc. Natl. Acad. Sci. U.S.A.*, 92, S749, 1995.
159. Wang, C. C. and Simashkevich, P. M., Purine metabolism in a protozoan parasite *Eimeria tenella*, *Proc. Natl. Acad. Sci. U.S.A.*, 78, 6618, 1981.
160. Mayberry, L. F. and Marquardt, W. C., Nucleic acid precursor incorporation by *Eimeria nieschulzi* (Protozoa: Apicomplexa) and jejunal villus epithelium, *J. Protozool.*, 21, 599, 1974.
161. Morgan, K. and Canning, E. U., Incorporation of ^{3}H-thymidine and ^{3}H-adenosine by *Eimeria tenella* grown in chick embryos, *J. Parasitol.*, 60, 364, 1974.

162. Ouellette, C. A., Strout, R. G., and McDougald, L. R., Incorporation of radioactive pyrimidine nucleosides into DNA and RNA of *Eimeria tenella* (coccidia) cultured *in vitro*, *J. Protozool.*, 20, 150, 1973.
163. Ouellette, C. A., Strout, R. G., and McDougald, L. R., Thymidylic acid synthesis of *Eimeria tenella* (coccidia) cultured *in vitro*, *J. Protozool.*, 21, 398, 1974.
164. Perrotto, J., Keister, D. B., and Gelderman, A. H., Incorporation of precursors into *Toxoplasma* DNA, *J. Protozool.*, 18, 470, 1971.
165. Roberts, W. L., Elsner, Y. Y., Shigematsu, A., and Hammond, D. M., Lack of incorporation of ^{3}H-thymidine into *Eimeria callospermophili* in cell cultures, *J. Parasitol.*, 56, 833, 1970.
166. Iltzsch, M. H., Pyrimidine salvage pathways of *Toxoplasma gondii*, *J. Euk. Microbiol.*, 40, 24, 1993.
167. Pfefferkorn, E. R. and Pfefferkorn, L. C., *Toxoplasma gondii*: specific labeling of nucleic acids of intracellular parasites in Lesch-Nyhan cells, *Exp. Parasitol.*, 41, 95, 1977.
168. Schwartzman, J. D. and Pfefferkorn, E. R., Pyrimidine synthesis by intracellular *Toxoplasma gondii*, *J. Parasitol.*, 67, 150, 1981.
169. Cama, V. A., Ortega, Y. R., and Sterling, C. R., *In vitro* evaluation of clarithromycin, 14-OH clarithromycin and paromomycin against *Cryptosporidium parvum* infectivity, in *Proc. 95th Ann. Meet., Am. Soc. Microbiol.*, Washington, D.C., 21–25 May, 1995, American Society for Microbiology, Abstract A-120, 164, 1995.
170. Doyle, P. S., Crabb, J., and Petersen, C., Anti-*Cryptosporidium parvum* antibodies inhibit infectivity *in vitro* and *in vivo*, *Infect. Immun.*, 61, 4079, 1993.
171. Gut, J., Peterson, C., Nelson, R. G., and Leech, J., *Cryptosporidium parvum*: *in vitro* cultivation in Madin-Darby canine kidney cells, in *Proc. 44th Ann. Meet., Soc. Protozool.*, Montana State University, 28 June–2 July, 1991, Society of Protozoologists, Abstract WS40, Allen Press, Lawrence, KS, 1991, 52.
172. Marshall, R. J. and Flanigan, T. P., Paromomycin inhibits *Cryptosporidium* infection of a human enterocyte cell line, *J. Infect. Dis.*, 165, 772, 1992.
173. Martinez, F., Mascaro, C., Rosales, M. J., Diaz, J., Cifuentes, J., and Osuna, A., *In vitro* multiplication of *Cryptosporidium parvum* in mouse peritoneal macrophages, *Vet. Parasitol.*, 42, 27, 1992.
174. Naciri, M., Yvore, P., de Boissieu, C., and Esnault, E., Multiplication de *Cryptosporidium muris* (Tyzzer 1907) *in vitro* entretien d'une souche sur oeufs embryonnés, *Rec. Méd. Vét.*, 162, 51, 1986.
175. Rosales, M. J., Cifuentes, J., and Mascaró, C., *Cryptosporidium parvum*: culture in MCDK cells, *Exp. Parasitol.*, 76, 209, 1993.
176. Rosales, M. J., Lazcano, C. M., Arnedo, T., and Castilla, J. J., Isolation and identification of *Cryptosporidium parvum* oocysts with continuous Percoll® gradients and combined alcian blue giensa staining, *Acta. Trop.*, 56, 371, 1994.
177. Favennec, L., Egraz-Bernard, M., Comby, E., Lemeteil, D., Ballet, J. J., and Brasseur, P., Immunoflourescence detection of *Cryptosporidium parvum* in Caco-2 cells: a new screening method for anticryptosporidial agents, *J. Euk. Microbiol.*, 41, 39s, 1994.
178. Griffiths, J. K., Moore, R., Dooley, S., Keusch, G. T., and Tzipori, S., *Cryptosporidium parvum* infection in Caco-2 cell monolayers induces an apical monolayer defect, selectively increases transmonolayer permeability, and causes epithelial cell death, *Infect. Immun.*, 62, 4506, 1994.
179. Gut, J. and Nelson, R. G., *Cryptosporidium parvum* sporozoites deposit trails of 11A5 antigen during gliding locomotion and shed 11A5 antigen during invasion of MDCK cells *in vitro*, *J. Euk. Microbiol.*, 41, 42s, 1994.
180. Tilley, M., Upton, S. J., Fayer, R., Barta, J. R., Chrisp, C. E., Freed, P. S., Blagburn, B. L., Anderson, B. C., and Barnard, S. M., Identification of a 15-kilodalton surface glycoprotein on sporozoites of *Cryptosporidium parvum*, *Infect. Immun.*, 59, 1002, 1991
181. Aji, T., Flanigan, T., Marshall, R., Kaetzel, C., and Aikawa, M., Ultrastructural study of asexual development of *Cryptosporidium parvum* in a human intestinal cell line, *J. Protozool.*, 38, 82s, 1991
182. Buraud, M., Forget, E., Favennec, L., Bizet, J., Gobert, J., and Deluol, A., Sexual stage development of cryptosporidia in the Caco-2 cell line, *Infect. Immun.*, 59, 4610, 1991.
183. Flanigan, T. P., Aji, T., Marshall, R., Aikawa, M., and Kaetzel, C., *Cryptosporidium* development from sporozoite to schizont *in vitro*, in *Proc. 44th Ann. Meet., Soc. Protozool.*, Montana State University, 28 June–2 July, 1991, Society of Protozoologists, Abstract WS44, Allen Press, Lawrence, KS, 1991, 53.
184. Lumb, R., Smith, K., O'Donoghue, P. J., and Lanser, J. A., Ultrastructure of the attachment of *Cryptosporidium* sporozoites to tissue culture cells, *Parasitol. Res.*, 74, 531, 1988.
185. Upton, S. J., Tilley, M., Mitschler, R. R., and Oppert, B. S., Incorporation of exogenous uracil by *Cryptosporidium parvum in vitro*, *J. Clin. Microbiol.*, 29, 1062, 1991.
186. Wagner, E. D. and Prabhu Das, M., *Cryptosporidium* in cell culture, *Jpn. J. Parasitol.*, 35, 253, 1986.
187. Pfefferkorn, E. R., *Toxoplasma gondii*: the enzymatic defect of a mutant resistant to 5-fluorodeoxyuridine, *Exp. Parasitol.*, 44, 26, 1978.
188. Pfefferkorn, E. R. and Pfefferkorn, L. C., *Toxoplasma gondii*: characterization of a mutant resistant to 5-fluorodeoxyuridine, *Exp. Parasitol.*, 42, 44, 1977.
189. Pfefferkorn, E. R. and Pfefferkorn, L. C., Specific labeling of intracellular *Toxoplasma gondii* with uracil, *J. Protozool.*, 24, 449, 1977.
190. Schmatz, D. M., Crane, M. S. J., and Murray, P. K., *Eimeria tenella*: parasite-specific incorporation of ^{3}H-uracil as a quantitative measure of intracellular development, *J. Protozool.*, 33, 109, 1986.
191. Schmatz, D. M., personnal communication.

192. Mitchell, A. and Finch, L. R., Pathways of nucleotide biosynthesis in *Mycoplasma mycoides* subsp. *mycoides*, *J. Bacteriol.*, 130, 1047, 1977.
193. Olson, J. A., *In situ* enzyme-linked immunosorbent assay to quantitate *in vitro* development of *Eimeria tenella*, *Antimicrob. Agents Chemother.*, 34, 1435, 1990.
194. Merli, A., Canessa, A., and Melioli, G., Enzyme immunoassay for evaluation of *Toxoplasma gondii* growth in tissue culture, *J. Clin. Microbiol.*, 21, 88, 1985.
195. Woods, K. M., Nesterenko, M. V., and Upton, S. J., Development of a microtiter ELISA for rapid screening of pharmaceuticals against *Cryptosporidium in vitro*, in *Proc. 70th Ann. Meet., Am. Soc. Parasitol.* and *40th Ann. Meet., Am. Assoc. Vet. Parasitol.*, 6–10 July, 1995, Pittsburgh, PA, Abstract 92, Allen Press, Lawrence, KS, 1995, 86.
196. Long, P. L., Some factors affecting the severity of infection with *Eimeria tenella* in chicken embryos, *Parasitology*, 60, 435, 1970
197. Current, W. L., *Cryptosporidium* sp. in chickens: parasite life cycle and aspects of acquired immunity, in *Research in Avian Coccidiosis, Proc. Georgia Coccidiosis Conf.,* McDougald, L. R., Joyner, L. P., and Long, P. L., Eds., University of Georgia, Athens, 1986, 124.
198. Current, W. L. and Long, P. L., Development of human and calf *Cryptosporidium* in chicken embryos, *J. Infect. Dis.*, 148, 1108, 1983.
199. Lindsay, D. S., Sundermann, C. A., and Blagburn, B. L., Cultivation of *Cryptosporidium baileyi*: studies with cell cultures, avian embryos, and pathogenicity of chicken embryo-passaged oocysts, *J. Parasitol.*, 74, 288, 1988.
200. Garza, D. H., Fedorak, R. N., and Soave, R., Enterotoxin-like activity in cultured cryptosporidia: role in diarrhea, *Gastroenterology*, 90 (Abstr.), 1424, 1986.
201. Datry, A., Danis, M., and Gentilini, M., Développement complet de *Cryptosporidium* en culture cellulaire: applications, *Méd./Sci.*, 5, 762, 1989.
202. Bonnin, A., Salimbeni, I., Dubremetz, J. F., Harly, G., Chavanet, P., and Camerlynck, P., Mise au point d'une modéle expérimental de culture *in vitro* des stades asexués de *Cryptosporidium* sp., *Ann. Parasit. Hum. Comp.*, 65, 41, 1990.
203. Cifuentes, J., Rosales, M. J., Diaz, J., Osuna, A., Boy, M., and Mascaro, C., Au sujet de la culture *in vitro* de *Cryptosporidium parvum*, *J. Protozool.*, 38 (Abstr. 141), 24A, 1991
204. Flanigan, T. P., Marshall, R., Ungar, B., and Kaetzel, C., *In vitro* screening of immunologic and pharmacologic therapies against *Cryptosporidium*, in *Proc. 44th Ann. Meet., Soc. Protozool.*, Montana State University, 28 June–2 July, 1991, Society of Protozoologists, Abstract WS106, Allen Press, Lawrence, KS, 1991, 63.
205. Kuhls, T. L., Mosier, D. A., and Crawford, D. L., Effects of carbohydrates (CHOs) and lectins (Lecs) on cryptosporidial sporozoite penetration of cultured cell monolayers, in *Proc. 44th Ann. Meet., Soc. Protozool.*, Montana State University, 28 June–2 July, 1991, Society of Protozoologists, Abstract WS41, Allen Press, Lawrence, KS, 1991, 52.
206. Lyagoubi, M., Malet, I., Datry, A., Danis, M., and Gentilini, M., Optimisation of a *Cryptosporidium parvum* cell cultures model, in *Proc. 44th Ann. Meet., Soc. Protozool.*, Montana State University, 28 June–2 July, 1991, Society of Protozoologists, Abstract WS42, Allen Press, Lawrence, KS, 1991, 52.
207. Joe, A., Hamer, D. H., Kelley, M. A., Pereira, M. E. A., Keusch, G. T., Tzipori, S., and Ward, H. D., Role of a gal/galNAc-specific sporozoite surface lectin in *Cryptosporidium parvum*-host cell interaction, *J. Euk. Microbiol.*, 41, 44s, 1994.
208. Oretega, Y. R. and Sterling, C. R., Improvements to the *in vitro* cultivation of *Cryptosporidium parvum*, in *Proc. 47th Ann. Meet., Soc. Protozool.*, Cleveland State University, 24–29 June, 1994, Society of Protozoologists, Abstract C66, Allen Press, Lawrence, KS, 1994, 79.
209. Thulin, J. D., Kuhlenschmidt, M. S., Rolsma, M. D., Current, W. L., and Gelberg, H. B., An intestinal xenograft model for *Cryptosporidium parvum* infection, *Infect. Immun.*, 62, 329, 1994.
210. Joe, A, Pereira, M. E. A., Keutsch, G. T., Tzipori, S., and Ward, H. D., Role of a galactose-specific *Cryptosporidium parvum* lectin in attachment of sporozoites to host cells, in *Proc. 95th Ann. Meet., Am. Soc. Microbiol.*, Washington, D.C., 21–25 May, 1995, American Society for Microbiology, Abstract B-472, 247, 1995.
211. Yang, S., Healey, M. C., Du, C., and Zhang, J., Complete development of *Cryptosporidium parvum* in bovine fallopian tube epithelial cells, in *Proc. 70th Ann. Meet., Am. Soc. Parasitol. and 40th Ann. Meet., Am. Assoc. Vet. Parasitol.*, 6–10 July, 1995, Pittsburgh, PA, Abstract 137, Allen Press, Lawrence, KS, 1995, 99.

Chapter 9

Laboratory Models of Cryptosporidiosis

David S. Lindsay

CONTENTS

0-8493-7695-5/97/$0.00+$.50

I. INTRODUCTION

Cryptosporidium parvum[1] is the species most commonly used in laboratory studies because it is a human and domestic animal pathogen, it is infectious for a wide variety of laboratory mammals, and large numbers of oocysts can be obtained, usually from young ruminants. Most of this chapter will concern *C. parvum* and mammalian hosts. *Cryptosporidium wrairi*,[2] confined primarily to guinea pigs, is discussed in reference to transmission and as a model for pathology. *Cryptosporidium muris*[3,4] is briefly discussed with reference to transmission studies.

Cryptosporidium baileyi[5] is most often used in studies of avian cryptosporidiosis because it produces more oocysts than *Cryptosporidium meleagridis*[6,7] in infected hosts and it infects a wide range of avian hosts.[8] Avian models are discussed briefly in this chapter.

Several structural types of *Cryptosporidium* oocysts occur in reptiles and all are presently called *C. serpentis*, whereas *C. nasorum* is the only species described from fish.[9] The reptile and fish species of *Cryptosporidium* are difficult to work with and little has been done experimentally with these parasites. These species will not be considered further in this chapter.

II. EXPERIMENTAL BIOLOGY OF *CRYPTOSPORIDIUM PARVUM* IN LABORATORY ANIMALS

A. INFECTIVE DOSE 50

The 50% infectious dose (ID_{50}) for 5-day-old Swiss-Webster mice (*Mus musculus*) was reported to be between 100 and 500 oocysts;[10] for suckling BALB/c mice, it was 60[11] or 1000 oocysts;[12] for 3- to 5-day-old C57BL/6J mice, it was 600 oocysts;[13] and for 4-day-old CD-1 mice, it is was 79 oocysts.[14] Ten oocysts were infectious for two of two pigtailed macaques (*Macaca nemestrina*).[15]

Most researchers use between 1×10^5 and 1×10^7 oocysts per animal to induce experimental infections. The number of oocysts inoculated can influence such parameters as prepatent period, site of intestinal colonization, and the number of oocysts excreted by infected hosts.[16] Little difference in clinical signs is usually noted with different doses of inoculum. The influence of parasite strain on the above-mentioned parameters is not well understood.

B. ROUTE OF INOCULATION

Patent intestinal tract infections have been established in immunosuppressed C57BL/6N mice by intraperitoneal (IP) and intravenous (IV) injection but not by subcutaneous (SC) or intramuscular (IM) injection of 1×10^6 oocysts.[17,18] Patent infections were delayed by several days in some IP- and SC-inoculated mice. The uterus was colonized in some IV-inoculated mice. Direct inoculation of 2×10^5 oocysts into the uterus produced intrauterine infections in adult BALB/c mice.[19] Direct inoculation through the body wall into the stomach (bypassing the oral cavity and esophagus) of suckling 4- to 5-day-old hamster pups produced infections identical to orally induced infections in hamster pups.[20] Patent infections were produced in suckling mice by direct inoculation of oocysts into the colon.[12] Direct intra-tracheal inoculation of 1×10^6 oocysts produced respiratory cryptosporidiosis in immunosuppressed but not in immunocompetent rats.[21] Developmental stages were demonstrated on both the respiratory and intestinal epithelium of these rats.

III. MOUSE MODELS OF *CRYPTOSPORIDIUM PARVUM*

A. BACKGROUND

The mouse is the most commonly used model of *C. parvum* infections. Soon after *C. parvum* was recognized as a serious pathogen, scientists began developing mouse models for the parasite, and it was determined that most outbred strains of weaned or adult mice were resistant to experimental infection but suckling animals could be readily infected.[22,23]

B. IMMUNOCOMPETENT MICE

1. Suckling Mice

Many researchers have used the suckling mouse system to examine a variety of the aspects of the biology of *C. parvum* (Table 1). The susceptibility of eight strains of suckling and 21-day-old mice to *C. parvum* was examined.[23] Random and inbred Porton, Schneider Swiss white, BALB/c, CBA and CBA nude, HR/HR-ADR hairless, and C57 black were much more susceptible to infection than 21-day-old mice

Table 1 Experimental *Cryptosporidium parvum* Infections in Various Strains of Nursing Mice

Mouse Strain	Result	Ref.
Random Porton	1- to 4-day-old susceptible	23
Inbred Porton	1- to 4-day-old susceptible	23
Schneider Swiss White	1- to 4-day-old susceptible	23
BALB/c	1- to 4-day-old susceptible	23
BALB/c	14-day-old resistant	24
BALB/c nude	6-day-old develop persistent infection	57
CBA nude	1- to 4-day-old susceptible	23
CB-17 SCID	2- to 5-day-old develop persistent infection	58
HR/HR-ADR hairless	1- to 4-day-old susceptible	23
C57 black	1- to 4-day-old susceptible	23
CBA	1- to 4-day-old susceptible	23
C57BL/6J (MHC Class I-deficient)	3- to 5-day susceptible	13
C57BL/6J (MHC Class II-deficient)	3- to 5-day susceptible	13
WBB6F_1/J (W/W^v, mast cell-deficient)	7-week-old susceptible	56

of the same strain. Of Schneider Swiss white and HR/HR-ADR hairless mice, 21-day-old mice were completely resistant.

The susceptibility dynamics of *C. parvum* in various ages of suckling BALB/c mice revealed that mice were susceptible from birth to 14 days of age.[24] Younger mice had more parasites in their tissues than did older mice. Mice infected at 4 days of age remained infected for about 3 weeks. These results were verified, and it was determined that the prepatent period varied with the numbers of oocysts inoculated and the age at which the mice were inoculated.[25]

a. Utility of Suckling Mouse Models

Several researchers have used neonatal mice to examine many aspects of the biology of *C. parvum,* including developmental biology,[26-28a] the efficacy of anticryptosporidial agents,[29-35] maternal and other immune mechanisms,[36-39] interactions with other infectious agents,[40] and viability of treated oocysts.[11,41-46]

An outbred, ICR strain of suckling mice was developed as a model for chemotherapy studies.[32] Pups are cross-fostered based on weight prior to allocation of litters to treatment groups. This produced litters with pups of approximately equal weights prior to initiation of the study. Treatment was begun when the mice in the groups were 3 g, and mice were orally inoculated with 1×10^5 oocysts 1 day later. Mice were treated 2 hours after inoculation, then daily for 6 days. Mice were weighed daily and medicated via a microdoser and syringe pump equipped with a tuberculin syringe fitted with microbore tubing. The volume of treatment compound was adjusted daily for the weight of the mouse pup, thereby keeping the dose constant. This daily adjustment prevented treatment of the fast-growing pups with too little medication relative to weight. Mice were killed 7 days postinoculation (PI), and the intestinal tract from the pylorus to the rectum was removed and homogenized in 10 ml of 2.5% (w/v) potassium dichromate solution. The number of oocysts were determined by a hemocytometer. Comparisons of oocyst production in treated and control mice were used to determine efficacy of anticryptosporidial agents.

A neonatal SCID (severe combined immune deficient) mouse model was used to determine the efficacy of anticryptosporidial agents (atovaquone) in chronically infected mice.[47] Five-day-old mice were orally inoculated with 1×10^6 oocysts, and treatment was initiated 14 days PI. In this system, mice died within 7 weeks when the treatment was not effective.

A sporozoite neutralization assay was developed in 7-day-old BALB/c mice.[12] Sporozoites obtained from oocysts were exposed to test sera. Following exposure, 2×10^5 sporozoites were inoculated into the colon by inserting a 2-cm tube through the anus. A rubber patch, glued over the anus to prevent defecation immediately after inoculation, was removed 5 hours PI. Mice were killed 4 days PI, and the intestinal tract was processed for histological examination. A numerical score was given based on the numbers of developmental stages present.

2. Adult Mice

Conventional adult BALB/c and CD1 mice are infrequently susceptible to even light *C. parvum* infection, but germfree adults of the same strains are readily susceptible to infections. No clinical signs occur in germfree or conventional mice of these strains.[48]

Of 19 strains of adult mice (Table 2) tested for susceptibility to *C. parvum* infection, C57BL/6J-bgJ beige mice were the most susceptible.[49] They were, however, much less susceptible than neonatal mice.

Cryptosporidium parvum infection became chronic in 6-week-old, female C57BL/6 mice inoculated with 1×10^7 oocysts.[50] Small foci of infection were identified in the epithelium covering the Peyer's patches of the jejunum and ileum, but oocysts could not be detected in the feces.

C. ARTIFICIALLY IMMUNOSUPPRESSED MICE

1. Suckling Mice

The effects of administration of anti-CD4, anti-CD8, and a combination of anti-CD4/anti-CD8 monoclonal antibodies were tested in BALB/c mice orally inoculated with 1×10^7 oocysts (divided on days of age 3 and 5).[51] Mice that received anti-CD4 or the combination of anti-CD4 and anti-CD8 antibodies developed chronic infections that lasted as long as the immunosuppressive antibodies were administered. Some mice developed severe cryptosporidiosis similar to that observed in outbred adult nude mice (see below).

2. Adult Mice

The effects of chemical immunosuppression with dexamethasone (DEX) was determined by giving a dose equivalent to 0.25 μg of DEX per gram of body weight in the drinking water for 14 days before mice were inoculated with 1×10^6 oocysts.[52] Chronic infection occurred in C57BL/6N but not DBA/2N, C3H/HeN, or BALB/cAnN mice. When these four strains of mice were given 125 μg DEX IP they shed oocysts earlier and for a longer period of time, with the exception of DBA/2N mice that all died by day 14 PI. These findings were extended in C57BL/6N mice and when alcohol-soluble DEX and water-soluble (phosphated) DEX were compared in the drinking water at 10, 33, and 100 μg/ml to IP inoculation of 125 μg per mouse per day.[53] The DEX water treatment was found comparable to IP DEX treatment, and the water treatment was simpler and required less time.

When 0.125 mg DEX was administered IP for 3 days following infection of 6-week-old, female C57BL/6 mice with 1×10^7 oocysts, foci of infection were heaviest in the epithelium covering the Peyer's patches of the jejunum and ileum.[50] Oocysts were detected in the feces of these mice. If the DEX treatment was delayed for 2 weeks following inoculation, infections were still demonstratable in the intestines but oocysts were not present in the feces.

Intramuscular administration of 50 mg/kg methylprednisolone acetate (MPA, 4 times at 3-day intervals) was used to immunosuppress 7-week-old ICR female mice prior to inoculation with *C. parvum*.[54] Mice were given 1×10^5 oocysts, and they excreted large numbers of oocysts on days 4 through 10 PI.

Two of 10 C57BL/6N mice maintained on an 8% protein diet (low protein diet) for 2 months prior to oral inoculation with 1×10^6 oocysts shed oocysts after inoculation.[52] None of the mice had stages in their intestines on day 7 PI.

The effects of exogenous anti-CD4, anti-CD8, anti-IL-2 and combinations of anti-CD4/anti-CD8, anti-CD4/anti-IFN-γ monoclonal antibodies on the development of *C. parvum* were studied in 6-week-old BALB/c mice orally inoculated with 1×10^7 oocysts.[55] Mice treated with anti-CD4 or the combination of anti-CD4/anti-CD8 antibodies developed chronic infections which lasted as long as the antibody treatment was continued. Mice treated with anti-CD4/anti-IFN-γ monoclonal antibody had greatly enhanced oocyst excretion and were chronically infected as long as the treatment was continued. Mice treated with anti-IL-2 monoclonal antibody alone or in combination with anti-IFN-γ did not develop infections.

D. GENETICALLY IMMUNODEFICIENT MICE

1. Suckling Mice

One-week-old, mast cell deficient $WBB6F_1$/J (W/W^v) mice reacted similarly to *C. parvum* infection, as did their mast cell sufficient littermates when orally inoculated with 1×10^5 oocysts.[56] Intestinal infection was marked 1 week PI but declined 2 weeks PI and was cleared by 5 weeks PI.

Six-day-old nude BALB/c mice inoculated with 1×10^5 oocysts developed persistent cryptosporidial infection that occasionally resulted in death.[57] Mice developed diarrhea about 30 days PI and shed oocysts for at least 56 days. Microscopic lesions of villous atrophy and crypt hyperplasia were present in the small intestines. Cystic hyperplasia was present in the colonic mucosa of some mice.

Table 2 Experimental *Cryptosporidium parvum* Infections in Various Strains of Adult Mice

Mouse Strain	Treatment[a]	Result	Ref.
A/J	None	Low susceptibility	49
AKR/J	None	No infection detected	49
BALB/c	None	No infection detected	49
BALB/c	None	Low susceptibility	48, 57
BALB/c	Germfree	High susceptibility	48
BALB/c	Anti-CD8	Low susceptibility	55
BALB/c	Anti-CD4	Low-level chronic infection	55
BALB/c	Anti-CD4/CD8	Low-level chronic infection	55
BALB/c	Anti-IL-2	No infection detected	55
BALB/c	Anti-IL-2/anti-IFN-γ	No infection detected	55
BALB/c nude	Anti-IFN-γ	Increased oocyst production	55
BALB/c	Anti-CD4/IFN-γ	Chronic infection	55
BALB/c nude	None	Low susceptibility	57
BALB/cAnN	DEX-W	Oocysts excreted on days 6 to 12 PI	52
BALB/cAnN	DEX-IP	Oocysts excreted on days 4 to 21 PI	52
B10.D2/J	None	Low susceptibility	49
B10.M/J	None	Low susceptibility	49
CB-17 SCID	None	High susceptibility	16
CBA	DEX-W	All died before infection	52
CBA	DEX-IP	All died before infection	52
CBA/NJ	None	No infection detected	49
CD1	None	Low susceptibility	48
CD1	Germfree	High susceptibility	48
C3H/Hen	DEX-W	Oocysts excreted on days 6 to 10 PI	52
C3H/Hen	DEX-IP	Oocysts excreted on days 4 to 14 PI	52
C3H/HeJ	None	No infection detected	49
C3H/HeJ/beige	Low-protein diet	Oocysts excreted for 5 days	52
C57BL/6	None	Prolonged infection but no oocysts in feces	50
C57BL/6	DEX-IP	Prolonged infection	50
C57BL/6N	DEX-W	Oocysts excreted on days 4 to ≥28 PI	52
C57BL/6N	DEX-IP	Oocysts excreted on days 4 to ≥28 PI	52
C57BL/6N	Low-protein diet	Oocysts excreted for 6 days	52
C57BL/6N	None	Ocysts excreted on days 4 to 7 PI	52
C57BL/6J	None	No infection detected	49
C57BL/6J	MHC Class I-deficient	Low susceptibility	13
C57BL/6J	MHC Class II-deficient	High susceptibility	13
C57BL/6J-bg^J	None	Low susceptibility	49
DBA/1J	None	Low susceptibility	49
DBA/2J	None	Low susceptibility	49
DBA/2N	DEX-W	Oocysts excreted on days 6–21 PI	52
DBA/2N	DEX-IP	Oocysts excreted on day 4; all died by day 14	52
S/J	None	Low susceptibility	49
HTG/J	None	Low susceptibility	49
ICR	MPA-IM	Oocysts excreted 4 to 10 days PI	54
NZB/B1NJ	None	No infection detected	49
NZW	None	No infection detected	49
P/J	None	No infection detected	49
RIII/J	None	Low susceptibility	49
SJL/J	None	Low susceptibility	49
SWR/J	None	No infection detected	49
WB/Re	None	Low susceptibility	49
WBB6F$_1$/J (W/W^v)	Mast cell-deficient	High susceptibility	56
Outbred nude	None	High susceptibility	51

[a] DEX-W = dexamethasone given in water; DEX-IP = dexamethasone given intraperitoneally; MPA = methylprednisolone acetate given intramuscularly.

The susceptibility of suckling C57BL6/6J mice made MHC Class I- or MHC Class II-deficient to *C. parvum* infection were compared to conventional C57BL6/6J mice.[13] Mice were orally inoculated with 1×10^7 oocysts. The MHC Class I-deficient mice responded in a manner similar to conventional C57BL6/6J mice to infection. The MHC II-deficient mice had much more abundant stages in tissue sections and persisted longer than conventional C57BL6/6J mice.

Two- to 5-day-old SCID mice inoculated with 1×10^6 *C. parvum* oocysts developed intermittent diarrhea, weight loss, and severe wasting by 4 to 6 weeks PI.[58] Lesions consistent with cryptosporidiosis were present in the ileum and colon. The infections terminated in death by week 7 PI. Interestingly, no hepatic involvement was observed in these mice infected as suckling pups. Hepatic cryptosporidiosis has been a consistent finding in SCID mice infected as adults (see below).

2. Adult Mice

Adult WBB6F$_1$J (W/W^v, mast cell deficient) mice were more susceptible to *C. parvum* infection than their mast cell sufficient littermates but cleared the infection quickly.[56] Inoculation of 1×10^6 *C. parvum* oocysts resulted in heavy infections 1 week PI that cleared by 2 weeks PI. Mice that cleared the infection were resistant to reinfection.

Three of 10 C3H/HeJ/beige (lack natural killer cell activity) maintained on an 8% protein diet for 2 months prior to oral inoculation with 1×10^6 oocysts excreted oocysts after inoculation. None of the mice had stages in their intestines on day 7 PI.[52]

The susceptibility of 5- to 6-week-old C57BL6/6J mice that were made MHC Class I- or MHC Class II-deficient to experimental *C. parvum* infection was examined.[13] Mice were orally inoculated with 1×10^7 oocysts. The MHC Class I-deficient mice responded in a manner similar to conventional C57BL6/6J mice to infection. The MHC II-deficient mice had infections that were much more abundant and persisted for at least 4 weeks.

Nude athymic BALB/c mice infected at 42 days with 1×10^5 oocysts developed only mild infections.[57] No clinical signs were present, and no microscopic lesions were observed in their tissues.

An athymic outbred nude mouse model was developed to study chronic cryptosporidiosis.[51] Six- to 10-week-old mice, orally inoculated with 1×10^7 oocysts appeared normal for the first 3 weeks, after which they developed clinical signs of cryptosporidiosis. Few oocysts were excreted during the first 3 weeks, but numerous oocysts were excreted thereafter. Clinical signs included dehydration, weight loss, intermittent diarrhea, and steatorrhea. Mice appeared to improve but then recrudesced. Three distinct patterns developed in the mice. The first was characterized by exacerbation of clinical disease that progressed to death in 2 to 3 days. The second form was characterized by gradual worsening of condition over a 4-week period followed by death. The third form was characterized by chronic cryptosporidiosis that lasted for at least 16 weeks. Mice that became icteric also were emaciated with abdominal distension and had hepatomegaly with dilated biliary tracts. Some mice also had gastric ulcers. Lesions in the intestines were characterized by degeneration, sloughing, and regeneration of both surface and glandular epithelial cells and dilated crypts and glands. Lesions in the livers of mice with hepatic cryptosporidiosis were inflammation of the bile ducts, portal and periportal areas, and hepatocellular necrosis. Choledochitis, cholecystitis, and cholangitis were indicated by hyperplasia, dilation, chronic-active inflammation, and occasional erosions. Some mice had pancreatitis with necrosis, atrophy, infiltration by inflammatory cells, and ductal hyperplasia. Mice with chronic stable infections had lesions mainly in the intestines. Cryptosporidia were identified on the epithelial surfaces of the pyloric ring, small and large intestine, cecum, gall bladder and biliary tree, and pancreatic ducts of some mice.

Nude 6-week-old BALB/c mice inoculated with 1×10^7 oocysts and treated with anti-IFN-γ monoclonal antibodies developed self-limited infections. Oocyst production was enhanced 75-fold in treated mice compared to nontreated nude BALB/c mice.[55]

A splenectomized oubred nude mouse model was developed[59] that was similar to the intact outbred nude model[51] but the splenectomized mice could be infected with a smaller number of oocysts and had higher oocyst excretion than did intact nude mice. This model was used to examine the efficacy of anticryptosporidial agents.

Many groups of researchers have used adult SCID mice for studies with *C. parvum*.[13,16,58,60-63] The infection dynamics and clinical features of cryptosporidiosis were critically examined in 6- to 8 week-old SCID mice experimentally inoculated with 1×10^3, 1×10^4, 1×10^5, 1×10^6, or 1×10^7 oocysts.[16] Mice inoculated with the lower numbers of oocysts had longer prepatent periods. All mice developed chronic infections. Hepatic involvement was present in all inoculation groups. Of mice given 1×10^3 to 1×10^5 oocysts, 30 to 40% had hepatic involvement, while 80 to 90% of the mice inoculated with 1×10^7

oocysts had hepatic cryptosporidiosis. Some of these mice died or were euthanatized due to hepatic cryptosporidiosis.

E. UTILITY OF CHEMICALLY OR NATURALLY IMMUNOSUPPRESSED ADULT MOUSE MODELS

The chemically or naturally immunosuppressed mouse models have been used to examine immunological parameters,[63,64] anticryptosporidial agents,[54,59,65-69] and the importance of intestinal microflora.[62]

IV. RAT MODELS OF *CRYPTOSPORIDIUM PARVUM*

A. BACKGROUND

Several groups of researchers presently use rats to study *C. parvum* infections. Most studies involve chemically or naturally (athymic) immunosuppressed animals. Clinical cryptosporidiosis has been observed in naturally infected suckling Rapp hypertensive rat (*Rattus norvegicus*) pups in which 32 of 34 pups died within 3 weeks of birth.[70] Pups had diarrhea that led to fecal staining of the fur. They continued to suckle. Microscopically, there was villous blunting and fusion in the small intestine associated with cryptosporidia. No other pathogens were isolated.

B. IMMUNOCOMPETENT RATS

Suckling rats (SPF and conventional) were susceptible to infection but did not develop clinical cryptosporidiosis. The prepatent period was 5 to 6 days, and oocysts were excreted for 16 days.[22,71] Female, 200- to 250-g Sprague-Dawley rats were resistant to *C. parvum* infections.[21,72]

C. CHEMICALLY IMMUNOSUPPRESSED RATS

Several groups of researchers have used chemically immunosuppressed rats to examine *C. parvum* infections. The agent used to induce immunosuppression and the diet fed are notable differences between the models.

Also developed was a cyclophosphamide-treated Sprague-Dawley rat model in which the cyclophosphamide was given in the drinking water to 200- to 250-g females fed a standard laboratory animal diet.[72] Daily 50-mg/kg cyclophosphamide for 14 days before inoculation with *C. parvum* was optimal for producing infections. The optimum age of rats was 12 to 14 weeks. Oocyst doses greater than 1×10^4 produced consistent infections. The prepatent period was 5 to 6 days, and infections peaked on day 18 after inoculation. Parasites were present in the terminal 24 cm on the small intestine, and infections persisted as long as cyclophosphamide was given. Approximately 50% of the rats had mild to moderate villous atrophy, but the degree of atrophy did not correlate with the numbers of cryptosporidia present. Rats cleared the infection 5 to 10 days after immunosuppressive treatment was stopped.

A DEX-treated Sprague-Dawley rat model also was developed; DEX was given in drinking water to 200- to 250-g females fed a standard laboratory diet. The optimal dose of 0.25 mg/kg for 7 days before inoculation was sufficient to induce susceptibility to infection. Rats given 0.25 mg/kg of DEX and not infected had a 12% mortality. There was no increase in mortality with cryptosporidial inoculation. No significant differences in infection occurred between doses of 1×10^3 to 1×10^6 oocysts. The prepatent period was 4 to 10 days in most rats. The number of endogenous stages was highest 7 days PI. Rapid clearance of parasites was observed after DEX treatment was stopped. An interesting finding of the study was that some rats relapsed when DEX treatment was reinitiated, even up to 10 weeks later.

A hydrocortisone acetate-treated, low-protein diet, Sprague-Dawley rat model was developed.[74] Male 200- to 250-g rats were injected with 25 mg of hydrocortisone acetate SC twice weekly for 5 weeks before and 3 weeks after inoculation with *C. parvum* and fed a 7% protein diet. The minimal infective does was 1×10^4. Rats were infected with 1×10^5 oocysts given in the drinking water. The prepatent period was 2 to 9 days, and oocyst production peaked on days 7 to 23 PI. Oocyst production was detected as long as the rats were fed a low-protein diet and given hydrocortisone acetate. Seven of 32 rats died. No differences in oocyst production were noted between rats that survived and those that died.

An MPA-treated rat model was developed to study respiratory cryptosporidiosis.[21] Female, 120- to 140-g, Lewis rats were immunosuppressed by SC injection with 4 mg MPA per 100 g body weight, then were given another 2 mg MPA per 100 g 1 week later. Respiratory infections were established by direct tracheal inoculations with 1×10^6 oocysts. Clinical signs of respiratory cryptosporidiosis appeared 4 days PI and consisted of labored breathing, lethargy, and weight loss. No infections were found in inoculated immunocompetent rats. Cryptosporidial stages were present in the respiratory and intestinal tract of immunosuppressed rats examined at necropsy. Lesions in the respiratory tract consisted of exfoliative necrosis of respiratory epithelium and accumulation of large amounts of mucocellular exudate in the airways. About 50% of the immunosuppressed rats died from respiratory cryptosporidiosis.

D. GENETICALLY IMMUNODEFICIENT RATS

Suckling nude rats and their heterozygous littermates were inoculated with 2.5×10^6 to 6×10^7 *C. parvum* oocysts when less than 24 hours old.[75] Nude rats excreted oocysts 5 days PI and oocysts were still being excreted in one rat 52 days PI. Diarrhea and death were observed in some inoculated rats. Inoculated nude rats gained less weight than noninoculated nude rats. Heterozygous rats began excreting oocysts within 10 days, and no animals excreted oocysts beyond 23 days PI. The heterozygous rats developed diarrhea about 10 days PI that lasted for less than 4 days. Weight gains of heterozygous rats were not affected. Administration of 12.5 or 25.0 mg/kg cyclophosphamide had little effect on the responses of suckling nude or heterozygous rats to *C. parvum* infections.

E. UTILITY OF RAT MODELS

The DEX-treated rat model has been used to examine the effects of immuno-potentiators (dehydroepiandrosterone),[76-78] and chemotherapeutic agents.[79-85] The hydrocortisone acetate-treated rat model has been used to examine anticryptosporidial drugs.[86] The hydrocortisone acetate-treated/low-protein diet rat model has been used to examine anticryptosporidial agents.[87,88]

V. GUINEA PIG MODELS OF *CRYPTOSPORIDIUM WRAIRI* AND *CRYPTOSPORIDIUM PARVUM*

A. BACKGROUND

Natural cases of clinical and subclinical cryptosporidiosis in guinea pigs have been documented.[89-92] Deaths due to cryptosporidiosis in guinea pigs have been reported in both naturally[91,93] and experimentally infected guinea pigs.[90,93] In one retrospective study, 81 (3.4%) of 2348 guinea pigs examined had cryptosporidial infections. The most common clinical signs were failure to gain weight, weight loss (61%), and diarrhea (29%).[91] Infected guinea pigs often had distended abdomens and rough, greasy hair coats. Juveniles were most often infected, but adults were occasionally infected also. Cryptosporidia were distributed throughout the small intestine, cecum, and colon but not in other locations. Microscopic lesions, most prominent in the small intestines, consisted of villous atrophy and bridging in chronic infections and erosions and crypt hyperplasia in acute infections. A mild to moderate infiltration of inflammatory cells was present in the lamina propria of acutely infected guinea pigs.

B. *CRYPTOSPORIDIUM WRAIRI*

Cryptosporidium wrairi was first found in naturally infected guinea pigs from the Laboratory Animal colony at the Walter Reed Army Institute of Research (WRAIR).[2] Developmental stages were similar to those described by Tyzzer for *C. parvum*,[1,94] but *C. wrairi* in scrapings from the ilium of infected guinea pigs was not infectious for mice.[2] However, others have shown that *C. wrairi* is moderately infectious for mice,[90,95-97] SPF lambs,[90] and calves.[97] It is presently accepted that *C. wrairi* is a valid species that has adapted to the guinea pig host. The outer oocyst wall proteins of *C. wrairi* are strikingly different from those of *C. parvum*;[95] however, an anti-*C. parvum* monoclonal antibodies reacted with oocysts of *C. wrairi*.[93,97] Production of *C. wrairi* oocysts by infected mice was 100-fold less than mice inoculated with equivalent numbers of *C. parvum* oocysts.

Although transmission of *C. wrairi* was studied in guinea pigs, little quantitative data were obtained except that guinea pigs as old as 16 weeks could be infected.[90] Mortality was observed, but it was unclear whether *C. wrairi* was the sole cause of deaths.

Morphometric studies comparing villous height, crypt depth, and villous height/crypt depth ratios indicated that both adults (8- to 10-week-old) and juvenile (2- to 3-week-old) guinea pigs were equally susceptible to infection with *C. wrairi* oocysts.[93] Sixteen-week-old animals could be infected, but the severity was lessened. Developmental stages were present in the villi of the duodenum, jejunum, and ileum. Doses of as low as 325 oocysts produced infections in recipient animals. Infections were patent for 2 weeks in most animals and cleared by 3 to 4 weeks PI. Two guinea pigs died after inoculation, but the cause of death was not clearly determined.[93] Apparently most inoculated animals did not develop severe clinical signs.

C. *CRYPTOSPORIDIUM PARVUM*

Cryptosporidium parvum was transmitted to 1-day-old SPF guinea pigs.[22] The prepatent period was 7 days, and the patent period was 4 days. No clinical signs were present. Others were unable to transmit *C. parvum* (calf-origin *Cryptosporidium* sp.) to guinea pigs.[90]

D. UTILITY OF GUINEA PIG MODELS

The *C. wrairi*-guinea pig model was used to test the efficacy of hyperimmune bovine colostrum raised against *C. parvum* for treatment and prevention of *C. wrairi* infections.[98] The *C. wrairi*-guinea pig model was used to demonstrate a specific humoral immune response in infected guinea pigs, and inoculated animals were resistant to reinfection.[93]

VI. HAMSTER MODELS OF *CRYPTOSPORIDIUM PARVUM*

A. BACKGROUND

In natural cases of cryptosporidiosis, a weanling male Golden hamster[99] and two 27-day-old hamsters were infected.[100] Both reports presented similar findings. Cryptosporidia were observed in the ilea and colons of infected animals, and concurrent lesions of proliferative enteritis and *Campylobacter*-like bacteria were present. The species responsible for producing the infection was not determined but probably was *C. parvum.*

B. IMMUNOCOMPETENT HAMSTERS

1. Suckling and Recently Weaned Hamsters

A model of cryptosporidiosis used suckling Syrian golden hamster pups, 4 to 5 days old, that were inoculated orally or by a novel method that involved direct inoculation through the body wall into the stomach.[17] The prepatent period was 2 days and oocysts were detected for 26 days in one animal. Oocyst numbers peaked 6 days PI. None of the inoculated hamsters developed diarrhea. The ileum was the only tissue examined and microscopic lesions were minimal. Recently weaned, 12-day-old hamsters were susceptible to infection, but fecal oocyst production and the number of parasites in the ileum were less than those observed in suckling hamsters.[101]

2. Adult and Aged Hamsters

Experimental *C. parvum* infections in young adult (2 to 3 months old) and aged (20 to 24 months old) male hamsters were compared.[102] All hamsters were orally inoculated with 1×10^6 oocysts. Young adults exhibited a prepatent period of 4 days, and oocyst numbers peaked on this day but oocyst production was minimal. The aged hamsters had a prepatent period of 3 to 4 days, and oocyst numbers peaked 6 days PI. Oocyst production was high until 9 days PI. Although neither the young adult nor aged hamsters developed diarrhea, feces of the aged hamsters was softer and more moist than the young adult hamsters.

No lesions or stages of *C. parvum* were observed in any tissue sections from young adult hamsters. Lesions consisting of villous atrophy and crypt hyperplasia and *C. parvum* stages were observed in the small intestines of the aged hamsters. No stages were observed in the stomach, liver, colon, or large intestine of any hamster.

3. Immunosuppressed Hamsters

Cryptosporidium parvum infection was studied in immunosuppressed adult hamsters.[103] Female outbred white hamsters weighing 80 to 100 g were immunosuppressed by s.c. injection of 8 or 10 mg hydrocortisone acetate on days –7, –3, –2, 0, 1, and 4 PI. Additionally, hamsters were irradiated with X-rays (500 rads) on the day of inoculation. Hamsters were orally inoculated with 5.0×10^4 (Group 1) or 1.0×10^5 (Group 2) oocysts. The prepatent period was 3 to 5 days. Oocyst production peaked on day 9, followed by a second lower peak on day 13. Mean oocyst production was 3.28×10^7 in Group 1 and 5.4×10^7 in Group 2 on day 9 PI. The mean total oocyst production (days 3 through 15 PI) was 1.58×10^8 and 2.37×10^8 oocysts for Groups 1 and 2, respectively. Oocysts were still present in the feces 25 days PI, but the patent period was not determined.

Diarrhea was not observed in any hamsters. Two hamsters in Group 1 and three hamsters in Group 2 died after 9 days PI. The cause of death was not reported. Neither the location of the parasite within the hamsters' intestinal tract nor extraintestinal involvement was determined.

4. Utility of Hamster Models

Neonatal hamster and 12-day-old hamster models were used to evaluate the efficacy of arprinocid against experimental *C. parvum* infections.[101] Adult Syrian hamsters were used to examine the effects of the immunomodulator dehydroepiandrosterone on experimental *C. parvum* infections.[104] Treatment resulted in a marked reduction in infection.

VII. PRIMATE MODEL OF *CRYPTOSPORIDIUM PARVUM*

Four newly weaned pigtailed macaques were used as primate models.[15] Two macaques received oral inoculations of 2×10^5 oocysts and clinical cryptosporidiosis was seen in both by day 7 PI. Loose watery stools, lethargy, anorexia but not fever or weight loss were the clinical signs associated with infection in these animals. No other pathogens were present. The animals excreted oocysts for 32 days and had clinical signs up until day 16 PI. The remaining two macaques, orally inoculated with 10 oocysts, responded similarly. Large numbers of oocysts were excreted 8 days PI, and enteritis was observed 10 days PI. Both macaques were clinically ill for 2 weeks, and oocyst excretion continued until day 56 PI. Although three of the four excreted oocysts after re-inoculation, all were resistant to clinical disease.

The effects of immunosuppression on a primary infection was examined in three macaques (20 to 40 days old) given 10 mg/kg MPA IM on days –3, 4, and 11 and 5 mg/kg on days 18, 25, 32, 39, 46, and 53 PI. The animals were orally inoculated with 100 oocysts. This course of immunosuppression did not alter the clinical or parasitological responses of the animals.

Disseminated *C. parvum* infections has been observed in Rhesus monkeys (*Macaca mulatta*) experimentally infected with Simian immunodeficiency virus delta.[105] This suggests a potential experimental primate model of cryptosporidiosis following viral induced immunosuppression.

VIII. MISCELLANEOUS MAMMALIAN HOST MODELS OF *CRYPTOSPORIDIUM PARVUM*

Young calves, lambs, goat kids, and pigs are susceptible to experimental *C. parvum* infections. Because of their large size, high susceptibility for natural *C. parvum* infections, and high cost of housing, they are not often used in controlled laboratory studies.

Dogs and cats are susceptible to human *C. parvum* isolates;[106] however, little attention has been given these hosts. Chronic infections were produced in cats using oocysts resembling *C. pravum* obtained from naturally infected cats. These same oocysts were not infectious for 4- to 6-day-old BALB/c mice, 3-week-old Wistar rats, 1- to 9-day-old JY-10 strain guinea pigs, or dogs. The prepatent period in four cats inoculated with 6×10^4 to 2×10^6 oocysts was 8 to 10 days. The cats did not develop clinical signs, and oocysts were excreted for 69 to 203 days. If cats that had stopped excreting oocysts were given 10 mg/kg prednisolone for 4 to 9 days, then oocyst excretion would commence once again. Cryptosporidiosis has been observed in a cat with feline leukemia virus (FeLV).[108] This suggests that a FeLV-induced immunosuppression model of *C. parvum* cryptosporidiosis is possible.

Cryptosporidium parvum infections were studied in suckling opossums (*Didelphis virginiana*) orally inoculated with 5×10^6 oocysts.[109] Mild diarrhea was observed in four of seven infected opossums, and cryptosporidia were observed in the ileum, cecum and colons of infected animals. Adult gerbils (*Meriones unguiculatus*) were found to be resistant to experimental *C. parvum* infections.[49]

In an intestinal xenograft model for *C. parvum* infections, paired segments of near-term fetal New Zealand white rabbit small intestines were transplanted subcutaneously into male outbred athymic nude mice.[110] Grafts were implanted for 5 weeks then inoculated with sporozoites. Xenografts were infected and demonstrated lesions typical of intestinal cryptosporidiosis.

IX. LABORATORY MODELS OF *CRYPTOSPORIDIUM MURIS*

Although originally described from mice, many field isolates of *C. muris* do not readily infect laboratory mice. In one study, a *C. muris* isolate from a cow was not infectious for mice but an isolate from a camel (*Camelus bactrianus*) produced infections in mice inoculated at 2 or 20 days of age.[111] The prepatent period was about 12 days, and mice remained infected for at least 27 days.

Cryptosporidium muris oocysts were found in 3 of 61 rats (*Rattus norvegicus*) examined in Osaka City, Japan.[112] This RN 66 strain of *C. muris* was successfully transmitted from rats to 3-week-old male ICR mice; 3-week-old SPF, male, Hartley strain guinea pigs; 6-week-old New Zealand white rabbits;

1- to 2-month-old dogs; and 1- to 2-month-old cats following oral inoculation of 1×10^6 oocysts into each animal.[113] None of the inoculated animals developed clinical cryptosporidiosis. The prepatent period in mice was 5 days, and many oocysts were present 14 through 21 days PI. The patent period was from 34 to 75 days. This *C. muris* isolate was confined in distribution to the microvilli of surface mucosal cells in the gastric glands of the glandular portion of the stomach.[114,115] The prepatent period in guinea pigs was 10 to 17 days, and the patent period was 11 to 17 days. No clinical signs were observed. None of the animals were euthanatized to determine lesions or site of development. Rabbits excreted moderate amounts of oocysts on days 11 to 43 PI. *Cryptosporidium muris* was observed in the stomach of the a rabbit examined 25 days PI. The prepatent period in dogs was 11 to 14 days, and only few oocysts were excreted for 2 to 10 days. The cats were heavily infected. The prepatent period was 15 to 21 days. Some cats excreted oocysts for 140 days. *Cryptsporidium muris* oocysts were observed in 10 of 58 feral house mice examined on a college farm in Warwickshire, U.K.[116] No transmission studies were done.

X. MODELS OF AVIAN CRYPTOSPORIDIOSIS

Small intestinal cryptosporidiosis can be a serious production problem in commercially raised bobwhite quail (*Colinus virginianus*) and turkeys (*Meleagris gallopavo*).[8] Respiratory disease is a common manifestation of cryptosporidiosis in turkeys.[117,118] Good model systems are not available for studying these diseases. *Cryptosporidium meleagridis* and *Cryptosporidium* sp. infecting bobwhite quail produce few oocysts, making experimental studies difficult.

Cryptosporidium baileyi has been associated with bursal and respiratory cryptosporidiosis in chickens (*Gallus gallus*). Bursa infections are asymptomatic, and lesions consist of epithelial hyperplasia and hypertrophy of the bursal epithelium with underlying inflammatory responses. Bursal follicle atrophy may be present. Respiratory disease is characterized by rales, coughing, sneezing, and dyspnea. Excess mucous may be present in the trachea and nasal cavities. Airsacculitis is common. Microscopic lesions consist of hypertrophy and hyperplasia of infected epithelial surfaces. Infiltrates of macrophages, heterophills, lymphocytes, and plasma cells usually are present.

Respiratory cryptosporidiosis can be produced in a variety of avian species by direct intratracheal inoculation of 2.5×10^5 to 1×10^6 *C. baileyi* oocysts.[8] Chickens were infected with oocysts by intranasal,[7] intracloacal,[119] ocular,[120] intra-abdominal,[121] and direct inoculation into the gall bladder.[122] Respiratory cryptosporidiosis resulted when birds were inoculated intra-abdominally but not by other modes. Intravenous inoculation of oocysts did not produce patent infections.[121]

There is an age-related susceptibility to *C. baileyi* infection in chickens.[123] When groups of chickens were intratracheally inoculated with oocysts at 2, 14, 28, or 42 days of age, only the birds inoculated at 2 and 14 days developed respiratory cryptosporidiosis. These birds excreted more oocysts and had more stages in their tissues at necropsy.

The *C. baileyi*-respiratory cryptosporidiosis model has been used to examine chemoprophylaxis,[124] to examine potential attenuated vaccines,[125] and to examine the effects of existing infections on response to vaccinations.[126] The *C. baileyi* bursal cryptosporidiosis model has been used for the above situations and to examine the effects of coninfection with viruses.[127,128]

REFERENCES

1. Tyzzer, E. E., *Cryptosporidium parvum* (sp. nov.): a coccidium found in the small intestine of the common mouse, *Arch. Protistenkd.*, 26, 394, 1912.
2. Vetterling, J. M., Jervis, H. R., Merril, T. G., and Sprinz, H., *Cryptosporidium wrairi* sp. n. from the guinea pig *Cavia porcellus*, with an emendation of the genus, *J. Protozool.*, 18, 243, 1971.
3. Tyzzer, E. E., A sporozoan found in the peptic glands of the common mouse, *Proc. Soc. Exp. Med.*, 5, 12, 1907.
4. Tyzzer, E. E., A extracellular coccidium, *Cryptosporidium muris* (gen. et sp. nov.), of the gastric glands of the common mouse, *J. Med. Res.*, 23, 487, 1910.
5. Current, W. L., Upton, S. J., and Haynes, T. B., The life cycle of *Cryptosporidium baileyi* infecting n. sp. (Apicomplexa: Cryptosporidiidae) infecting chickens, *J. Protozool.*, 33, 289, 1986.
6. Tyzzer, E. E., Coccidiosis in gallinaceous birds, *Am. J. Hyg.*, 10, 269, 1929.
7. Slavin, D., *Cryptosporidium meleagridis* (sp. nov.), *J. Comp. Pathol.*, 65, 262, 1955.
8. Lindsay, D. S. and Blagburn, B. L., Cryptosporidiosis in birds, in *Cryptosporidiosis of Man and Animals*, Dubey, J. P., Speer, C. A., and Fayer, R., Eds., CRC Press, Boca Raton, FL, 1990, 133.
9. Upton, S. J., *Cryptosporidium* spp. in lower vertebrates, in *Cryptosporidiosis of Man and Animals*, Dubey, J. F., Speer, C. A., and Fayer, R., Eds., CRC Press, Boca Raton, FL, 1990, 149.

10. Ernest, J. A., Blagburn, B. L., Lindsay, D. S., and Current, W. L., Infection dynamics of *Cryptosporidium parvum* in neonatal mice (*Mus musculus*), *J. Parasitol.*, 72, 796, 1986.
11. Korich, D. G., Mead, J. R., Madore, M. S., Sinclar, N. A., Sterling, C. R., Effects of ozone, chlorine dioxide, chlorine, and monochlororamine on *Cryptosporidium parvum* oocyst viability, *Appl. Environ. Microbiol.*, 56, 1423, 1990.
12. Riggs, M. W. and Perryman, L. E., Infectivity and neutralization of *Cryptosporidium parvum* sporozoites, *Infect. Immun.*, 55, 2081, 1987.
13. Aguirre, S. A., Masson, P. H., and Perryman, L. E., Susceptibility of major histocompatibility complex (MHC) Class I- and Class II-deficient mice to *Cryptosporidium parvum* infection, *Infect. Immun.*, 62, 697, 1994.
14. Finch, G. R., Daniels, C. W., Black, E. K., Schaefer, F. W., and Belosevic, M., Dose response of *Cryptosporidium parvum* in outbred neonatal CD-1 mice, *Infect. Immun.*, 59, 3661, 1993.
15. Miller, R. A., Bronsdon, M. A., and Morton, W. R., Experimental cryptosporidiosis in a primate model, *J. Infect. Dis.*, 161, 312, 1989.
16. Mead, J. R., Ilksoy, N., You, X., Belenkaya, Y., Arrowood, M. J., Fallon, M. T., and Schinazi, R. F., Infection dynamics and clinical features of cryptosporidiosis in SCID mice, *Infect. Immun.*, 62, 1691, 1994.
17. Yang, S. and Healey, M. C., Patent gut infections in immunosuppressed adult C57BL/6N mice following intraperitoneal injection of *Cryptosporidium parvum* oocysts, *J. Parasitol.*, 80, 338, 1994.
18. Yang, S. and Healey, M. C., Development of patent gut infections in immunosuppressed adult C57BL/6N mice following intravenous inoculations of *Cryptosporidium parvum* oocysts, *J. Euk. Microbiol.*, 41, 67S, 1994.
19. Liebler, E. M., Pohlenz, J. F., and Woodmansee, D. B., Experimental intrauterine infection of adult BALB/c mice with *Cryptosporidium* sp., *Infect. Immun.*, 54, 255, 1986.
20. Kim, C. W., *Cryptosporidium sp.*: experimental infection in Syrian golden hamsters, *Exp. Parasitol.*, 63, 243, 1987.
21. Meulbroek, J. A., Novilla, M. N., and Current, W. L., An immunosuppressed rat model of respiratory cryptosporidiosis, *J. Protozool.*, 38, 113S, 1991.
22. Tzipori, S., Angus, K. W., Campbell, I., and Gray, E. W., *Cryptosporidium*: evidence for a single-species genus, *Infect. Immun.*, 30, 884, 1980.
23. Sherwood, D., Angus, K. W., Snodgrass, D. R., and Tzipori, S., Experimental cryptosporidiosis in laboratory mice, *Infect. Immun.*, 38, 471, 1982.
24. Novak, S. M. and Sterling, C. R., Susceptibility dynamics in neonatal BALB/c mice infected with *Cryptosporidium parvum, J. Protozool.*, 38, 102S, 1991.
25. Vitovec, J. and Koudela, B., Location and pathogenicity of *Cryptosporidium parvum* in experimentally infected mice, *J. Vet. Med. B,* 35, 515, 1988.
26. Current, W. L. and Reese, N. C., A comparison of endogenous development of three isolates of *Cryptosporidium* in suckling mice, *J. Protozool.*, 33, 98, 1986.
27. Scaglia, M., Bruno, A., Chichino, C., Atzori, C., Cevini, C., and Gatti, S., *Cryptosporidium parvum* life cycle in suckling mice: a Normarski interference-contrast study of a human-derived strain, *J. Protozool.*, 538, 118S, 1991.
28. Tilley, M., Upton, S. J., and Freed, P. S., A comparative study on the biology of *Cryptosporidium serpentis* and *Cryptosporidium parvum* (Apicomplexa: Cryptosporidiidae), *J. Zoo. Wildl. Med.*, 21, 463, 1990.
28a. Upton, S. J. and Gillock, H. H., Infection dynamics of *Cryptosporidium parvum* in ICR outbred suckling mice, *Folia Parasitol.,* 43, 101, 1996.
29. Tzipori, S. R., Campbell, I., and Angus, K., The therapeutic effects of 16 antimicrobial agents on *Cryptosporidium* infection in mice, *Aust. J. Exp. Biol. Med. Sci.,* 60, 187, 1982.
30. Angus, K. W., Hutchison, G., Campbell, I., and Snodgrass, D. R., Prophylactic effects of anticoccidial drugs in experimental murine cryptosporidiosis, *Vet. Rec.*, 114, 166, 1984.
31. Arrowood, M. J., Mead, J. R., Mahrt, J. L., and Sterling, C. R., Effects of immune colostrum and orally administered antibodies on the outcome of *Cryptosporidium parvum* infections in neonatal mice, *Infect. Immun.*, 57, 2283, 1989.
32. Blagburn, B. L., Sundermann, C. A., Lindsay, D. S., Hall, J. E., and Tidwell, R. R., Inhibition of *Cryptosporidium parvum* in neonatal Hsd:(ICR)BR Swiss mice by polyether ionophores and aromatic amidines, *Antimicrob. Agents Chemother.*, 35, 1520, 1991.
33. Fayer, R. and Ellis, W., Glycoside antibiotics alone and combined with tetracyclines for prophylaxis of experimental cryptosporidiosis in neonatal BALB/c mice, *J. Parasitol.*, 79, 533, 1993.
34. Fayer, R. and Ellis, W., Qinghaosu (Artemisinin) and derivatives fail to protect neonatal BALB/c mice against *Cryptosporidium parvum* (CP) infection, *J. Euk. Microbiol.*, 41, 41S, 1944.
35. Cama, V. A., Marshall, M. M., Shubitz, L. F., Ortega, Y. R., and Sterling, C. R., Treatment of acute and chronic *Cryptosporidium parvum* infections in mice using clarithromycin and 14-OH clarithromycin, *J. Euk. Microbiol.*, 41, 25S, 1994.
36. Campbell, P. N., and Current, W. L., Demonstration of serum antibodies to *Cryptosporidium* sp. in normal and immunodeficient humans with confirmed infections, *J. Clin. Microbiol.*, 18, 165, 1983.
37. Moon, H. W., Woodmansee, D. B., Harp, J. A., Abel, S., and Ungar, B. L. P., Lacteal immunity to enteric cryptosporidiosis in mice: immune dams do not protect their suckling pups, *Infect. Immun.*, 56, 649, 1998.
38. Harp, J. A. and Whitmire, W. M., *Cryptosporidium parvum* infection in mice: inability of lymphoid cells or culture supernatants to transfer protection from resistant adults to susceptible infants, *J. Parasitol.*, 77, 170, 1991.
39. Uhl, E. W., O'Connor, R. M., Perryman, L. C., and Riggs, M. W., Neutralization-sensitive epitopes are conserved among geographically diverse isolates of *Cryptosporidium parvum, Infect. Immun.*, 60, 1703, 1992.

40. Vitovec, J., Koudela, B., Vladik, P., and Hausner, O., Interaction of *Cryptosporidium parvum* and *Campylobacter jejuni* in experimentally infected neonatal mice, *Zbl. Bakt.*, 274, 548, 1991.
41. Anderson, B. C., Effect of drying on the infectivity of cryptosporidia-laden calf feces for 3- to 7-day-old mice, *Am. J. Vet. Res.*, 47, 2272, 1986.
42. Upton, S. J., Tilley, M. E., Marchin, G. L., and Fine, L. R., Efficacy of a pentaiodide resin disinfectant on *Cryptosporidium parvum* (Apicomplexa: Cryptosporidiidae) oocysts *in vitro*, *J. Parasitol.*, 74, 719, 1988.
43. Fayer, R., Nerad, T., Rall., W., Lindsay, D. S., and Blagburn, B. L., Studies on the cyropreservation of *Cryptosporidium parvum*, *J. Parasitol.*, 77, 357, 1991.
44. Villacorta-Martinez, I., Ares-Mazas, M., Duran-Oreiro, D., and Lorenzo-Lorenzo, M. J., Efficacy of activated sludge in removing *Cryptosporidium parvum* oocysts from sewage, *Appl. Environ. Microbiol.*, 58, 3514, 1992.
45. Fayer, R., Effect of high temperature on infectivity of *Cryptosporidium parvum* oocysts in water, *Appl. Environ. Microbiol.*, 60, 2732, 1994.
46. Fayer, R., Effect of sodium hypochlorite exposure on infectivity of *Cryptosporidium parvum* oocysts for neonatal BALB/c mice, *Appl. Environ. Microbiol.*, 61, 844, 1995.
47. Rohlman, V. C., Kuhls, T. L., Mosier, D. A., Crawford, D. L., Hawkins, D. R., Abrams, V. L., and Greenfield, R. A., Therapy with atovaquone for *Cryptosporidium parvum* infection in neonatal severe combined immunodeficiency mice, *J. Infect. Dis.*, 168, 258, 1993.
48. Harp, J. A., Wannemuehler, M. W., Woodmansee, D. B., and Moon, H. W., Susceptibility of germfree and antibiotic-treated adult mice to *Cryptosporidium parvum*, *Infect. Immun.*, 56, 2006, 1988.
49. Enriquez, F. J. and Sterling, C. R., *Cryptosporidium* infections in inbred strains of mice, *J. Protozool.*, 38, 100S-102S, 1991.
50. Johansen, G. A. and Sterling, S. R., Detection of a prolonged *C. parvum* infection in immunocompetent adult C57BL/6 mice, *J. Euk. Microbiol.*, 41, 45S, 1994.
51. Ungar, B. L. P., Burris, J. A., Quinn, C. A., and Finkelman, F. D., New mouse models for chronic *Cryptosporidium* infection in immunodeficient hosts, *Infect. Immun.*, 58, 961, 1990.
52. Rasmussen, K. R. and Healey, M. C., Experimental *Cryptosporidium parvum* infections in immunosuppressed adult mice, *Infect. Immun.*, 60, 1648, 1992.
53. Yang, S. and Healey, M. C., The immunosuppressive effects of dexamethasone administered in drinking water to C57BL/6N mice infected with *Cryptosporidium parvum*, *J. Parasitol.*, 79, 626, 1993.
54. Kimata, I., Shigehiko, U., and Iseki, M., Chemotherapeutic effect of azithromycin and lasalocid on *Cryptosporidium* infection in mice, *J. Protozool.*, 38, 232S, 1991.
55. Ungar, B. L. P., Kao, T. C., Burris, J. A., and Finkelman, F. D., *Cryptosporidium* infection in an adult mouse model: independent roles of INF-γ and CD4+ lymphocytes in protective immunity, *J. Immunol.*, 147, 1014–1022, 1991.
56. Harp, J. A. and Moon, H.W., Susceptibility of mast cell-deficient W/W^v mice to *Cryptosporidium parvum*, *Infect. Immun.*, 59, 718, 1991.
57. Heine, J., Moon, H. W., and Woodmansee, D. B., Persistent *Cryptosporidium* infection in congenitally athymic (nude) mice, *Infect. Immun.*, 43, 856, 1984.
58. Kuhls, T. L., Greenfield, R. A., Mosier, D. A., Crawford, D. L., and Joyce, W. A., Cryptosporidiosis in adult and neonatal mice with severe combined immunodeficiency, *J. Comp. Pathol.*, 106, 399, 1992.
59. Leitch, G. J. and He, Q., Putative anticryptosporidial agents tested in an immunodeficient mouse model, *Antimicrob. Agents Chemother.*, 38, 865, 1994.
60. Mead, J. R., Arrowood, M. J., Healey, M. C., and Sidwell, R. W., Cryptosporidial infections in SCID mice reconstituted with human or murine lymphocytes, *J. Protozool.*, 38, 59S, 1991.
61. Mead, J. R., Arrowood, M. J., Healey, M. C., and Sidwell, R. W., Chronic *Cryptosporidium parvum* infections in congenitally immunodeficient SCID and nude mice, *J. Infect. Dis.*, 163, 1297, 1991.
62. Harp, J. A., Chen, W., and Harmsen, A. G., Resistance to severe combined immunodeficient mice to infection with *Cryptosporidium parvum*: the importance of intestinal microflora, *Infect. Immun.*, 60, 3509, 1992.
63. Kuhls, T. L., Mosier, D. A., Abrams, V. L., Crawford, D. L., and Greenfield, R. A., Inability of interferon-gamma and aminoguanidine to alter *Cryptosporidium parvum* infection in mice with severe combined immunodeficiency, *J. Parasitol.*, 80, 480, 1994.
64. Rohlman, V. C., Kuhls, T. L., Mosier, D. A., Crawford, D. L., and Greenfield, R. A., *Cryptosporidium parvum* infection after abrogation of natural killer cell activity in normal and severe combined immunodeficiency mice, *J. Parasitol.*, 79, 295, 1993.
65. Bjorneby, J. M., Hunsaker, B. D., Riggs, M. W., and Perryman, L. C., Monoclonal antibody immunotherapy in nude mice persistently infected with *Cryptosporidium parvum*, *Infect. Immun.*, 59, 1172, 1991.
66. Mead, J. R., You, X., Pharr, J. E., Belenkaya, Y., Arrowood, M. J., Fallon, M. T., and Schinazi, R. F., Evaluation of maduramicin and alborixin in a SCID mouse model of chronic cryptosporidiosis, *Antimicrob. Agents Chemother.*, 39, 854, 1995.
67. Riggs, M. W., Cama, V. A., Leary, H. L., and Sterling, C. R., Bovine antibody against *Cryptosporidium parvum* elicits a cicumsporozoite precipitate-like reaction and has immunotherapeutic effect against persistent cryptosporidiosis in SCID mice, *Infect. Immun.*, 62, 1927, 1994.
68. Healey, M. C., Yang, S., Rasmussen, K. R., Jackson, M. K., and Du, C., Therapeutic efficacy of paromomycin in immunosuppressed adult mice infected with *Cryptosporidium parvum*, *J. Parasitol.*, 81, 114, 1995.

69. Rasmussen, K. R., Healey, M. C., Cheng, L., and Yang, S., Effects of dehydroepiandrosterone in immunosuppressed adult mice infected with *Cryptosporidium parvum, J. Parasitol.*, 81, 429, 1995.
70. Moody, K. D., Brownstein, D. G., and Johnson, E. A., Cryptosporidiosis in suckling laboratory rats, *Lab. Anim. Sci.*, 41, 625, 1991.
71. Reese, N. C., Current, W. L., Ernst, J. V., and Bailey, W. S., Cryptosporidiosis of man and calf: a case report and results of experimental infections in mice and rats, *Am. J. Trop. Med. Hyg.*, 31, 226, 1982.
72. Rehg, J. E., Hancock, M. L., and Woodmansee, D. B., Characterization of cyclophosphamide-rat model of cryptosporidiosis, *Infect. Immun.*, 55, 2669, 1987.
73. Rehg, J. E., Hancock, M. L., and Woodmansee, D. B., Characterization of dexamethasone-treated rat model of cryptosporidial infection, *J. Infect. Dis.*, 158, 1406, 1988.
74. Brasseur, P., Lemeteil, D., and Ballet, J. J., Rat model for human cryptosporidiosis, *J. Clin. Microbiol.*, 26, 1037, 1988.
75. Gardner, A. L., Roche, J. K., Weikel, C. S., and Guerrant, R. L., Intestinal cryptosporidiosis: pathophysiologic alterations and specific cellular and humoral immune responses in RNU/+ and RNU/RNU (athymic) rats, *Am. J. Trop. Med. Hyg.*, 44, 49, 1991.
76. Rasmussen, K. R., Martin, E. G., and Healey, M. C., Effects of dehydroepiandrosterone in immunosuppressed rats infected with *Cryptosporidium parvum, J. Parasitol.*, 79, 364, 1993.
77. Rasmussen, K. R., Arrowood, M. J., and Healey, M. C., Effectiveness of dehydroepiandrosterone in reduction of cryptosporidial activity in immunosuppressed rats, *Antimicrob. Agents Chemother.*, 36, 220, 1992.
78. Rasmussen, K. R., Martin, E. G., Arrowood, M. J., and Healey, M. C., Effects of dexamethasone and dehydroepiandrosterone in immunosuppressed rats infected with *Cryptosporidium parvum, J. Protozool.*, 38, 157S, 1991.
79. Rehg, J. E., Hancock, M. L., and Woodmansee, D. B., Anticryptosporidial activity of sulfadimethoxine, *Antimicrob. Agents Chemother.*, 32, 1907, 1988.
80. Rehg, J. E., Anticryptosporidial activity is associated with specific sulfonamides in immunosuppressed rats, *J. Parasitol.*, 77, 238, 1991.
81. Rehg, J. E., Anti-cryptosporidial activity of macrolides in immunosuppressed rats, *J. Protozool.*, 38, 228S, 1991.
82. Rehg, J. E., Anticryptosporidial activity of lasalocid and other ionophorous antibiotics in immunosuppressed rats, *J. Infect Dis.*, 168, 1293, 1993.
83. Rehg, J. E., A comparison of anticryptosporidial activity of paromomycin with that of other aminoglycosides and azirthromycin in immunosuppressed rats, *J. Infect. Dis.*, 170, 934, 1994.
84. Rehg, J. E., The activity of halofuginone in immunosuppressed rats infected with *Cryptosporidium parvum, J. Antimicrob. Chem.*, 35, 391, 1995.
85. Verdon, R., Polianski, J., Gaudebout, C., Marche, C., Garry, L., and Pocidalo, J. J., Evaluation of curative anticryptosporidial activity of paromomycin in a dexamethasone-treated rat model, *Antimicrob. Agents Chemother.*, 38, 1681, 1994.
86. Lemeteil, D., Roussel, F., Favennec, L., Ballet, J. J., and Brasseur, P., Assessment of anticryptosporidial agents in an immunosuppressed rat model, *J. Infect. Dis.*, 167, 766, 1993.
87. Brasseur, P., Lemeteil, D., and Ballet, J. J., Anti-cryptosporidial drug activity screened with an immunosuppressed rat model, *J. Protozool.*, 38, 230S, 1991.
88. Brasseur, P., Lemeteil, D., and Ballet, J. J., Curative and preventive anticryptosporidium activities of sinefungin in an immunosuppressed adult rat model, *Antimicrob. Agents Chemother.*, 37, 889, 1993.
89. Jervis, H. R., Merrill, T. G., and Sprinz, H., Coccidiosis in the guinea pig small intestine due to *Cryptosporidium, Am. J. Vet. Res.*, 27, 408, 1966.
90. Angus, K. W., Hutchison, G., and Munro H. M. C., Infectivity of a strain of *Cryptosporidium* found in the guinea-pig (*Cavia porcellus*) for guinea-pigs, mice, and lambs, *J. Comp. Pathol.*, 95, 151, 1985.
91. Gibson, S. V. and Wanger, J. E., Cryptosporidiosis in guinea pigs: a retrospective study, *J. Am. Vet. Med. Assoc.*, 189, 1033, 1986.
92. Marcial, M. A. and Madara, J. L., *Cryptosporidium*: cellular localization, structural analysis of absorptive cell-parasite membrane-membrane interactions in guinea pigs, and suggestion of protozoan transport by M cells, *Gastroenterology*, 90, 583, 1986.
93. Chrisp, C. E., Reid, W. C., Rush, H. G., Suckow, M. A., Bush, A., and Thomann, M. J., Cryptosporidiosis in guinea pigs: an animal model, *Infect. Immun.*, 58, 674, 1990.
94. Vetterling, J.M., Takeuchi, A., and Madden, P.A., Ultrastructure of *Cryptosporidium wrairi* from the guinea pig, *J. Protozool.* 18, 248, 1971.
95. Tilley, M., Upton, S. J., and Chrisp, C. E., A comparative study on the biology of *Cryptosporidium* sp. from guinea pigs and *Cryptosporidium parvum* (Apicomplexa), *Can. J. Microbiol.*, 37, 949, 1991.
96. Blewett, D. A., Quantitative techniques in *Cryptosporidium* research, in *Cryptosporidiosis,* Angus, K. W. and Blewett, D.A., Eds., Moredun Research Institute, 1989, 85.
97. Chrisp, C. E., Suckow, M. A., Fayer, R., Arrowood, M. J., Bealey, M. C., and Sterling, C. R., Comparison of the host ranges and antigenicity of *Cryptosporidium parvum* and *Cryptosporidium wrairi* from guinea pigs, *J. Protozool.*, 39, 406, 1992.
98. Hoskins, D., Chrisp, C. E., Suckow, M. A., and Fayer, R., Effect of hyperimmune bovine colostrum raised against *Cryptosporidium parvum* on infection of guinea pigs by *Cryptosporidium wrairi, J. Protozool.*, 38, 185S, 1991.
99. Davis, A. J. and Jenkins, S. J., Cryptosporidiosis and proliferative ileitis in a hamster, *Vet. Pathol.*, 23, 632, 1986.

100. Orr, J. P., *Cryptosporidium* infection associated with proliferative enteritis in Syrian hamsters, *Can. Vet. J.*, 29, 843, 1988.
101. Kim, C. W., Chemotherapeutic effect of arprinocid in experimental cryptosporidiosis, *J. Parasitol.*, 73, 663, 1987.
102. Rasmussen, K. R. and Healey, M. C., *Cryptosporidium parvum*: experimental infections in aged Syrian golden hamsters, *J. Infect. Dis.*, 165, 769, 1992.
103. Rossi, P., Pozio, E., Gomez-Morales, M. A., and La Rosa, G., Experimental cryptosporidiosis in hamsters, *J. Clin. Microbiol.*, 28, 356, 1990.
104. Rasmussen, K. R. and Healey, M. C., Dehydroepiandrosterone-induced reduction of *Cryptosporidium parvum* infections in aged Syrian golden hamsters, *J. Parasitol.*, 78, 554, 1992.
105. Blanchard, J. L., Baskin, G. B., Murphey-Corb, and Martin, L. N., Disseminated cryptosporidiosis in Simian immunodeficiency virus/delta-infected monkeys, *Vet. Pathol.*, 24, 454, 1987.
106. Current, W. L., Reese, N. C., Ernst, J. V., Bailey, W. S., Heyman, M. B., and Weinstein, W. M., Human cryptosporidiosis in immunocompetent and immunodeficient persons: studies on an outbreak and experimental transmission, *N. Engl. J. Med.*, 308, 1252, 1983.
107. Asahi, H., Koyama, T., Arai, H., Funakoshi, Y., Yamaura, H., Shirasaka, R., and Okutomi, K., Biological nature of *Cryptosporidium* sp. isolated from a cat, *Parasitol. Res.*, 77, 237–240, 1991.
108. Monticello, T. M., Levy, M. G., Bunch, S. E., and Fairley, R. A., Cryptosporidiosis in a feline leukemia virus-positive cat, *J. Am. Vet. Med. Assoc.*, 191, 705, 1987.
109. Lindsay, D. S., Hendrix, C. M., and Blagburn, B. L., Experimental *Cryptosporidium parvum* infections in opossums (*Didelphis virginiana*), *J. Wild. Dis.*, 24, 157, 1988.
110. Thulin, J. D., Kuhlenschmidt, M. S., Rolsma, M. D., Current, W. L., and Gelberg, H. B., An intestinal xenograft model for *Cryptosporidium parvum* infection, *Infect. Immun.*, 62, 329–331, 1994.
111. Anderson, B. C., Experimental infection in mice of *Cryptosporidium muris* isolated from a camel, *J. Protozool.*, 38, 16S, 1991.
112. Iseki, M., Two species *Cryptosporidium* naturally infecting house rats, *Rattus norvegicus, Jpn. J. Parasitol.*, 35, 521, 1986.
113. Iseki, M., Maekawa, T., Moriya, K., Uni, S., and Takada, S., Infectivity of *Cryptosporidium muris* (strain RN 66) in various laboratory animals, *Parasitol. Res.*, 75, 218, 1989.
114. Uni, S., Iseki, M., Maekawa, T., Moriya, K., and Takada, S., Ultrastructure of *Cryptosporidium muris* (strain RN 66) parasitizing the murine stomach, *Parasitol. Res.*, 74, 123, 1987.
115. Yoshikawa, H. and Iseki, M. Freeze-fracture study of the site of attachment of *Cryptosporidium muris* in gastric glands, *J. Protozool.*, 39, 539–544, 1992.
116. Chalmers, R. M., Sturdee, A. P., Casemore, D. P., Currey, A., Miller, A., Parker, N. D., and Richmond, T. M., *Cryptosporidium muris* in wild house mice (*Mus musculus*): first report in the U.K., *Eur. J. Protistol.*, 30, 151, 1994.
117. Hoerr, F. J., Ranck, F. M., and Hastings, T. F., Respiratory cryptosporidiosis in turkeys, *J. Am. Vet. Med. Assoc.*, 173, 1591, 1978.
118. Glisson, J. R., Brown, T. P., Brugh, M., Page, R. K., Kleven, S. H., and Davis, R. B., Sinusitis in turkeys associated with respiratory cryptosporidiosis, *Avian Dis.*, 28, 783, 1984.
119. Lindsay, D. S. and Blagburn, B. L., *Cryptosporidium sp.* infections in chickens produced by intra-cloacal inoculation of oocysts, *J. Parasitol.*, 74, 615, 1986.
120. Lindsay, D. S., Blagburn, B. L., Hoerr, F. J., and Giambrone, J. J., Experimental *Cryptosporidium baileyi* infections in chickens and turkeys produced by ocular inoculation of oocysts, *Avian Dis.*, 31, 355, 1987.
121. Lindsay, D. S., Blagburn, B. L., Sundermann, C. A., Hoerr, F. J., and Giambrone, J. J., *Cryptosporidium baileyi*: effects of intra-abdominal and intravenous inoculation of oocysts on infectivity and site of development in broiler chickens, *Avian Dis.*, 31, 841, 1987.
122. Hatkin, J. M., Lindsay, D. S., Giambrone, J. J., Hoerr, F. J., and Blagburn, B. L., Experimental biliary cryptosporidiosis in chickens, *Avian Dis.*, 34, 454, 1990.
123. Lindsay, D. S., Blagburn, B. L., Sundermann, C. A., and Giambrone, J. J., Effect of broiler chicken age on susceptibility to experimentally induced *Cryptosporidium baileyi* infection, *Am. J. Vet. Res.*, 49, 1412, 1988.
124. Lindsay, D. S., Blagburn, B. L., Sundermann, C. A., and Ernest, J. A., Chemoprophylaxis of cryptosporidiosis in chickens using halofuginone, salinomycin, lasalocid, or monensin, *Am. J. Vet. Res.*, 48, 354, 1987.
125. Lindsay, D. S., Sundermann, C. A., and Blagburn, B. L., Cultivation of *Cryptosporidium baileyi*: studies with cell cultures, avian embryos, and pathogenicity of chicken embryo-passaged oocysts, *J. Parasitol.*, 74, 288, 1988.
126. Blagburn, B. L., Lindsay, D. S., Giambrone, J. J., Sundermann, C. A., and Hoerr, F. J., Experimental cryptosporidiosis in broiler chickens, *Poult. Sci.*, 66, 442, 1987.
127. Guy, J. S., Levy, M. G., Ley, D. H., Barnes, H. J., and Gerig, T. M., Interaction of reovirus and *Cryptosporidium baileyi* in experimentally infected chickens, *Avian Dis.*, 32, 381, 1988.
128. Levy, M. G., Ley, D. H., Barnes, H. J., Gerig, T. M., and Corbett, W. T., Experimental cryptosporidiosis and infectious bursal disease virus infection of specific-pathogen-free chickens, *Avian Dis.*, 32, 803, 1988.

Chapter 10

Molecular Biology of *Cryptosporidium*

Mark C. Jenkins and Carolyn Petersen

CONTENTS

I. INTRODUCTION

Research on the molecular biology of *Cryptosporidium* is in its infancy. Advances in our understanding of the basic biology of *Cryptosporidium* and greater insight into host immunity that develops after cryptosporidial infection have spawned a number of studies on the molecular biology of the parasite. Recent waterborne outbreaks of cryptosporidiosis in North America and elsewhere have stimulated an effort to develop improved methods for detecting *Cryptosporidium* in environmental samples. One detection method being evaluated in a number of laboratories is amplification of *Cryptosporidium* DNA using polymerase chain reaction (PCR) technology. Also, the preliminary success in ameliorating this parasitic disease in humans and animals by passive administration of *C. parvum*-specific hyperimmune colostrum has prompted research on the antigenic and molecular nature of *Cryptosporidium*. The goal of this work is to develop recombinant antigens that may elicit high titer colostrum and serum for passive immunotherapy of acute infection. In general, the research reviewed in this chapter has been conducted for the purpose of developing a rapid and sensitive detection method and identifying genes and gene products that might serve as targets of chemotherapy or immunotherapy.

II. CHARACTERISTICS OF *CRYPTOSPORIDIUM* NUCLEIC ACID

A. CHROMOSOMES

The actual number and size of *Cryptosporidium* chromosomes is controversial. Initial studies using field-inversion gel electrophoresis (FIGE) showed *C. parvum* to contain five chromosomes ranging in size from 1.4 to 3.3 mb and *C. baileyi* to contain six chromosomes ranging in size from 1.4 to over 3.3 mb.[1] Subsequent studies using a similar separation method, orthogonal field alteration gel electrophoresis (OFAGE), showed *C. parvum* to contain at least five chromosomes between 0.9 and 1.4 mb.[2] Based on different intensities of the chromosomal bands, these authors speculated that more than five chromosomes are present. A recent study indicates that by using a modified contour-clamped homogeneous electric fields (CHEF) at least seven chromosomes between 0.945 and 2.2 mb can be resolved and that an eighth chromosome at about 1.3 mb may be present.[3] Although a variety of estimates have been proposed, about 10 to 20 mb of chromosomal DNA appears to be present in *Cryptosporidium*.

0-8493-7695-5/97/$0.00+$.50

Table 1 DNA Sequences of *Cryptosporidium* Available in GenBank

Accession Number	*Cryptosporidium* Species	Description	% AT (CDS/NC)[a]
X64340	*parvum*	Ribosomal RNA gene for 18s rRNA	NA/62
X64341	*parvum*	Ribosomal RNA gene for 18s rRNA	NA/62
Z22537	*parvum*	Precursor of oocyst wall protein	60/71
Z17386	*parvum*	190-kDa wall protein	61/NA
M59419	*parvum*	DNA segment A for PCR-based detection	NA/67
M59420	*parvum*	DNA segment B for PCR-based detection	NA/67
M59421	*parvum*	DNA segment C for PCR-based detection	NA/61
L08612	*parvum*	15-kDa/60-kDa cross-reactive sporozoite antigens	64/76
M86241	*parvum*	Actin gene, complete CDS	61/73
L31806	*parvum*	Beta-tubulin gene, partial CDS	60/NA
L01269	*parvum*	DNA fragment for PCR-based detection	NA/51
M95743	*parvum*	190-kDa oocyst protein	61/NA
L25642	*parvum*	18S ribosomal RNA gene	NA/62
L16996	*parvum*	18S ribosomal RNA gene	NA/61
L16997	*parvum*	18S ribosomal RNA gene	NA/61
L20049	*parvum*	5S ribosomal RNA genes	NA/74
L19068	*baileyi*	18S ribosomal RNA gene	NA/59
L19069	*muris*	18S ribosomal RNA gene	NA/58
S74588	*parvum*	DNA fragment for PCR-based detection	NA/67
S71380	*parvum*	18S ribosomal RNA	NA/57
L34568	*parvum*	15-kDa sporozoite protein	58/69
U11440	*wrairi*	Small subunit ribosomal RNA gene	NA/61
X77586	*parvum*	TRAP-C2 gene for putative thrombosondin-related adhesive protein	59/60
X77587	*parvum*	TRAP-C1 gene for thrombosondin-related adhesive protein	63/68
X64342	*muris*	Ribosomal RNA gene for 18S rRNA	NA/62
X64343	*muris*	Ribosomal RNA gene for 18S rRNA	NA/62
U11761	*parvum*	70-kDa heat shock protein	59/75
U24082	*parvum*	Acetyl CoA synthetase	56/77
U22892	*parvum*	CP15 sporozoite protein	64/74

[a] Percent AT in coding region (CDS) and noncoding regions (NC). NA = not applicable.

B. CODONS

A notable feature of *Cryptosporidium* is the high AT percentage of the genome (60 to 70% AT). Although this estimate is based mostly on sequencing of genes encoding protein and ribosomal RNA, Southern blot hybridization of *Cryptosporidium* DNA digested with restriction enzymes that have either AT- or GC-rich recognition sequences support this finding.[4-9] The protein encoding regions of most *Cryptosporidium* genes are composed of about 60% AT, while noncoding regions have an even higher percentage of AT, in some instances comprising over 75% of the DNA sequence. Analysis of the protein coding sequences in Table 1 (M95743, Z17386, L34568, M86241, Z22537, L08612, L31806, X77587, U11761, X77586, U24082, U22892) also showed a marked codon bias. In general, codons containing an A or T in the third position occurred in greater frequency than those containing a G or C in this position.[9a] This codon usage may impact attempts to identify recombinant *Cryptosporidium* clones in *Escherichia coli* phage or plasmid expression libraries. It is possible that many *Cryptosporidium* DNA sequences may not be translated at detectable levels in *Escherichia coli* because the tRNAs necessary for efficient expression are present at too low of a concentration. For instance, the codon AGA, which codes for the amino acid arginine, is utilized by *C. parvum* at a frequency of 81% during expression of acetyl CoA synthetase (Accession No. U24082). Conversely, AGA is utilized at a frequency of only 3.5% by *E. coli,* which uses the codons CGT or CGC instead at a frequency of over 80% to produce arginine. Another example is the usage of GGA (for glycine) by *C. parvum* to produce CP15/60 protein (Accession No. L08612). This codon is utilized by the parasite at a frequency of 75%, compared to *E. coli* which uses the other three codons (GGT, GGC, and GGG) at a combined frequency of over 90% (see Table 2). One interesting example is the exclusive use of the codon CCA (for proline) by *C. parvum* to express heat shock protein 70 (HSP 70, Accession No. U11761). *E. coli* uses CCA at a frequency of under 20%,

Table 2 Comparison of Average Codon Usage for *Cryptosporidium parvum* (CP), *Escherichia coli* (EC), *Saccharomyces cerevisae* (yeast, SC), and *Autographa californica* Nuclear Polyhedrosis Virus (Baculovirus, ACNPV)[a]

	Percent Codon Used by Organism				
Amino Acid	**Codon**	**CP**	**EC**	**SC**	**ACNPV**
Phe	TTT	63	50	46	73
	TTC	37	50	54	27
Leu	TTA	35	10	27	21
	TTG	23	12	35	37
	CTT	22	10	11	8
	CTC	9	10	5	9
	CTA	9	3	13	10
	CTG	2	55	9	15
Ile	ATT	60	47	49	49
	ATC	22	46	30	24
	ATA	18	7	21	27
Val	GTT	54	28	44	26
	GTC	8	21	24	16
	GTA	34	16	16	18
	GTG	4	35	16	40
Ser	TCT	34	18	31	15
	TCC	6	17	18	11
	TCA	33	11	19	11
	TCG	4	14	9	21
	AGT	17	13	14	17
	AGC	6	27	9	25
Pro	CCT	30	15	29	19
	CCC	6	10	13	37
	CCA	62	19	48	13
	CCG	2	56	10	31
Thr	ACT	47	19	38	27
	ACC	10	46	25	21
	ACA	42	12	26	21
	ACG	1	23	11	31
Ala	GCT	55	19	44	19
	GCC	8	25	24	27
	GCA	35	21	34	21
	GCG	2	35	8	32
Tyr	TAT	73	53	50	45
	TAC	27	47	50	55
His	CAT	76	51	60	50
	CAC	24	49	40	50
Gln	CAA	80	30	73	73
	CAG	20	70	27	27
Asn	AAT	68	40	55	44
	AAC	32	60	45	56
Lys	AAA	56	24	55	73
	AAG	44	76	45	27
Asp	GAT	82	59	62	40
	GAC	18	41	38	60
Glu	GAA	66	69	74	68
	GAG	34	31	26	32
Cys	TGT	67	43	34	50
	TGC	33	57	66	50
Arg	CGT	17	43	17	20
	CGC	2	38	5	31
	CGA	2	5	5	15
	CGG	0	8	3	4
	AGA	70	4	54	20
	AGG	9	2	16	10

Table 2 (continued) Comparison of Average Codon Usage for *Cryptosporidium parvum* (CP), *Escherichia coli* (EC), *Saccharomyces cerevisae* (yeast, SC), and *Autographa californica* Nuclear Polyhedrosis Virus (Baculovirus, ACNPV)[a]

		Percent Codon Used by Organism			
Amino Acid	**Codon**	**CP**	**EC**	**SC**	**ACNPV**
Gly	GGT	50	37	60	29
	GGC	6	41	9	43
	GGA	37	9	16	21
	GGG	7	13	15	7

Note: Values for *C. parvum* derived from sequences listed in Table 1; values for other organisms derived from Wada et al.[32] Values for Met and Trp are not shown since only one codon is used for these amino acids.

[a] Numbers refer to percentage a codon is utilized for a particular amino acid averaged for all protein-encoding genes available in Genbank sequence database.

instead using CCG at a frequency greater than 50%. These data exemplify the potential difficulty in expressing *Cryptosporidium* genes in *E. coli.* Other expression systems, such as yeast or baculovirus, that may have codon usages more similar to *Cryptosporidium* may be required for producing recombinant protein. Some caution must be exercised in interpreting codon usage data, however, because the average codon frequency as shown may not reflect the actual intracellular pool of each tRNA available to the organism. Several authors have shown that highly expressed genes utilize codons that correspond to abundant tRNAs.[10-11] In general, DNA sequences encoding proteins that are expressed at low to moderate levels do not show the marked codon bias as observed in highly expressed genes. Thus, parasite genes that might appear difficult to express in *E. coli* based solely on codon usage may nevertheless be expressed at high levels in this prokaryote. It is suggested that a particular clone be tested for expression in *E. coli* first, and then, if the level of recombinant protein produced is low, other systems could be evaluated using codon frequency as one criterion. This information may also be useful in designing degenerate oligonucleotides for identifying genes encoding a specific *Cryptosporidium* protein by hybridization screening of phage or plasmid libraries. Certain codons, those ending in A or T for instance, could be used at a higher frequency than those ending in C or G.

C. DNA FROM OOCYSTS

To isolate *Cryptosporidium* DNA, most researchers have treated either purified oocysts or excysted sporozoites with 10 to 50 m*M* Tris, pH 7.5 to 8.0, 10 to 500 m*M* EDTA containing 1% sodium dodecyl sulfate; conducted repeated freeze-thaw cycles; and then digested with 0.05 to 2.0 mg/ml proteinase K. DNA extractions have been carried out at elevated temperatures (42 to 60°C) for 1 to 48 hours. The DNA solutions were then extracted with phenol, phenol-chloroform, and chloroform, followed by ethanol precipitation, and yields of 10 to 20 μg purified DNA per 10^9 oocysts were obtained. Greater yields have been reported based on optical density readings at 260 nm, but several workers have found an amorphous substance (lipid, glycogen) to absorb at O.D. 260, thus causing overestimates of DNA yield (Jenkins, unpublished observations; Lally, personal communication).

D. RNA FROM OOCYSTS

Cryptosporidium oocyst RNA has been prepared for direct RNA sequencing, *in vitro* translation, and production of cDNA libraries. In general, sporozoites or oocysts have been frozen in liquid nitrogen and then treated with either guanidine hydrochloride containing sodium acetate and 2-mercaptoethanol [4,12,13] or guanidinium thiocyanate containing phenol, sodium citrate, sodium acetate, and 2-mercaptoethanol.[9,14] After extraction, the RNA was precipitated with ethanol and stored in DEPC-treated H_2O or 70% ethanol at –70°C. Using the latter procedure, 100 to 150 μg total RNA per 10^9 oocysts can be obtained. The proportion of mRNA in these samples is quite low, representing less than 2% of total RNA, which may reflect an inactive state of sporozoites in the oocysts. Ancedotal evidence suggests that DNA and RNA yields decrease with the length of time oocysts are stored. An accepted procedure for purifying oocysts from fecal material has been centrifugation over cesium chloride[15] followed by freezing in extraction buffer in liquid nitrogen or immediate processing for nucleic acid isolation.

III. CLONING OF SPECIFIC *CRYPTOSPORIDIUM* GENE SEQUENCES

A. PHYLOGENETICS

Numerous genes have been cloned and characterized for the purpose of studying phylogeny of *Cryptosporidium*, detecting the parasite by PCR, or identifying targets of chemotherapy or immunotherapy. At present, there are 30 gene sequences in the GenBank database (Table 1). Numerous authors have obtained the complete sequence of the 18S small subunit ribosomal RNA and used these sequences to infer phylogenetic relationships between *Cryptosporidium* and other apicomplexan species. In one study,[4] the semi-conserved regions of small subunit ribosomal RNA were compared using both parsimony (DNA-PARS) and distance (FITCH) analyses. This work provided some evidence that *Cryptosporidium* was not closely related to other apicomplexan parasites. Others have employed phylogenetic analysis using parsimony (PAUP) to compare sequences of semiconserved and variable regions of small subunit rRNA from a variety of apicomplexans. As opposed to prior work, these authors showed that the Apicomplexa are a monophyletic group containing *Cryptosporidium*.[16] In this study, *Cryptosporidium* appeared most closely related to *Plasmodium*. The genes for 18S ribosomal RNA of *C. parvum* and *C. muris*[17,18] and for 5S ribosomal RNA of *C. parvum*[19] have been cloned by PCR and complete sequences derived for both. The 18S ribosomal RNA sequences of *C. parvum* and *C. muris* appeared more than 99% identical to one another.[17] Restriction enzyme analysis of cloned small subunit rRNA genes of three *Cryptosporidium* species showed that *C. parvum* could be distinguished from *C. baileyi* and *C. muris*.[20a]

B. DETECTION

Information derived from cloned gene sequences have been used to identify *Cryptosporidium* oocysts in a variety of environmental samples. As little as 300 fg of purified oocyst DNA was detected by PCR using oligonucleotide primers based on a cloned gene sequence.[6] This procedure has been extended to detect *C. parvum* in paraffin-embedded human intestinal tissue[21] and to obtain PCR primers specific for *C. wrairi*.[21a] Using the same oligonucleotide primers after boiling oocysts and exposing them to DNase prior to sporozoite excystation, PCR enabled investigators to distinguish live vs. dead *C. parvum* oocysts.[22] Similar to other detection methods (e.g., immunofluorescence), PCR cannot distinguish infectious vs. noninfectious oocysts. However, using a combined excystation-proteinase digestion procedure, one research group has produced a PCR-based technique that distinguishes oocysts that harbor viable sporozoites vs. oocysts that cannot be excysted.[22a] A recent study showed specific detection of DNA extracted from 2×10^3 purified *C. parvum* oocysts using PCR and oligonucleotide primers based on a repetitive DNA sequence.[23] Hybridization of PCR products impregnated on a membrane was found to improve sensitivity of detection by 10- to 100-fold.[21,23] A number of recent studies have compared conventional detection methods (e.g., dye staining, immunofluorescence, ELISA) against a combined PCR-chemiluminescence procedure. Most authors have found the latter to be three to four orders of magnitude more sensitive for detecting pure oocysts in solution and about two orders of magnitude more sensitive for detecting oocysts in fecal matter.[23a-d] The PCR-based method of random amplified polymorphic DNA (RAPD) also has been used to differentiate species and isolates of *Cryptosporidium*[24-24b] but is limited by the need to purify *Cryptosporidium* oocysts from contaminating fecal material. The difficulty in adapting PCR for detection of parasites in environmental samples was exemplified in a recent study.[25] PCR detected 20 to 100 oocysts in highly purified samples, but the sensitivity of the assay decreased 100- to 1000-fold after oocysts were mixed with filtered tap water. However, recent advances in oocyst purification methods, such as immunomagnetic separation (IMS),[23b] should decrease sample handling times and remove inhibitors from the oocyst preparations.

C. ANTIGEN-ENCODING GENES

Several strategies have been employed to obtain protein-encoding genes of *Cryptosporidium*. For cloning a thymidylate synthase (TS) gene fragment, oligonucleotide primers based on conserved DNA sequences were used to amplify TS genes directly from *C. parvum* genomic DNA.[26] Unless the DNA sequence shares significant homology with *Cryptosporidium* DNA, as in the cloning of *C. parvum* actin,[2,27] or alpha- and beta-tubulin,[27] or elongation factor-2[27a] genes, bacteriophage or plasmid expression libraries must be constructed and screened with antigen-specific sera. Although cDNA libraries have been prepared from *C. parvum* oocysts,[9] this effort may be unnecessary because introns are absent or occur in low frequency in genes of *Cryptosporidium*.[7,8,28] In fact, only one small intron has been identified thus far in the gene for beta-tubulin (Nelson, personal communication). Thus, *Cryptosporidium* genomic DNA may be the preferred source of genetic material for hybridization and expression screening, because all sequences are likely to be represented. Most genomic DNA libraries have been derived from isolated oocyst DNA that was subjected to partial or complete digestion with moderate- to high-frequency cutting restriction enzymes.

Several *C. parvum* antigen-encoding genes obtained by immunoscreening have been subjected to molecular and immunological characterization.[5] Partial DNA sequences encoding epitopes of *C. parvum* sporozoite proteins of 15/35, 23, 45, 68/95, and >500 kDa have been identified by screening with hyperimmune sera prepared against unfractionated oocyst/sporozoite protein.[29] Elution of recombinant antigen-specific antibodies from these clones and probing air-dried *C. parvum* sporozoites showed either a diffuse or a localized immunofluorescence pattern associated with several surface or internal components of the parasite. One encodes a portion of a 900-kDa microneme protein which contains mucin-like domains (Petersen, unpublished observations). A gene fragment encoding an epitope associated with a 190 kDa oocyst-specific antigen was obtained by screening a *C. parvum* genomic DNA library with anti-sporozoite/oocyst serum.[7] The predicted protein sequence contains two highly repetitive regions; the first region is composed of 13 repeats of the amino acids CPXG(7X)C which are followed by either 9 or 11 residues, and the second region is composed of three cysteine-rich, 53-amino-acid repeats. A larger genomic clone of this sequence was obtained, and both antisera and monoclonal antibodies prepared against the encoded recombinant protein confirmed that the antigen was associated with the oocyst wall.[8] A gene encoding a 70-kDa heat shock protein of *C. parvum* was produced by screening a genomic DNA library with anti-oocyst/sporozoite serum.[28] The gene appears to lack introns and shares homology at the amino acid level with known cytoplasmic HSP70s from a variety of sources including *Plasmodium falciparum*.

Immunoscreening cDNA libraries with polyclonal anti-oocyst/sporozoite sera has been used to identify several DNA sequences encoding epitopes associated with *C. parvum* sporozoite surface proteins[9,30] or enzymes.[30a] This technique was employed to produce recombinant 15-kDa antigen, which was used to stimulate production of hyperimmune bovine colostrum for immunotherapy of cryptosporidiosis. One clone, designated CP15/60, encoded epitopes shared by native *C. parvum* sporozoite 15- and 60-kDa antigens.[9] A genomic DNA clone identical to this cDNA sequence was identified and corroborated the immunological reactivity with native CP15 (Upton, personal communication). A second cDNA clone, designated CP15, encoded an epitope specific for native CP15 which, by immunofluorescence staining, appears to be associated with both sporozoites and oocysts.[30] CP15 is unrelated at both the nucleic acid and protein level to CP15/60. Subtractive cDNA libraries from *C. parvum*-infected Madin-Darby bovine kidney cells have been prepared.[31] Differential screening with subtracted cDNA identified several clones that were either host cell or parasite derived. This approach and techniques such as PCR-based differential display have identified parasite and host cell genes that are regulated during *Cryptosporidium* infection (Abrahamsen, personal communication). A genomic *C. parvum* DNA clone encoding a protein which has hemolytic activity has been isolated.[32] The encoded hemolytic protein may play some role in host-parasite interaction since the gene encoding this protein appears to be transcribed in both excysted sporozoites and in intestinal tissue obtained from *C. parvum*-infected mice.[32]

Several groups have been successful in protecting humans and animals against cryptosporidiosis by passive administration of monoclonal antibodies or hyperimmune serum or colostrum against native *C. parvum* antigens (see Chapter 5). Molecular cloning may produce epitopes of specific *Cryptosporidium* proteins for immunization studies, whereas large-scale preparation and purification of a distinct native protein is impractical. In addition to cloning and expressing genes for antigenic cryptosporidial proteins, direct DNA injection technology has been used to elicit antigen-specific immunoglobulin responses.[33] The titer and duration of CP15/60 antigen-specific responses were dependent on DNA immunization dose and site of injection, suggesting that induction of appropriate immune responses may depend on host cell processing of translated protein.

IV. FUTURE DIRECTIONS

Based on initial success in cloning and expressing specific gene sequences of *Cryptosporidium*, progress in the next few years should be rapid. Several areas should be emphasized if the goal of controlling cryptosporidiosis is to be realized. One such effort is to evaluate prokaryotic vs. eukaryotic expression of *Cryptosporidium* genes. Biochemical studies have shown that many *Cryptosporidium* proteins are post-translationally modified and that these moieties are important for immune cell recognition. Expression of *C. parvum* DNA in *E. coli* often does not reproduce epitopes that are recognized by immune serum and colostrum from *C. parvum*-infected animals.[30] However, others have expressed recombinant

Cryptosporidium antigens in *E. coli* that are recognized by neutralizing monoclonal antibodies (Petersen, unpublished observations). One bifunctional enzyme, thimidylate synthase-dihyrofolate reductase, has been expressed in *E. coli* in a functional form (Petersen, personal communication). In choosing an expression system, one must consider a number of factors, including codon usage of *Cryptosporidium* and the microorganism harboring the recombinant vector, the biochemical nature of the native protein (e.g., are post-translational moieties critical for epitope formation or enzyme function), and the effect of vector-encoded fusion protein on folding of the cryptosporidial protein. Another area of study should be a search for novel biochemical pathways, similar to the mannitol cycle in *Eimeria* that cryptosporidial parasites utilize during intracellular development. These studies will require biochemical and molecular approaches to characterizing such pathways and the particular enzymes involved. Insight may be gained on how the parasite interacts with the host cell during development by studying regulation of parasite and host gene transcription and translation. Methods such as PCR-based differential display and two-dimensional gel electrophoresis of infected and uninfected host cells may provide such information. An accurate estimate of the number and size of *Cryptosporidium* chromosomes is also needed. A panel of cloned DNA sequences could be used to localize genes on specific chromosomes. Merging of DNA technology with biochemistry, drug design, and immunology of *Cryptosporidium* offers exciting opportunities for designing therapies against cryptosporidiosis.

REFERENCES

1. Mead, J. R., Arrowood, M. J., Current, W. L., and Sterling, C. R., Field inversion gel electrophoretic separation of *Cryptosporidium* spp. chromosome-sized DNA, *J. Parasitol.*, 74, 366, 1988.
2. Kim, K., Gooze, L., Petersen, C., Gut, J., and Nelson, R. G., Isolation, sequence and molecular karyotype analysis of the actin gene of *Cryptosporidium parvum*, *Mol. Biochem. Parasitol.*, 50, 105, 1992.
3. Hays, M. P., Mosier, D. A., and Oberst, R. D., Enhanced karyotype resolution of *Cryptosporidium parvum* by contour-clamped homogeneous electric fields, *Vet. Parasitol.,* 58, 273, 1995.
4. Johnson, A. M., Fielke, R., Lumb, R., and Baverstock, P. R., Phylogenetic relationships of *Cryptosporidium* determined by ribosomal RNA sequence comparison, *Int. J. Parasitol.*, 20, 141, 1990.
5. Dykstra, C. C., Blagburn, B. L., and Tidwell, R. R. Construction of genomic libraries of *Cryptosporidium parvum* and identification of antigen-encoding genes, *J. Protozool.*, 38, 76, 1991.
6. Laxer, M. A., Timblin, B. K., and Patel, R., DNA sequences for the specific detection of *Cryptosporidium parvum* by the polymerase chain reaction, *Am. J. Trop. Med. Hyg.*, 45, 688, 1991.
7. Lally, N. C., Baird, G. D., McQuay, S. J., Wright, F., and Oliver, J. J., A 2359-base pair DNA fragment from *Cryptosporidium parvum* encoding a repetitive oocyst protein, *Mol. Biochem. Parasitol.*, 56, 69, 1992.
8. Ranucci, L., Muller, H.-M., La Rosa, G., Reckmann, I., Angeles, M., Morales, G., Spano, F., Pozio, E., and Crisanti, A., Characterization and immunolocalization of a *Cryptosporidium* protein containing repeated amino acid motifs, *Infect. Immun.*, 61, 2347, 1993.
9. Jenkins, M., Fayer, R. Tilley, M., and Upton, S. J., Cloning and expression of a cDNA encoding epitopes shared by 15- and 60-kilodalton proteins of *Cryptosporidium parvum* sporozites, *Infect. Immun.*, 61, 2377, 1993.

9a. Char, S., Kelly, P., Naeem, A., and Farthing, J. G., Codon usage in *Cryptosporidium parvum* differs from that in other Eimeriorina, *Parasitol.,* 112, 357, 1996.

10. Ikemura, T., Codon usage and tRNA content in unicellular and multicellular organisms, *Mol. Biol. Evol.*, 2, 13, 1985.
11. Shields, D. C., and Sharp, P. M., Synonymous codon usage in *Bacillus subtilis*, *Nucl. Acids Res.,* 15, 8023, 1987.
12. Johnson, A. M., McDonald, P. J., and Illana, S., Characterization and *in vitro* translation of *Toxoplasma gondii* ribonucleic acid, *Mol. Biochem. Parasitol.*, 18, 313, 1986.
13. Brooker, J. D., May, B. K., and Neoh, S. H., Synthesis of gamma-amino-laevulinate synthase *in vitro* using hepatic mRNA from chick embryos with induced porphyria, *Eur. J. Biochem.*, 106, 17, 1980.
14. Xie, W., and Rothblum, L.I., Rapid, small-scale RNA isolation from tissue culture cells, *Biotechniques*, 11, 324, 1991.
15. Taghi-Kilani, R. and Sekal, L., Purification of *Cryptosporidium parvum* oocysts and sporozoites by cesium chloride and Percoll® gradients, *Am. J. Trop. Med. Hyg.*, 36, 505, 1987.
16. Barta, J. R., Jenkins, M. C., and Danforth, H. D., Evolutionary relationships of avian *Eimeria* species among other apicomplexan protozoa: monophyly of the Apicomplexa is supported, *Mol. Biol. Evol.*, 345, 8, 1990.
17. Cai, J., Collins, M. D. McDonald, V., and Thompson, D. E., PCR cloning and nucleotide sequence determination of the 18S rRNA genes and internal transcribed spacer 1 of the protozoan parasites *Cryptosporidium parvum* and *Cryptosporidium muris*, *Biochim. Biophys. Acta*, 1131, 317, 1992.
18. Kilani, R. T. and Fayer, R., Geographical variation in 18S rRNA gene sequence of *Cryptosporidium parvum*, *Int. J. Parasitol.*, 24, 303, 1994.
19. Taghi-Kilani, R., Remacha-Moreno, M., and Wenman, W. M., Three tandemly repeated 5S ribosomal RNA-encoding genes identified, cloned and characterized from *Cryptosporidium parvum*, *Gene*, 142, 253, 1994.

20. Awad-El-Kariem, F. M., Warhurst, D., and McDonald, V., Detection and species identification of *Cryptosporidium* oocysts using a system based on PCR and endonuclease restriction, *Parasitology*, 109, 19, 1994.
20a. Leng, X., Mosier, D. A., and Oberst, R. D., Differentiation of *Cryptosporidium parvum, C. muris,* and *C. baileyi* by PCR-RFLP analysis of the 18s rRNA gene, *Vet. Parasitol.,* 62, 1, 1996.
21. Laxer, M. A., D'Nicuola, M. E., and Patel, R., Detection of *Cryptosporidium parvum* DNA in fixed, paraffin-embedded tissue by the polymerase chain reaction, *Am. J. Trop. Med. Hyg.*, 47, 450, 1992.
21a. Chrisp, C. E. and LeGendre, M., Similarities and differences between DNA of *Cryptosporidium parvum* and *C. wrairi* detected by the polymerase chain reaction, *Folia Parasitol.,* 41, 97, 1994.
22. Filkorn, R., A. Wiedenmann, A., and Botzenhart, K., Selective detection of viable *Cryptosporidium* oocysts by PCR, *Zbl. Hyg*., 195, 489, 1994.
22a. Wagner-Wiening, C. and Kimmig, P., Detection of viable *Cryptosporidium parvum* oocysts by PCR, *Appl. Environ. Microbiol.,* 61, 4514, 1995.
23. Webster, K. A., Pow, J.D., Giles, M., Catchpole, J., and Woodward, M. J., Detection of *Cryptosporidium parvum* using a specific polymerase chain reaction, *Vet. Parasitol.*, 50, 35, 1993.
23a. Johnson, D. W., Pieniazek, N. J., Griffin, D. W., Misener, L., and Rose, J. B., Development of a PCR protocol for sensitive detection of *Cryptosporidium* oocysts in water samples, *Appl. Environ. Microbiol.*, 61, 3849, 1995.
23b. Webster, K. A., Smith, H. V., Giles, M., Dawson, L., and Robertson, L. J., Detection of *Cryptosporidium parvum* oocysts in faeces: comparison of conventional coproscopical methods and the polymerase chain reaction, *Vet. Parasitol.,* 61, 5, 1996.
23c. Leng, X., Mosier, D. A., Oberst, R. D., Simplified method for recovery and PCR detection of *Cryptosporidium* DNA from bovine feces, *Appl. Environ. Microbiol.,* 62, 643, 1996.
23d. Mayer, C. L. and Palmer, C. J., Evaluation of PCR, nested PCR, and fluorescent antibodies for detection of *Giardia* and *Cryptosporidium* species in wastewater, *Appl. Environ. Microbiol.,* 62, 2081, 1996.
24. Morgan, U. M., Constantine, C. C., O'Donoghue, P., Meloni, B. P., O'Brien, P. A., and Thompson, R.C.A., Molecular characterisation of *Cryptosporidium* isolates from humans and other animals using RAPD (random amplified polymorphic DNA) analysis, *Am. J. Trop. Med.* Hyg., 52, 559, 1995.
24a. Morgan, U. M., O'Brien, P. A., and Thompson, R. C. A., The development of PCR primers for *Cryptosporidium* using RAPD-PCR, *Mol. Biochem. Parasitol.,* 77, 103, 1996.
24b. Carraway, M., Tzipori, S., and Widmer, G., Identification of genetic heterogeneity in the *Cryptosporidium parvum* ribosomal repeat, *Appl. Environ. Microbiol.,* 62, 712, 1996.
25. Johnson, D. W., Pienzack, N., and Rose, J. B., DNA probe hybridization and PCR detection of *Cryptosporidium* compared to immunofluorescence assay, *Water Sci. Technol.*, 27, 77, 1993.
26. Gooze, L., Kim, K., Petersen, C., Gut, J., and Nelson, R. G., Amplification of a *Cryptosporidium parvum* gene fragment encoding thymidylate synthase, *J. Protozool.*, 38, 56, 1991.
27. Nelson, R. G., Kim, K., Gooze, L., Petersen, C., and Gut, J., Identification and isolation of *Cryptosporidium parvum* genes encoding microtubule and microfilament proteins, *J. Protozool.*, 38, 52, 1991.
27a. Jones, D. E., Tu, T. D., Mathur, S., Sweeney, R. W., and Clark, D. P., Molecular cloning and characterization of *Cryptosporidium parvum* elongation factor-2 gene, *Mol. Biochem. Parasitol.*, 71, 143, 1995.
28. Khramtsov, N. K., Tilley, M., Blunt, D. S., Montelone, B. A., and Upton, S. J., Cloning and analysis of a *Cryptosporidium parvum* gene encoding a protein with homology to cytoplasmic Hsp70, *J. Euk. Micro*., 42, 416, 1995.
29. Petersen, C., Gut, J., Leech, J. H., and Nelson, R. G., Identification and initial characterization of five *Cryptosporidium parvum* sporozoite antigen genes, *Infect. Immun*., 60, 2343, 1992.
30. Jenkins, M. C. and Fayer, R., Cloning and expression of cDNA encoding an antigenic *Cryptosporidium parvum* protein, *Mol. Biochem. Parasitol*., 71, 149, 1995.
30a. Khramtsov, N. V., Blunt, D. S., Montelone, B. A., and Upton, S. J., The putative acetyl-CoA synthetase gene of *Cryptosporidium parvum* and a new conserved protein motif in acetyl-CoA synthetases, *J. Parasitol,* 82, 423, 1996.
31. Steele, M.I., Kuhls, T.L., Nida, K., Meka, C.S.R., Halabi, I.M., Mosier, D.A., Elliot, W., Crawford, D.L., and Greenfield, R.A., A *Cryptosporidium parvum* genomic region encoding hemolytic activity, *Infect. Immun.,* 63, 3840, 1995.
32. Jones, D. E., Tu, T. D., Sweeney, R. W., and Clark, D. P., Isolation of *Cryptosporidium* and bovine cDNA clones from a *Cryptosporidium*-infected MDBK cell line subtraction library, *J. Euk. Micro*., 41, 46, 1994.
33. Jenkins, M.C., Kerr, D., Fayer, R., and Wall, R., Serum and colostrum antibody responses induced by jet-injection of sheep with DNA encoding a *Cryptosporidium parvum* antigen, *Vaccine*, 1996.
34. Wada, K., Wada, Y., Doi, H., Ishibashi, F., Gojobori, T., and Ikemura, T., Codon usage tabulated from the GenBank genetic sequence data, *Nucl. Acids Res*., 19, S1981–S1987, 1991.

INDEX

B

C

D

E

F

G

H

I

J

N

O

P

Q

R

S

T

U

V

W

X

Y

Z